FISTS FULL OF GOLD

A COMPLETE GUIDE TO THE ART OF PROSPECTING

BY CHRIS RALPH

GOLDSTONE SCHOOL OF PROSPECTING

Goldstone Publishing,
A Division of Speaking Stones, Reno, Nevada

 This book is published by the author. For additional copies of this book, or to contact the author, he can be reached at PO Box 3104, Reno, Nevada 89502.
The Author's Website is at: **http://nevada-outback-gems.com/prospect/chris_prospect.htm**

International Standard Book Number (ISBN): **978-0-9842692-0-4**
Library of Congress Control Number: **2009911066**
First Printing: January 2010; Second Printing: May 2011; Third Printing: August 2013
Fourth Printing: July 2020; Fifth Printing: Feb. 2021; Sixth printing; August 2025

Cardinal GRAPHICS & DESIGN

Cover Art, Layout and Graphics design by **Cardinal Graphics**
For more information, Contact John O'Dell at
cardinalgd@aol.com

Considerable amounts of artwork and illustrations from a number of century old mining related texts now out of copyright protection and in the public domain appear in this book. Most have been specially restored, shaded or annotated for use in this book. For a listing of these works and further information about them, see Part 2 of the "Recommended Books" section in the appendix.

SPECIAL NOTE AND DISCLAIMER: This publication contains the opinions and ideas of the author. It is intended to provide helpful and informative material on the subject matter covered. It is sold with the understanding that the author is not engaged in rendering professional services in this book. If the reader requires personal assistance or advice, a competent professional should be consulted for that specific advice. The author specifically disclaims any responsibility for any liability, loss or risk, personal or otherwise, which is incurred as a consequence, directly or indirectly of the use and application of any of the contents of this book. Like many outdoor activities, there are potential risks and dangers associated with prospecting. Slips, rock falls, bad air underground, wild animal attacks and vehicle breakdowns are only a few of the many potential risks. All readers are strongly advised to stay out of underground mines. When venturing into the field, the reader must always have sufficient materials on hand to deal with potential problems and dangers, including at a minimum extra water, food and clothing, first aid medical gear, vehicle repair and maintenance items, maps, and communication equipment. The reader is also responsible to comply with all related legal requirements, which includes the evaluation of all applicable laws and standards which may apply to the prospecting activities of the reader. Although the author has taken due diligence to present some of the most basic laws which apply to prospecting, the details of these laws vary around the country, and do change over time. It is incumbent upon the reader to determine the specific laws and requirements that will apply at the location and time of activity, and to exercise all necessary diligence in complying with them. Nothing in this book should be construed as legal advice, and if such advice is needed, the reader should consult a competent attorney.

TABLE OF CONTENTS

Early Day Hydraulic Gold Mining

Thanks and Acknowledgements: *First, I want to thank my wife for her encouragement, support and understanding while I was writing this book. I also appreciate the support and understanding of my children. Steve Herschbach, Mike Furness and Barbara Ralph (my mom the English teacher) contributed significantly to the review, preparation and editing of the text, and I want to express my appreciation for their assistance.*

FOREWORD

FORWARD - FIRST THROUGH SIXTH EDITION

Yes, it's true that you can prospect for and find your own gold – it's still out there! For years, the pastime of small scale prospecting and mining for gold has been like a well-kept secret among a closed fraternity of friends. The necessary knowledge and skills have often been passed down among enthusiasts almost like a trade-apprenticeship type of operation. However, things have changed in recent years and the increase in the price of the yellow metal has created a new interest from many who would like to find their own gold. Individual prospecting is most popular in the USA and Australia, but there is growing interest from a number of other locations including Canada, Europe and other localities.

While there are many excellent books that specialize in how to operate various types of gold recovery equipment, including the gold pan, suction dredges, dry washers or metal detectors, there is little information on the geology of gold deposits and how to prospect for them. Most "how to" books offer a little basic prospecting advice, but it is very difficult to find good information that goes beyond the most simple beginner's level. There are also books that tell where gold was found in the past, but they don't tell much about how to prospect for and find it once you get there. Finally, there are some advanced textbooks on gold deposits, but these are directed at professional geologists and require formal, university training in geology to really make much use of them. This book takes a unique and different approach in teaching the "trade skills" of prospecting – not just equipment, but the knowledge you really need to find locations with recoverable gold. This information is important because in the final analysis, no matter how well you operate your dredge, metal detector or other prospecting equipment, unless you can find the deposits where the gold is located, your equipment cannot help you recover it.

I want to help you to understand how gold deposits form, how to gather information on an area and how to prospect efficiently so you can find gold. I will address prospecting for both new deposits as well as finding additional deposits within or adjacent to known districts. This book puts together information from a wide variety of sources into a format that is useful for the individual who has discovered the excitement of prospecting. The goal is to present a detailed but plain language explanation of the geologic processes that form gold deposits, and give helpful suggestions to those hoping to get involved in this unique and special field. While some may consider this book a bit too technical, I am hopeful it will open doors for those who wish to move to a higher level of knowledge about the geology of prospecting.

There is a large amount of information presented in this book, including many advanced topics. I have tried to make it as easy to understand as possible, but probably much of it will be new to many readers. I would encourage you to take your time in reading it. Re-read those sections you may be a bit hazy on. It is important to get a firm grip on the topic under discussion before moving on. Many of the concepts are interdependent and combine to give the full understanding needed for the independent prospector. A number of the topics covered briefly in this book could easily be whole books on their own. I have given the coverage to them that would be useful for the individual, independent prospector to know. For those that wish to learn more, there are many books and web pages listed in the appendixes which offer that opportunity. If you are interested in seeking more detail, I encourage you to go out and explore those worlds. After all, no one knows it all, and prospecting is about learning more – the learning never stops.

Chris W. Ralph
Reno, Nevada, USA - June, 2025

Part I - The Lure and Fun of Prospecting

WHY GO PROSPECTING?

Prospecting for gold (and silver) is an exciting pastime that more and more folks are discovering each day. It's certainly been enjoyable for me - I have been prospecting and mining for gold both as a hobby and as an occupation for over 30 years and in my opinion it's a blast! From the deep green forests to the rolling sagebrush hills, few people see as much of America's wide-open spaces as I do. I kick around kooky little old towns in the middle of nowhere. I visit historic sites where the pioneers of the west toiled for years to extract precious metals from the ground. As fun as that is though, finding your own gold, either as a nugget or in solid hard rock is a special experience that's hard to equal.

School kids in California learn how James Marshall accidentally discovered gold nuggets while constructing a water-powered sawmill in the Sierra foothills. The excitement resulting from Marshall's discovery was a fire that ignited gold and silver rushes all across the western US. Almost as well known is the story of O'Reiley and McLaughlin who accidentally discovered the Comstock Lode silver bonanza while working a small deposit of placer gold, tossing away a blue-black waste that later turned out to be rich silver ore. A century ago, Jim Butler, while traveling from his ranch in central Nevada, noticed some quartz vein material. Being a good prospector, he collected a sample, but he thought so little of his find that it sat on his porch for months before it was tested. That sample became the first of many rich discoveries at Tonopah. I could write a whole book telling the stories of these individual prospectors who, whether intentionally or by accident, found rich deposits of gold and other valuable ores. These finds have had no small impact on the development of our country - historically millions upon millions of ounces of gold have been recovered from deposits found by individual prospectors.

Though everyone knows that literally tons of gold were recovered by individual miners in the pioneer days, we also know that times have changed since those rugged single blanket prospectors trod the mountains of the western US with their gold pans and trusty mules. Most people believe that such rich discoveries are a thing of the past – the long past. After all, the great finds I mentioned above were all made 100 or more years ago. Today giant corporations explore all over the world for gold. What can an individual possibly hope to accomplish prospecting in the present day? He is competing with multi-national firms that possess millions of dollars in resources and a staff full of Ph.D. geologists. Most people would believe that surely today, the individual is at such a disadvantage that there is no chance that he can expect to find anything.

You might easily believe that, but it's not necessarily true. Just consider the following bit of history. The most important gold discovery on this planet within the last 100 years was made in the hills of Nevada north of the small town of Carlin. It led to the discovery of a whole new type of gold deposit, one not previously recognized by geologists or the old time prospectors – the "Carlin type" sedimentary-hosted gold deposit. The old timers could not find it because the gold was too small to see - it couldn't be recovered by crushing the ore to a powder and panning to see the result. This unusual discovery has made Nevada one of the richest gold bearing regions on earth. The Carlin trend district has produced more than 60 million ounces of gold since its discovery in 1961, making it the most productive gold district in the United States. The first samples of low-grade gold ore were taken from the Carlin deposit by an individual prospector by the name of Frank Mahoney in the 1950's. Unfortunately, Mr. Mahoney did not recognize the potential of his find, but he showed some geologists what he had found, and they did understand the significance of his discovery. A short distance to the south, on the nearby Rain section of the Carlin trend, located about 8 miles southeast of the town of Carlin, favorable outcroppings were initially recognized and staked by Price Montrose, a local independent prospector. In 1979, Mr. Montrose recognized a geologic similarity between that area and the main Carlin mining areas, which are northwest of town. Subsequent discoveries on the Rain property have produced over one million ounces of gold.

In the early 1980's FMC mining discovered a rich gold deposit at Paradise Peak near Gabbs in central Nevada. The ore body contained over a million ounces of gold and nearly 40 million ounces of silver. The FMC geologists were initially led to the area by the husband-and-wife prospecting team of Bill and Ardith Arnell. The Arnells had discovered low grade gold mineralization in a wide spread area of altered volcanic rocks. Further work in that area lead to the discovery of the rich Paradise Peak ore body.

So, even though most of the latest commercial discoveries are being made by professional geologists, a significant number of finds are still being made by independent prospectors, even among the large and rich hard rock discoveries of Nevada. In spite of this, most individual prospectors focus all of their energy on finding small gold flakes and nuggets, called placer gold. Even so, there is significant potential in seeking and discovering gold found in hard rock veins and similar types of "in place" gold deposits as well. These modern day individual prospectors are also finding some very impressive gold nuggets from time to time.

Prospecting can be a fun family activity. Here the author and his young nephew do a little dredging for gold in Northern California.

I won't give the details, but roughly about a decade ago, a prospector was hunting for gold with a metal detector in the Mother lode country of California. He got a loud signal that he thought was a big piece of junk - perhaps a horseshoe or a tin can. But he'd found some decent gold nearby, so he figured he'd dig it because big and loud trash can hide smaller bits of gold in the same area. When he got that target out of the hole, it was a 75-ounce nugget, the largest found in California in several decades. There is still very nice gold to be found by those who keep their eyes open and search for it. So, I'll give you the same admonition, when you are out in the gold country, keep your eyes open!

The gold prospecting world is basically divided into two halves. They are placer gold and hard rock gold. Hard rock gold is mineral which remains in the original host rock in which it formed. Northern Nevada is rich in gold, mostly as these primary hard rock type deposits. The hard rock, open pit mines of Nevada have produced nearly 100 million ounces since large scale open pit mining became prominent in 1960s. Although a few small operations still exist, hard rock mining is usually done on a large scale. The main problem for individuals interested in hard rock gold deposits is high capital costs for the equipment to crush and process hard rock ore in order to extract the gold from its solid rock enclosure.

Where water was available, early day miners used high pressure water to wash gravels into their sluice boxes. This was a low-cost method to recover the gold, but it did a lot of environmental damage. This process was known as hydraulic mining.

Because of this, the prospectors who look for hard rock gold often seek to sell their finds to large companies that possess the resources to develop them. I'll talk a lot more about that in a later chapter of this book.

Any gold that has weathered out of its original rock matrix, be it a quartz vein or another source, is called placer gold. Once it is freed from the rock, any accumulation of that free gold is called a placer deposit. This includes things such as gold flakes and nuggets, and this is the type of gold which is of the greatest interest to most individual prospectors. There are several different kinds of placers depending on how far the gold traveled, its origin, etc. The four most common types of placer deposits are: 1) **Residual** - where the original vein has weathered, but the placer gold remains more or less "in place" and still within a few feet of the original source; 2) **Eluvial** - where the gold has traveled a short distance down from the source, but has not made it into streams and other drainages - these are often called hillside placers; 3) **Alluvial** - Where the gold has made it into area streams and rivers. These placers are sorted by running water and usually the gold lies mostly on or near the bedrock; 4) **Beach** placers occur where small gold particles make it all the way down river to the ocean. Wave action can concentrate the heavier fraction of the sand, producing black sand layers containing fine gold.

Depending on the mining region being considered, the portion of the gold mined as placer vs. gold in the solid hard rock lode varies quite a bit. In California, over 64% of the gold produced has come from placer deposits, while in Alaska the proportion of placer was a much higher percentage. However, in Nevada and Colorado, most of the gold has come from deposits of gold where the metal is still within the matrix of its original hard rock formations. Initially, the rich, easily discovered, surface and river placers were mined throughout the western US by individual miners using small scale methods like panning and sluicing from the 1850s through the 1860s. Discoveries in Australia began in about that same time frame. While we no longer have the huge number of gold "rushes" we experienced across the globe in the second half of the 1800s, there are still occasional situations where thousands rush to an area because of new gold finds. Such floods of miners have occurred in recent decades in remote locations of places such as Brazil and Mongolia.

Where water was available, the early miners developed large hydraulic mines, using powerful water cannons to wash down whole hillsides. Placer miners have also gone underground; drift mining for placer gold deposits in the deep lead gravels of ancient channels and bench deposits.

Light colored cliffs like these mark the site of an old hydraulic mine. Old mining sites such as this one still yield gold to modern prospectors.

Because of the comparative ease of recovering gold from placer deposits, most individual prospectors start out seeking placer gold nuggets and flakes. Some later progress to an interest in hard rock deposits, but nearly all still start out looking for flakes and nuggets of free placer gold. Hard rock deposits usually require blasting to excavate the ore, crushing and often the step of chemical leaching to extract it from the rock. A lone prospector with a gold pan can quickly verify the presence of placer gold in a gravel deposit in a matter of minutes. Small placers are also relatively easy to mine, as the gravels usually require far less processing than ore from lode mines. Once you find your first gold, you won't have much trouble seeing the excitement that kept those old pioneer prospectors going under such rugged conditions. Additionally, the modern prospector has many equipment options and possibilities for finding and recovering gold that were simply not available to the old time prospector of the last century.

It's always great when you come up with your own gold, and the excitement is real. There is no doubt in my mind that gold fever is a condition that actually exists. In my experience, staring too closely at gold nuggets or thinking too much about the quest to find them often causes it. Luckily, it's an enjoyable condition with few, if any, harmful side effects. No doubt prospecting for gold is a hobby that's easy to fall into.

Gravity based separation methods are still the most widely used recovery methods to extract placer gold and the basis of most placer mining equipment. Gravity methods are all based on the fact that gold is dense and heavy, far denser than the rocks, sands and gravels associated with it in the typical placer deposit. Gravity recovery equipment, including things like gold pans, sluice boxes, suction dredges, and more commercial equipment such as shaking tables, jigs, and others, have been used since the times of the great 1800s gold rushes, and many present day operations still employ the same types of equipment. The major flaw of the gravity separation method is that a significant percentage of the very fine gold, often referred to as flour, flood, or gold dust, is lost in processing. Early miners sometimes recovered no more than 60% of assayed gold values when large amounts of fines were present. Because of this, many of the remaining placer deposits left by the old timers have a higher percentage of fine gold than the placers worked during the early days.

With the current high price of gold, it is understandable, then, that today more care is given to the recovery of that fine gold. In recent times a number of changes and new designs in gravity separation equipment have been developed. Today, many types of equipment exist for the efficient recovery of fine placer gold. On the other hand, for many small scale miners, getting the very last little bit of gold dust can be costly and time consuming to the point that it is not really worth pursuing. The recovery of placer gold in most types of gravity based separation operations involves a series of steps. The first step is concentration where the valuable material is separated from the large bulk of valueless waste. The material from this bulk concentration process, whether it is sluicing, dredging, dry washing or something else, is cleaned and further concentrated and eventually smelted or otherwise refined into a final bullion product.

Early day miners worked with simple tools, like picks, pans and shovels. They learned about finding gold as they went along. While today's prospectors have many options, it is still often best to start out small and grow as you learn.

Many types of efficient placer gold recovery equipment are available to the individual today. For the small operator, the choice is usually determined based on the environment he desires to prospect in, as well as his spending budget. In desert environments, dry washers and metal detectors are the chief choices. Other water conserving devices may also be a possibility. For those with access to water that is close to their placer deposits, options like sluice boxes and small scale suction dredges can also be considered. The versatile gold pan gets used in almost all environments, although in dry environments one has to bring water and the tub to hold it in order to use the pan effectively. We will take a look at all of these devices and how they are used by the individual prospector and miner to recover gold flakes, dust and nuggets.

So in the deserts, on the hillsides and in the mountain streams, there is still gold to be found in many places. The old timers didn't get it all, in many places all they did was skim off the easy rich deposits before they moved on looking for greener pastures. Every year there are still great finds being made, mostly in the western states of the USA and various parts of Australia (but sometimes in other locations as well). Even large nuggets over an ounce in size are still being recovered - they are not all that common, but they are still found. Flakes and small chunks of gold are far more common finds. It's not easy, and not every day will be a home run, especially for the beginner. Patience is required and some days just have to be chalked up to the quest to gain knowledge and experience. Even for most of the experienced participants it's a pastime, and you need to consider the benefits of other things like good exercise, getting away in the great outdoors along with the value of the gold recovered. Very few individuals find enough gold to make a difference in their lifestyle.

It doesn't necessarily cost a mint to get into prospecting. It can be as simple as purchasing a gold pan for $10 and grabbing a bucket and the garden spade from the garage. On the other hand, a top-of-the-line metal detector designed to find gold may cost over $5000. A large suction dredge may cost up to $10,000. Other prospecting equipment covers all the levels in between.

The author is shown panning for gold on the North Fork of the Yuba River in California. The Yuba River has produced some exceptional large gold nuggets, and several clubs have claims on this river to provide their members with access to this historically productive mining area.

It is not necessary for the new prospector to start with the most expensive items of equipment. It is both sensible and reasonable to start off small, purchase some books to learn about how to prospect and have your equipment grow with your knowledge and abilities. So my suggestion is to give it a try, start out small and see how you like it. Learn about the various types of mining equipment that are available on the market. It's highly unlikely you will strike it rich, but if you work at it, you will find some gold and have a lot of fun along the way.

Some types of equipment on the market allow the modern prospector to accomplish things no old timer could ever dream of. From metal detectors, to portable suction dredges, to dry placer machines and other gold recovery devices of all types, many significant improvements have been made in small-scale prospecting and mining equipment. There certainly is no problem finding ways to spend as much money on good equipment as you would like - lots of great stuff is available. Most individuals start off small and purchase more advanced equipment as they get more involved in the hobby.

Club meetings and other gatherings such as the AZO outing at Rich Hill pictured here, represent a significant opportunity to learn more, meet new friends and share experiences.

Although it's said that "even a blind squirrel trips over an acorn once in a while", there is some knowledge required to be regularly successful at prospecting. The individual well-armed with a thorough knowledge of prospecting techniques, gold geology and placer deposits will tend to find more gold over the long haul. Learning how and where to find gold is an educational experience that really never ends. There is always more to learn about gold deposits, and even the most experienced prospectors find there is always some new and useful information to be gained. No doubt that it does take some time and effort to "learn the ropes", but there are many excellent "how to" books available on operating various types of gold recovery equipment, including the gold pan, suction dredges, dry washers, metal detectors, and others. There are also many "gold location" books that tell where gold has been found and recovered in the past and what the deposits look like in those districts. A good basic college geology textbook can be helpful in understanding the geologic conditions a prospector may find in the field. YouTube, on my channel "Chris Ralph, Professional Prospector," I publish many good basic reference videos on a variety of prospecting related topics. Books and magazines offer a considerable amount of useful reference information. Following me on YouTube will allow you to keep up to date on all that is going on in the prospecting world.

Another way to learn more about prospecting is to join one of the many clubs dedicated to the individual prospector. A few have been around for a while, but over the last 10 years quite a number have sprung up around the country. This one fact alone has had a profound effect on spreading interest in prospecting and brought hundreds of new folks into the fold. A number of prospecting related clubs have their own websites and can be found searching with Google on the internet. One of the most pressing problems that commonly presents itself to new prospectors is finding a place to prospect. Most clubs own mining claims, which solves that problem for their club members. They range in size from a very large, nationwide group like the Gold Prospectors Association of America (GPAA), to regionally oriented clubs that focus only on a certain part of the country, like the Comstock Gold Prospectors Club in the Reno, Nevada area. Clubs provide their members not only a place to prospect but also a great opportunity to meet other prospectors. Many clubs make a special effort to reach out to new prospectors and help them learn the basics of finding gold. Group outings by the club can provide the new prospector a chance to see various types of prospecting equipment in action.

The earliest prospectors and miners always had their pans close by. Was the gravel at a certain place rich? The gold pan could answer that question with a quick and easy test.

One of the most innovative changes now available to the individual prospector is the wealth of information available over the Internet. If you already have a computer, it's a great way to learn more about gold mining. A number of web sites and forums dedicated to various aspects of prospecting for gold have sprung up in recent years. Most of these web sites fall into two broad categories - web pages and forums. The web pages are sites with specific information and reports on the experiences of others, sources for prospecting research, and pages for manufacturers that have useful equipment information. There are even sites with topographic maps and aerial photos that may be useful in researching places to prospect. The second source is forums where questions can be asked or topics discussed - most are open to anyone who is interested. These forums offer a very unique chance for prospectors worldwide to converse about their favorite topic. From Australia and New Zealand, to Alaska and the western states of the "lower 48", prospectors from around the globe can now share their insight and knowledge on finding the elusive yellow metal. Many of the web and forum sites focus on prospecting in certain regions like the Arizona desert or the Rocky Mountain region - you may be able to find one that specializes in your area. Other websites specialize their focus in a certain type of prospecting, putting most of their attention on prospecting by metal detecting, and others giving more of their effort on water related methods like panning and dredging. I have a list of them in the appendix, but a simple way to find the URL address locations for these sites is simply by using a search engine like Google. Many of the sites have links to other prospecting sites, so explore around and take a look. There is a tremendous amount of useful information available. This is a great way to spend time during the winter months when the weather doesn't allow prospecting in the field.

So, whether it's searching for the next million ounce ore deposit or just finding a small gold nugget you can call your own, rest assured, it is still possible. For those who enjoy hunting, hiking, fishing, off road exploring or any of the multitude of outdoor hobbies so many folks participate in, prospecting may be something you would be interested in. For just about any outdoor enthusiast, it's worthwhile to know a little about gold deposits - because the next big find may be yours!

In some places, historic old mines still yield beautiful gold specimens to individual prospectors. However, interesting they may seem, these sites can be dangerous and may be located on private property. Obtain permission where needed, stay out of underground workings and take all due safety precautions. This particular mine is located at Goldfield, Nevada, site of a historic and productive gold rush in the early years of the 1900s. The district produced over four million ounces of gold, essentially all if it from underground mines working hard rock lode deposits.

BEGINNER'S CORNER: BASIC QUESTIONS AND ANSWERS

Out in the western gold country, the traffic of new gold prospectors up in the hills appears to be increasing at a rapid rate. While a lot of different factors are contributing to this, probably the biggest force attracting these new folks is the continuing increase in the price of gold. Many experts think inflation is around the corner and gold is headed higher for the long term. The number of new prospectors out in the field will surely reflect this increase in the coming years.

While this book is written for both beginners and more experienced hands, at this point I thought I'd take a few pages to just go over some of the basic questions beginners often ask and try to provide some useful information before we get on with building foundations toward some more advanced topics. Here are some of those common questions:

1. Can I really find gold for myself?

The simple answer is yes. While there is no doubt that the biggest richest placer districts were found by the old prospectors in the 1800s, they missed some significant deposits. Not many folks know that some important placer districts in Nevada (Rye Patch, Sawtooth) and Arizona (Gold Basin, Lost Basin) were not discovered until the 1930s when a new boom of prospectors spread across the west spurred on by the joblessness of the great depression, and an increase in the price of gold to $35 per ounce. In many locations, the old time prospectors took what they could get easily and moved on. Not enough attention has been paid to smaller deposits or deposits on the fringe of well-known districts and these can present a significant opportunity for the modern day prospector. Many times, if the old timers were not able to operate a profitable mine or make a living from their diggings at $20 dollars per ounce, they moved on. They left a lot of gold bearing material behind. With a few exceptions, we are not trying to make a living off our findings, and so our actual take per day is not as critical.

Finds like this quarter ounce of Arizona gold nuggets are still made regularly in the Western US by experienced prospectors.

In addition, we have new equipment that the old timers could never dream of. These include things like portable suction dredges (with Hookah diving gear) and metal detectors that allow us to find and recover gold the early day pioneers could not easily reach. This book will take a good look at that equipment and how it can be used to its best advantage. We'll talk more about this, but while your chances of striking it rich are poor, your chances of finding some good gold are excellent. The bulk of this book will be focused on helping you find those places with good, recoverable gold.

2. How will I be sure that my finds really are gold?

One of the questions most commonly asked by new prospectors is how to recognize real gold nuggets. Some of the other minerals found in streams, in the deserts and around old mines, can initially confuse those who have not recovered gold before. Fortunately, gold is usually fairly easy to identify once you have seen and handled a good amount of it. Few minerals have as intense a metallic luster, or form in the same types of shapes. It is heavy and soft, and deforms like a soft metal when struck or cut. The two minerals that are most commonly confused with gold by beginners are pyrite and mica. Mica breaks into

Small cubic crystals covering this brassy colored specimen indicate it is made of the mineral pyrite, often called "fools gold".

sparkling little sheets when rubbed between the fingers and just does not have a true metallic luster. Mica is also fairly light - it does not have the density of gold, and it will not be found in the heavy concentrates when panning or dredging. Pyrite, when fresh and unoxidized, does have a nice metallic gold color. Although it's not as heavy as gold, it is heavier than most other rock forming minerals, so that it will be found in the heavy concentrates. That is how it got tagged with the name of "fool's gold". It often forms angular, cubic shaped crystals, which are easily recognizable. Luckily, pyrite also rapidly oxidizes in the air and the surface changes to a rust brown mineral, usually limonite. In this form it is not often mistaken for gold. I always tell folks that once you've seen enough real gold, it's hard to mistake mica or pyrite for the real thing. If you have a friend with prospecting experience, ask him. My YouTube channel (**Chris Ralph, Professional Prospector**) has videos on how to identify natural gold, so you can take a look there too. On unglazed porcelain tile, pyrite leaves a dark streak, while gold leaves a light gold one. If you are still not sure, the last test is to hit the suspected gold with a hammer. Pyrite shatters because it is brittle, Mica breaks into sparking little sheets looking something like glitter, but gold will smash down and flatten out in a manner similar to lead. Try not to do the hammer test - it ruins the nugget value of the piece. It's still valuable as gold but it's no longer a pretty nugget.

At times, some prospectors may find a silvery gray, heavy substance in with their concentrates. If it is in the form of small silvery liquid globs - it is mercury. If it is a slushy liquid-solid mix, it is most likely amalgam, a mixture of mercury and gold. Old timers used mercury to help capture their gold, and small amounts sometimes spilled out. Silver-gray solids are usually lead, but on rare occasions they can be platinum nuggets; they are found along with gold in some placers. While it may be hard to acquire, dilute nitric acid will dissolve lead, while platinum will not dissolve. Nitric also dissolves mercury away from gold.

3. Do I need to spend a mint buying expensive equipment?

Well, you can spend just about as much or as little as you want - remember that the bottom line goal in all of this process, beyond finding some gold, is to have fun. You can get a decent plastic pan for around $10, and with a screwdriver, a bucket, and other tools from home, you are ready to begin your quest for gold. On the other hand, a top of the line metal detector with a few normal accessories can be over $5000 and a large dredge with diving equipment can cost far more. There are also lots of possible alternatives between these two extremes.

An inexpensive pan, rock pick and small shovel are really all you need to get started in the world of gold prospecting.

It is normal that new prospectors start slow and buy a few inexpensive items. Try it out and see if you enjoy it. There are even some tools and equipment that it's possible to make for yourself if you have fabrication and welding experience. For the beginner, a gold pan and a sluice box is a good level to start at if you are working streams and rivers. This kind of starter kit can be put together for less than $200. Don't start by spending a mint on equipment even if you can afford it. Buy a gold pan and perhaps a sluice, then spend the money to join a club and learn about prospecting. Even if you are in the desert you can still pan for gold. Dig and screen the gravel and then pan the material out in a tub with water you have brought to the site. Start out small and build up - many long time prospectors acquire quite a load of various types of prospecting equipment (you should take a look at my garage!). Find out what types of prospecting you enjoy first. If you take your time, you can learn what works in your area before you spend too much of your hard-earned money.

Electronic prospecting is one of the most difficult forms of prospecting to master, and one of the most expensive when you buy all your equipment from the top makers. The author believes it's worth taking some time and learning about the concepts of prospecting and trying it out before you go out to make a very expensive purchase.

Finally, there are some types of equipment one can build at home. Simple sluice boxes come to my mind, but I have seen well-built high bankers, dry washers and even dredges. If you are going to attempt something like this, you ought to have good welding and metal fabrication skills, access to the necessary equipment, and a very good understanding of how the device works. This is because I have also seen homemade equipment doomed to fail because the maker either cut corners or did not fully understand the nature of the device. Examining closely the commercial units that are for sale can tell you lots about how you ought to construct any homemade prospecting equipment.

4. I've never prospected for gold before, but I want to learn. What should I do?

Well, you made a good start in buying this book because that's just what it is all about! The first thing you can do is read this book - it has a lot of useful information, and it will help you. Prospecting is, in a lot of ways, like a trade. There is a lot to learn and it takes time to learn what to do in different situations, how to read a river, what to see in the quartz on the surface of the desert floor, the color of the soil or a dozen other indicators that might point out the best places to start looking for gold. There are procedures to learn which are time-tested to yield good results. The more you know and the more experience you have, in general, the more gold you will recover, and like a trade, it is a skill that takes time and effort to learn.

As a beginner, make plans to attend club meetings, then go out on club field trips and learn how to find gold and prospect. Talk to the local guys in the club and see what special tricks they know about finding gold in your area. While reading books, searching the Internet and other forms of armchair learning are great, nothing substitutes for getting out in the field and really doing it. Spend some time in the field and see how you like prospecting, then if you are still gung-ho about it you will have learned lots more in the meantime and be able to make a well-informed decision on what kinds of equipment you might want to buy.

5. I've been out and poked around and I didn't find any gold (or not much). What should I do?

That's pretty much what happened to me when I first started prospecting. I bought a pan and some books on where to find gold in southern California (which was where I lived in those days), I built myself a sluice box out of wood, and read some other related books from the library. I went out to a place the "Where to Find Gold" book said gold had been found. I spent a day digging around and ran some material through my sluice box. I found nothing - not a single speck. I was unwilling to give up, so I tried a couple other places mentioned in the book, but still not a bit of gold. It wasn't until a few months later, when I took a trip with my family up to the mother lode country on the North Fork of the Yuba River, that I found a little 3/16 inch flake in my pan. It was not very large, but plenty large enough to see it was clearly gold. After that, I learned more about gold and kept exploring different locations closer to my home in the LA area and ended up doing better, finding spots where I always got at least some color and sometimes a few very small nuggets (little pickers). One of the places at which I had initially failed to find even a single speck early on, later yielded pretty well as I learned more about how to prospect and to be more selective about the exact spots I chose for digging. The bottom line is: Don't give up - prospecting is a skill that takes time to learn. Successful prospectors have a lot of tenacity and the commitment to stick with it.

Within a couple years I bought a small puffer type dry washer and a 3-inch dredge (both from Keene Engineering) and I started processing more material and getting a little more gold. In those days (30 years ago) metal detectors were not nearly as powerful as the current models, and could not do what they do now in finding nuggets, so I did not own one. Joining a prospecting club is a great idea, especially for a beginner - you will learn loads. The more you learn, in general the better you will do. Reading books like this one is another great way to learn and gain knowledge.

6. Where should I go to prospect for gold?

This is not just a beginner's question - everyone who is looking for gold asks this of himself or herself. Even experienced prospectors are always on the lookout for better places where they might be able to get more gold. Much of the answer to this question depends on where you live. In the USA, the western states are a much better place to look. Though there are places in the east where gold was mined, the private property closed off to the public and other issues limit prospecting opportunities for those folks living east of the Rockies. While there are still some decent locations in the eastern US, most are "pay to prospect" fee locations or spots operated by clubs.

A good idea for starting on this is to buy some books describing where gold has been found in your area. They can help you get information about specific locations. While this book has only a limited amount of information on deposits and locations within specific states, it does have some maps and a lot of information about how you would go about researching potential locations to explore and prospect around. It usually takes some significant time and exploration in the field to ferret out good spots to work.

This can be a big problem for a new guy from out of state who wants to travel and prospect while on vacation. If you live in the area, then you have time such that you can go round and try different areas until you find someplace good. It may take a while - like months of weekends. On the other hand, if you are from out of state, and only have a week or ten days, then you can't afford to just bum around and explore until you find a good spot. Doing your own research to find places with gold is a whole art in and of itself.

The fast answer is to join one or two prospecting clubs in the area you are interested. The cost is comparatively inexpensive, and then you'll have ground to prospect that does have gold. You still will need to do hunting around on the claims, but at least you'll have a place to start and perhaps a few club compadres that can help answer a few questions for you.

7. It seems like there are claims everywhere in the gold fields. How can I find where the claims cover and where there are no claims?

US law allows folks to stake mineral claims on certain types of government land. The claim owner owns all rights to any gold and other valuable minerals there. There are literally millions of valid claims and finding out where one can and cannot legally prospect can sometimes be a problem that takes a bit of research to solve. In the Appendix of this book there is a good summary of mining claim law, but for now I'll give a brief answer to this question because it's one you will hear often.

Some claims are well marked in the field, but actually it's just as important to stay off claims that appear unmarked as well. Any valid claim is one you should stay off - and that's the hard part. The problem is that people steal or remove claim markers all the time (even though its illegal to do so). Sometimes, they just get knocked down in the wind and rain. Either way they become no longer easily visible. In forests, it can be really hard to find markers for all the trees and brush. So even if no markers are visible, the claim may still be legally valid. How does one determine which claim is a valid claim and what lands does it cover? Heck, there are whole books written on answering the details of that question! That's another good reason why I recommended new prospectors join a club - to give them a chance to learn to walk before they run. Claims are located using the federal public land survey (PLSS), which marks out land based on sections, townships and ranges (more on this in the mining claim portion of the Appendix). The County recorder's office where the claim may be located keeps records, but Counties don't distribute maps showing the locations of all valid mining claims - and neither does the federal government. There is no such thing because the status is constantly changing.

In 2021, the BLM came out with a new mineral and mining claim records system intended to update and replace the old LR2000 system they had used for years. The new system is called the MLRS (Mineral and Land Records System). If you are going out to prospect, it's good to know the land status of the location where you are searching.

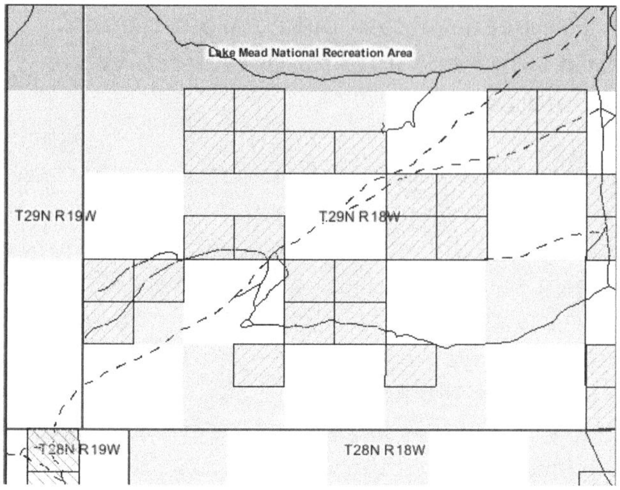

This is an example of one of the land status products which can be generated with the BLM's internet based MLRS system. It shows a portion of the Gold Basin Placer District in Arizona. Hashed portions are existing claims.

The website portal can be found at https://mlrs.blm.gov/s/ or you can just Google MLRS. The new system has a map-based research tool. If you are logged in or not, you can click on the research tab on the MLRS home page to access the map. You then zoom in to the areas you are interested in. Claims are not put onto the MLRS system for a month or more after they are submitted; so, loads of perfectly valid claims just don't show - at least for a while. So, if an area seems open and unclaimed on the MLRS, there is still a chance it is claimed. The claims viewer research tool can also be set up to display the grid of the Public Land Survey, and the underlying layers can be changed to topographic maps, satellite images or many other useful map bases. This map-based viewer is great for rough checking of an area for claims. It's a high-speed way to check over an area to see where any current or closed claims might be located. I highly recommend it. I also recommend generating a written report on the township you are interested in. They have additional detail that I think is easier to review in report format than on the maps. The map will show you that there is at least one claim in a quarter section, but there may be one or many. Another possibility is to visit your state's main BLM office, they keep records there and you can request a look at the maps in the claim files for the claims which are located in your state.

8. Can I make a living prospecting for gold?

Don't let the high value of an ounce of gold mess with your head. Lots of guys get attracted to the concept of being self-employed out in the wilds, making a living off the land, but it's not really something a new hand should consider. Even at $3000 an ounce it's really tough to score an average over that long haul that pays for bean and gas money, let alone all the other things we need to survive. A lot of the guys who prospect regularly have other jobs that allow this or they are retired and have enough savings and pension to cover their needs so that the gold is just extra. Most of the small group who do live only on prospecting, do it by living in the field with a very low cost, hand to mouth kind of existence. Although they may have some very productive days, their wages are usually less than minimum wage overall. A very few experienced operators are able to make a reasonable living, but this takes lots of skill (and certainly not the skill level of a brand-new prospector).

This 156 ouncer is perhaps the largest all-gold nugget recovered with a metal detector in the "lower 48". It is a very rare and unusual find.

Although some days are more productive and some are less, a typical average take for an experienced prospector in the US might be around 2 grams per day over the long term. Even at $3,000 per ounce that is less than $200. Yes, by camping outdoors and watching every dollar one can eek out a survival on that kind of money - but it is certainly not a life of luxury. Bad weather and other problems mean no gold and no income. Inexperienced new prospectors should expect far less, perhaps less than half a gram per day average over the long haul (roughly $50/day at $3,000/oz). The professional living off his finds must constantly go out and find new places to work - all locations eventually become worked out and yield so poorly that one must move on. What began as an exciting odyssey questing for free gold becomes in time a more mundane grind - a hunt for gold to survive. The gold you find cannot be saved as a trophy; it must be sold quickly to pay the bills - sometimes it must be sold so quickly that you cannot wait to get top dollar for it.

If you are a new prospector, don't quit your day job. Learn to walk before you try to run - try prospecting on weekends and see how you do. Don't count your chickens before they hatch - I cannot tell you how many newbies that I have seen quit their jobs go out to try prospecting for a living that have failed. It is a huge and important thing to note that the vast majority of the guys saying they are going to go out and live off the land prospecting are folks with virtually no prospecting experience whatsoever (the flip side of which is that the guys who know what prospecting is about and how it works know better). Remember that 99% of the old timers with their trusty mules lived a rough hand to mouth existence, barely making enough gold to survive. The ones who discovered a new spot or hit a big bonanza were very much the exception. This is just as true today (in fact, big finds are even rarer today). When you start prospecting, keep a tally of what you get on your weekends and vacation time - figure that as dollars and see if it's a wage you would want to live on, especially considering that there are bad weather periods and other situations when one simply cannot be out prospecting. On the other hand, there are many retired folks who spend a few weeks to several months and even more prospecting each year and thereby successfully supplement their income. They have other sources of income so the gold is not as crucial and they can wait and get a good price for their finds.

Part II- Modern Gold Mining Equipment

Tools Of The Modern Prospector

The image of the old-time prospector, leading his trusty mule loaded down with all his food and tools, is indelibly etched in the minds of many people. Movies and TV have done their full share to reinforce this historic stereotype. However, things have changed in the last century, and the modern prospector has a whole lot of options in the way of tools and equipment that the old timer could never have even imagined. From RVs and ATVs to metal detectors and so many others, there is no question that modern prospector has it pretty good by comparison. In this section of the book, we are going to take a look at the equipment the modern prospector uses to find gold, chiefly placer gold. This includes the gold pan, the sluice, suction dredges, high bankers, dry washers, metal detectors and even a few others. First we will take a quick look at some of the ancillary equipment that many prospectors use, and then focus a lot more time and attention on the main types of placer gold mining equipment used by modern prospectors. So to start, let's consider a few handy accessory items that most prospectors own and use for good reasons. Now these items are not strictly required to find gold, but if you want to get serious about your prospecting, they are handy to have. These subjects could each be their own full book topic, but we will have space to only take a brief look just to consider the basics.

RVs, ATVs, Quads and Dirt Bikes

Because worthwhile gold deposits usually lie out away from the city, the old-time prospector hiked or rode a horse, and slept on a blanket out under the stars. While that may be picturesque, things have changed - I drive a big suburban and haul a trailer full of gear. Now whole books have been written about selecting RV's and ATVs, but for our purposes we will just make a very cursory consideration of them, because while many prospectors own them, they are not the main topic I want to cover. Owning a nice RV is actually very handy, because when you are working hard at your prospecting - either digging out a deposit or searching for a new one, it's very nice to have a warm, soft bed to sleep on. I used to do a lot of tent camping, but no more. Call me an old cream puff if you will, but I won't give up the comforts of my trailer. Prospecting is hard work and takes its toll on your body. Getting a good night's sleep in the evening is important. Feeling good and well rested often means you get more gold the next day. Especially on those longer trips, you have to take care of yourself or you will get hurt, tired and burned out. As a result, you will be quicker to give up, quit and head home - yet we know that hanging in there and persevering is one of the keys to getting good gold.

The author's suburban and travel trailer here comfortably set up at a campground in the heart of the mother lode gold country of California.

Some pretty fancy campers and trailers are available out there. This camper features pull out sections which expand it in two directions.

My trailer has also kept me dry when the weather turned stormy, and warm when it has turned cold and windy. It's an older unit, so I don't feel bad when I haul it out to the middle of nowhere on dirt roads. While sometimes I can haul it right out to the mine, more often I set it up at a central location in or near the mining region and work outward from there. Another great advantage of the trailer is that it can haul a lot of the basic stuff I take on trips. In the summer time, my trailer is always loaded up and ready to go - all I need to do is put in my groceries for the trip, charge up the batteries, top off the water and I am ready to hit the hills. There are all sorts of RVs on the market, from big motor homes that are too large to fit in some campgrounds, to tent trailers and even campers that fit into the bed of a pickup. Toy haulers are a special type of trailer set up to haul all the gear and equipment you need when out prospecting, while still offering a nice living space. Each type has its own purpose, but all are handy to have, and a large number of prospectors own an RV of one type or another.

While there may be a few who still use horses, the ATV, also known as the quad, and the dirt bike type motorcycle have given the modern prospector great access into the field. Even a good 4WD sport utility vehicle like my suburban can only go so far out into the hills. In addition, most of the old mining roads in both the desert and mountains were constructed many decades ago (in some cases well over a century) and they are not maintained - many are no longer passable to even a good 4 WD off road vehicle. All-terrain vehicles can get you in there where a car cannot. While one can always hike in to remote spots, the motorized vehicle will get you in quicker and allow you to cover more ground. Most folks will only hike so far, and these vehicles can get you into prospect at places few others would attempt. More and more it's those remote locations where the best finds are being made, so there is a real advantage to these vehicles in some situations. Most of these ATVs get far better gas mileage than my suburban does, and with high gas prices being the way they are, that's a worthwhile consideration as well.

Quads are available in sizes suitable for all members of the family, from Dad all the way down to the younger kids. They can make a big difference in providing access to the backcountry.

Quads have a much greater carrying capacity as far as additional gear and equipment, but a trailer or open full sized pick up bed is needed to move it around. Many operators set up their quads with various equipment boxes to carry their gear - and the quads have a good capacity for that. Bikes are easier to transport, but can carry far less when it comes to equipment. There is a balance and choice to be made between the two. A few manufacturers used to make 3-wheeled all-terrain vehicles, but they tend to tip when they are not on flat surfaces, and new units of this type are no longer made. For the prospector thinking about purchasing an RV, a toy hauler type trailer that can provide a nice living space and still allow plenty of room for an

ATV quad or a dirt bike might well be a good option.

This toy hauler has a drive up ramp that allows the prospector to simply drive his off road vehicle right inside. The deck above the vehicle storage area drops down and makes into a bed.

GPS Hand Held Locators

One of the most hi-tech items available to the modern prospector is the Global Positioning System device, better known as the GPS. While some larger devices that are designed to be used in a car can be purchased, smaller hand held GPS devices intended for hikers and other outdoor pursuits are the type prospectors are most interested in. They are comparatively inexpensive and can be picked up at "Big Box" sporting goods stores and a variety of other outdoor equipment retailers. These smaller, hand held units, are lightweight and easy to carry around.

The GPS device works by receiving a signal from a series of orbiting satellites that are constantly revolving slowly around the Earth. Our hand held devices are receivers, and they receive information from the chosen satellites and use radio waves to calculate their exact location. The receiver simply notes the time a signal was received by a satellite in comparison to the time it was sent from a ground-based station. Although the radio waves travel at the speed of light, the small difference in time that is measured allows the receiver to determine how far away the satellite is in space. A series of these satellites provide additional distance measurements, so by comparing and averaging all of the results, the exact position of the GPS device is then geometrically calculated and displayed to the user. The concept and technology that is used in Global Positioning Satellite systems was originally developed by the US Department of Defense for military purposes, but it has now been made available to the general public. They are now employed for a variety of uses in many types of industries.

The GPS provides the prospector a precise way of knowing just where something (like a mine) is located. There are many thousands of lost mine stories out there, some real, and some made up, but the use of a GPS could well have made the difference in locating and not losing them (at least for the real ones). Unfortunately, GPS technology was simply not available in the 1800s, but here in the present they are pretty handy for modern prospectors.

Perhaps a little example story could tell about the usefulness of such a device. The first hand held GPS I purchased almost 10 years ago made a big difference its first time out in the field. There was a little old mining prospect I was looking for, and I had been out in the field unsuccessfully once before trying to find it. I had a topographic map that showed its location, but there were no nearby roads and the area was heavily forested. You could easily come within one hundred feet of this little mine and never see it for all the trees and brush. In addition, once you were into the dense forest, it was difficult to see the nearby mountains and other landmarks that one would normally use as a guide. The hand held GPS seemed to offer just the right solution, but it was a technology I had never used before.

I was able to determine the GPS coordinates of the little mine from a topographic map, and armed with my new GPS, a compass and my map, I was determined to locate this deposit. I set the first coordinate for my hike where I parked the car as a test to see if my GPS device was working properly, and when I checked the results against the map, the numbers seemed right to me. I set off into the forest. I hiked for over an hour, over hills, up some small ravines and through the forest and brush. I followed the indicated directions with my compass and found the little prospect, just where the GPS device said it was supposed to be. I was amazed that I had found this mine, because I could see that there was no way I would have ever found it without the GPS. I had only found it because the device had indicated the coordinates where it was located and the directions I needed to travel in order to get to the mine.

I sampled and explored the area and eventually it was time to return to the car. I was deep in the forest and could not see the landmarks and mountains around me - I could not see the faintest indication of where my car was. However, I had taken a GPS reading at the car, and I figured the GPS could lead me back to the car just as easily as it had taken me to the little prospect I was looking for. I headed off, again following the GPS and my compass, and although I was not going exactly the way I had come in, I could still see by the sun's angle that I was going in generally the right direction toward the car (remember I had explored around the general area, so I did not start my return trip from the mine itself).

Eventually, after a long hike back, the GPS said the car was only a couple hundred feet ahead. Rather than make me feel better, this concerned me because while the GPS said it was only a couple hundred feet away, I couldn't see the slightest sign of the car or even the paved road it was parked along, and I thought I should have seen at least some sign of it through the brush. Additionally, there was not a sound of any traffic from the road, either - and from a couple hundred feet away in a quiet forest, I should be able to hear the other cars easily. Was the GPS going to fail me? Had I made a mistake in how I was using the GPS? I hiked on in the direction indicated, because I really didn't have any better information or any other good alternatives. After pushing through some scrub brush and trees for almost 200 feet, I stepped into a clearing and looked up. There was my car, only about 30 feet in front of me, parked right beside the road, just exactly where I left it. I hadn't heard anything because there just hadn't been any traffic for a few minutes - it was not a busy road. I was real impressed how my GPS had led me right back to the car even when I could not see it at all for the heavy brush. I have used my GPS unit many times since that day for a variety of mining and prospecting related purposes. The GPS really has proven itself as a handy tool, so let's consider some of the prospecting and mining related possibilities for it.

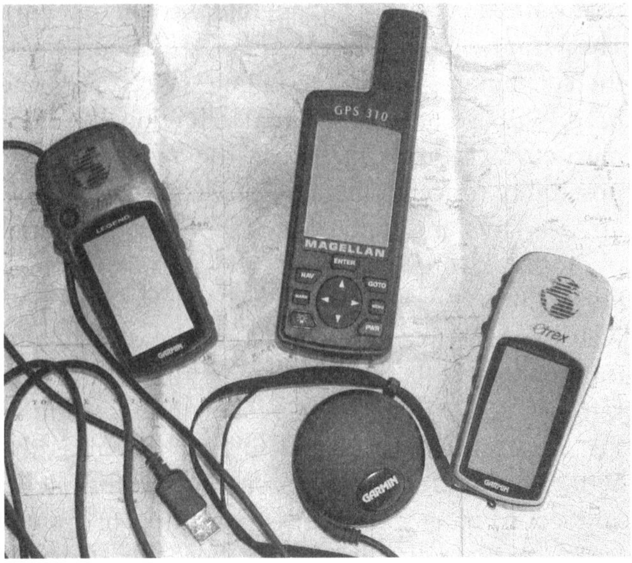

Here are four different GPS units – three hand held units with screens, and a GPS sensor with a USB output plug made to be mounted on your car roof and used with your laptop computer.

The first and most obvious use is finding old mines. As my story points out, it's real handy in forested areas, but I have used it in the desert as well. I remember a trip looking for some little high grade gold mines in the Nevada desert. You could see for miles, and I wasn't using my GPS, but had it along with me. I was going down a road that went along the foot of some mountains, looking for a road that took off to my left. Eventually, I got to a point where I knew I had driven much too far, and I had missed my turn off. I had been looking for it all the way and was very confused. I took some readings with my GPS and sure enough, I had driven past the point where I was supposed to turn up into the mountains. I had the coordinates for the turn off point, so I kept heading back on the road and taking readings until I got to the point where the turn off was supposed to be. I looked hard and - there it was - a very faint track off the main road, almost impossible to see even when I knew where to look and I was standing right there looking at it. I had driven right past it because it was so hard to see. I turned down this faint track and followed it. Sure enough, a couple miles later I was at the mines I was looking for.

There are a number of ways to obtain the coordinates for old gold mines or deposits you might want to explore (as well as the locations of roads and intersections that will take you to them). They can be obtained using topographic maps as the coordinates are printed along the edge of the map, with some maps actually having the lines marked as a grid all across the map. Another possibility for obtaining coordinates from maps is to use the internet. Topographic and aerial photographs are available at websites like Google Earth (http://earth.google.com). These apps and websites will calculate and show the coordinates for whatever spot you place your curser over. You can also research and obtain coordinates through the USGS mines database - better known as the MRDS system. This database gives information and coordinates for many thousands of mines, large and small, all across the USA. Further discussion of this database and how to use it appears in the chapter about doing research to find places to prospect.

Another great use of the GPS device is in the staking of mining claims. With the high cost of claims paperwork and the 10 claims small miners limit, you want to get your claims posted correctly and the corners laid down right the first time. You want the claim to cover the full area that you are allowed by law. The GPS allows me to draw the claim exactly as I want it on a topographic map, then get the coordinates for each corner, and go out in the field and put the stakes in the ground exactly at the correct locations. The typical hand held GPS device is accurate to about plus or minus 15 feet under most conditions, and for most mining claim staking that is a sufficient level of accuracy.

Some prospectors use a GPS to keep track of the locations of their finds, plotting on a map the exact coordinates of their finds. This can be very handy for those who specialize in electronic prospecting with a metal detector as finds can be spread over a large area and seeing the distribution of the nuggets already found can help lead to additional finds.

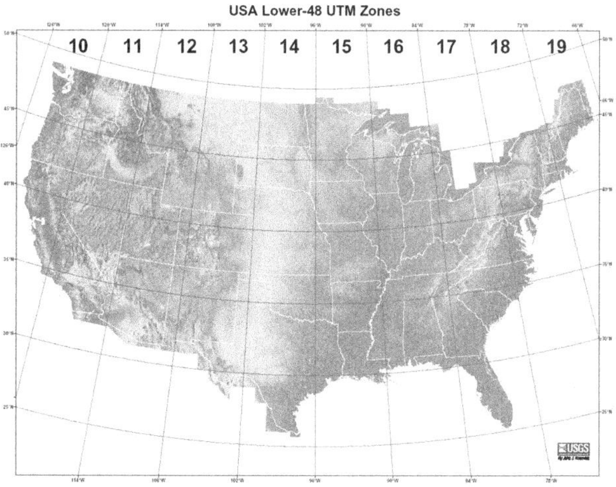

The UTM coordinate system divides the US into a number of zones, as shown on the map above. There are other zone numbers (up to 60) that cover the entire globe.

Throughout this chapter I have mentioned coordinate systems, and I think it would be worthwhile to consider them, as they are important in understanding and using your GPS. Think of these coordinate systems as a giant grid system that has been laid across all the earth. Most hand held GPS devices are capable of using a variety of systems, but the most popular in the US are the Latitude - Longitude system and the UTM or Universal Transverse Mercator system. There are simple computer programs available to convert the coordinates of one grid system to another. Generally most prospectors use the Latitude - Longitude system, sometimes called lat-long for short. Latitude lines run parallel to the equator, with the equator being zero and the poles at 90 degrees. Longitude lines run from the north to the south pole, going in a full circle of 360 degrees. Degrees of latitude and longitude are then further subdivided into minutes and seconds:

there are 60 minutes per degree (they are marked with a '), and 60 seconds per minute (seconds are marked with a ").

Also commonly used is the UTM system - it's a particular favorite of mine. In 1947, military cartographers adopted a special grid for use throughout the world called the Universal Transverse Mercator (UTM) grid. In this grid, the world is divided into zones, each covering a strip 6° wide in longitude. In each zone, the coordinates are measured on a grid numbered north and east in meters. I like the UTM system because when I am out in the field if the GPS tells me the point I want to be at is 200 meters to the west, I know I have to go a little more than 200 yards in a westerly direction. If I am using the lat-long system, and my GPS says something is one minute, 17 seconds to the west, I have no intrinsic idea what that distance actually is. Such things can be calculated, but one would not normally be able to make that sort of calculation in your head. So, I like the UTM system because it gives you a plain and easy feel for the distances you are evaluating because the numbers are in meters.

There are some limitations, and in order to work, your GPS must be able to see the sky and the satellites in it. While they can see through glass windows and clouds, they cannot see through thick layers of rock. As a result, a GPS device cannot be used underground in mines. Newer units can see through obstructions better, but with my older GPS I have had problems a few times in very dense forests where the canopy of trees and branches overhead did not allow it to see enough satellites to take a reading. Even with the older units, usually stepping to a clearing in the forest will resolve the problem. The fewer satellites your GPS can see due to obstructions of any type, the less accurate your readings will be - each satellite that is used in the calculation improves the reliability of the result. The various GPS units currently available offer various features including maps, two-way radios for communication, waterproof units, etc. Most of the pre-loaded maps are larger scale, and really designed for traveling within cities. However, there are some units which allow the user to self load maps onto the GPS device. There are also GPS sensors that come with software and a USB type connection, which are designed to be used with laptop computers. There are a number of makers of GPS units, but prospectors are most interested in the makers of hand held GPS devices. In this field, Garmin and Magellan are leaders, but there are other makers as well. Although some prospectors are intimidated by such high tech devices, they really are not difficult and can be very helpful in a number of situations.

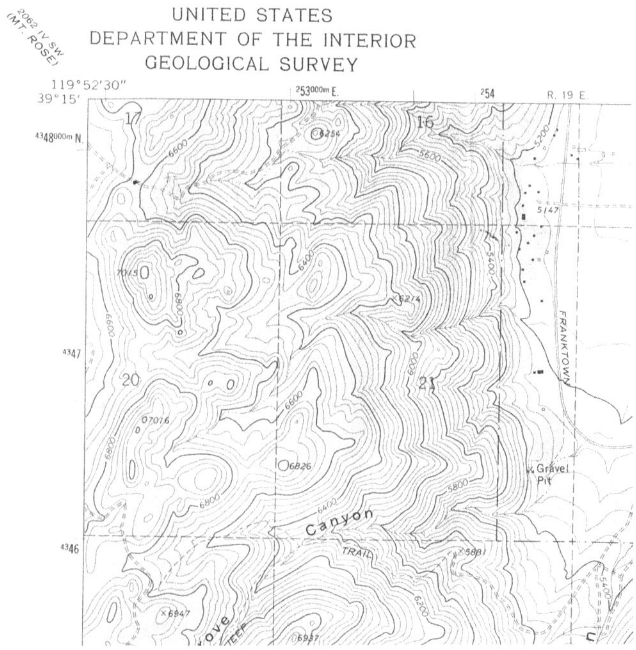

The information along the edges of a topographic map can provide lots of information – they give the coordinates in both the UTM and latitude – longitude systems.

Simple Hand Tools: Gold Panning

Perhaps the oldest and most widely used gold concentration device is the gold pan - originally called a "gold dish" by the 49ers. Although pans are available in various shapes and sizes, the standard American gold pan is 12 to 18 inches in diameter at the top and 2 to 2 1/2, inches in depth, with the sides sloping at roughly 30 to 45 degrees. They can be purchased at prospecting stores, rock shops and even some hardware stores for around $10 to $20. Gold pans are constructed of metal or plastic and both are used in prospecting for gold, for cleaning gold-bearing concentrates, and rarely, for hand working of rich, isolated deposits. Plastic pans are recommended for most folks, because they have some significant advantages over the older steel types - including the fact that they are lighter, don't rust, and don't conflict with the use of a magnet. Around the world miners use a wide variety of containers in place of the pan. In South America, prospectors have long used a cone shaped wooden bowl called a batea to process the gravels. In Mongolia, prospectors are currently using large plastic tubs. The 49ers of California made their first "pans" out of cattle horn, and they were shaped like large spoons. The exact shape of the device is not all that critical, its point is to allow the operator to shake the gravels so the gold settles downward, and then wash the lighter material off the top. Eventually, after all the lighter materials have been removed, only the heavy concentrates will remain in the pan. Hopefully, this will include some gold.

The simple pan and the individual prospector go hand-in-hand -- it is a tradition. The inexpensive pan has long been the standard piece of prospecting equipment carried by old timers since the first great gold rushes in the United States. It's been used to look for hard rock gold and placers in mountain streams and in deserts far and wide. In the past, most pans were made of metal and on long exposure to water they had a big tendency to rust. You can still buy metal pans, and I do own a couple of them myself, but the pans I use normally are all made of plastic. Plastic pans are cheap, durable and they don't rust. Unless you drop a big boulder on one, they will last for many years.

For most new prospectors a plastic pan should be the first piece of prospecting equipment they purchase, and panning should be the first prospecting technique they learn to master. The basic idea of panning is that by shaking and moving the pan around, the gold which is so much heavier than the sand, will settle out downward to the bottom of the pan. Lighter sand and gravel can then be washed off the surface and out of the pan. The process of shaking and washing off the lighter material continues until only a small concentrate of heavy materials remain.

It doesn't take loads of water to process a pan full of gravel – just a little puddle full of water will do the job quite nicely.

The basic operation of a pan is actually fairly simple, but experience and skill are needed to process large amounts of material and achieve maximum recovery. Panning is best learned by watching an experienced operator work, but the general principles and steps are outlined below. With a little practice, you can learn to pan gold for yourself. Just follow the instructions in this chapter and you can teach yourself. When you have read these pages and think you understand the panning process, do a little practice run and try it for yourself. For your first trial attempts, take a few pieces of small lead buckshot, mix it with some sand and gravel, follow the directions of this chapter and see if you can pan away the gravel and keep all the lead shot in your pan. If you don't have buckshot, cut some lead (real lead metal not pencil lead) into small pieces less than 1/8 th of an inch in size. The smaller the lead piece, the more its capture is a test of your skills. For maximum gold recovery out in the field, the material to be panned should be as uniform in size as possible - it's a good idea to remove big pebbles and stones larger than about 1 inch from your pan before you even begin the process. Take care to wash these stones so that the sand and clay attached to them stays in your pan. Some studies have shown that the fine materials attached to the larger rocks actually contain a lot more fine gold than one would expect. Panning is best done in a tub or pool of still, clear water, a spot where you can get good balance or even sit on a rock. You don't want to lose your balance and slip and fall into the river along with all the gold in your pan (I have done this a few times, and it's not pretty). The advantage of using a tub, especially if you are new to panning, is that you always have the opportunity to re-pan the material in the tub.

At the Begining of the Panning Process, After the First shaking, the gold has moved to the Bottom of the Pan.

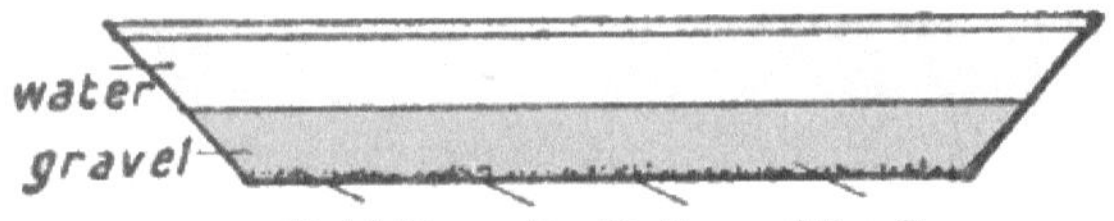

The first step in panning is to fill your pan then shake it back and forth a good number of times in order to allow the gold (or at least the most of it) to settle down to the bottom of the pan.

First, fill the pan one-half to three fourths full of gravel or concentrate - be sure to leave some space at the top especially if you are beginning panner. Add water to the pan or carefully hold the pan under water and mix and knead the material by hand, carefully breaking up lumps of clay and washing any rocks present. As much as possible, wash the clay out from the pan. Re-fill the pan with water (if not held underwater) and carefully remove rocks and pebbles, checking them before discarding - occasionally you may find a piece of quartz laced with gold. I will tell you that this is a very rare occurrence in most locations. Tilt the pan slightly away from you and shake with a back-and-forth motion from side to side. You have to do it hard enough that the gravel material shakes and moves around.

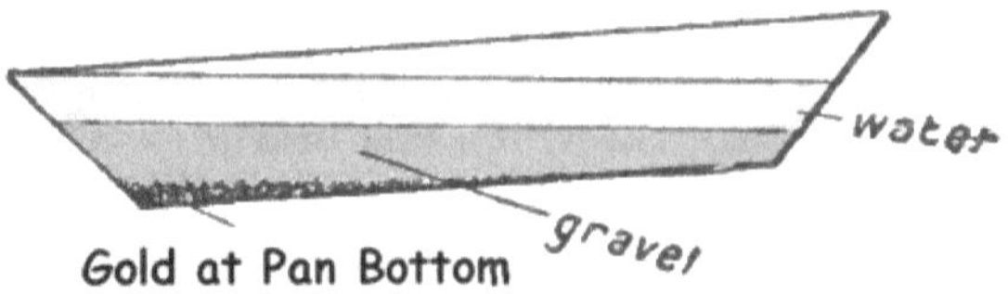

As You Wash out the Lighter Materials off the Top of the Gravels, Slowly Tilt the Pan Forward

During the panning process, carefully wash (don't just pour) the lighter material off the top of the gravels, shaking again after every few sweeps you make to wash off the lighter material.

The point of the shaking is to allow heavy materials to settle and work their way downward toward the bottom of the pan. Gold is extremely dense and heavy - it is actually much heavier than lead. So as you shake the pan it naturally wants to move downward through that sand toward the bottom of your pan. As you shake, the heavies move down, and this leaves the light materials on top. The tilt of your pan should be such that even when shaking back and forth, material does not shake out of the pan. Many short little shakes from side to side do the job better than fewer big jerking motions.

The next step is to wash the light material now on the top layer of your gravel out of the pan. With a gentle forward and back motion while holding it just below the surface of the water, lift the pan and allow a gentle flow of water to wash the light material over the edge of the pan and out of it. Removal of lighter material is facilitated by gently raising and lowering the lip of the pan in and out of the water. After these lighter materials have been washed off the top (typically you have taken off the top quarter inch of material), you will then repeat the process of shaking the pan from side to side with the same type of motion used originally to concentrate the heaviest materials on the bottom. After a bit of shaking, when the heavies have settled downward, you repeat the second process of washing the lighter material off the top. You have to repeat this over and over because gold will not always work its way right down to the bottom of the pan - sometimes it gets stuck along the way down. Large pebbles should be periodically removed by hand as they come to the top or are revealed during the removal of the light materials.

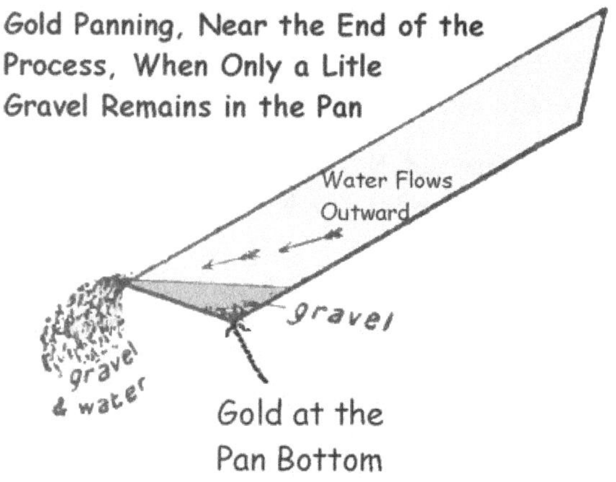

Gently wash (don't pour) the lighter materials over the lip of the pan. The heavier materials, including gold will remain at the bottom of the pan. If you see any gold show itself in the gravel as you are panning, shake the gravels to allow the gold to settle downward.

As you continue to repeat these steps, the tilt on the pan will slightly increase. Keep on repeating this two-step process of shaking down and then washing off the lighter material until you've reduced your material down to a couple of tablespoons full, when only the heaviest material remains. Usually (but not always) there will be some significant amount of black sand. Any gold that is present may be observed by gently swirling the concentrate into a crescent on the bottom rim of the pan. Coarse nuggets are removed by hand, while finer grained gold may be carefully picked out with tweezers or a snuffer bottle. The remaining black sands may be saved for further processing later. If you have a magnet, and a plastic pan, you can use the magnet to pull some of the black sand out of your concentrates. Take only a little at a time or you will pull gold up with the black sand. A small part of the gold from some locations is wrapped around magnetite and will lift with the magnet. It's worthwhile to check and see if this is a problem with your concentrates. Larger versions of the new, more powerful rare earth (Neodymium) magnets will pull not only the magnetite, but most of the hematite out of your concentrates as well (see the part of this chapter on processing black sands for more information and detail).

An experienced panner can process about one cubic yard in 10 hours. However, this does not count the time it takes to dig material out of the ground or river, and usually finding the right spot and then digging the material takes a lot longer than the time to actually pan the material. Panning was widely used as a primary recovery method in the early days of mining. Snipers and backpackers still use it in rivers today that are difficult to reach and may not have been heavily worked. Many prospectors use the pan as a means of processing concentrates from Dredging, dry washing, sluicing, etc. However, the process of panning is extremely limited, as only small amounts of gravel can be processed, even by the most experienced panners.

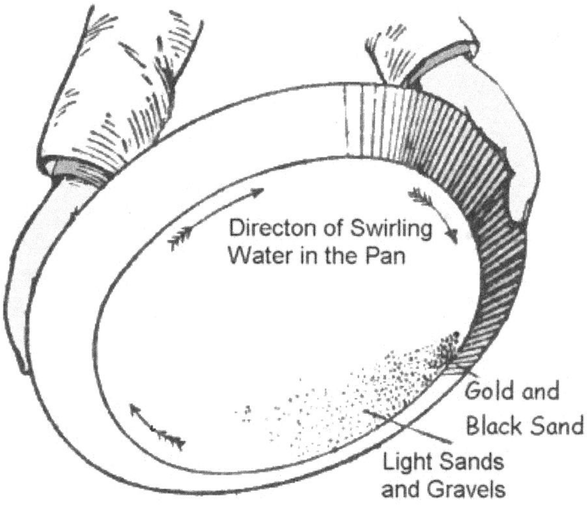

When you have panned off the majority of the material and are down to the last bit, you can swirl the water around and see what you have found. Heavy black sand and gold will resist the water flow and be at the head of the material.

Today, the gold pan is most commonly used for prospecting or for cleaning concentrates. Its low price, immediate availability, and portability make it an essential tool for the individual prospector or miner - virtually all individual prospectors own one or more gold pans. My instructions here really make panning for gold sound a lot more complex than it really is, and once you've practiced a bit with it you will see that panning is pretty easy. After a while, it gets to be a "muscle memory" kind of thing that you can do in your sleep.

One of the things that tend to confuse both new and old prospectors alike is the dizzying array of pans of various types. You can purchase pans which are big or little, deep or shallow, metal or plastic, with grooves or pockets of various types, rough or smooth surfaces, square, hexagonal or round - the variations seem nearly endless. Some foreign variations such as the batea or Lortok are actually made out of wood. So which is best? A 14 inch green or black round plastic pan is a good place to start, but in the end, it's a call you'll have to make for yourself as each has its strengths and weaknesses. The truth is that in experienced hands, all the variations can be made to capture gold efficiently. Here are some of the advantages and tradeoffs of various types of pans: Metal pans can rust, but they can take more of some kinds of abuse. Wet concentrates can be dried in a metal pan by setting it over a stove - a few have even used metal pans for cooking. Copper pans are used in mercury amalgamation, as mercury sticks to copper, but not steel. Plastic pans don't rust or corrode, but sometimes under really hot conditions they will melt. They are lighter and generally cheaper than steel pans. Because of injection molding methods, they can be formed to all sorts of odd shapes and textures, with groves, riffles or pockets wherever you want them. Pans with corners, like the hex pan, are easier to pour out of, as you have better control. Which of these best meet your needs is something you'll have to decide for yourself, or you can do as I have done, and just own a whole variety of them to suit whatever purpose you might have.

In time, you will see that the real difficult part is not so much in the learning how to operate a pan or any other type of mining equipment. The difficult part is in finding the material which carries significant gold! Gravel is everywhere, but gravel with some good gold nuggets in it is a whole lot harder to find. It's much better to dig half a bucket of naturally concentrated material holding an ounce of gold than 100 buckets full with .01 ounces in each! You get the same gold but at a far less amount of work. There is a lot to be learned as far as how to hunt for gold and how to find it even when you are in the right regions and locations. The later chapters of this book which talk about placer geology and digging for gold will cover that most important topic.

A variety of different Gold Pan designs in metal and plastic from the author's collection. Each one of these has a place, and from time to time, I use them all.

Prospecting With Only a Gold Pan

Although it's simple and inexpensive, a lot of prospecting can be done with just a gold pan. Let's take a look at some of the prospecting you can do with just a pan. There is no question that it is possible to find good gold with just a gold pan and simple hand tools - I've done it. Even if you are in the desert you can dig, screen and pan the material out in a tub with water you have brought to the site. This is the smallest scale form of prospecting - work you can do with just a simple gold pan and tools to dig in the bedrock. Many folks call this form of prospecting, sniping. When you are prospecting with just a pan, you advantage is portability. There is nothing lighter and more portable than a pan, and when you are working to find gold with a pan, you need to take advantage of that fact. The simple pan allows you to cover a lot of ground and take a lot of samples. If you follow proper procedures and look in the right types of places, you should be able to make some decent gold finds.

When you are looking for a place to pan, you are often looking for cracks and crevices in the bedrock of a stream, because these are some of the most common places for gold to get caught. A crevice is simply a narrow crack or opening; like a fissure or cleft in the rock. A pan is not a high volume piece of processing equipment, and so you are looking for small quantities of richer material, where the river or stream has already done most of the concentration for you. Since gold works its way downward in most cases, cracks and crevices become catch holds for gold to accumulate in. Even within the crevices, gold has a strong tendency to work its way right to the very bottom. Bedrock is the solid rock that sits under all sand and gravel, but not all types are equal. Various types of bedrock have different tendencies to form those gold catching crevices. As an example, granitic rocks tend to form smooth, almost polished surfaces, while limestone has a tendency to form very deep pocket like crevices, with slate and many other rocks lying in between those two extremes.

Gold has a tendency to follow a line within a river - a path of least resistance and minimum flow. Prospectors and miners call this line the paystreak. We will talk loads more about this in later chapters of this book, but it suffices to know for now that gold tends to follow a path within a stream. As a prospector working in a river environment, you are looking for that line, and where gold will get hung up or concentrated along that line. Finding gold's main "line of travel" in any given stream takes sampling and testing because it is not always obvious to the eye. There will be plenty of information later on in this book about reading a stream and how to find those natural accumulations of gold, but this is the beginning of the matter. We are just starting here with some basic information about using a pan to sample and prospect. The details of how to do this may be old hat to some, but all new to others.

Crevices in the bedrock like the one shown here are natural catches for gold, and good dig spots to test with a pan.

In time you will learn all of this, but it takes time. You won't gain the decades of experience the old-timers possess in just a few days or weeks.

The pan is the tool of choice for testing out new locations where bedrock is shallow - and it is important to remember that not enough testing is done by most prospectors! One of the most important lessons for old and new prospectors alike is that it is hard to be successful without lots of sampling on a regular basis. Rarely does gold occur in consistent and even concentrations throughout an area. Normally gold is spotty and can be rich in one place and only a few feet away be virtually barren. This is the reason why lots of sampling has to be done because there are a whole lot more low-grade spots than high grade ones. You want to keep looking until you find those good spots! Don't hunker down and just keep digging in a spot where there is little or no gold - if you don't find it, you've got to move on! If you find a spot with a fairly large amount of gravel that yields well you might want to consider bringing in a sluice box or other processing equipment.

In many ways, sniping and prospecting with simple hand tools has almost become lost art. This is partly because there are so many types of great equipment that are available, dredges, metal detectors, dry washers and the like. It just becomes too easy to get caught up in all the great opportunities that these equipment items offer to

Exposed bedrock at the edge of the stream like this can be a great gold spot, but so can the bedrock that crops out below the water line.

us. However, it's important to remember that simply breaking open small cracks and crevices in exposed bedrock can yield some fairly nice nuggets. Crevicing and sniping can be done even in the desert simply by following exposed, shallow bedrock and blowing or brooming off the surface of the rock and collecting material in the cracks. Any crevices in the bedrock which are exposed by doing this should be carefully cleaned out. Small nuggets can be exposed or materials from inside those cracks can be collected for later panning. A small mirror is often a handy tool to have when you are cleaning out crevices, as a mirror can easily be used to reflect the sunlight down into deep crevices where it is otherwise difficult to see. Some prospectors will also run a metal detector down as deep into a crevice as possible, as sometimes gold gets wedged into places where it is impossible to see.

Another possible sampling technique that is often over looked is mossing for gold. The moss that grows on rocks near the edge of a river often has a surprising capacity to catch and hold gold. The gold gets caught up in the tendrils and netting of the moss, and while the gold recovered from mossing is normally fine sized, sometimes there is a fairly good amount of it. In many ways, the moss acts like the spaghetti weave liner on the bottom of a sluice box to catch and hold fine gold. Every once in a while you might get a nice flake or two. The moss must be taken from a spot close to the stream - at a level where, when the river is in flood stage, the moss will be underwater. One way of quickly sampling an area to evaluate its potential is by carefully pealing the moss off the rock, collecting and saving all the sand and silt caught up in the moss. The moss has to be torn up and shaken out to extract the sandy materials it contains. The material recovered from the moss is then panned out by the normal methods to recover the gold.

While most people will do the bulk of their prospecting above the water line or in dry areas, it's important to remember that crevicing can be done both above the water line and below. Underwater crevicing is basically an art in itself -- there are prospectors who specialize in this type of work because it is often overlooked and can be very productive. Really the only tools necessary are wetsuit, mask, snorkel, a gold pan and tools like a pry bar for cleaning out crevices. A pair of tweezers and a specimen bottle to hold your gold is also worthwhile. Other than the wetsuit, all these are inexpensive items. The advantage of underwater crevices is that they get a whole lot less attention than their more easily reached brethren above the water line. During a recent summer while dredging at a spot that serves as the local swimming hole for the town youth, I picked out a quarter ounce nugget exposed to sunlight but about 18 inches below the surface of the water. I guarantee that quarter ounce nuggets above the water line which can be seen by the eye get picked up rather quickly. That little spot of bedrock below the water line had just never really been thoroughly explored in spite of the fact of the local kids come there nearly every day in the summer time. That's the big advantage of working under water it's a whole area you may have all to yourself. The underwater sniper commonly floats downstream going along with the current, looking for attractive crevices below the water line. When he finds a likely spot, he stops and removes the sand that commonly fills crevices in the bedrock by a method called fanning.

Fanning is the process where the diver moves his hands quickly to fan away the sand and other light debris thereby exposing the bottom of the crack and any gold that it may contain. The most common technique uses both hands in a motion that is much like "dog paddling" to scour the sand out of the crevices. Larger rocks are removed by hand and those that are really wedged into the crevice may require a pry bar. Tightly packed gravel is often a sign that that particular crevice has not been cleaned out in a long time - this is true both above and below the water line. Once exposed to the sun, the underwater gold just seems to glow, its metallic reflection of the light really makes it stand out. Smaller pieces of gold are removed with tweezers. Gravels from within the crevice can be processed by panning to recover the smaller flakes and pieces. There are guys who are very successful using this technique and can even out produce guys with dredges when they're working in the right kind of area. You need a good amount of exposed bedrock to make this kind of prospecting work well -- is not a good technique for areas with deep gravel and little or no exposed bedrock. Often smaller streams and tributaries up in the high country fit the bill perfectly since they are steeper, less gravel accumulates in them and more bedrock is exposed for easy work.

Finally, about dry panning: in theory, it is possible to pan material without water -- but in reality dry panning doesn't work very well in most situations. Water makes the process of separating waste gravel from gold much, much easier, helping the gold and heavy materials settle downward. There is no doubt that dry panning is a difficult art to master. Illustrations and directions on dry panning can be found on page 67, at the end of the dry washing chapter. In many places where heavy clay is present, it can be virtually impossible to effectively dry pan placer gravels. Gold gets locked up in clods with the clay, and won't settle to the bottom of the pan the way it should. Any real acceptable success at dry panning requires that the material be screened, very dry and contain little or no clay. This is why it is much more common for desert prospectors who are sampling and sniping around to bring in water in buckets or other containers and pan collected material within a tub. This allows the prospector to get reasonably good effectiveness and recover any fine gold that is present in the gravels which is collected. Bringing along water in a tub can be a logistical pain in the behind but it does work. Screening your material can also be very helpful. All the material you dig must be carried to water and panned. Screening out the coarser gravel and rocks is much faster than picking them out by hand and allows the prospector to focus in on smaller gravels and sands which are usually more likely to contain gold. Except when heavily coated with clay and iron staining most nuggets that are too large to go through the screen will be easily visible and the average prospector won't miss seeing them.

Early day panning for gold during the great Klondike rush of 1898. Everyone took an interest, and there were far more women present than during some of the earlier western US gold rushes of the mid 1800s.

So the next time you attend a show or read a magazine, and see all the wonderful equipment that is available for recovering gold, don't get stuck with the idea that the only way to be successful in prospecting is to spend a mint on equipment. Sure, that stuff is very useful and great to have, but the skillful prospector can really do quite a bit with nothing more than just a simple plastic gold pan.

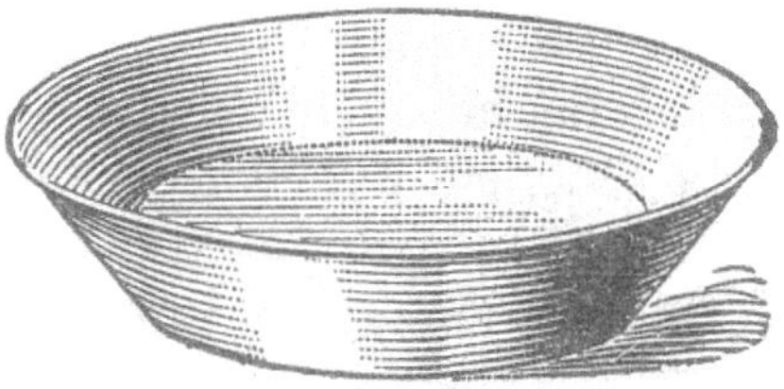

Finding Gold With A Simple River Powered Sluice Box

A sluice box lined with riffles is one of the oldest gold separation devices still in use today. They are simple to operate and have been used all across the world for thousands of years. The size of a sluice box can range from small, portable aluminum models used for prospecting all the way up to large units hundreds of feet long, such as those used in commercial operations. The sluice is a normal next step up for a new prospector who has purchased a gold pan and is interested in increasing the amount of material he can process. This chapter will focus on hand sluices that are run directly by the flow of river water. Sluice boxes that are run on powered flows from motorized pumps will be discussed directly at the end of this chapter once we finish with river flow boxes. Sluice boxes can be made out of wood, aluminum, plastic or steel. A sluice is really nothing more than an artificial channel lined with devices to catch gold. Water flows through it, moving the lighter materials such as clay, sands and gravels out of the sluice. The heavier materials (including gold) remain behind, trapped by the riffles. The huge advantage that the early prospectors found in sluice boxes was that they could process the gold bearing gravels much faster than they could with a gold pan. An experienced prospector might be able to pan one cubic yard of gravel per day, but he could run a cubic yard of gravel through a sluice box in less than an hour. This allowed those miners to produce a lot more gold, and it still does the same thing for us today. The prospector armed with a sluice can try to tackle gold deposits with a good bit more gravel volume than one might try to take on with just a pan.

Modern sluices are often built as one unit although sluices formed in sections are still used especially for larger commercial systems. Sluice boxes designed for individual prospectors are small and light, designed for maximum portability. As a rule, a long narrow sluice is more efficient than a short wide one. The sluice should have a slight slope, usually in the range of about 1/2 to 2 inches per foot, depending on the amount of available water, the size of material processed, and the size of the gold particles. Larger flows of water and steeper slopes are required to move larger rocks out of the sluice box. Unfortunately, steep slopes and large flows of water also result in poor recovery of fine gold. The riffles in a sluice retard the material flowing in the water, and form a sand bed which also traps heavy particles and creates turbulence. This turbulence causes heavy particles to tumble, and helps flush lighter material out of the box. Hungarian or Lazy L riffles are one of the most popular riffle designs, have an overhanging lip on each riffle, which increases the turbulence, and forms a small eddy zone behind the riffle, helping to trap fine gold. Spaghetti weave mats, as well as some types of carpet, screens, and other liners are also used today to cover the bottom surface of the sluice, aiding further to trap the fine gold. The Nomad ® or similar types of spaghetti weave liners do a lot to help capture the small sized gold. Some also suggest a flat "slick sheet" section placed ahead of the riffles. Except in areas where the majority of the gold is very fine, these are usually not necessary for river-operated hand sluices as the normal water flow is slow enough that the extra sheet is not required.

Prospecting with a simple hand sluice is easy. Light modern aluminum units are easy to carry and work well. Molded plastic sluices also work well. This item of equipment is a good second step for the prospector who wants to prospect in areas with flowing streams.

Riffles can be made of wood, rocks, rubber, iron or steel, and while historically woolen sheep hides, wood or rock were common, wood and wool are not durable and rock is too heavy and cannot be formed to specific shapes easily. Most riffle systems today are made from some form of mild steel. They can take a number of shapes, slats set at 45 degrees, angle iron, and the well-known Hungarian style riffle. Riffles range from one half to about 2 inches high depending on the size of rock being processed through the sluice, with larger riffles being used on larger rock. Generally, each riffle is placed several inches apart, one after the other. A set of riffles is commonly fastened to a rack that is wedged or bolted onto the sluice so that they can be easily removed for recovering the concentrates when the sluice is to be cleaned out.

The water flowing through the sluice should be just fast enough to move the material through the sluice, without forming white water conditions which quickly blow the material on through.

The sluice box is not hard to operate - it is fairly forgiving of the surges of gravel that come in with each shovel full we add. To perform efficiently, a sluice needs a good amount of water. This is my simple rule of sluice box operations: Just enough water should be going in with the feed gravels to move the material out of the sluice in a reasonable time, allowing only the heavier materials to build up behind the riffles. Too little water and the incoming material simply piles up as the rocks stop and get stuck in the sluice box. If the flow is too fast, fine gold will be blown out with the materials. Optimum water flow is just what it takes to keep things clear and no more. The riffles should not become fully buried, nor should the entire bed of the sluice a blanketed over with gravel. For maximum recovery, the flow should be turbulent, yet not frothy in any way. Excessive turbulence results in poor fine gold recovery.

Setting up a sluice box in a stream actually takes a little effort – it's important to find just the right spot, where you can get just the right flow speed. Areas of a river that are pond-like with a very slow flow won't do, and a whitewater kind of flow is too fast. You want to select a spot with moderate flow. The typical pitch of the sluice is normally between 1/2 and 2 inches of fall per foot of length. A little more or less fall is OK as long as you are getting the right amount of water flow. I find that in some places selecting a good spot is easy, but in others it is difficult. I also find in many of the good spots where I place my sluice, it is necessary to put a rock on top of the sluice to hold it down or else wedge it in place with rocks on the side of the sluice. This is because those places with sufficient water flow will have a tendency to lift and move the sluice around, and you want to keep it anchored firmly in place while you are feeding it. On the other hand, sometimes it's necessary to increase the flow by diverting a small part of the stream. I am not talking about building a dam and making a major diversion, but just putting a few rocks near the head of the sluice to direct a bit more of the flow into the sluice. This is why some boxes have a "Y" shaped entrance to increase the flow through the box. These few rocks can make the difference in getting just the right flow into the box – I consider them "flow adjusters". Once you have set up and run the box a few times you will be able to quickly set it up without any difficulty. Once set up with the right flow, operating the sluice is simply a matter of feeding the material into the box. One of the important parts of operating the sluice is to not overload it. If material is shoveled in too quickly, the riffles get buried, which can cut down on fine gold recovery. Any rocks that are too large or otherwise get stuck in the sluice should be picked out and removed. Usually, it is only necessary to clean up a river fed sluice about once each day, except where you are testing and need to clean up after each test.

For cleanup, clear water is run through the sluice until the riffles are clear of excess gravel. A container is placed at the discharge end to prevent loss of concentrate. I lift my sluice out of the water at this point, and then the riffles are removed and carefully washed into the bucket. Any matt or bottom liner is removed and washed as well. Any large pieces of gold should be removed by hand during the sluice cleaning process, as the concentrate is being washed out of the sluice into a suitable container. The concentrates are further processed by panning, tabling, or another method, depending on the volume of concentrates that are obtained from your operation. For most individual operations, a pan is sufficient. After cleanup, the sluice is reassembled and is once again ready for more material to be processed.

The sluice box is really just a simple trough lined with riffles which are perpendicular to the flow.

Gold recovery with sluices can vary depending on a number of factors. Heavy clay in the gravels may cut down yields when the gold is caught in the clay and the clay clods do not break down as they travel through the sluice. Careful washing of rock covered in clay may be very worthwhile in some rich areas. To a certain extent, fine gold losses can be minimized by cleaning up more frequently, but excessive cleanup results in loss of time processing new material. It is better to examine the operating factors such as the flow of water, the slope pitch of the sluice, the size of the gravel feed, and the riffles and liners being used in the sluice. The addition of a liner such as the spaghetti weave material to the bottom of the sluice really helps trap that fine gold. A much more detailed discussion of optimizing your sluice for fine gold recovery is found right after the chapter on suction dredges. Overall, sluices are widely used today due to their low cost and ease of manufacture. Most individual prospectors own a sluice because they are lightweight and portable. They require very little supervision or maintenance and can tolerate large fluctuations in feed volume. Even commercial units are portable with the help of the right equipment. When properly operated, in many locations sluices can obtain a gold recovery of better than 90%, and they also require only a minimal initial investment to purchase and set up.

The biggest disadvantage for commercial operations is that very fine particles of gold are not effectively recovered. With coarse gold pieces, recoveries of better than 90% are possible, but with fine gold less than 20 mesh, the recovery can be much less. In addition, sluices cannot operate when they are being cleaned which requires a certain amount of downtime. Fairly large volumes of wash water are also needed to operate them – however in many circumstances much of the water can be recirculated. Generally, these issues are not a big concern for the individual prospector. They are available in various configurations as pre-made units from a variety of manufacturers. However, many sluice boxes (especially in commercial operations) are simply fabricated by the user for their specific operations, as their construction is not particularly difficult. Building a sluice box is a comparatively easy project to complete, and is commonly done by new prospectors - I have a full section on sluices in my chapter on building you own equipment. Early day miners used high pressure flowing water and sluice boxes to wash away huge hillsides in a process known as hydraulic mining. Smaller scale, lower pressure hydraulic operations were known as ground sluicing. Over the years, a wide variety of variations to the standard sluice box have been developed. Let's take a look at a couple of old time varieties of sluice boxes. The long tom and the dip box / rocker are really very rarely used, and almost never seen in the field. The rocker / dip box set up is used once in a while by prospectors working in the field with very limited water in puddles or pools - not enough water to use a pump with.

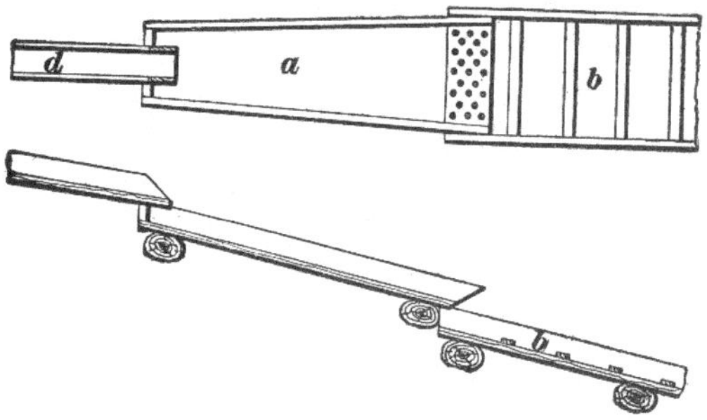

The old style miner's Long Tom sluice

The Long Tom Sluice:
Among the many variants of the sluice, the old time long tom was the type used by the 49ers and other early day prospectors. It is actually a small sluice that uses less water than a regular sluice. It consists of a sloping trough about 12 feet long, 15 to 20 inches wide at the upper end, flaring to 24 to 30 inches at the lower end. The lower end of the box was set at a 45 degree angle and is covered with a perforated plate or screening. Below this screen a second long tom sluice box was set up containing additional riffles; it was wider and usually shorter and set at a shallower slope than the first box. It was common to set these up in a long series of sluices were multiple operators to be shoveling in to the box at various points. The long tom uses much less water than a sluice but requires more labor. Material is fed into the upper box and then washed through, with water. An operator breaks up the material, removes boulders, and works material through the screen. Larger rocks were often removed and tossed aside using a pitchfork. Coarse gold settles in the upper box and finer gold in the lower. The typical capacity of a long tom is estimated at about 5 to 10 yards per day. This type of sluice was widely used with ground sluicing operations in the early days. Other than using slightly less water, advantages and disadvantages are the same as for other sluices.

The Ground Sluice:
A ground sluice is a variety of mining operation where the sluice box is dug as a trough into the ground. In general, it will use up 6 times as much water as a standard sluice box to process the same amount of material, however the flow is often used to partly wash materials off the slope and into the sluice. Ground sluicing is essentially a variety of hydraulic mining, and in most locations is probably illegal. The sluice trough in the ground is formed partly by diverting the stream through it, and assisted by loosening the earth with a pick; when the trough is made, the pay dirt is either washed into it by the stream itself or carried by the miners. If the miner is fortunate enough to have a hard and uneven bed rock for the bottom of his ground sluice, the rough bedrock floor may be enough, in itself to hold the gold even without riffles. Usually some sections of wooden sluice boxes were used in the trough to better catch the gold, but sometimes the old timers would just put boulders and heavy stones, too large to be moved by the water, into the ground trough in a haphazard manner on the bottom of the ground sluice to partly take the place of normal riffle-bars. Ground sluicing is noted here because of its historical importance and because these old sites can make good spots for metal detecting and other prospecting.

The Rocker, cradle or dip-box:
The rocker or dip-box is another modification of the sluice that was used where water is scarce and the grade is too low for an ordinary sluice. It is also commonly called a cradle. The dip box is a close cousin to the rocker box, and may be considered as a variety of a rocker. This device is simply a short sluice with a bottom of one by 12 inch lumber, with 6-inch-high sides and a 1 to 1-1/2 inch end piece. To catch gold, the bottom of the box is covered with burlap, canvas, carpet, Astro-Turf, spaghetti weave mat or other suitable materials. The box is set with the feed end about 30 inches high and the discharge end 6 to 12 inches lower. Material is fed, one dipper at a time, into the back of the box. Water is poured gently over it from the dipper, bucket, or hose until the water and gravel are washed out over the lower end. When the hopper becomes filled with stones, and they are all washed clean, they are then tipped out and carefully examined for any gold nuggets that may be mixed up with them. A certain proportion of very fine gold dust will be caught and held by the hairs and fibers of the cloth in the apron, but most of the larger particles of gold will collect behind the riffle-bars on the bed of the cradle. These devices are little used by modern prospectors.

The old time miner's rocker box is little used today because there are so many better alternative methods available for processing your gravels.

Two or three times a day, depending on the nature and richness of the pay dirt, the rocker must be cleaned up. First, the hopper is taken out so that the apron can be withdrawn. The apron is then washed off in a bucket of clean water. This will dislodge the gold dust held in the fibers of the apron material, and this gold can then be recovered from the bottom of the bucket. The gold and other gravels which have been caught by the riffle-bars are scraped out with a steel spoon. The collected gravel is put in a pan, and the gold then is panned out. As a bucket full of water normally weighs more than the same volume of pay dirt, the gravel is generally is brought to the water, when it is not possible to pump or let the water flow to the pay dirt by gravity.

The dip box or rocker uses much less water than a sluice box, however fine gold is not effectively recovered with this type of device. Rocking is neither economical nor efficient, which why it is rarely used now. Much fine gold is lost by its use, but the rocker is cheap, requires very little water and is portable. Rocking or cradling is mentioned here mostly because of its historical interest.

Early Day Gold Sluicing Scene

Highbanker (Power Sluice) Operations:

A highbanker is basically a sluice box that operates away from the river by using a motorized pump to provide the water necessary to operate it. This allows the operator to shovel directly into the box and not have to carry the gravels to the water. Of course, it can't be too far from the water source as the pump must be able to get the water up to the sluice. The highbanker, which is also known as a power sluice or sometimes a hydraulic concentrator, consists of a feed hopper and a sluice box. The feed box portion has a water spray and some sort of screen to scalp off the coarser material. The fines that pass through the screen then go into the sluice box section where the gold is captured. Like a regular sluice, you still have to shovel the material to get it into your hopper. Most of the units have a water adjustment so that the flow of water entering the hopper can be adjusted. As with other sluices, set the water stream to the minimum rate it takes to keep the gravels that fall through the screen moving down the sluice box section without any rocks getting hung up in it.

Some of these models are designed so that the sluice section could be separated from the unit and operated in a river without the pump, but most are not. Other models are set up so they can be run as a dredge, sucking materials out of the stream so it does not have to be shoveled. These latter units are known as highbanker & dredge combinations.

The dredge-highbanker combination is a compromise piece of equipment, and as a result it's not perfect, but it does offer some flexibility and may save you a bit of money. The dredge part of the high-banker & dredge combos are generally smaller dredges, with most being in the 2 to 3 inch hose size. I do not know of any 4 inch or larger sized dredge / power sluice combinations. Most highbanker - dredge units have a suction nozzle type of intake (one maker uses a jet). This type of equipment is best for shallow, small streams and backcountry tributaries. These systems are designed to be set up alongside the stream's edge, and have adjustable legs rather than a float system as would normally be seen in a traditional suction dredge.

Highbanking for gold at the Moore Creek Mine in Alaska. The mine was a pay to prospect operation near McGrath in the Iditarod district. The author dug some nice gold at this location.

The advantage of a highbanker is found in working materials that are deposited in a location away from a river. Moving the water to the sluice with a pump saves the operator the work of carrying the material down to the stream. If you are processing a lot of material, this makes a big difference. One of the disadvantages of these types of systems is that you need a whole lot more equipment to run a highbanker system than a river flow hand sluice. If you are taking it to a remote location, you need to pack in a motor and pump, hoses, fuel, and other gear for the highbanker to work. Another issue with these types of units is the fine gold recovery. The pitch of the sluice box must be set steep enough that rock and other material will move on out. Generally, the sluice on a highbanker drops about 3 to 4 inches per foot of length, an angle much steeper than a standard hand sluice set up in a river. This is needed to keep the small rocks moving through the sluice box. As a result, the rate of fine gold recovery will usually be a little less than a river run sluice. This is not to say that highbankers don't recover any fine gold, it's just that the rate of recovery with most designs is a little less than that of a well set up hand sluicebecause the power sluice is running faster water and set up at a steeper pitch angle. Some

The modern highbanker / power sluice unit showing how it can be set up for dredging by attaching the dredge hose to the feed hopper box in the front.

special units made for working beach sand get good fine gold recovery. They use expanded metal riffles, spaghetti weave liners, slow water and pre-screened sand less than ¼ inch. The optimal place to use a highbanker is a spot where there are lots of bench gravels that have been left high and dry by erosion. They are also good where there is no moving water (like ponds) and a sluice won't work. There has to be sufficient water close by – at least within a couple hundred feet so the hose can reach to it. In some locations, there are good opportunities for using a highbanker and in other locations there are not. I have used a highbanker, but I do not own one myself, I just don't consider it an essential item for me. That's just my personal opinion based on conditions in the areas I normally prospect. For the few times I have material away from water that I want to process, just use a hand sluice in the stream and carry the material down to it with buckets. There is no question that using buckets is much slower, and if I did a lot of that, I'd get a highbanker. Whether a highbanker is right for you just depends on how much you might use it in your area.

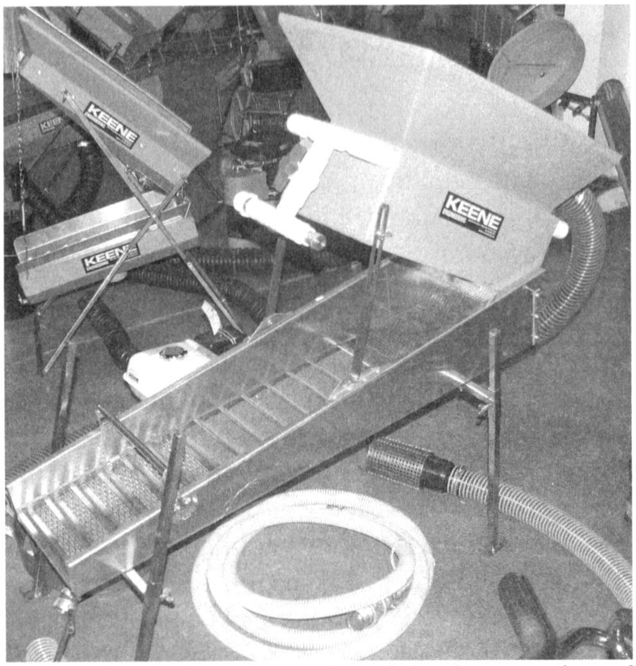

A shot that better shows the sluice box section of the highbanker unit.

There are some environmental issues with the water draining from a highbanker, as most bench materials have a large amount of clay in them. Too much dirty water draining directly into a stream may cause issues of water pollution. I recommend you have your water drain into a settling pond or pit that is not directly connected to the stream (so the clay and mud can remain there). Always make sure there are no legal issues with how you are operating especially if you are putting that mud into a flowing stream. Right now few states require permits for high bankers, but that can change in the future, so be sure to check the current legal conditions. If you are interested in building a highbanker set up or perhaps even converting a river sluice to run with a pump away from the stream as a highbanker unit, I have some thoughts on that topic later on in the "Build Your Own Mining Equipment" chapter.

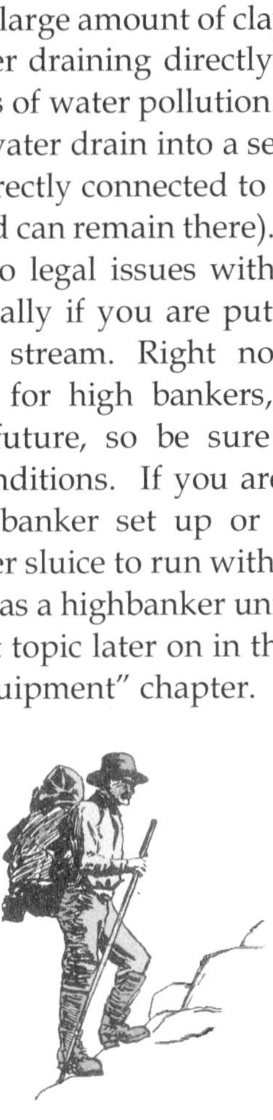

Suction Dredging For Gold

Diving for nuggets in the streams of the gold country has been a popular activity for many years, but the suction dredge as we know it has only been around a bit more than 50 years. A small amount of diving was actually done in California by the 49ers during the days of the gold rush. Their equipment did not allow them to conduct continuous underwater operations, but the old timers knew there was gold in the bottoms of the streams - a place they couldn't easily reach. Sometimes they could build dams, divert the stream and harvest its bed, but often the canyon walls were such that they had no place to divert it to. Bucket line dredges which used long chains of large steel buckets to dig gravel from stream beds were first operated successfully in the 1890s, but these were large scale affairs operated by big mining companies. It wasn't until the 1950s, with the advent of lightweight portable "lawn mower" type engines, that suction dredges became small enough that they could be easily transported, set up and run by a single operator. Roughly at that same time, scuba and similar diving equipment also became easily available. During these early days, there were no big manufacturers that sold suction gold dredges. They were all handmade in garages and small shops by men like Ernie Keene. Today, a number of makers produce a variety of gold dredges in all sizes, including the family of Ernie Keene.

Dredging can be a lot of fun and produce some nice gold. Here the author is working with a four inch dredge in the Yuba River of California.

I bought my first dredge from Keene Engineering way back in 1978, and over those decades, I have sucked up a lot of gravel with the dredges I have owned, and I've had a lot of fun doing it. The lightweight, portable suction dredge allows the individual prospector the ability to explore a world of gold deposits that the old timers had great difficulty reaching. The old timers could build dams and divert the rivers in some locations, but today's prospector simply dons his wetsuit, fires up the motor on his dredge and dives under water. The great advantage of the modern dredge is that it can be easily moved around and taken to new locations for prospecting new areas.

What a Dredge is and How it Works:

A suction dredge is basically a device that is designed to suck gravel from the bottom of a river, pull it up through a hose and run it over a sluice box. In the sluice box, any gold which is present becomes trapped and the lighter materials like sand and gravel move down and out of the sluice box and are dumped back into the river. The operator guides the nozzle of the hose to suck up the rocks and gravel which he desires to process off the river bottom, the advantage being that he does not have to actually "dig" the material.

Normally, a small "lawn mower" type engine is used to pump the high-pressure water that creates the suction which pulls sand and gravel up through a hose and into the sluice box. The suction in the hose is actually created through a Venturi effect by pumping high pressure water through a side jet. The fast-moving water creates the suction in the main dredge tube. This way the sand and gravel does not actually go through the pump which would quickly wear it down. In addition to pumping water, the small engine can also be used to produce compressed air for the diver to breathe while he is working underwater. In cases where the water is shallow, the gold diver may simply use a snorkel instead of compressed air. The engine, pump and sluice box are all mounted on a flotation system that keeps them floating above the water line. There are some exceptions where the sluice is actually mounted under water and these types of designs will be discussed shortly.

There are two main types of Venturi suction jets currently being used on modern dredges. These are: 1) The suction nozzle where the jet is up at the nozzle point; and 2) The power jet where one or more jets are located up near the entrance to the sluice box. The advantage of the suction nozzle is that it can be picked up above the water line without losing suction. If you allow much air into the nozzle of a power jet, it loses its suction and the hose must be refilled with water before suction can be re-established. Because of this, the suction nozzle is better for shallow water conditions. On the other hand, power jets have a greater level of efficiency and create a greater amount of suction for the amount of water coming out of the pump. As a result, most medium and large dredges are equipped with power jets, while small dredges are often equipped with suction nozzles. A few dredges use a third type, the ring jet. Ring jets are touchy to operate, much less efficient and require larger motors and more gas to operate them, so few dredges are made using them.

Sometimes gold gets hung up among the boulders in fast moving waters. Some care for safety is required when working in fast water.

Most modern dredges are made to float on the surface of the water, allowing the operator the greatest level of flexibility to move from place to place and work small deposits. Most flotation systems are made of rigid plastic pontoons, but there are still a number of units in use that employ other flotation systems such as truck inner tubes (truck tubes were popular for dredge flotation years ago). Today's suction dredges have come a long way from those early days - they are far more efficient, lightweight and reliable than their early predecessors. They come in a wide variety of designs and intake sizes. Suction dredges are commonly available "off-the-shelf" in sizes ranging from an intake of 2 inches up to an intake of 10 inches. The larger the intake size, the more material will be processed through the dredge in each hour of operation. However, the larger the intake, the heavier the dredge will be, and with increased weight there is decreased portability. In addition, dredges with intake sizes larger than 5 inches can be difficult for a single operator to manage. Quite a bit of improvement has also been made in the capture efficiency of the dredge sluice boxes, and many designs support undercurrent sluice systems to increase fine gold recovery.

The first step in underwater prospecting is a careful examination of the stream itself and the adjacent terrain to find a spot where gold has already been concentrated by the flowing water. Natural features that should be observed are the topography, the nature of the bedrock including any joint and crevices systems, stream currents, bends in the streams, visible underwater features such as large boulders, or any other type of barriers that might act as a trap for gold. Anywhere the natural features cause some change in the volume of water flow which results in a sudden decrease in stream velocity, normally those are the conditions favorable for gold deposition. The inside portions of river bends and the downstream area behind boulders and rock outcrops are common places to search for concentrations of gold. Deep narrow crevices and cracks, especially those that occur in steeply dipping rocks whose strike or trend is roughly perpendicular to the stream flow are particularly favorable for the occurrence of gold. A series of parallel deep narrow cracks or crevices at right angles to the direction of the stream are especially good because they form natural riffles that catch and retain gold. Such traps are most likely to be found in rocks like slate, schist, gneiss and other heavily jointed forms of bedrock. The gold is almost always concentrated at or near the very bottom of such natural traps where it may occur as flat or rounded grains and flakes ranging from microscopic sized particles, up to nuggets several inches long. The underwater dredger, armed with the right equipment, has a special opportunity to uncover and work these cracks and crevices and discover the gold they contain. The particles of gold found in the streams originally came from veins, mineralized zones, and older placer deposits from which they were released by weathering and disintegration of the rock. The efficient prospector will evaluate locations being considered for dredging and test them by sampling before bringing in his dredge. Even once the dredge is brought in, the dredger needs to sample around in the river to find good gold bearing deposits and pay streaks that can be worked profitably. In many cases, the first step in dredging is to remove the layer of overburden gravels which often cover the bedrock. When removing overburden, it is best to plan ahead and start at the downstream end of your planned work area and move upstream so as not to force yourself to re-dredge your own tailings. Any deep and narrow crevices which are found under the overburden must be completely cleaned out to the deepest level because it is here that most of the gold is to be found. Gold has an amazing capacity to work its way down to the very bottom of a crevice in bedrock. Sometimes it will make its way down into hair line fissures to the point where even though it appears that the crevice has been completely cleaned out, gold may still be hiding underneath. Fully cleaning out the bedrock crevices is very important and where possible, it's best to break up the rock in and on the bottom of these crevices especially ones that you've already found gold in. Bedrock slide pounders and rugged crevicing tools will be most helpful in accomplishing this task. High-pressure flusher hoses and nozzles reduced down in size so that they fit into deep crevices are other important tools for crevice cleaning.

For those prospectors interested in building their own suction gold dredge, there is a section of this book on building a dredge, and the efforts involved to accomplish that task. I also have some information about this on my website. Sometimes you can modify your existing dredge to operate better or help improve your gold recovery. This includes things like adding an undercurrent section to your sluice box if you don't already have one. This type of information can also be found in the dredge portion of the "build your own mining equipment" chapter. Summer and fall are usually the best times of the year to dredge for gold because the water in the streams is low and clear. However, there are some areas that are best worked at other times of the year, and on some streams in which the water is only seasonal, one must work during the time when the water is flowing. Under those conditions, the dredger may be forced to work in murky water. Clear water allows the operator to see both the gold in the river as well as any hazards that may be present. There may also be legal restrictions set by your state's Fish and Game Department that limit the times of the year in which dredging may be conducted on some streams. Different states have different laws, so it is important to find out which rules apply in your location. Small scale suction dredging is an exciting and interesting pastime. Even though it's very unlikely that you will strike it rich or make a life-changing gold find, a good amount of gold is found by those miners

It doesn't take long to excavate a decent sized work hole even with a smaller dredge like this very portable 2-inch model. These small dredges can be packed into remote locations with better gold.

who know what they're doing and where to look. Many prospectors spend considerable amount of time dredging each year.

Small dredges like this 2 inch backpack unit can be carried into the remote backcountry where the bedrock is exposed and gold may be plentiful.

Dredge Sizes and Types:

There are a large number of different types and sizes of dredges available on the market. They include small, lightweight and "backpack" dredges that have a high portability factor and a minimal cost; there are also medium sized dredges that represent a compromise between small size and big production capacity; and finally, there are larger production dredges. These may be a bit heavy and costly, but they can really process huge quantities of gravel. If you are considering a dredge purchase, it's worthwhile to take a look at the various possibilities, so let's get started with that.

We will start with the small dredges - those with an intake hose having an inside diameter of 2.5 inches and less. The advantage of these units is their high portability factor and minimal purchase cost. They are also often a good first time operator's gold dredge, especially the 2.5 inch dredge. They are made for shallow waters - and usually don't have compressed air for diving. Most folks who own these dredges use a snorkel for dredging, or simply keep their heads above the water line. Many are designed with a suction nozzle type set up which is actually better for working in very shallow water conditions. They can be packed into the high country tributaries that are little worked, and may contain some good gold. The problem I find with these is that every cobble that can't be sucked up the intake has to be moved by hand, and there are a whole huge number of little cobbles that can't go up these tiny nozzles (remember that the nozzle intake is smaller than the hose diameter, so a 2 inch rock is about the largest piece you can suck up with a 2.5 inch dredge). Tossing every rock in a river that is larger than 2 inches can be a real pain, and it will slow your production down to a crawl. As a result, you need to work places with very little gravel on top of the bedrock. Some of the manufacturers formerly made a 1.5 inch dredge, but these have mostly been replaced by the 2 inch, as the 2 was only slightly heavier, but had a much greater capacity.

Next, let's consider the medium sized dredges - those with an intake hose inside diameter of 3 inches to 6 inches. These can be handled by a single operator and give a good compromise on cost and portability vs. increased production capacity and weight. The cost is definitely more, but still not in the league of a larger, commercial, production capacity dredge. As a result, this is the size of dredge owned by a lot of individual prospectors who enjoy dredging. Many are equipped with compressed air so they can be used with Hookah diving setups. They normally are set up with power jet suction systems, and offer enough suction to pull easily from depths of 15 to 20 feet. A medium sized dredge offers a lot of flexibility. They are light enough to be moved around easily and with some effort, can be packed into fairly remote locations, yet with time they can be used to punch some fairly large dig holes. A four-inch dredge is a good compromise machine, and is the dredge size I have used the most over the years. As with all dredges, there is still the issue of moving all the cobbles that won't fit into the nozzle. With my 4 inch dredge I still have to toss a good number of cobbles, but there are a whole lot fewer rocks that won't fit into a 4 inch nozzle than the number that won't fit in a 2.5 inch dredge nozzle.

Finally, let's consider those large production dredges - those with an intake hose having an inside diameter of 8 inches and larger. These are the high production, high cost units that can really process a lot of gravel. The downside is that they are expensive and heavy, but once you get them set up, you can really get things done. All of the dredges of this class which I have seen feature power jet systems and compressed air for diving. This is the type of unit you need to take on big gravel bar paystreaks and other large and deep dredging projects. The 6 inch dredge also gets to about the limit of what one man can handle alone. I know that at my age (I'm 49 as I write this) running my 6-inch subsurface dredge really works me to death over the course of a day. For just that reason, it's actually been several years since I ran it - I have stuck with the 4 in more recent years. I can't imagine running an 8-inch as a single man operation. These larger units, because of their larger water flows can have problems with fine gold recovery. The better-quality large dredge units feature some sort of screen classification and undercurrent type sluice box to improve the catch of fine gold. Another minor downside issue with these large dredges is that they can consume a fair amount of fuel, which can add up in cost, especially with high gasoline prices.

One of the things that it is important to remember is that the production capacity numbers which are described in the dredge sales flyers and catalogs are not based on the typical operating conditions found in the field during a normal dredging experience. They are based on the maximum amount of screened, light sand one could suck through the dredge under the most perfect of circumstances. Real conditions in the field are much, much different, and always far more difficult than optimal perfection. As an example, sales flyers may state that a 4 inch dredge could process "up to" about 5 cubic yards of material each hour. In real life, my experience is that in 6 hours of dredging, I can normally dig about 4 to 6 cubic yards of material. Even less if I have to do a lot of bolder moving or the gravel is compacted. This means that when you have to take the time to toss a bunch of rocks, resolve system plug ups, refill the gas tank from time to time and deal with other issues - all while functioning as a one man operation - then you will probably realize less than 1 cubic yard of gravel actually processed each hour as opposed to the 5 claimed by a sales flyer.

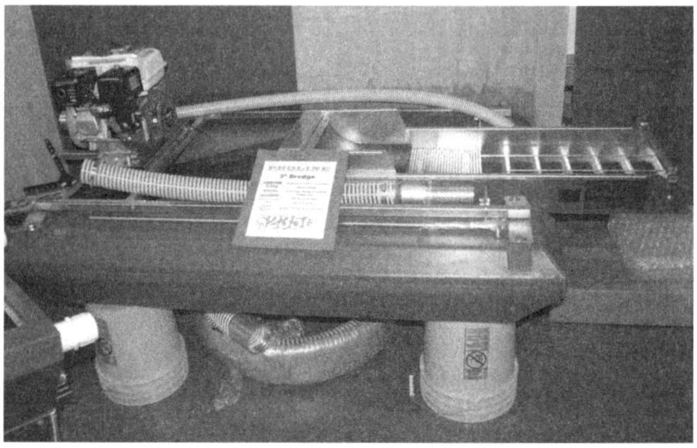

There are a number of different dredge builders. This dredge was built by Proline, a well known maker of suction dredge equipment.

While most dredges have their sluice box on the surface, some have their sluice below it. Because they do not have to lift the water and gravels above the surface, it takes less pump and motor power to run them. Some of the earliest small dredges consisted of a single solid suction nozzle unit with a tube and small sluice box attached - these are known as "submersible" type dredges. I have seen pictures of these, but have never seen one myself in operation. They have the distinct disadvantage that the gravels are dumped out only 6 to 8 feet behind the nozzle. They are also famous for poor recovery of fine gold. There is another, similar class of dredge known as "sub-surface" type. These are set up to hang just below the flotation, and have longer, better sluice boxes and while they do not equal the recovery of a surface sluice for fine gold recovery, the increased throughput often results in more gold recovered per hour worked, even though a certain amount of the fine values are lost. The other disadvantage of these sub-surface systems is that they cannot be used in extremely shallow water conditions. They are best with four or more feet of water and impractical where the stream is less than about two feet deep.

There is a lot to think about when considering the project limitations for different sized dredges - specifically, what kinds of gold deposits you can take on. If you are running a small dredge, it's senseless to sample and test a big gravel bar 20 feet wide and 100 feet long - it's a project that's too big for you to tackle. That would be a big project even for a 6 inch dredge (but it would be a doable project as long as you had the time). With a small dredge, you want an area with shallow or even exposed bedrock - those are the types of deposits you should be sampling or looking at. So when you are considering what type of dredge you want to purchase, be sure to also consider the type of dredge project you are likely to be working on. Additionally, with a really large dredge, you may well need a partner and if you don't have one or if you like working alone, then you may have a problem with the requirements of a large dredge. All these factors need to be considered in your purchase.

Way back in the 1980s, there used to be a large number of dredge makers, but as a result of the poor market for dredges in the 1990s when gold prices were low, there are far fewer today. Perhaps if the price stays high in the future, new manufacturers will come on line. Keene Engineering is perhaps the best known of the different makers, but Proline, Dahlke and others are also well known. Both Proline and Keene dredges are manufactured in California. Various folks have their own personal opinions about one maker vs. another, but they all make good products. Probably more important than the maker is the quality of the materials and construction. This is true even for homemade dredges - quality is what counts. Dredges get knocked around a lot going in and out of the water as well as while they are being used. So durability matters a lot in the long run. That is why I dislike the dredge designs with plastic sluice boxes - I just think the dredge sluice is someplace where maximum durability is needed. When you are considering buying a dredge, take a hard look at how they are built with an eye toward durability and how they will stand up to being knocked around.

Buying Used Dredges:

Let's face it. Dredges are not cheap, and sometimes a good bargain on a used dredge will come along. As I noted above, dredges are subject to abuse, and if your used dredge is beat up junk, it may not be a bargain at all. Plastic materials that are left in the sun will oxidize, becoming brittle and subject to cracking. This is especially common in the flotation pontoons that crack and then leak. It is possible to temporarily repair them, but the brittle plastic will soon crack in another location, so you really ought to consider brittle, cracked plastic as junk that needs replacement before the dredge is used. Hoses set in the sun can also be a problem. Another common thing to see in used dredges is homemade modifications or even entire homemade dredge units. Check carefully with these - some are for sale because they never worked properly, and you don't want to pay good money to buy someone else's mistake. Be sure the price is worth the project you may be taking on. Building or modifying your dredge can be an interesting and useful project, and further information along those lines can be found in this book's chapter on building your own dredge.

Setting Up Your Dredge:

There are a number of important considerations in setting up and operating a dredge. Other than the smallest dredge units, carrying in and setting up a dredge takes time. Take the time to get it set up right the first time. You don't want things vibrating and falling off later in the day, so get all the nuts wrenched on tight, and be sure to use those lock washers or other systems to keep everything held together firmly. Things that fall off into the water have a real high tendency to get lost in the sand, so this is definitely one of those times where you want to get it right the first time.

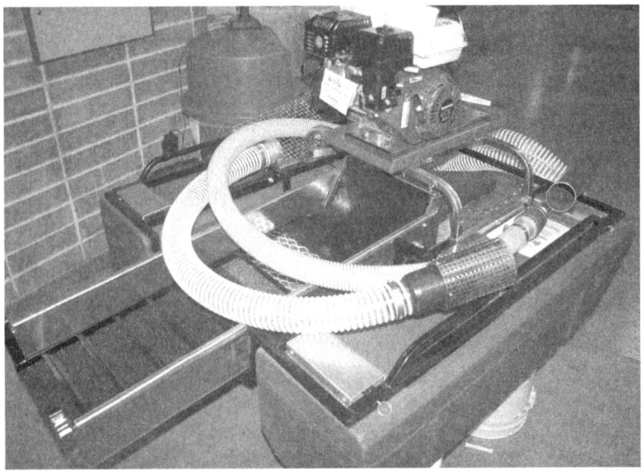

This dredge is built by yet another manufacturer, with the engine mounted up higher up on top of the sluice box.

Properly tying off your dredge is also important. Normally to get the dredge pointed the way you want, it will have to be tied off to nearby trees or rocks. Dredges also have a tendency to motorboat - the water coming out the back has a tendency to push it forward. This is another reason to anchor it to the side of the river. Finally, there is the issue of sudden or unexpected flooding. One time I set up my dredge and tied it down with a several ropes - just in the way I normally used to do, which was not very carefully. I only tied my dredge down enough just to keep it in place while I was dredging. That afternoon it unexpectedly started to rain. It came down heavy and I kept thinking it would end, but it poured all night. By the next morning, the river had risen about a foot and was flowing high and fast. When I got back down to my dredge, the branches on which two of the ropes were tied had broken. One little quarter inch diameter rope tied off on a big boulder was all that was holding my dredge from breaking loose and flying downstream to its destruction. I had expected at the most a light summer shower that afternoon, and the heavy downpour was very unexpected. I was lucky, but I have had friends who have lost their dredges to sudden, unexpected storms. Be sure your dredge is well tied down to large, sturdy trees or big boulders. Unexpected large rains, and even big summer time water releases from upstream dams can raise the river and destroy your dredge - be prepared and don't let it happen to you!

Also, if you are on a larger river where folks use kayaks and other boats to travel up and down in the water, this can sometimes present a problem. You have to keep any ropes that cross the river high enough above the water to avoid problems with the boaters. A typical way to do this is to tie off your rope about 10 feet above the river using two trees that are more or less opposite each other on the sides of the river. A guy rope is then dropped down from this crossing rope to the dredge to tie it off.

Sometimes you will need to set your dredge up in fast moving water, and keeping your dredge stable will be import. Tying your dredge off properly is extra important when operating in fast water. Sometimes dredgers will build special outriggers to act as stabilizers for fast water. These are extra floats that attach to the sides of the dredge and keep it from moving too much in a side-to-side direction (too much side-to-side rocking can hurt your gold recovery). There are

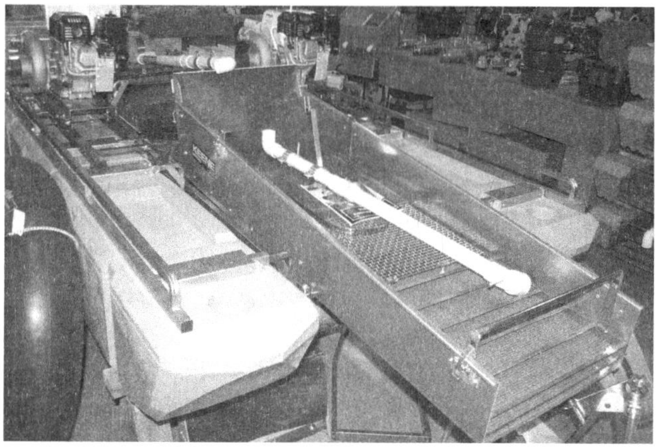

A dual motor 6 inch dredge like this may be a tough beast to handle in a one man operation, but it can really suck up large quantities of gravel.

also situations where your dredge may get tilted to one side during normal operations. If you suck too much gravel too fast the rock may lodge on one side of the sluice causing a tilt even further, and under the most extreme situations it is possible to tip the dredge over in fast water by overloading the sluice box. As you work, keep an eye on how your dredge is operating to make sure it doesn't overload on one side. Even if your dredge does not tip over, severe tilting can lead to very poor gold recovery in your sluice box.

The foot valve of your dredge must be set so that it does not draw in sand. Your pump is precisely machined, and the normal dredge pump is not made to pass a significant amount of sand. Where the water is deep enough, the intake will hang above the gravel bed of the river and so that it will not take up sand. In shallow water, it will often be necessary to take precautions to prevent sand from entering the pump. Normally the best method is to dig a shallow hole and set a plastic bucket in it. The foot valve for the intake is placed in the bucket so that no sand is drawn in. Additionally, if the stream has significant amounts of roots, leaves and other organic material, you will need to inspect the foot valve and remove these organic materials as necessary. They have a tendency to clog up the screen on the intake and will eventually limit the amount of water going into the pump. If not cleaned off, in time this material will clog the intake and prevent the pump from supplying the necessary flow of water to your dredge.

Once you have fully set up your dredge, you will need to prime your water pump. Normally this is done by quickly moving the foot valve up and down until water appears in the pump outflow hose (on suction dredges, these foot valves have a one-way flap). With larger dredges there is a spout set up on the pump intake for pouring water into the pump. The next step in getting your dredge running is establishing suction in your dredge hose. With suction nozzle systems, this is no problem; you just put the hose intake below the water line. However, with power jet systems, you need to make sure the hose is submerged and filled with water before starting the engine. With the pump primed, and the intake hose full of water, you are ready to start the engine (so long as it is full of fuel and oil).

This dredge is made of the "triple sluice" design with an undercurrent that directs the smaller gravel to slower flowing side sluices.

Dealing with Plugs in the System:

When you finally get everything together, and you've donned your wet suit, you'll probably be eager to jump in and get started digging for that gold! Enthusiasm is great, but don't overdo it and just shove that nozzle into the nearest pile of underwater gravel. It is important to regulate the speed of the materials you are taking into the dredge. If you suck up too much material too fast, your dredge will quickly plug up - and that's something you want to avoid. There is a balance of sucking material fast and sucking material too slow; you want the happy medium. Too slow and it will take forever to dig that hole in the gravel, too fast and you'll spend half your time unplugging the hose and it will still take forever to dig that hole. Let the suction of the nozzle pull the gravel into your hose - don't plunge the nozzle into the gravel. On the other hand, don't be afraid to suck up the gravel - you have to find that balance where you can process material at a good rate without excessive plugging.

Keep your eye out for flat rocks just a bit smaller than the intake and long narrow rocks. Both of these have a tendency to lead to plug ups. If you are able to grab them and toss them aside before they go up the nozzle, you will save yourself some plugged hose headaches. Part of seeing these problem rocks and getting the right flow going up the nozzle takes experience, its just something that you will get a feel for in time. If you are doing well, in most situations you should not get more than about 3 or 4 plugs per 8-hour day of dredging. Some of the new dredges with features designed to minimize plugs will have even fewer.

However, in spite of your best efforts, there will be times that you will face plug up situations and you will be forced to deal with them. Plug-ups come in several varieties, and the resolution of each depends on the variety. In suction jet dredges, the hose plug is the most common. When this happens, you will notice a significant drop in suction. As soon as you notice that drop in suction, stop sucking additional material. Filling up the hose with more gravel only makes the problem worse. If you are underwater, listen to the gravel going through the jet - under normal conditions you can hear the sound of the gravel as it goes through the metal tube of the jet. If you don't hear the sound of the gravel moving, you probably have a plug. With smaller and medium sized dredges the first thing you want to do is to slap the nozzle opening several times with the palm of your hand. This works great with 4 inch and smaller dredges, as the stopping and restarting of the flow creates pulses that will reopen the plug about half the time. You can hear when the plug is released - you will hear the sound of a surge of gravel going through the metal power jet tube. If the plug is too tight to be released by the pulse of slapping the nozzle opening, or you have a 5-inch or larger dredge, you will have to locate the plug and free it. The first step in this process is to look at the hose. Normally, you will be able to see the dark plug of rock in the translucent hose plastic (this is why dredge hose is clear). Take a nice smooth river rock a bit larger than your fist, and give the hose a gentle whack in the spot where the plug is located. If the plug frees, you will see the gravel take off in the hose and then hear it go through the metal jet tube. You are not trying to hit it so hard you break it, but your hose can take a moderate bump with a rock. Occasionally, really bad plugs will require the operator to shut down the dredge to stop all water flow, as the force of the flow sometimes holds the plug in place. A few times I have even had to partly disassemble the dredge, taking the hose off the jet and shake the plug cause out of the hose, etc. One time I even had a rusty nail get pushed into the plastic wall of the hose and puncture it. Until I figured out the problem 15 minutes later, it caused one plug after another.

If you cannot see the plug inside the hose, it may be located in the jet tube. This is the reason for the long piece of conduit or similar pipe that came with your dredge. You must go "up top" and use the pole to loosen the plug from the sluice box side. Since you are not underwater you will not hear the gravel move through the jet, but normally you will see a surge of gravel go through the sluice box when the plug is loosed. I mentioned that you will notice most plugs in your suction jet type dredge as a drop in suction. Occasionally, you will find the water in the dredge hose has reversed flow and in now flowing out the nozzle of your hose. This situation is caused by a plug in the short section between the point where the pressurized water enters the power jet, and the jet's exit point. These are fixed using the same long plug pole as previously described.

With suction nozzle type dredges, most plugs cause a reverse flow. The most common plugs occur in the bend of the nozzle, where the gravel makes a sharp turn. Plugs in a suction nozzle type system are resolved in a manner similar to those used on the hoses of power jet dredges. If you run your dredge right, recognize and avoid the "wrong" types of rocks, you will have very few problems with plugs. However, even in the best of situations, you will still have a few and it's just a part of dredging to deal with them.

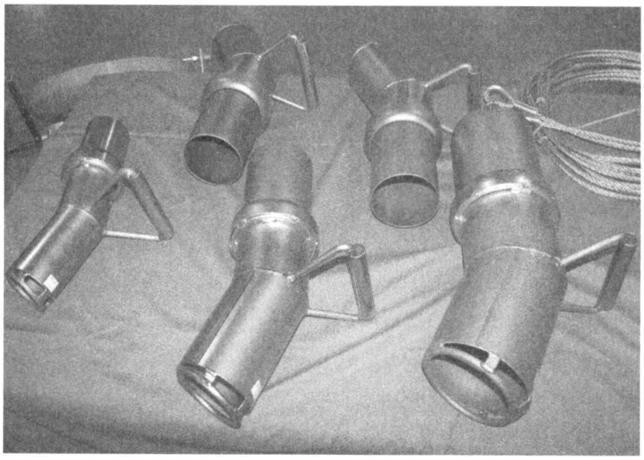

Swivel nozzles like these shown here, where the operator can rotate the intake from side to side may be a very handy accessory for your dredge, especially on larger sized intake nozzles.

When you first get started digging at a new location, make a plan for the work you want to accomplish. Planning stepwise operations at the start can save lots of time later. Where will you toss your cobbles? Make sure it's not someplace you might want to dredge later! Will your sluice tailings cover over a place you will want to dredge later? Where will you actually start the hole? Sometimes starting a hole is one of the more difficult parts of the process, especially when working in faster waters. When you are opening up a deep hole be sure to leave gentle slopes on the sides. If you make the walls too steep, they will cave, perhaps bringing down a boulder on top of you or at least making a mess of your hole. Unexpected boulders rolling out of the wall of your hole can be extremely dangerous, so you need to be aware of the big rocks in the wall of your dredge hole and keep track of their potential to pop out and fall. Good safety practice is to leave gentle slopes on your hole. Really deep holes are often taken downward in lifts - 4 or 5 feet at a time.

Occasionally, when the dredger is working down through the gravel searching for bedrock, it is necessary to punch a quick hole down to the bottom to see how deep it really is. If you expected to see bedrock at 2 feet deep and you are now down 5 feet, it can be a problem! Until you check, you just don't know how deep it is, so sometimes a small hole is dug to see the depth to bedrock or a productive flood layer. When punching a steep sided sampling hole, be careful and make sure you are not going to create a safety issue, because shifting boulders can be extremely dangerous.

There are some specific issues when the operator is dredging tributaries in the high country. The water in these places is often shallow, access is poor, and steep climbs to move your dredge in and out of the area are the norm. In these types of places, little dredges with suction nozzle jet systems are the most common; often the amount of gravel is limited and the bedrock is shallow. They rarely contain large paystreaks, and the operator is often hunting for small catches of gold behind boulders, in crevices and potholes. Now don't get me wrong, a lot of gold can be found in these kinds of places, but the volume of the gravel is often small. This is the type of location where the portability of the small "back pack" type dredge is very handy.

Here the author is dredging a crevice in shallow bedrock along the edge of a stream.

Dredging in the larger streams and rivers usually allows for better access and the use of bigger dredges. There is more water, and the paystreaks are often larger and more consistent. However, these locations also have much more gravel and often a significant overburden on top of the bedrock. Medium to large dredges, with the capability of making air for diving are the normal first choice for working these deeper, more extensive locations.

Desert or limited water dredging also presents its own special set of circumstances. The normal situation with this type of work is that you have a very limited amount of water which is recirculated, keeping the flow of your dredge within the same pool you are working. This results in muddy water dredging, where the operator cannot see what is in front of him. Dredging in these circumstances is done by feel, as the operator is in the blind. Where the material and depth to bedrock is shallow, there are no large boulders, and you keep your head above the water, the risk is small. Under those circumstances, the operator will work the dredge by feel and what he can see above the water line. However, there is a definite safety issue with dredging deeper gravels in muddy pools when the work must be done blind and the dredger is operating below the water line. I simply don't recommend deep, blind dredging, because it can be very dangerous to life and limb. If you attempt such a thing against my recommendations, be sure to take every safety precaution, have very gently sloping sides in your hole, feel around carefully for obstructions and take a partner with you for safety assistance. Small to medium sized dredges are the ones most commonly used in these situations.

Useful Accessory Equipment:

There are a lot of useful accessories and tools for the dredge operator. Crevice cleaning tools and equipment are critical to get the gold out of the deepest crevices. Gold tends to work its way down to the very bottom of the cracks and crevices in the bedrock, and you need tools to open them up. There are many types of crevice cleaning tools on the market. When I go out into the field, I have quite an assortment of crevice cleaning tools, but I mostly use an Estwing gad / pry bar, a rock pick, screwdrivers and some homemade hand scraping tools. Chisels can be useful to help break open crevices that hide gold down deep, or need to be opened up so you can suck them clean with your dredge. I also have a narrow intake nozzle extender that will fit down into narrow crevices and suck out the material. I welded up the nozzle extender that I use myself, and made it to fit both my 4 and 6 inch dredges, but you can also buy them commercially. There are also other handy tools that many guys use for the same purposes, including slide hammers for breaking open crevices and pressure hoses that divert some of the water from the pump to a nozzle that can be used to flush materials out of a crevice underwater. These accessory items can be critical in getting the gold that hides at the very bottom of crevices.

Another important set of accessories is the equipment for moving boulders. Most streams have large rocks, and in many situations, the dredger will need to be moving some of these around to get the gold which is often found underneath them. There is a large range of equipment available for this, but in general they fall into the categories of: 1) rock slings; 2) snatch blocks and related items; 3) large pry bars and 4) winches. Personally, I use only a large pry bar and a simple hand operated pulley winch for pulling boulders. The bar is sufficient for rolling rocks away when they are not too large. However, much fancier and more powerful systems employing gas or electric powered winches are also available. These more powerful units are needed for moving those larger boulders.

Boulder moving in progress – Although this boulder seemed to be well gripped, just a few seconds after this photo, the sling slipped off. Luckily, the dredger was keeping a safe distance.

When pulling rocks out of a large, deep hole you will need to create some sort of ramp to draw the boulder up and out of the hole, so be sure to leave a place for that in your work. It is important to have a secure hold on boulders, as they are often smooth and slippery and can come undone from your rock sling. The rock then usually rolls back to the hole where it comes to rest, often causing problems and maybe damage when it does. Slings are made out of steel cable, truck chains or sometimes straps. In either case, a secure hold on the boulder is very important. Additionally, it's important not to overload your equipment. Cables, straps and chains can only take so much force. When using more powerful equipment like power winches, be sure about the ratings of your equipment. When moving boulders, be sure you, your tools and dredge parts are all out of the way, so if something should suddenly go wrong, no one will be hurt and none of your equipment will be damaged. If you are moving a boulder downhill where it could roll, don't stand too close! Slings sometimes do come off the boulder and those are the times you want to be sure you give yourself some distance between you and that big rock. The few deaths I know of related to dredging all were the result of large boulders shifting unexpectedly either when they were still in place or while they were intentionally being moved. Good safety procedure is necessary for boulder moving.

Although it may seem odd to think about, the use of snatch blocks and pulleys can greatly reduce the force that your winch must exert to move that boulder. Believe it or not, when properly set up, pulleys can effectively double your winch's pulling power. Snatch blocks and pulleys are available from various sources, so you may want to take a look at these for use in your winching operations. They can be a whole lot cheaper than buying a larger winch.

I kind of have a policy with boulders of not doing any more work with them than is really necessary. Yes, you don't want your boulders to pile up until they bind you in, but sometimes merely rolling or flipping boulders around will allow you to accomplish the work you need to do with your dredge and save a lot of time. If you get all the rocks and cobbles out of your way, you may have room to roll the boulders aside and still accomplish the dredging work you intended to do. Sometimes when the overburden is not real thick, it is possible to simply work around big rocks. On the other hand, undercutting materials is something to be strictly avoided, so plan your operations carefully leave room for your work, and deal with those big boulders safely.

Keeping Warm When Working Underwater:
Working underwater requires its own set of equipment and tools. Except in a few unusual warm water conditions, the use of a wet or dry suit is necessary to keep the dredger warm. If you are working and get to the point that you are getting

the shakes from being too cold, your body is giving you a message, telling you to get out of the water quickly and do something to get warm. Uncontrollable shivering means that your core body temperature has dropped too low, and you need to do something about it in short order. If you continue working in the water, with your body continuing to get colder, you will risk much more serious health problems. The rivers and streams of the gold country in most regions are often cool even in the summer and its impossible to dredge for long without some sort of protection. Water temperatures do vary quite a bit and depending the location and time of year, to stay warm you will likely need some sort of wet or dry suit if you are in the water for any significant period of time. Dredging in the summer time in California, I have always just used a wet suit, and this has always been sufficient for me. However, if one wishes to dredge in Alaska or California streams in winter, you will need a dry suit.

Wet suits are made of neoprene rubber lined with a smooth cloth liner, usually made of nylon. They are made to allow just a little water in between you and the suit, which is warmed by your body heat. The insulating effects of the neoprene rubber foam keeps you warm in the cool water. They come in several different thicknesses, the thicker the neoprene, the warmer the suit will keep you, but the less flexible it will be as well. I use a fairly thin suit because I am a warm water dredger. Dry suits are made to be watertight and keep the wearer fully dry - they fit tightly at the wrists, neck and ankles (or they have built in feet). They can be worn comfortably in water temperatures much cooler than can be endured in a normal wet suit. In addition, insulating garments like a sweatshirt and similar items can be worn underneath them to help keep you warm. They are little bit more expensive but if you will be dredging in very cold water, this is the type of suit you need.

In order to see underwater, you will need a diving mask. Masks come in a variety of sizes and designs, but the one you want as a dredger is the type with a one-way air valve built into it. This allows you to blow out any water that leaks into the mask without going to the surface. In the old days the walls the mask used to be made of rubber, but all the new ones are made of silicon and are much more durable and last far longer. A good mask is a worthwhile investment. To keep your view to the mask crystal clear, there are some "anti-fogging" chemicals available on the market, but a little spit applied to the inside surface of the mask and spread around with your fingers will help quite a bit to keep it from fogging up during use. It's always worked for me and I'm told it is an old diver's trick. Learn how to work with your mask underwater and deal with problems should it get bumped or knocked off your face while dredging.

A wet suit like the author is shown wearing here is needed even in moderate temperature waters, as eventually your core body temperature will drop too low without one.

In shallow water it is possible to do a lot of dredging with just a snorkel as your air supply. In fact, over the years nearly all of my dredging has been shallow enough that I have used a snorkel. Using a snorkel and what is known among dredgers as the "long arm" technique, I have found it possible to dredge to depths of almost 5 feet below the water surface. I basically use the suction of the hose to pick up rocks too large to enter the hose and pull them up to a point where I can grab them. With rocks too large for this or small "problem" rocks I don't want to see enter the hose, I simply dive down and get them, and blow the water out of the snorkel when I get back to the surface. It's not very efficient and you can't do a lot of this and be truly productive, but it works for quick samples. In case you are considering building a custom "long" snorkel, there are some serious safety issues you want to think about and avoid. The air in the length of the snorkel is rebreathed during its use. Extra long snorkel lengths can lead to oxygen deficit issues from re-breathing too much of your own air. Standard commercial snorkels are a good length, but if you add to that, you may be asking for trouble.

For diving to deeper depths, you will need an independent compressed air source. While most folks are familiar with or have seen SCUBA diving on TV, far fewer have seen or heard of hookah diving. SCUBA stands for "self-contained underwater breathing apparatus", but it is not normally used in dredging, as your time underwater is then limited to the amount of air stored in your tanks. As your dredge has an engine, it can be used with an air compressor to generate an unlimited supply of compressed air for diving, and this type of set up is known as hookah diving.

Diving Safety, Snorkels and Hookah Units:
It is far beyond the scope of this book to get into a full discussion of all the hazards of diving and working underwater. However I think it reasonable to touch on some of these issues as they apply to the dredger. Care and maintenance of your equipment is critical. To really get the full amount of information you need for safe diving, it is recommended that you take a diving class through the local dive shop. Another hazard of diving that every dredger ought to be aware of is diving embolism. This is the formation of tiny bubbles in your bloodstream. It is recommended that you never ascend faster than your air bubbles. If you are diving to depths of more than 30 feet, you need to stop on the way up and allow your system to catch up. There are diving tables that specify safe ways to come up from deeper depths, but most dredging is done at depths of less than 30 feet. When diving, always use the correct type of airlines - sometimes, cheap compressed air hose that is intended for use with tools and equipment powered with compressed air is available. Don't use tool line air hose to breathe through. There is a difference - the equipment lines are intended to be used with and lubricated by a spray of oil. The equipment lines often contain oil and you don't want to be breathing that. In all cases you need to use some sort of a compression storage tank. They insulate you from the hot air of the compressor by cooling and provide a source of emergency air for a safe ascent should the engine suddenly fail. There are units made for this, but I have also seen stainless steel soda syrup tanks (thoroughly cleaned out first of course) that have been re-plumbed with quick release fittings to be used as a compression storage tank.

Dredging with only a snorkel for air can be done in many situations where the gravels and river waters are shallow.

Because of the air they trap or contain, both wet and dry suits will make you float like a cork. For working underwater, it is necessary to wear lead to keep yourself in the dredge hole. An average sized guy may need 50 or more pounds of lead to accomplish this. In fast waters you will need an even larger amount to hold yourself firmly to the bottom. Weights are arranged on belts and worn by the dredger. Make sure to use the right kind of belt that is designed for this purpose - they are equipped with a quick release so that the dredger can make a quick move to the surface if necessary in case of an emergency.
Common safety warnings tell us to stay away from bodies of water during lightning storms, and rivers are no exception. If an active lightning storm is pounding the area you are working, it's definitely best to get out of the water until the lightning passes. This doesn't happen often but is good common sense. Heavy storms (with or without lightning) also have another effect: they muddy the water. Being able to see what you are doing while working underwater is essential for safe operations. Working blind - by feel only - while under water is just plain dangerous, and I don't recommend it. The deeper the water, the greater the risk is to the diver when working blind.

When I am working underwater and I can't see at least 3 feet, I normally get out, especially when conditions are getting worse as time goes on (such as right after a storm).

This large boulder sits on an inside bend, rests on a thin layer of gravel, just above bedrock. The gravel underneath yielded 4 ounces of gold, but the dredger had to tunnel in underneath it – a very dangerous effort that is not recommended.

In all the work you do underwater remember not to undercut the gravels - it is tempting once you are down to bedrock to just dig the bedrock layer and the materials immediately above it, because they generally have more gold. However, while the gravels may stand a short time while you are digging, they can also collapse suddenly dropping large rocks onto the diver. Under the worst-case conditions you could be pinned underwater and drown, if you are lucky, maybe you will be hit by a 200-pound rock and only break a couple bones. Respect the rocks in the gravel, because no amount of gold is worth your life. Many of the most serious accidents related to dredging happen due to falling or shifting boulders. I have seen where guys tunneled under boulders the size of a commercial delivery van chasing nuggets to the point where their head and shoulders were completely underneath the giant rock. They were lucky and the rock did not come down, but they certainly didn't know for certain that it would not come down while they were still digging. They were getting good gold, and gold is tempting. Yet in spite of the fact that some nice gold came out of that spot, a human life is worth a lot more than a few ounces of gold - use your head and play it safe around boulders. I'll say it again because it's so important: No amount of gold is worth dying for.

Preparing For the Field:

There are legal aspects of suction dredging - some states require special permits, some do not. On some lands administered by the Forest Service, they require a plan of operations be filed and approved before dredging can start. The local forest manager decides if this is necessary for the national forest lands they oversee. At the time I write this, most do not require it, but because it's a determination of the local manager, that can change. As the legal requirements vary from location to location and can change over time it is difficult in a book like this to spell them out, because what is OK today may not be tomorrow. As I was initially finishing up this book, some California politicians succeeded in passing a bill making suction dredging illegal across the entire state, enacting a ban on all suction dredging. Supposedly certain salmon are threatened, but only a very few streams contain them and dredging was banned in all of California. It was challenged in the courts and as of this time all dredging is still prohibited (I'm updating this in 2025). Some folks are suggesting some legal workarounds, but things change from time to time, and it's difficult to say what will happen in the future. For now, it is allowed in some states like Alaska, but prohibited in California and other locations. I recommend that you include in your research efforts some work on determining all the current legal requirements in your area. Local prospecting clubs or prospecting equipment shops are a great place to start your legal research, but also contact the local office of the land administration agency like the Forest Service.

Once you have selected an area you are interested in and are ready to make plans, you need to consider creating a packing list of parts and accessories you will need to take. I take a lot of different items with me when I dredge. In addition to all my mining hand tools for opening up crevices, I take some stuff to repair and maintain my dredge, because breaking down in the field can be a real pain. Unlike my home, up in the mountains there is usually not a hardware store just a mile or two away. I take a number of spare parts for my dredge - extra nuts and bolts, an extra pump seal, motor oil, etc. I take some basic hand tools for making minor repairs - a battery powered hand drill / screwdrivers, files and of course the most famous fix-all: duct tape. Being stuck out in the field without critical items can easily shut you down, so be sure to make up a good packing list that covers everything you might need when you are out dredging.

Selecting a Site and Sampling:

The most important thing about successful dredging is to find those natural concentrations of gold. If you are working a spot in the river which contains very little gold, no matter how much work you do or how well your dredge is built, you can't recover the gold if it isn't there to begin with. That's why sampling to find paystreaks and other gold concentrations in the river is so very critically important. Before you even bring your dredge down to the water, take a look around and select a series of spots you want to sample. Sometimes, when I know I am going to dredge an area in the future, I'll actually take pictures from different angles and look at them later to get ideas on where to sample. Don't just figure one or two places to try, pick 5 or more different locations. Take some samples with your pan and see what you can find. Get in the frame of mind that when you first bring in your dredge you are sampling for a good deposit, and you need to move around until you find it. Look around closely and think about how the river would flow during flood conditions. Consider possible gold traps behind boulders, or bedrock outcrops; consider gravel bars on the inside bend of the river, places with shallow bedrock and other possibilities. If you need to review the concepts of reading a river for its potential gold deposits and the kinds of locations you might want to be looking for, take a look at this book's chapter on placer geology - there is a lot more detail there. Select spots that are within your ability to work with the dredge size and time you have available.

Don't forget that your primary objective is to find gold. That is your first task; do not get bogged down with focusing on moving gravel and getting into production until you have found a decent gold deposit where the values have been concentrated in the river. Far too many dredgers pick one likely spot, set up their equipment and then if they don't find much at that location, just keep on digging in that same spot hoping that by enlarging their hole somehow, eventually they will get into some good gold. I can tell you from experience that such changes in fortune are not very common. So don't get into that mindset - be mobile, make full use of the portability of your dredge and keep moving around until you find a good spot. If your first spot doesn't work out, continue onto the next and next and next until you get into some good gold. Sometimes it takes a while and you need to remember the old adage that if at first you don't succeed, try, try again.

Behind and underneath large boulders in the river can be good spots for gold to accumulate. Because of this, they make good spots to sample.

Once you find some decent gold with your sampling efforts, start planning ahead for your production operations. Think about the location of your hole - try to drop back to the beginning of the deposit and work upstream as you go. Make a plan as to where the rocks will go and how you will get the job done. Set up your dredge so that you will dump the tailings in an area that you will not be working later. This can make a huge difference in your productivity because it is important to save time by not burying your paystreak with tailings that have to be removed later. Think also about how you will be tying off your dredge so that the ropes do not interfere with others, yet they keep your dredge properly secured and effectively anchored down.

Flood (False Bedrock) Layers:

While conducting both sampling and production dredging, it is extremely important to keep an eye on the materials you are working and be aware of where the gold you are finding is coming from within them. Are you getting your gold from a bedrock crevice, or is it coming from behind an outcrop or bolder? Knowing where the gold is deposited can make all the difference in the world as to how you operate your dredge and what materials you choose to dig. Many streams have packed flood layers that act as a false bedrock, and it is very common that it will be more profitable to work down to that flood layer, and no farther. These layers are formed when the river fails to scour out the gravel all the way down to bedrock during flooding, leaving behind layers of gravel. Gold then gets caught up on the surface of this layer, much like the case with normal bedrock.

Recognizing these layer changes and false bedrock situations is an important skill for he prospector. Often these flood layers are identifiable by their high clay content and hard packed nature. Sometimes they have a different color as well. Flood layers may be deep or shallow, so you should try to punch down through the flood layer to see how far down bedrock is located. If the packed flood layer is shallow, and bedrock rich, then you need to work through it, but if the depth below the flood layer is significant, and the gold on top of it is good, it may be most profitable just to work down to the top of the flood layer and no farther. It is important to see where your best gold is coming from and make a determination about only digging the materials that get you the most gold. This allows you to be the most efficient with your dredge time, produce the most gold and achieve the most profitable results.

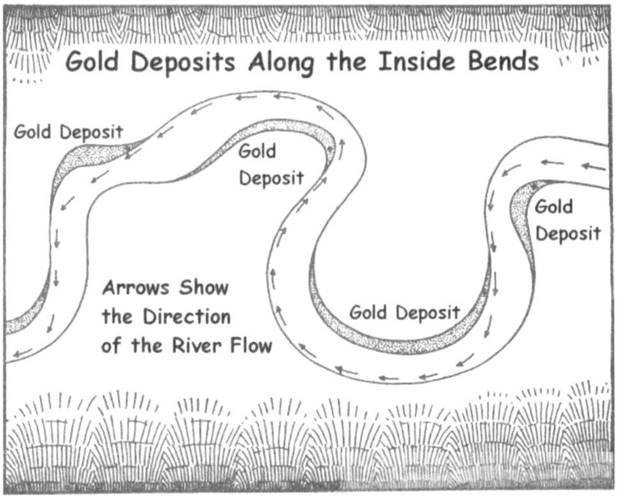

This Illustration shows the gold concentrations that form along the inside bends of a river. Considerable coverage of paystreaks and how to read a river for its gold deposits is provided in the chapter on placer geology.

Paystreaks and Gold Deposits:

The gravels within any gold bearing river are anything but uniform - some portions are rich in gold, and most other parts are nearly barren. Your search is to find the richer portions! This huge variance in stream richness is due to a stream's changing width, steepness and bends as well as any obstructions that are present. Unfortunately, in any given stream there are a whole lot more places where there is only a little gold present compared to the number of rich deposits. Gold is deposited in various catches along the stream and in larger deposits known as paystreaks. If you want to succeed at dredging, your job is to locate and find those gold catches and paystreaks and then suck them up with your dredge. Finding those rich spots is the real skill of dredging, and the true secret of successful operators. I have extensively covered the subject of gold catches and paystreaks elsewhere in this book, so for more information about sampling for them, how they occur and how you can read a river to locate them, see the chapter of this book on placer geology.

Commercial Production Operations:

For a good number of prospectors there will come a time when they decide to get very serious about prospecting and consider turning it into a profession, quitting their job and prospecting for a living. I do not wish to discourage anyone from trying this great adventure, but I do very much encourage you to be realistic. If you are a brand new prospector or have less than a year's experience, I think you should continue learning and practicing on weekends before you quit your day job. "Excessive exuberance" is sometimes a hallmark of new prospectors, so be realistic in your expectations. Still, many experienced prospectors, with true and realistic expectations, will want to try their hand at full time dredging.

For many guys, the switch to professional prospecting eventually turns the quest and fun of prospecting into the grind of a job. Not only do you "need" to go dig gold, if you don't find much, you may be going hungry or otherwise going without stuff you'd really like to have. During the winter, there will be periods of time where the water is running high and muddy when it is simply not safe to go in the water. Be sure you have a good hefty bankroll to start in getting yourself established. It is of course, easier if you are a single guy with limited or no family obligations. You'll need to learn to live cheaply and carefully squirrel away your money. You can't go off blowing money like a typical miner who makes a strike and goes out to celebrate. Making and setting aside money really is the key to success - also see the chapter of this book on selling your finds and getting top dollar for your gold. One alternative to becoming a full year 'round dredging pro is partial year prospecting - one could work for a school district, ski resort or some other seasonal operation and have summers free for prospecting and dredging. This type of plan has the benefit of allowing the prospector to build up a nice grubstake bankroll during the period he is working, which can then see him through any difficulties encountered during the prospecting season. Another popular alternative along those same lines is snow birding - dredging in the summer and doing a different alternative winter prospecting technique like metal detecting or dry washing.

Care and Maintenance of Your Dredge:
The engine and pump assembly are normally among of the most expensive parts of your dredge and properly caring for them is important if you want them to last. New dredge engines are shipped from the factory without oil in their crankcase. If you just bought a new dredge, be sure to fill it with oil. I once saw a guy bring in a dredge and set it up about 100 yards from me. I did not know at that time that it was brand new. After about half an hour I could see he was having serious problems. I went over and saw that his motor has seized. In talking with him I found the engine was new and there was no oil in the crankcase. There was nothing that could be done besides purchasing a new engine. If your dredge has sat for a while, always check the oil before you try to start it up. Sometimes the oil will have leaked out unexpectedly, perhaps while in storage. Many new modern engines of this type have a low oil level shut off switch that is automatic. However not all of them do, so always check your oil.

Here is my new Briggs & Stratton engine with the old Keene pump installed on it. This engine, with its aluminum block and overhead valves, is much lighter than the old one it replaced.

Maintaining fresh engine oil in your dredge motor is important. Your dredge is put through a lot more work than a normal lawnmower engine. One important key making it last for years is to change and replace that oil regularly. Don't let it get gooey and dark - that will cost you life on your engine. Air and fuel filters are some other maintenance items, although most dredging environments usually have little if any dust. Modern small engines also do better if you store them over the winter without fuel, which is known as dry storage. This is because our modern gasoline has a tendency to degrade and deposit gum and other compounds inside the engine, and can lead to expensive repairs. I suggest that at the end of the season you empty the fuel from the gas tank, and then run the engine for a minute or so to remove the remaining fuel in the lines (don't forget to put a bit of water in the pump to wet the seals). Alternatively, there are gasoline stabilizer chemicals on the market that can be added to the fuel, but the dry storage is best.

With good maintenance, the small engines that power these dredges will last for many, many years, but they do not last forever. Recently, I had to replace the engine on my 4 inch dredge. It was more than 15 years old, and just plain old worn out. I did some research and found that there were a lot of choices on the market for various engines in the correct size. While many users like the small engines produced by Honda, the American makers have done a lot to improve the quality of their products in recent years. I selected an engine by Briggs & Stratton and have been very happy with it so far.

The old yellow plastic pontoons as shown on this dredge are prone to becoming brittle from long exposure to the sun. Cracks and holes can be filled but the damage cannot be undone.

Dredge parts can be large and inconvenient to store during the off season. One thing I have seen is that folks often want to store some of them outside - especially the flotation pontoons. There is a problem with this, in that over time the sunlight causes serious damage to the plastic materials. On long exposure to the sun, they become brittle and prone to cracking. A sign of sun damage is that the surface becomes covered with small cracks. Many folks wonder about repairing sun damaged plastics, and while some try to re-melt the material or seal the cracks with resins and glues, these are normally only temporary "band-aid" types of solutions. The problem with plastic materials is that once they become sun damaged, the damage is done and there really is no process by which it can somehow be "de-brittled" and restored to its original condition. The plastic only becomes more brittle over time, and new cracks will continue to appear with regularity. The cracks let in water, which will eventually sink your dredge. If your dredge sinks suddenly while in operation, not only will your sluice box concentrates be lost, but your engine may well be destroyed also. The answer isn't much better should your dredge sink slowly over night while you sleep (but you might save the engine if you are careful). The real answer is to replace those sun damaged pontoons. This is a costly choice but worth it to prevent sinking and more serious damage.

The seals on your water pump are cooled and lubricated by the water in the pump. If you run the pump dry for more than half a minute or so, you may do damage to the seals. They are replaceable but not cheap, and doing so is a pain in the neck, especially when damaging them is easily avoided. Always put at least some water into the pump when you want to start your engine, even if it is only for short term testing purposes.

Dredging by Yourself or With Others:
There are considerations for dredging by yourself in a lone wolf type of operation, or working in a partnership. With smaller and medium sized dredges, it is very possible to operate them as an individual, but with the larger sized dredges it can be difficult to move them around and operate them at their full capacity when you are alone. There can be safety issues as well, and having a partner certainly is a safer proposition. If you can find the right partner, having someone with you can be both an encouragement and a help, but the wrong partner can lead to major problems - so choose wisely. It is very important that both partners be clear about their expectations and plans. One guy may be out there mostly to have fun, while the other is focused on working hard and digging gold. Miscommunication and a general lack of communication are commonly at the root of many partnership problems. I will admit that most of the dredging I have done, I have done as an individual - in many aspects it's just easier that way. When considering a partner, it's important to remember that gold has split up a lot of friendships. Gold just has an odd sort of effect on some people, bringing out the greed and worst parts of their personalities. It is easy for one guy to start feeling like his partner is not putting in his share of the work - other types of misunderstandings are also common. Sometimes in those situations, if both guys have equipment, it is best to allow each worker to keep the gold he digs, rather than splitting the total take. If you enter into an agreement with a partner, be sure to

choose carefully and make sure both parties are very clear on their expectations and how various situations will be handled.

Moving Your Dredge Around:

Successful dredging often means taking advantage of your dredge's mobility to sample and find valuable gold concentrations. It is important to take advantage of that portability factor when you're sampling around. Moving the smaller dredges is no big deal, but moving the larger dredges requires much more effort, especially in faster water situations. Make sure everything is firmly tied down and secured on board. Two guys, one on each side of the river, controlling the dredge with ropes can direct the dredge as it moves downstream. Usually it is a lot easier to move a dredge downstream with the current helping along than it is to move it upstream fighting the flow, so plan where you put into the water with that in mind. Always do what is necessary to keep the dredge under control, as things can get out of hand very quickly, especially in fast water.

The other part of moving your dredge around is getting it down to the river from your car for the original access to the area you want to work. For short distances, I just carry the pieces, but I have also used various wheeled devices to carry it in for longer distances. One of these is like the device used to carry a deer or other large game by hunters, sort of a platform with hand rails that rides on top of a single bicycle sized wheel. Another takes the form of a hand truck with both wheels much larger than normal. Over the years I've used both these types of devices and they can be very handy in moving your dredge around over a trail and down to the water.

Cleaning Out the Sluice Box of Your Dredge:

Cleaning out your sluice to recover your concentrates is one of the most important parts of operating your dredge. Do it wrong and you may well loose a fair part of the gold you've worked so hard to capture. Many guys ask how often a dredge sluice ought to be cleaned out. I clean out my sluice at the end of each day, though I make a quick check of the box each time I come up out of the water for any nuggets or pickers that I can remove. This tells me about what I am producing

Moving a large dredge around is a project that's really better done with at least two guys.

and how well I am doing in my work - and multiple pickers and a few visible nuggets in the box is always a good sign. Now my sluice is a three-stage sluice, with a short head section followed by a screen and an undercurrent sluice section. I clean out the head portion of the sluice daily, but the undercurrent part is only cleaned weekly, including the section of the sluice that handles the material that is too large to go through the screen. For sluices that have no undercurrent section, my recommendation is daily cleanup of the full length of the box. For sub-surface sluices, it's a different matter and because of their smaller, shorter sluice box sections, I'd recommend cleanup 2 or 3 times daily. They are also easier to clean out as they normally have a smaller detachable tray that can be easily dumped into a bucket. The first step in doing a sluice cleanup is to make sure your dredge is in an area where you can easily access your dredge. This means that in some fast water areas you may need to move your dredge into quiet water in order to safely do a clean-up. The last thing you want to do is slip or stumble in fast water and dump your concentrates back into the river. River rocks can be plenty slippery! I know this well from personal experience. You'll need a good sized tub, or bucket, one large enough to cover the end of your sluice. Lift or remove your riffles, and carefully roll up the carpet, and place it in your tub. Most of the concentrates should be on the carpet. When I clean out just the head of the sluice, as much of the remaining concentrates as possible is picked up and collected by hand. When cleaning out the whole sluice, what remains in the sluice is then washed down into the tub or bucket. Very often the material under the matting in the undercurrent section of the sluice has a good amount of fine gold. Once you have captured all your concentrates into a tub, it is time to begin the initial rough processing of them in the field. Select a nice calm part of the stream where you can easily access the water to process your concentrates. I often use a couple pans and two buckets in completing this process. With my concentrates, I screen them into a coarse and fine sized material. The first screen I use has holes about ¼ inch in size. All nuggets and picker sized gold pieces are removed off the screen and placed in a bottle. The coarse materials are panned, and any nuggets that were missed are captured. Usually this goes fairly quickly. The smaller material is then screened again with a 20-mesh screen. This material is also panned and it also goes quickly, with the gold being picked out and placed in the bottle. The fine material is panned down until all that remains is mostly pure black sand with fine gold. With the final black sand material containing all the fine sized gold, I place that material in a container to be taken home and processed later. Usually that amounts to less than a pound of black sand materials generated each day. At home I have equipment and a special procedure for processing all my black sands, no matter if it comes from sluicing, dredging or dry washing. For a full description of the procedure I use, see the chapter in this book on the final clean up and dealing with black sands.

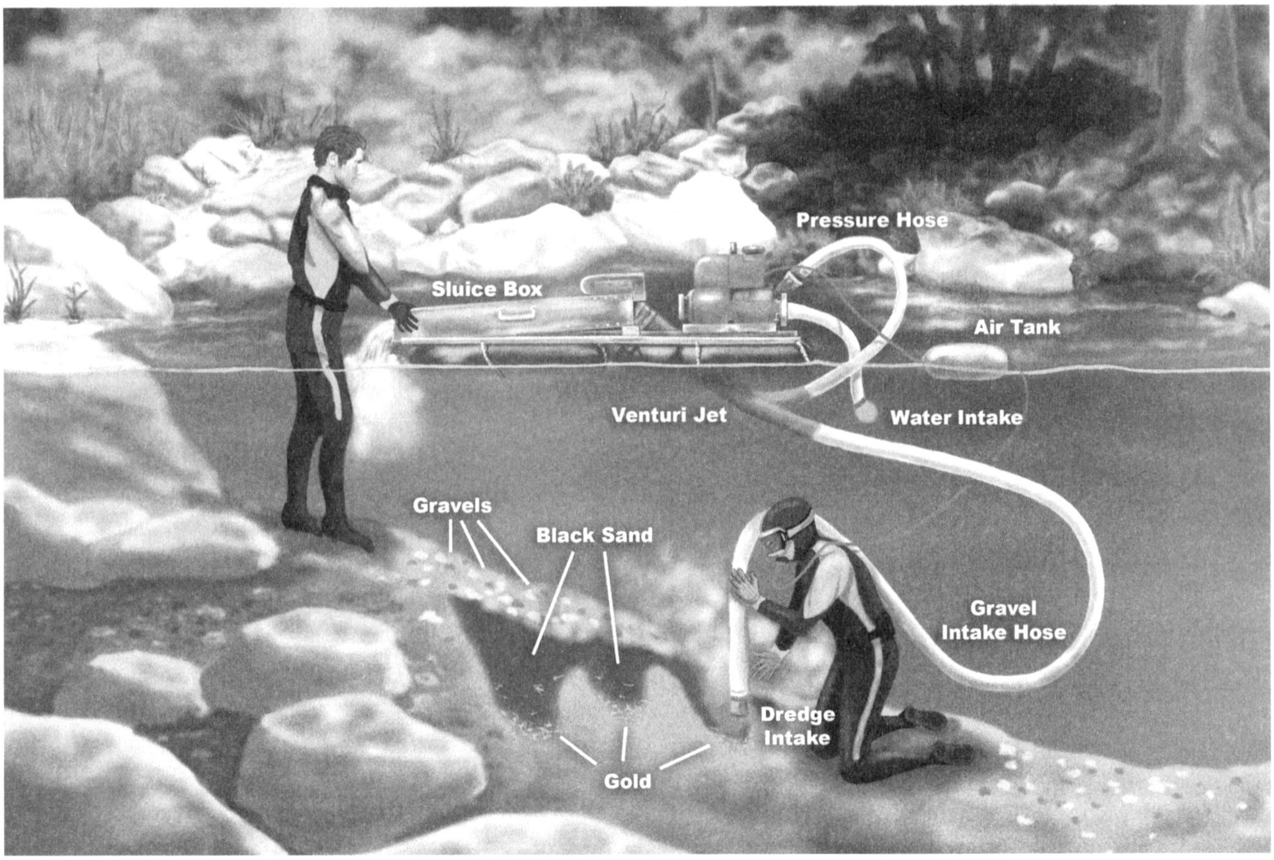

The diagram above shows the basics of suction dredging operations. Gold that is otherwise inaccessible in crevices under water can be captured with a suction dredge. Unfortunately, many environmental extremists have decided to oppose and target small scale suction dredging, resulting in it becoming illegal to operate a suction dredge in those jurisdictions. Be sure to check first for the latest legal status of dredging in your jurisdiction.

Maximizing Your Recovery: Getting The Most from Your Sluice Box

The sluice box is probably the most common gold recovery device after the pan. For centuries, the sluice has been one of the world's most popular devices for primary placer gold recovery. Every dredge and every highbanker, has one. It is used by very small-scale hand miners all the way up to commercial scale placer operators that run several hundred tons per hour. They are also very reliable, inexpensive to build and simple to operate. They can recovery a very high percentage of gold coarser than 50 mesh, and properly designed and operated, can even recover a decent percentage of the gold larger than 150 mesh in size. Historically there have been a huge number of designs, with parallel riffles, cross riffles, stone lined bottoms - all the way to the more modern Hungarian riffle types. Sluice boxes can provide much higher concentration ratios than most other gravity type concentrators with very good overall gold recoveries if properly designed and operated. The sluice box is a fairly forgiving piece of equipment. However, depending on how it is operated, sluice efficiency can range from very good to very poor, and no miner wants to be one of the operators that get poor recovery from their sluice box.

The flow of water in your sluice should not be too turbulent. The flow rate pictured here looks to be just about right.

There is no question a book like this one needs a section to fully discuss sluice box operations, and I could have put a bit of this discussion under several headings, but I have decided to consolidate them all here in a separate section, rather than repeat myself under sluice boxes, highbankers, commercial placer operations, etc. Because of this, this section on sluice box operations will combine aspects and concepts applicable to all sluice operations. This works out OK, since while there are a few differences, there are many similarities in all sluice box designs.

In some ways, a sluice box is much like the bed of a river - some areas of a river have just the right conditions and concentrate considerable gold. Other areas in a river have the wrong conditions and so retain very little gold. If you run your sluice box correctly, it will have the right conditions and concentrate lots of gold - if you do not run it right, it will allow much of the gold to run right through, just like those sections of river that are nearly barren. So it's worthwhile to take a little time when you are setting up to make sure your sluice box is operating properly and getting the reliable recovery you want to see. The toughest gold to recover in a sluice is the fines - they are the most likely to fail to settle in the sluice, and the most likely to be scoured out by turbulent waters. If you can run your box in a way that it reliably catches 50 mesh gold, you won't have problems with the pennyweight coarse stuff (50 mesh means 0.0117 of an inch or roughly three tenths of a millimeter). The actual shape of gold particles in the placer deposit can greatly affect recovery - flat particles tend to lay down and resist the water's flow, while round particles of equal weight would have a tendency to roll. However, the placer miner cannot choose the shape and size of the particles in his deposit. He must accept the hand he has been dealt.

Even if one does recover some fine sized gold, it does not mean you are getting the majority of the fines in the gravels you are mining. Sluice box operations are just not that simple. Even if most of your gold settles in the upper end of the box, it does not necessarily mean you are doing as well as you could. To really get a feel for what your sluice is losing, you need to sample the tailings of the box when you are processing good, gold bearing material. The tailings should be checked with some very efficient placer processing system such a shaker table or by careful panning. Only this type of testing really tells you what you are missing. Fire assay is not recommended; as it often yields misleading results because any gold that is contained in the rocks of the gravels that are being processed will be recovered by a fire assay. This gives a misleading result, since that hard rock gold cannot be captured in any normal placer system.

Reading the sluice and interpreting the results takes some understanding of how a sluice works, so let's take a look at how the gold is recovered. So how does a sluice box do its magic of separating gold and sand, sending the worthless sand on down the line while holding onto the heavy materials like gold or platinum? There is probably more than one mechanism at work in a sluice, and there have been a number of different opinions as to how these forces interact to capture the gold. In the past, the settling velocity of gold (and other high-density materials) was considered the chief factor, but more recent scientific research emphasizes the centrifugal force of the swirling water behind the riffles. Some of these modern studies of sluice box operations are actually available on the internet and have become a topic of discussion for prospectors trying to get the best performance out of their equipment.

Historically, riffles have been made from a number of different materials, and in a number of designs. The floors of sluice boxes have been paved with many different types of arrangements including large stones, railroad steel, wood poles or rods in various positions that are either vertical, crosswise or lengthwise to the sluice. These arrangements were often more the result of random guessing by the old time miners working with the materials that they had available, as opposed to the result of any actual scientific study. Many years of testing and experience has demonstrated that a removable ladder of steel obstructions set perpendicular to the flow of the water in the sluice is the best arrangement. Modern riffles are made in a variety of configurations, depending on the water flows and size of the materials being processed. Popular riffles include 90-degree angle iron, flat slats of steel set at a 45-degree angle to the bottom of the sluice, and Hungarian or "lazy L" type riffles. Hungarian riffles consist of flat steel set at a 45 degree angle to the base of the sluice box, but the top lip of the riffle is bent over to be roughly parallel to the floor of the sluice.

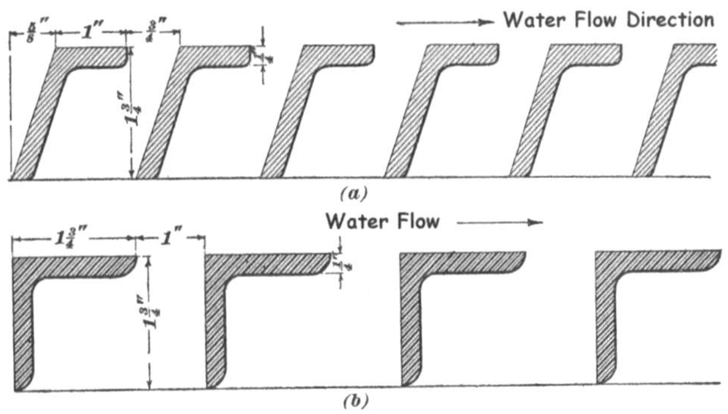

Older style Hungarian riffles above (a) and angle iron riffles (b) below. In both cases I would recommend leaving much more space between the riffles – every other one could be removed. If riffles are too close together the likelihood of material packing in behind them is greatly increased. Packed gravels between the riffles can lead to large gold losses as the gold moves through the sluice and finds no place to lodge.

Researchers have stated that riffles work because "turbulent eddies are formed in the slurry as it flows over and around the flow obstructions that comprise the riffles". The sand/water slurry velocity (momentum) provides the energy that powers the vortex behind the riffle. That energy is necessary. If the velocity of the slurry is reduced through overloading with solids, insufficient water flow, not enough tilt to the sluice box or some other reason, a full size vortex may not be created, leading to riffle packing and loss of gold. Pay gravel which contains a high proportion of high specific gravity minerals such as magnetite, or a high percentage of clay are more susceptible to riffle packing. These may require increased box tilt and/or water flow. On the other hand, if the sand/water slurry velocity is too high, extreme turbulence will occur and fine gold will not settle out. Even after settling it may be scoured out causing excessive gold losses. The goal then is to find that happy medium where good flow

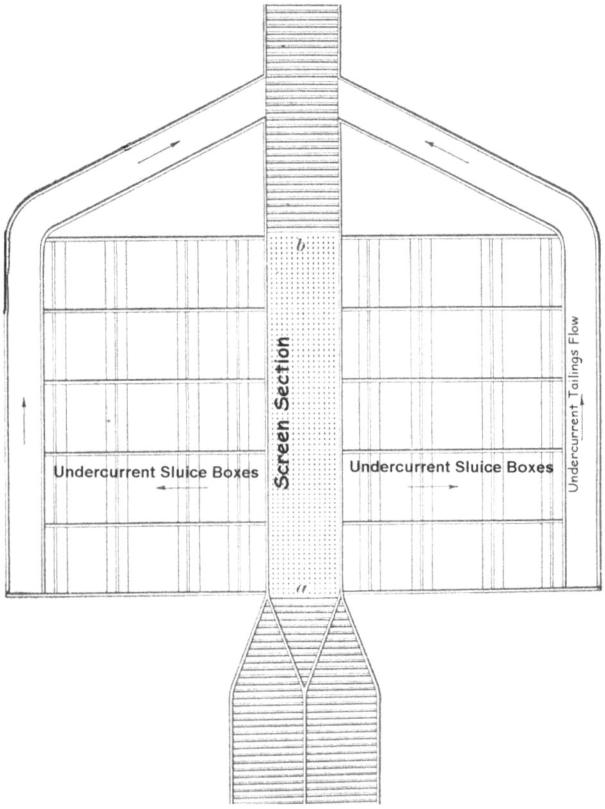

This old time drawing shows an undercurrent sluice section with screen and side undercurrent boxes to handle the fines at a much slower flow rate. Even the old miners knew that this type of system helps a lot to capture the fine gold.

keeps the gravel materials moving along efficiently through the box, but not to the point that the high velocity and turbulence simply blows out all the fine gold.

Scientists and engineers have long noted that turbulence is a function of water speed. The more water you try to push through even a smooth bottomed trough (like a sluice without riffles), the more turbulent that water flow will be. Now turbulence in the water flow of sluices is actually both good and bad, because a little is good and large amounts are bad. A little turbulence is good behind the riffles where the turbulence helps sort the material in the slurry and accumulate the gold while sending the lighter materials downstream. That's planned and limited turbulence exactly where you want it. On the other hand, turbulence is bad where it is large and unplanned - boiling white foam from excessively high water speeds, around large rocks that get lodged in the sluice, etc. Its bad because the excessive turbulence from those high water speeds can cause big losses in the smaller sized gold. Random turbulence (mostly caused by high water speeds) is the enemy of good fine gold recovery. Slowing down the water speed really is the core "secret" of improving recovery of fine gold at both small scale and commercial placer operations alike. Water speed control is important to remember when running many types of placer recovery systems used by individual prospectors.

For commercial operations, if you can screen out the big rocks and small boulders to say 1 inch minus, then you only need a water speed sufficient to move a one inch rock down the sluice. Without removing the big stuff, you then need a stream of water sufficient to move an 8 or 10 inch rock through the sluice. The water speed needed to move that 10 inch rock (and the inherent turbulence it causes) is also sufficient to blow much of the 20 or smaller mesh gold right on out the box as well. That's why frothing white water flying down a sluice box is a BAD thing. This is also why most highbanker units have a scalping screen built in to the feed box. With a dredge, you really have little ability to size the material in advance - other than stopping down the intake nozzle to a smaller size. However, this is not recommended as it means a lot of hand movement of waste rock. Undercurrent sluices, which screen the material and divert the fines are the way to go with suction dredges and similar systems. Undercurrent sluices recover fines by sending them to a slower moving sluice box where the fines can be caught. Commercial sluice systems may employ screening systems like trommels and what is called a nugget trap to capture any very coarse gold that ends up with the oversized material which will not pass through the screen.

So let's take a look at some of the possible steps that the prospector can make to improve the recovery in his sluice box. There are nine important considerations leading to adjustments or possible modifications to your equipment that when done properly will allow you to bring home more gold. The following checklist and discussion topics are important things to think about for improving your sluice box recovery:

1: Proper Box Angle and Water Flow;
2: Gold Holding Mattings;

3: Expanded Metal;
4: Material Sizing and Undercurrent Sluices;
5: Feed Rates;
6. Dampers;
7. Settling Plates;
8. Flared Sluice Entry on Suction Dredge Systems;
9. Flared Sluice Boxes;

1. Proper Angle and Water Flow:
These are the adjustments and settings that concern folks the most, because it is true that excessive water speed and box tilt will allow gold to be flushed out of the sluice box - especially the smaller stuff. However, it is equally true that water flows too slow and a tilt too flat will clog the riffles and also result in the loss of gold. So where is the happy medium where the recovery is at its best? Generally speaking, sluices should be run at as slow a speed as you can get by without accumulating rocks in the sluice and without packing the space between the riffles - the slowest speed that will give a good movement of the slurry down the box. That is my simple rule of sluice box operations. The minimum flow you need to get by and keep the box looking right. Over time you should accumulate some material in the box, but some matting should still show between the riffles. There is no one angle that sluices should always be run at. Sometimes new folks think that there is some magic single sluice slope which works in all cases.

Fast flowing white foamy water as shown in this sluice is a very bad thing – it will blow out most of the fine sizes of gold.

This is one of those rare situations where more power is NOT better. For guys that like real powerful, high suction dredges, I'd say that trying to increase suction at the nozzle until it is strong enough to suck the gloves off your hands is usually going to result in sluice box flows fast enough and high enough to lose fines. High suction often means high water flows, which equals poor fine gold recovery. Higher volume (larger sized) dredges often need an undercurrent sluice design to get decent fine gold recovery. High flows create turbulence that prevents small gold from settling out, keeping it in suspension until it eventually just goes out the back of the sluice. The smaller the gold size, the more easily and longer it is kept in suspension - and the more easily it is lost. This also goes for commercial operations - if the water just blows through your sluice box, you'll never have a rock hang up in it, but on the other hand, your fine gold recovery will stink. This is also the reason underwater subsurface dredges have narrow sluice boxes and an inherently lower fine gold recovery - you have to keep the flow speed fast enough to blow those big rocks on through the system.

On the other hand, this means that if you constantly need a box tender to pick out the rocks that get stuck in your sluice, you need either to reduce the size of the largest rocks being fed (as in a hand sluice or highbanker) or increase the flow and/or tilt of the box (as on a dredge). A tender should not be necessary. The same is true if your sluice is overloaded and you are accumulating so much material that there is no matting showing between the sluices. You need to have enough flow to form the wanted and desired turbulence behind the riffles to keep the lighter sand materials moving along. The higher the riffles are, the greater the minimum flow must be to form good centrifugal turbulence behind them. This is why hand sluices which run on river flow have low height riffles and expanded metal, while a large 6 inch suction dredge has riffles an inch tall or taller.

A balanced midpoint flow is the goal. If, after running a while you have very little material accumulating behind the riffle, slow the flow down or decrease the tilt. If you run a while and your box is accumulating sand and rock or burying the riffles, then you need more tilt or

A properly operating sluice has smooth, even flowing water with a minimum of froth and turbulence.

more water flow. A "normal" sluice box tilt will typically run between 1.5 and 2.5 inches per foot of length, though high banker sluices as well as some commercial set ups often having a bit higher tilt angles.

On my own dredge, I turn down the throttle and run my dredge as slow as possible, while still maintaining good suction at the nozzle without plugs and without rock accumulation in the box. If my sluice is retaining some material, but showing the matting for about half the space between riffles, I figure I have the system dialed in pretty good, and my experience is that I get pretty good fines recovery when that happens.

2. Gold Retaining Mattings:
What's underneath your riffles? One of the best sluice box improvements made in the last 20 years are the spaghetti loop mattings. They are known by various trade names such as Nomad® or wayfarer, and are used under the riffles to catch fine gold and prevent it from eventually being washed down the box and out. These mattings can capture and hold considerable amounts of gold. They work by allowing the gold to fall down into the space between the loops and get out of the main flows of water, permitting the fine gold to be captured and protected from the scouring effects of the water flow. Based on a number of tests, the unbacked versions are preferred and appear to be the best box matting now in common use. There are a number of benefits to using these mattings: 1) they do not interfere with vortex formation behind the riffle; 2) most of its volume is available for gold storage; 3) it does not easily release trapped gold particles during sluice box operations and 4) it is easy to clean. As a result, these mattings greatly improve the recovery of fine gold and are well worth the cost of adding to your sluice if you work in an area that has significant fine gold. After repeated use a few small pebble grains can get lodged in the loops, but usually a good stretching of the matting will allow these to pop out (if you try this be sure to do it over a tarp or container as the material popping out may well include gold).

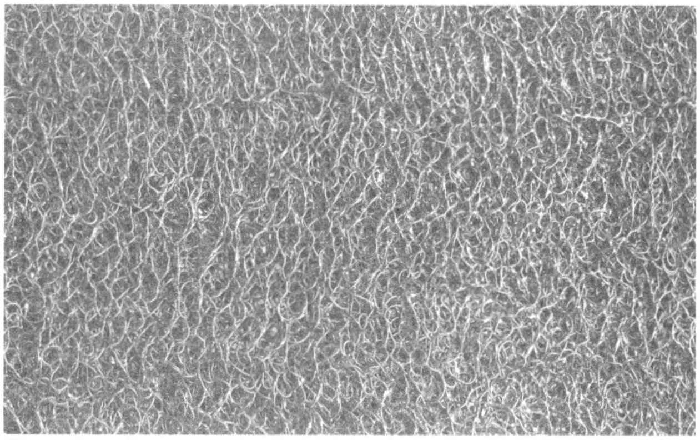

Spaghetti type matting as shown above is a great help in capturing fine gold within a sluice box.

3. Expanded Metal: Better Riffles for Fines
Testing has shown that expanded metal actually recovers fine gold a bit better than the standard Hungarian or angle type riffles. The greater height of the tall "standard" riffles makes more turbulence, which tends to flush out some of the smallest fine gold. The lower height of the expanded metal allows lower turbulence, therefore better recovery of fines. Researchers suggest that for best recovery, sluices should contain sections of short height expanded metal and sections of taller Hungarian riffles. If you are building or remodeling your sluice, you might want to keep those facts in mind.

4. Material Sizing and Undercurrent Sluices
For over a century, miners have known that separating out the fine gravels and processing them in a slower flow of water will result in a higher overall gold recovery. The design is called an undercurrent sluice and is also sometimes known as an over and under style sluice. The standard method is to pass the gravel slurry over a screening surface such as a punch plate or other

similar device, allowing the small gold and finer sands to pass downward where they are diverted to a separate sluice area where water flows are slower, and fine gold particles can be recovered. The size of these holes are in the 3/16th to 3/8ths size for most suction dredges. The trick in designing these systems is to allow sufficient screening area so that the smaller material can settle down and pass through the holes. Where the fines are held up in suspension in the slurry, they cannot drop down through the screen. These then go to the "coarse" area of the sluice where fines recovery is not as good. Some commercial set ups fully process the gravels by wet screening and only sluice the smaller material that passes through the screen. Another problem with these systems is that the screening unit, if it is a non-moving type such as a punch plate, will often plug after enough material has run over it - small rocks will lodge in the punch plate holes. So the system needs to be designed such that this plate can be removed and cleaned. Some dredge makers offer an aftermarket retrofit kit to turn a single deck sluice into an over and under type, but these can also be done by an individual with some fabrication skills. The conversions, if properly done, are well worth the effort.

5. Material Feed Rates

One very common problem with sluice operation that leads to low recovery is surge feeding as waves of material. This overloads the sluice, temporarily clogging the riffles. When the riffles get clogged the normal sluice processes are not working and gold may roll over the top and be flushed back out into the river. This can be especially common in some commercial operations that are directly fed by a dozer. Shoveling into highbankers, one can experience this as well. However, it can happen even with hand fed sluices run on river flow, if too much material is dumped in at once. There is no question that overloading your sluice with too much feed results in poor fine gold recovery. In most hand fed sluice operations, it's easy to avoid this condition and it is worthwhile to do so. It is also important to note that when there is no feed coming into the sluice, and only clear water is running, the water flows have a tendency to scour out the fine gold and push it out the box. So it is important not to leave a sluice with no feed for long periods of time. A good way to deal with this is with deep mattings underneath the riffles. The spaghetti loop type mattings also known as the "no mad" type material do a great job of letting the gold drop down into the matting and preventing scour during periods when only clear water is entering the sluice.

High feed rates, which produce high-density slurries, are different than surge feeding. High density slurries are often found at commercial operations. As long as the other parameters of the sluice are properly designed and operated, good recoveries can be achieved even with fairly high feed rates so long as the box is not overloaded.

The black plastic flap at the head of this sluice is a damper. It helps to even out flow in the box.

6. Dampers

Dampers are flat sheets of rubber that ride on top of the slurry entering the sluice box. They spread out the flow and even out distribution of the feed, helping to reduce excess turbulence. Dampers can drop any floating fine gold off the surface, but gold rarely floats in the normal turbulence of a sluice box. Dampers are well worth adding to a dredge sluice box that does not have them.

7. Settling Plates (Slick Plates)

Some research reports done on sluice operations recommend slick plates in front of the riffles. This is to allow the gold to settle and be riding along the bottom of the sluice when it hits the riffles. This is the optimum point where one would want to have the gold located when it enters the riffle area. Some of the research reports recommend slick plates of around 4 feet long for large commercial dozer fed operations, however for smaller scale individual operators with dredges or hand fed sluices, much shorter slick plates would suffice. In dredges with a flared type sluice entry system, the flare performs some of the function of a settling plate as it allows the gold to settle and move downward before it enters the riffle section of the sluice box.

8. Flared Entry on Suction Dredges and Similar Slurry Flow Systems:

The flared entry on suction dredges function in a manner similar to a slick plate on other sluice box systems. By slowing down the flow speed, they allow the gold to settle and be riding along the bottom surface when it enters the riffle section of the sluice. Suction dredges have a certain amount of induced turbulence and mixing of the gravels caused when the material goes through a power jet or crash box type header. This type of turbulent mixing keeps fine gold in suspension and is the enemy of good gold recovery. Flared entry on dredges is a modification that I think well worth making, as they do a lot to improve fine gold recovery. If you have the old crash box style head on your dredge's sluice, the extra turbulence it creates is definitely costing you some fine sized gold.

9. Flared Sluice Boxes:

Some sluice box systems flare out the width of the box, so that the end of the box is wider than the front. This type of system is not recommended. The width of the sluice affects the flow speed, so flaring the box width slows down the speed at the end of the box. However, the best operated sluice runs the minimum flow that will keep all the rocks and gravels moving down and out of the box. By having the speed adjusted so that the wide lower end is running at the minimum necessary speed, you will find the water speed at the narrow top of the sluice is too fast. If you adjust the flow so that it is right at the top of the box, then where it slows at the bottom, rocks will pile up. Flared sluices are not common, are harder to manufacture, and I think that in the end you simply are better off without them.

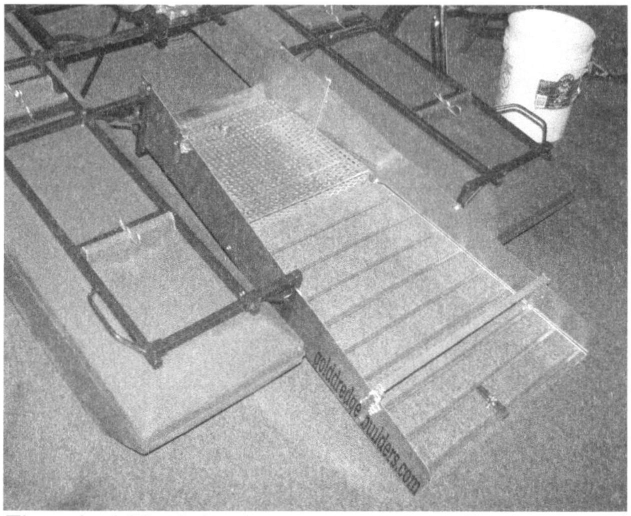

Flared width sluice boxes are generally not a good design. They make a proper water speed harder to achieve. It often achieves a sluice with too fast flow at the top and too slow at the bottom.

Testing Your Sluice Box:

It is important that you have confidence in your sluice. It makes a big difference. That's why I recommend testing your sluice box to make certain it is working properly. Perhaps the best way to do this is with small bird shot (not big buck shoot, use the tiny stuff). I take about 20 pieces of bird shot, whack each one once or twice with a hammer to flatten it, then paint them red. I add them to the sluice after I have run a good amount of material through, and then dump in the 20 pieces. I then run more material through the box. When I clean up, I count the number of shot I have recovered. Painting them red makes sure that I won't count any shot that I might have picked up in the placer gravels. If all the shot are recovered, then your box is operating reliably, and you can have confidence in it.

Reference Bibliography:

The following research papers are available for further reading. These and similar documents can be found on the internet. The scientific research done on sluice box operations has been done for commercial type, larger scale operations. The availability of these reports on the internet has resulted in some misconceptions because these commercial operations are very different from the smaller scale operations of the individual prospector. One of the things we dredgers need to consider in our evaluations of these research papers is the major differences between our equipment and typical heavy equipment fed commercial placer operation. They are running at a much higher feed rate and ratio of solids to water. They often run one inch or taller riffles, where we run lazy L type Hungarians that are much shorter in our dredges. Their boxes also run at a greater tilt. As a result, the differences between how they run their sluices and how our suction dredge sluices operate are significant, even if both are sluice boxes. Keep that in mind when reading the commercial research results. The following reports can be found on the internet with a little bit of searching.

I. Fine Gold Recovery of Selected Sluice Box Configurations
G. W. Poling and J. F. Hamilton
Department of Mining and Mineral Process Engineering, 1986
University of British Columbia, Vancouver , BC

II. Placer Gold Recovery Research: Final Summary
Randy Clarkson, PE, New Era Engineering
Prepared for the Klondike Placer Miners Assn., December 1990
Whitehorse, Yukon

This photo shows sluicing for gold more than a century ago at Nome, Alaska. These old wooden sluices could be easily made on site with local materials, but rarely got good recovery on fine gold. However, the deposits worked by the old timers were often rich enough that it was still very profitable to mine them even when the fine gold recovery was poor.

This photo shows the view from inside a wooden sluice box while preparing for clean up. Klondike mining district, Yukon, Canada.

These miners are sluicing for gold at number 2 below discovery on Bonanza Creek in 1898 at the Klondike district, Yukon Territory, Canada during the height of that great rush. Piles of rocks like those in the photo were taken out of the sluice because they blocked the flows. Rock piles like these mark the sites of historic gold mining operations all over the world.

Dry Washing For Gold Nuggets

Placer deposits have been mined in the desert regions of the American southwest, Australia Mongolia, and a number of other places across the globe where very little water is available. Since conventional wet methods like sluice boxes cannot be used to recover gold in these regions, various dry methods using flowing air have been devised. In some areas, where there is a seasonal choice to process materials wet or dry it is far better to process the placer gravels with water because a much higher percentage of the fine sized gold will be captured. Dry concentration techniques are much slower and inherently less efficient than wet concentration methods, and can only be used on placer materials consisting of small, dry particles that can be sorted by streams of blowing air.

Dry washers can be fed by direct shoveling or by bucket. Here the author feeds a 12V motor operated bellows style dry washer.

Winnowing was the original dry placer method - it was practiced in antiquity by the Romans and others. This process involves screening out all the coarse gravel, placing the sand and other fines on a blanket and tossing them in the air, preferably in a good wind. The lighter particles are blown away by the wind and the heavier, usually more valuable minerals fall back onto the blanket. The weave of the blanket tends to hold fine gold. Winnowing is a very primitive method and is not used by prospectors today. Basically, all this method really does is remove the light clay and dust particles from the placer gravel mix - there is no effective concentration beyond that.

Dry panning is partially a winnowing process, and is only slightly better than blanket winnowing as far as gold recovery. I do not recommend it, but it is sometimes used as a sampling method. A normal gold pan is used, but no water is added. The material must be dry and contain a very minimum of clay for this method to work. It also needs to be screened to materials smaller than about a quarter inch in order to work at all. The first step in dry panning is to use winnowing to remove the finer clay and dust materials. The gravel is flipped up into the air allowing the fine clay and dust to blow away. The second step uses rapid but gentle side to side shaking of the pan to allow the gold to settle in an action similar to a jig. Larger gold pieces will tend to move downward through the material as in normal wet panning, but it does so in a much less efficient manner, and a much higher percentage of the gold is lost. After shaking, the lighter material, which is then on top, can be removed by pushing or scraping it off by hand. In my opinion, if the gravel you are working is really rich, it is better to haul it home or hand pan it in a tub with water.

A far better and more successful method for processing larger amounts of dry material is dry washing, also called dry blowing in Australia. The dry washer allows for the processing of gravels at a much higher rate than possible by winnowing or dry panning, and is far more efficient at collecting the gold (but still not nearly as good as wet methods). I own and use a dry washer as I live near some gold bearing desert areas in the state of Nevada.

The dry washer is the most widely used piece of dry recovery equipment. For all intents, the dry washer is basically designed to be a short, waterless sluice. It separates gold from sand by pulsations or flows of air coming up through a porous cloth medium. The upward flow of air "fluidizes" the bed of material, allowing it to "flow" almost like water. The vibrations and the flow of air replace the function that water occupies in the normal sluice, allowing the gold to settle downward as it normally would. Because the air will only lift the smaller materials, dry washers usually screen the gravels to some size roughly around ½ inch. The oversize is dumped off the front. The smaller screened gravel passes down an inclined riffle box with cross riffles. The bottom of the box consists of a thin, light weave linen or some other fabric. Beneath the riffle box is the source of air. There are several types of air sources. One type is a bellows system which operates by pulling a rope which blows air in short, strong puffs through the canvas. This gives a combined shaking and classifying action to the material. There is also a bellows type which has a battery powered electric motor that operates the bellows for the operator. Other designs employ a motorized blower to provide a more continuous flow of air. The blower is usually set up with an eccentric device of one sort or another, which creates vibrations that help the gold to settle downward. With both types, the gold gravitates down to the canvas and is held by the riffles, which are set up in the direction opposite of a normal wet sluice. Unlike standard sluice boxes that are made to be used with water, the gold in a dry washer does not get caught behind the riffles, but instead the riffles are made such that the gold gets caught in front of the riffle. The waste passes over the top of the riffles and out of the end of the sluice section. A basic dry washer is composed of a frame in which a well braced, heavy support screen is covered with fine linen. Above this, riffles made of one-half to three-quarter-inch, L shaped steel or a quarter-round wood molding placed about 4 inches apart. The slope of the box varies from 3 to 6 inches per foot.

Dry washing, by its very nature, is an extremely dusty process and the operator is urged to plan out your setup so that you can stay up wind as much as possible. Keep motor air intakes out of the dust as well. Most operators should even consider wearing a dust mask, especially when the dust cannot be avoided. Over the longer term, dust particles can potentially do serious damage to the tissues of your lungs, and dry washing produces copious amounts of dust in most situations. There are even some viruses or other disease causing agents in the soil. Valley fever is perhaps the best known of these, but there are others, and some are quite serious. Overall, it's good to take reasonable precautions to minimize the dust you breathe as much as possible.

Here the author operates a pull string type bellows dry washer. The amount of dust here is minimal compared to the norm. Often dry washing is very dusty.

A power blower washer of the type described above can process up to 21 cubic feet (approximately 0.8 cubic yards) of loose material an hour. Hand-powered washers operated by two men can process 1 or more cubic yards per 8 hours, depending on the size of the material handled. For recovery of gold, the gravels to be processed must be completely dry and well disintegrated. This is the real key to getting the best recovery possible with your dry washer. Clumped or wet material just ruins the efficiency of the dry washer. It is critical that the gravels are fully dry because in even slightly damp materials, the clay will clump with the gold preventing good recovery. As a result, any gravel must be thoroughly dry before treatment. During the dry times of the year placer gravels may be naturally desiccated and dry enough to process, but after stormy periods drying may be required. For small-scale work, sun drying on tarps will usually dry the material sufficiently to do the trick. In addition, materials with a heavy clay content, or those which are bonded together with the caliche will clump up even when dry, so breaking up or crushing the clumps by hand or other methods may be necessary to get good recovery. Fully dry, sandy material gives by far the best recovery results in dry washing. Even so, rich gravels are often run a second time through the dry washer to recover any gold lost the first time around. During a dry washing operation, dry gravels are fed onto the scalping screen of the dry washer where the finer materials fall through the screen into the sluice section and are processed over the riffles. The oversize portion rolls off the top of the screen, and the processed gravels fall off the edge at the end of the sluice section, leaving an undersize and oversize. These paired piles of coarse and fine materials are the standard sign of past dry washing efforts when one is prospecting in desert areas.

The bellows type dry washer is operated by hand or powered by a small motor. The bellows should be operated at about 20 pulsations per minute with a stroke of about 3 inches. These figures will vary with the coarseness of processed material and the fineness of the gold. Operation continues until about one or two cubic yard of material has been processed, after which the sluice section is normally cleaned.

One type of operation that often goes hand in hand with dry washing is dry vacuuming. There are various "dry vac" models and some similar systems can be made by hand. They consist of a suction device that feeds the material into a collection bucket. The point of these units is that it's a lot faster to suck the material out of the cracks and crevices in bedrock than it is to broom them out and pick them up. The materials collected by the dry vacuum can then be dry washed to collect their gold content. These dry vac units are not strictly necessary, but if you are planning on doing a lot of dry washing they can be pretty handy. They really do a better job of efficiently cleaning off the bedrock.

Operating a dry washer is not terribly difficult. Reading this chapter will give you the basic information and instructions necessary to

Here the author cleans out the riffle tray of his bellows style dry washer. The concentrates are later panned out in a tub to recover the gold.

learn how to operate a one. The tilt of the oversize screen is not that critical, but should be steep enough to allow the loose rock oversize to run off, without being so steep that the dirt and undersize material you are seeking to process runs off the screen as well. The tilt of sluice box section of the dry washer is more critical. Like a wet sluice box, it should be set at an angle steep enough to allow most of the material to flow through the sluice box section within a reasonable period of time. This would be something like 20 or 30 seconds for material to work its way down from the top of the sluice down to the bottom and out on the tailings pile. The material should not bury the riffle section and you should be able to see the material "float along" as the air pushes through it. If you feed too much material too fast, you will not get a proper fluidizing of the material flowing through the sluice, and gold will simply go off the back of the sluice section. While it depends on the nature of the material, often the correct tilt for the sluice section of the dry washer is in the range of 6 to 8 inches drop per foot of length.

Many dry washers have an adjustable feed port control. These should be adjusted to prevent overloading of the sluice box section. Too much material, coming in too fast will bury the riffles and result in poor recovery. Most of the motorized blower types are really a two-man operation, or at least one man working very hard. They process a lot of gravels and a single operator will be working hard to keep the unit properly fed with material. Another important fact about blower type dry washer operations is to make sure that a reasonable amount of material is coming in at all times. Allowing the air to blow for a long period of time with no new material coming in will tend to scour the riffles and lead to loss of gold.

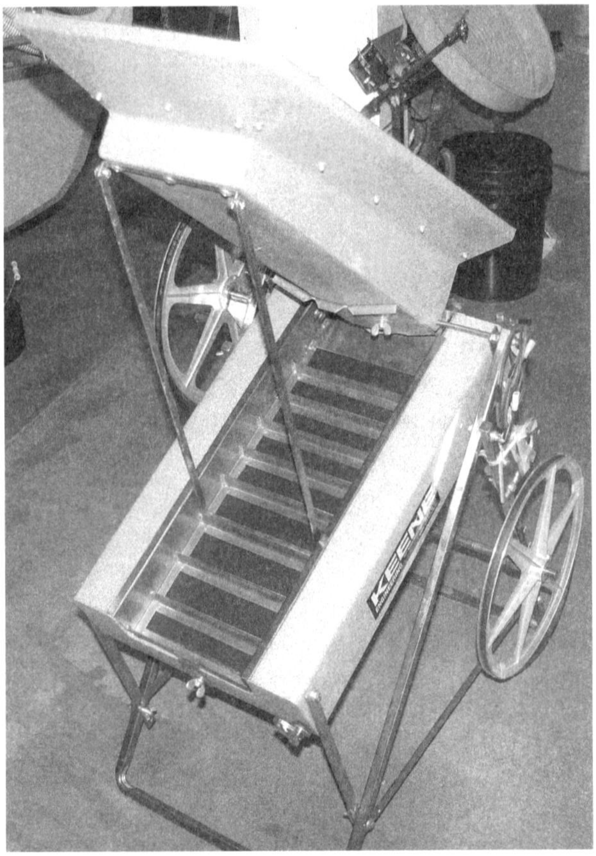

Here is a new bellows style dry washer made by Keene that is powered by an electric motor operated on a battery.

Because all dry placer methods are inefficient, it is normal that a certain percentage of the very fine gold is lost. As a result of this, it is common to reprocess the dry washer tailings and recover additional gold. This is especially true if you hit a good paystreak and are recovering significant amounts of gold - you will want to rerun your tailings, because it will almost always be well worth the effort. In fact, if you hit some really good gold it may be worthwhile to take home the fines and reprocess them by some sort of wet device like a recirculating sluice. Additionally, sometimes coarse gold is lost when it is too big for the screen on the dry washer feed box. The best way to deal with this is to go over your over size (coarse) gravel pile with a metal detector, even raking it down flat so the detector can see all the material. The exposed faces of the dig hole, the bedrock (if any is exposed) and even the dry washer fine material piles are often checked with the metal detector as well. A metal detector and the dry washer are two prospecting tools that go hand in hand.

Unlike the water powered wet sluice box, where you may go a full day between cleanups, regular cleanup of the dry washer sluice box is a must to prevent excessive loss of gold. Perhaps two hours is the very most you would want to run between cleanups, much better would be ninety minutes or even one hour - this equates to roughly 1 or 2 yards of material processed per clean up. During cleanup, the riffle box is lifted out and turned over into a large bucket or tub. Light tapping of the riffles gets all the material out and into the bucket. Usually the coarse and some fine gold are most abundant in the upper riffle section. The lower riffles may contain a few colors, but the majority of all the recovered gold is caught in the upper riffles.

The concentrates from the dry washer riffles are panned the same as any placer concentrates, and the gold removed. If water is very scarce, and you don't want to carry in big amounts of water, the concentrates may be saved and processed elsewhere. I normally bring along water or take my concentrates to a water source (often at the spot where I am camped). Sometimes if I have a large volume, I run them through a sluice. Any panning is done in a tub. In a procedure similar to handling the concentrates from dredging, I first screen my dry washer concentrates in a screen about ¼ inch in size. The coarse material is panned quickly and easily, as any gold larger than ¼ inch is a nice nugget. The fines are then screened a second time, through a 20 mesh screen. The ¼ to 20 mesh size is panned down and the nuggets and pickers captured out of that fraction. In the smallest fraction (the portion smaller than 20 mesh), I will pan it in the field until I get down to a mostly black sand concentrate. This contains all the very fine gold and I will usually take it home and process it with the rest of my black sand concentrates from dredging and other activities. For more information on dealing with black sand concentrates, see the chapter on black sand and final cleanup of the concentrates.

Dry washers are portable, comparatively inexpensive, and easy to use, but the dry washer is merely a tool for finding gold. You are the one who has to find a spot that has the productive gold bearing gravel. As with other prospecting tools, the dry washer is not that difficult to learn to operate, the difficulty is in finding paying deposits of gold bearing gravels. The same as with dredging and sluicing, the key to finding paying deposits of gold is an intelligent sampling program. Be sure to check and sample a number of areas that look good and see what amounts of gold they yield. Rather than using dry panning or another method to sample, I think it is best to do your sampling with the dry washer itself.

Many of the best places to dry wash fall into one of two categories. One is in drainage areas and the gravel benches adjoining to them, as placer gold becomes concentrated in the ravines and streams that drain gold bearing areas. In many ways these are similar to most wet placers in that the gold is normally concentrated on bedrock, behind obstructions, on inside bends. False bedrock layers are far more common in desert situations. Caliche is also a common desert phenomenon that cements gold bearing gravels into a solid mass. While desert stream paystreaks are typically very spotty and inconstant, they do exist. The other type of area is known as the hillside type of area. Long ago, prospectors discovered that there were placers located on some hillside areas which contained gold that was not associated with any stream or drainage. These are residual placers and may be associated with gold bearing veins, stock works of veins or gold bearing zones in schist. There are still some nice fine gold placers out there to be found. When checking out a new location, if you are looking for the good places to dry wash, take a look at where the old timers did their work. Look for those paired piles of coarse and fine material. Were they working along the drainages or up on the hillsides? Take your cue from them and try similar locations in that area. Sample around the old workings and see if you can't find a hot spot that they missed. If you are not getting much gold pick another spot until you start finding some good gold. I've done really well dry washing nugget patches I found with my metal detector. The detector will only see gold down to a certain size, and the dry washer will capture that smaller gold. I usually get almost double what I got in the original nugget patch. It's a great way to get a lot of gold the detector missed.

Another interesting thing about dry washers is that the bellows type can be easily made at home by prospectors. There is a long tradition of homemade dry washers that has been around for many years, and it makes an excellent do-it-yourself project. For those prospectors interested in building your own, I have a chapter on building a dry washer with more information.

Commercial Scale Dry Blower Systems:

In Australia and a few other locations, larger commercial scale dry washers have been built and successfully operated. The soil must be fairly rich, and not too cemented nor contain excessive clay. In this type of operation, larger quantities of material are picked up by front end loader and put into a hopper. These operate in a manner similar to the standard dry washer in that they screen the material and process it using a stream of air. This type of equipment is much more popular in Australia than the US. There is no question that in some parts of the US such equipment could probably be used economically.

This is a larger, gas engine blower type dry washer – it will normally process gravel faster than what two guys can shovel into it.

Four Basic Steps of Dry Panning

Step 1: *Free fall winnowing is used to remove the lighter clays and silt. Be sure to use a tarp around your target pan as materials can and do bounce. I also recommend dropping from a much shorter height than shown above. Pick out any big rocks.*

Step 3: *After the heavy materials have settled, gently scrape the lighter gravels off the top. Remember that gold won't settle as efficiently in the dry material. Any fine clay or silt will hinder the settling process and cause the loss of gold. Pick out the worthless pebbles and stones.*

Step 2: *Turn over the gravels and then shake them well with a side to side motion in order to allow all the heavy materials to settle down to the bottom of the pan. Good shaking is important.*

Step 4: *After repeating steps 2 and 3 several times you will be down to only the heaviest gravels. Carefully check your pan for gold – you can use gentle blowing to help remove the last of light materials. Dry panning loses gold, so it's best to use some water and the normal wet panning process whenever possible.*

Metal Detecting For Gold Nuggets

Metal detecting for gold is something that is comparatively new. The old timers, decades ago, never had any piece of equipment remotely like a metal detector. Modern technology however, has brought us this new prospecting tool - the metal detector. It is an electronic device which is able to sense the presence of metallic objects buried in the ground. Since gold is certainly a metallic material, metal detectors can be used to find gold nuggets. On the other hand, metallic trash left behind by old miners will also be detected. In this chapter I am going to take a look at the basic technology of the metal detector, answer the common questions asked by many beginners and others that are considering taking up detecting for gold, and then go through a practical, in depth look at the considerations of what it takes to be successful in metal detecting for gold nuggets.

Finds like the handful of beautiful nuggets shown here are what attract prospectors to the potential of searching for gold with a metal detector.

In the last 30 years, detecting for nuggets has become a very fast-growing aspect of the prospecting world. Several companies have developed metal detectors specifically optimized to find nuggets. Mind you, there is no such thing as a metal detector that finds only gold and no other types of metal, but there are some detector systems capable of distinguishing iron trash from gold in most cases, and the technology continues to improve. In fact, keeping up with the improving technology is important if you are interested in detector prospecting. When I was just getting into prospecting in the mid 1970's, I read a book about prospecting which discussed the possible opportunities of electronic prospecting with metal detectors. The capabilities of the detectors of that day were very limited and they had many drawbacks. The book was accurate about the weaknesses of the detectors of that day, and so for years I paid little attention to them. That was a major mistake of mine - because new, much improved units were released beginning in the mid-1980s, with greatly increased capabilities. Even better detectors were released later, but I was not paying attention back then because of what I'd read about the detectors of the 1970s. So take a lesson from my ignorance - pay attention to the increasing capabilities of the new units being released. Detector technology is constantly improving and new and better units are being released regularly. The detectors of tomorrow may well make the fine units of today look as poor as the ancient BFO technology detector units of the mid 1970s appear to us now.

How Metal Detectors Actually Work:

There are many varieties of metal detectors with technologies and features designed for certain characteristics and certain types of uses. All of the technologies currently in use are based on the scientific fact that any electric current flowing through a conductive material creates a magnetic field. Conversely, any conductor that moves within a magnetic field generates an electric current. While this is a simplified explanation, all metal detectors work by putting electrical energy into a transmitting coil, which temporarily creates a magnetic field. If there are any metallic objects in the ground as you swing your coil, the moving magnetic field from the transmit coil induces an electric current into the metallic object - also known as the target. The tiny current which is created in the target is known as an eddy current and decays down to nothing in a fraction of a second - however as the eddy current decays, it creates a tiny magnetic field of its own. Before this tiny magnetic field disappears, the receiving coil in the metal detector senses the induced magnetic field coming from the metallic target. In some cases, one coil of wire serves as both the transmit and receive coil, as power is put to it to transmit, then is turned off and the detector then looks to the same coil to see and receive the decaying magnetic field of the target. Because the received signal can be millions of times weaker than the original transmitted signal, a large amount of amplification is required to process the received signal that is returned back from the target. The magnetic iron minerals (like magnetite) that are naturally present in the ground also return back some magnetic response also. The detector must sort between the two, filtering out the ground noise from the iron bearing mineralization and correctly identifying the presence of a metallic target (in our case, hopefully a gold nugget). All of this requires some sophisticated electronics, which the design engineers of the various detector companies continue to develop. That is why the technology of metal detectors continues to improve from year to year.

Some detectors are designed for hunting deep caches and treasure, others for general purpose coin and jewelry hunting, and still others for underwater or beach hunting. While all of these can find gold if the piece is large enough and close enough to the surface, the types we are most interested in are those detectors designed specifically to prospect for gold nuggets. This is because most of the gold that is out there to be found is not huge, nor is it buried only a hair below the surface. Much of the gold is deep or small, which makes it harder to detect. The depth at which a nugget can be detected by a specific detector varies quite a bit with size, density and shape of the nugget, the mineral content of the soil, the technology of the detector, the size of the coil being used and a number of other factors as well. Even the presence of interference from nearby planes and airports can affect some units.

The current world of metal detectors which are designed to detect nuggets is dominated by two technologies: one is called VLF or induction balance, the other is called PI or pulse induction. With the VLF technology, the flow of electric energy into the transmit coil alternates back and forth as a sine wave (similar to AC electrical power), but energy is essentially always being transmitted. The receive coil is placed so that it is shielded from the transmit coil energy, and the receiver coil is always looking for responses. With PI technology, the electric energy is pulsed into the coil when it turns on, then it shuts off for a brief period of time, after which the receive coil looks for any eddy currents which may indicate the presence of a metallic object. In general, VLF technology is able to sense smaller gold and to better discriminate gold from most iron trash, but it is also more sensitive to negative effects from iron mineralization in the ground. VLF detectors are also generally less expensive. PI technology is able to sense large gold deeper and is much less sensitive to interference from hot rocks or other iron mineralization in the ground. Much of the iron ground mineralization produces an eddy current which lasts only a very short time, and so by switching off for a brief period, the PI detector is able to ignore the signal of the ground mineralization as the response eddy from the soil has mostly died out by the time the receive coil switches on to see it. Intense mineralization is a common problem in many gold mining districts, as it is often introduced into the rock by the same series of geologic processes that lead to the formation of gold nuggets. Because they are much less sensitive to ground mineralization, detectors using PI technology are the ones most commonly used to find gold nuggets in many parts of the Western US, Australia as well as other parts of the world. Still, having said this, there is also no

Metal detecting for gold is becoming increasingly popular, as prospectors take advantage of new technology.

question that VLF detectors still have an important place in the nugget detecting world, which I will discuss shortly.

Getting Into Metal Detecting for Gold: Purchasing the best quality equipment isn't going to be cheap. The cost of a top of the line, brand new Minelab PI technology GPX 7000 with the typical accessories and an extra coil currently runs well over $10,000, and that is no small chunk of change. To be sure there are other options much less expensive, yet in spite of the cost, there seems to be an unending stream of new prospectors - often retirees - that are willing to shell out some significant money to get involved in this exciting pastime. A surprising number are completely new prospectors. These are guys who have never panned, sluiced or dry washed a single flake of gold previously, but are attracted to the possibility that sophisticated detector technology can help them find some trophy-sized gold nuggets (I can say from experience that staring too long at those trophy nuggets will give you a bad case of gold fever faster than you can imagine!).

In my mind, nugget detecting is the toughest form of individual prospecting to consistently succeed at, so new electronic prospectors can really use as much additional information and training as they can get (and that's part of why I've written this book). It is certainly very possible for the average guy to be successful at nugget shooting, but it is not easy - you have to be willing to make the effort. Learning the basic techniques needed to operate your detector such as which settings to use, etc., may only take a few hours to complete - but learning to find gold is a much more difficult proposition. Consider it like a trade, because it is a skill that takes time to learn. You don't buy a wrench or a voltmeter and somehow become a journeyman plumber or an electrician overnight - it is the same with electronic gold prospecting. Learning a skill or trade takes a lot of time and practice. I believe it is much more difficult for novice prospector to consistently find nuggets with a metal detector than it is to consistently get a little "color" with a dry washer, sluice box or dredge. With a metal detector, it's easy to spend a whole day looking very hard and still get skunked. It is normal that it often takes a new detector owner many months to even more than a year to find his or her first nugget - even if you are out detecting for nuggets regularly. The attraction of the metal detector is the fact that the yields can be very much higher - when you are successful, you can find a significant amount of gold very quickly. It takes time, however to learn those skills and gain the experience needed to be consistently successful.

Locations (sometimes called patches) yielding multiple nuggets like this three ounces of coarse gold collected by the author are not common and should be appreciated. They are rare finds, but they are out there if you work hard.

There have been some really amazing finds made with metal detectors: Things like large multi-ounce nuggets, small patches of ground with dozens of ounces of gold, quartz veins laced with gold and other spectacular finds. They don't happen very often and there is a good deal of skill and a bit of luck involved in making some of those spectacular finds. Don't expect that somehow you can just walk out anyplace near an historical old gold district and start digging handfuls of gold nuggets. Please don't figure that you can make a load of money and get rich quickly with a metal detector. Unfortunately, it's just not that easy, and the prospect for getting rich quick is extremely remote. It has happened to a few, and I know a couple guys who have made these amazing finds, but there are thousands of guys armed with metal detectors in Australia and the Western US, and many of them have never found a speck of gold with their metal detector, in spite of working very hard at it.

I do think a lot of new guys sometimes look at sales brochures, YouTube videos, or the internet forums, see other guys posting photos of gold and get the wrong idea. Too many new guys expect that somehow you can just walk out anyplace near an historical old gold district and start digging bean-sized nuggets every hour or so with the occasional walnut sized nugget thrown in for good measure. I wish it was that easy, but the old timers have already skimmed off the really easy stuff. If you are willing to spend the money to get quality equipment, to put in the time it takes to learn to operate that equipment, then do the work to research potential areas and put in the time to explore around in places where detectable nuggets are found, then you'll be able to find some gold too, but you should never count on it being easy. Buying a great detector is a fine start, but as I have pointed out, owning a powerful metal detector no more makes one an experienced nugget shooter than owning a pipe wrench makes one a plumbing contractor. With both, there are lots of skills to learn, and it takes time to gain the needed knowledge and experience. Dogged, stick-to-it determination is a necessary asset for the prospector!

To be successful, it's important to learn what you can about how gold occurs - its geology and how it forms in the districts you will be prospecting in. Each district has its own special features and geology, and what works in one location to find gold may not be a good idea in another. All of this means that taking up the hobby of electronic prospecting can be very exciting and productive but it also means there's a lot of work to learn about how to use a metal detector to successfully find gold.

Get what information you can from the more experienced guys you might know. They may even show you a few good spots to detect. In the end however, the most important thing is to learn to be able to explore on your own and find your own gold patches. Learn to do your own research - it's like the old saying: "give a guy a fish and you've fed him for a day, but teach him how to fish and you've fed him for life". Read "location books" then go out in the field and explore for yourself. If you are a new prospector, I'd suggest you join a local prospecting club. They have claims you can legally prospect on that are within a reasonable drive, and it's nice to not have to worry if you're trespassing on someone else's claim. You can meet other guys in your area interested in prospecting and learn a lot from the other club members. It really is a good place to gain useful prospecting experience. Finding gold isn't easy and the knowledge and experience you will gain are the factors that make the difference between finding some nice gold and getting skunked. Additionally, there are always pay to detect locations in places like Alaska or Australia. There are some others as well. Even at these locations, it's still not easy, and there is no guarantee. The finds are anything but evenly distributed. Some folks will do very well and others not as well; this reflects various skill levels among the prospectors, the amount of work put in and some short-term luck.

So, with that introduction and some initial thoughts on getting started in field of metal detecting for gold, let's take a look at some of the common questions beginners and other folks considering metal detecting for gold typically are wondering about.

Some Things New Electronic Prospectors Should Know.

There is little question that electronic prospecting is one of the fastest growing aspects of this pastime, and many new comers to prospecting are focusing in on nugget detecting. This is true in spite of the fact that the equipment needed to pursue this possibility is not at all cheap. While a lot of factors are contributing to this renewed interest, probably the biggest force attracting these new folks is the continuing increase in the price of gold. The new technology has certainly led to some spectacular finds in recent years. The nugget photos posted on the forums and shown in magazines demonstrate that there is still plenty of gold out there, which is true. But it is also important for new prospectors to know that finding gold with a metal detector takes knowledge, skill, hard work, and sometimes a little luck. Too many new guys jump in thinking it's not that hard - but they soon get hit with the reality that it is not at all easy.

If you are seriously considering buying that expensive metal detector, commit in your mind to do what it takes and know success won't happen overnight. The true bottom line is that you are going to dig a whole lot of trash. Research is an important key to gaining knowledge - it allows the prospector to learn of new places to explore and what to look for when he reaches the target area. *Actually,* hitting the field and swinging your coil with the knowledge you found out in your research and listening to the suggestions of the more experienced guys will allow you to find your own patches of gold nuggets in time. If you are willing to study and learn, then work hard to find your own gold, in time, you can and certainly will succeed.

Many desert locations in Nevada and Arizona are best prospected using Pulse Induction technology.

New guys often have a lot of questions, and here is some basic information to help you get started: Answered below are a dozen common questions that many new electronic prospectors (or those seriously thinking of getting into electronic prospecting) frequently ask:

1. What Detector Should I Buy for Prospecting, or Which Detector is the Best?

That's probably the most common first question asked on the electronic prospecting forums by new guys, and there are a lot of opinions out there. I am not a dealer, nor am I allied with any detector manufacturer, so I have no axes to grind one way or the other in buying or selling metal detectors. Many folks considering the purchase of a metal detector will ask: "which one is best"? This is not a simple question and the answer depends on your finances and what you intend to do with your detector. It should be phrased "which one is best **for me**"?

Metal detectors can find a lot of gold fast, including big gold! But metal detectors designed for hunting gold are expensive, and you don't want to buy the wrong one - so think about these considerations before you buy. Think about what features to look for and how to get the most bang for your buck. Take your time before making your purchase. Research the different types of metal detectors out there and the features they offer. Think about it first, and decide on what you want to do with your detector. There is no detector that is the one perfect unit to rule them all. Every choice is full of tradeoffs. So, you need to know what you want to do to decide which tradeoffs will work best for you. Will you hunt coins, jewelry and gold? Or just hunt nuggets? What is your experience level running a metal detector? Decide on a budget for your purchase and buy the best you can reasonably afford. Save up if necessary - don't get over excited and rush. Take your time. Familiarize yourself with what is available on the market, because there are new ones that come to market all the time. Because there is no one best detector that is the best of all, they vary quite a bit in price. All of them have advantages and disadvantages, learn about them so you're not negatively surprised. Whatever you choose you get some aspect and loose another. Avoid Chinese knockoff detectors at all costs. They will not save you money a few of them work marginally and many hardly work at all. New is better but you should consider buying used as it can save money. Good used deals take a little more work to find. See it work properly and test it on small targets in person if buying used. Buy only from someone reputable - some offer training in the use of your new detector - it's worth it if you can get that.

When considering what detector to buy, it is important to consider what type of detecting you want to do. If you plan to detect more frequently for coins and jewelry and only detect for nuggets once in a while, you probably will be most interested in a general-purpose detector which can also be used for detecting coins and jewelry at schools, parks and other locations. If you intend to focus only on detecting for nuggets and have the cash, it is best to purchase a detector which is optimized for nugget detecting. If you are trying to figure out what to buy, you need to ask yourself some questions. First, what do I intend to do with my detector? Will I be using it every week? Some guys know they want to be able to go prospecting with their detector, but can only get out in the field

The author hunting for specimens on an old gold mine dump with his White's MXT metal detector (a multipurpose VLF type unit).

for a few days each year, and that makes a difference. Second, you need to figure out what you can afford, even if you have to spend some time saving up your money. There is no simple one "best" detector for all purposes - that's why they make so many models. Each has their strengths and weaknesses. Additionally, new detectors will be released after I write this book and technologies may change. When looking at the costs, many guys see how expensive the top-of-the-line prospecting detectors really are and want to cut corners and buy a less expensive detector. I hope I've convinced you that gold detecting isn't easy, and going cheap may not be a good idea. In all honesty, this is an expensive pastime to get into. You can sometimes get decent bargains on used equipment, but in the end you do get what you pay for. Commercially available gold nugget detectors use either VLF or PI (pulse induction) type technologies in their operation. Many folks considering a detector purchase want to compare the VLF and Pulse Induction technologies. Each has their place, but it is no coincidence that the guys doing the best in AZ and NV (and many other locations) are doing so with Minelab pulse detectors in the GP and SD series. In certain parts of the mother lode country of California and also in Alaska, there are places where the advantage of a PI detector is not really significant because the mineralization is mild and the bedrock shallow. However, unless you know that those are the only places you will be hunting, you may well be better off in the long run with that more expensive Minelab. Even in CA and AK, Pulse Induction type detectors have a big advantage in many parts of those states. In Australia, there are places where a VLF is acceptable, but the majority of the Aussie goldfields seem to have fairly high levels of mineralization, generally higher than the goldfields of the western USA. Minelab is an Australian company and the pulse induction detectors they make are designed and tested in the heavily mineralized Australian goldfields. Each of the technologies has advantages that can allow them to shine under different circumstances. Many dedicated nugget prospectors choose to own both technologies, to allow themselves the maximum amount of flexibility. If you have no problem with the money, having both is the way to go.

In general, metal detectors specifically designed for use as coin and jewelry shooting machines make very poor gold detectors, as they are just not sensitive enough to detect small or deep nuggets, nor can they handle the intense mineralization found in most gold fields. My honest opinion is that nugget detecting with a cheap, low-cost detector (often from Amazon) made to hunt coins is a waste of time. Gold nugget detecting really is quite a bit different from coin or jewelry hunting in local parks. On the other hand, there are some multipurpose VLFs like Fisher's Gold Bug Tracker and Minelab's Equinox 900 which can also be used for coin hunting and other detecting, but these were also designed with nugget hunting in mind. VLF detectors that are specifically designed to detect gold nuggets have an advantage over PI detectors with tiny gold, and have much better discrimination. They are generally less expensive than PI detectors, which is often a reason many new guys are attracted to them. Unfortunately, they can also have serious problems with hot rocks and mineralized soils which are all too common in many gold districts. In the many gold districts where mineralized soils are present, it is often necessary to turn down the sensitivity in order to use them. This is a very serious disadvantage and greatly limits how deeply they can detect. VLFs work best and have their greatest advantage over PI detectors in old mine dumps, locations where the gold is finely crystalline and not very solid, and areas with shallow surface exposures of bedrock. Because of this, I think there will always be a place for VLF detectors in prospecting - at least there will be for many years to come. However, in mineralized areas and other locations where PIs have an advantage, that advantage can be very large. If you are working one of those regions where the advantage to the PI units is large, it can be very tough to find gold using a VLF machine when you are working locations that have already been well prospected by a good PI operator.

The Fisher Gold Bug-2 is a well-known and very successful VLF detector that can find very small gold flakes and nuggets. This a new easier to use version with a ground balance read out to help in making mineralization setting adjustments.

Gold is heavy and it tends to work its way down deep, and here the PI detectors like the Minelabs have an advantage in seeing gold deeper, especially in heavily mineralized soils. Because they can ignore considerable soil mineralization, large coils can be used where deep detection is an advantage, and this provides even more depth capability. The disadvantage to most people is the high cost of purchasing a complete PI outfit, including accessories and a few aftermarket coils. The other disadvantage is in prospecting areas with large amounts of very deep trash, where your PI may have you digging crater sized holes for deeply buried trash - this is the downside of deep detecting. Gold is hard to find and you should not intentionally handicap yourself getting a detector that will be at a disadvantage in competing with other prospectors who have PI detectors. Many of the well-known nugget patch areas have been hit hard over the years and the remaining targets are deep and faint. The deeper your detector can see, the more places will be open to you in detecting. Many prospectors deal with the two different technologies by having both a PI and a VLF detector, but it is certainly not necessary to own two detectors to be successful in finding gold. I was plenty skeptical about the Minelab PI units when they first came out, and I wondered if they were significantly better in the gold fields beyond what I knew the VLFs could do. Only after seeing one produce gold in the field did I understand how much better they were, that they were indeed a significant improvement over the existing VLF models of the time. If you are going to detect for nuggets regularly in well mineralized areas, it will be worth it for you to purchase a pulse induction detector.

A big part of this decision does depend on the part of the world you live in - there is no one simple answer that is perfect for everyone. The USA has many different detecting sites, as does Australia and other locations. If you don't know what kind of prospecting sites are near you and which type of detector might be best for that location, get some information first and don't rush things. Take your time and learn before you invest your money - don't rush. Probably the very best advice I can offer is to take your time and get the very finest detector you can possibly afford, even if you have to save up for a bit. Many new guys, who have never prospected before, enthusiastically decide to rush out and buy a VLF (mostly because they are cheaper), but later regret it. Most of these guys would have been better off buying a $10 pan, and joining a prospecting club, taking their time and learning about prospecting for 6 months or a year while they save to buy a PI detector, even if they end up purchasing a used model. In the end, much of this decision comes down to what you plan to do with your metal detector, and what you can reasonably afford. If you can afford it, get both a VLF and a PI machine. I own a number of detectors. These include a Minelab Gold Monster 1000, and a Minelab SDC 2300, a GPZ 7000, and A Garrett Axiom. Each has their place and I use them all. I also have a multipurpose detector to hunt coins and jewelry, and between all of these detectors, I pretty much have all the bases covered. If you plan to mostly detect for coins and jewelry in local parks and the like, with only an occasional prospecting trip, then a multipurpose machine like the Minelab's Equinox 900 or one of the similar multipurpose detectors like the Fisher F19 or Garrett AT Gold, might be a good choice. They sell new for less than one thousand dollars, and a couple hundred less if you can find one that is used. There are also fine detectors by Notka and other makers. For detecting the smallest gold in areas with mild to moderate mineralization and shallow bedrock, I recommend the Gold Monster 1000 by Minelab or the Gold Bug 2 by Fisher. These detectors can see surprisingly small gold flakes. I have found gold with these detectors weighing less than one hundredth or a gram or less than a tenth of a grain. Of course, the depth that these tiny bits of gold can be found is very shallow, while larger pieces of gold can be seen at deeper depths.

Over the years, there have been man new detectors designed for gold prospecting come out on the market. I've been involved with testing a number of them. Because what is available changes from time to time, I suggest you familiarize yourself with what is available on the market. Different models and design types have various strengths and weaknesses. And there are new ones that come to market all the time. There is no one best detector that is the best of all, and they vary quite a bit in price. All of them have advantages and disadvantages, learn about them so you're not negatively surprised. As an example, the top-of-the-line GPZ 7000 punches deeper than any detector on the market, but it has no discrimination. If you are searching a very trashy area with loads of rusty iron junk, then the GPZ will be a poor choice. A detector with good discrimination will be more productive. Garret also makes a fine PI detector, the Axiom, which has been out for a few years at the time of this writing. The Axiom has a number of different sized accessory coils available in both mono and double D designs. It is a PI detector so it has good resistance to hot rocks and mineralized soils, and yet the discrimination is fairly accurate. I like the Axiom dual tone discrimination is a lot better than the blanking type discrimination on some of the Minelab PIs. It also has a wireless headphone set up that is nice. If you are looking for a mid-priced PI detector with a lot of options and possibilities, the Axiom is a good choice.

There are several other lesser-known detector models that are used in the field. These include the several detector models by Notka, XP, and Algoforce. There is always talk of new detectors coming on the horizon, and it's a topic of interest to many prospectors. The truth is that all the detector makers are always working on various new models and improvements to old models. Some of these machines eventually materialize on the market and some never do. I always recommend waiting a bit when a new one comes out to be sure a detector is going to be able to do the job, rather than buying the first one on the block. It's just a cautious way of doing things, since not all new detectors live up to the hype surrounding them.

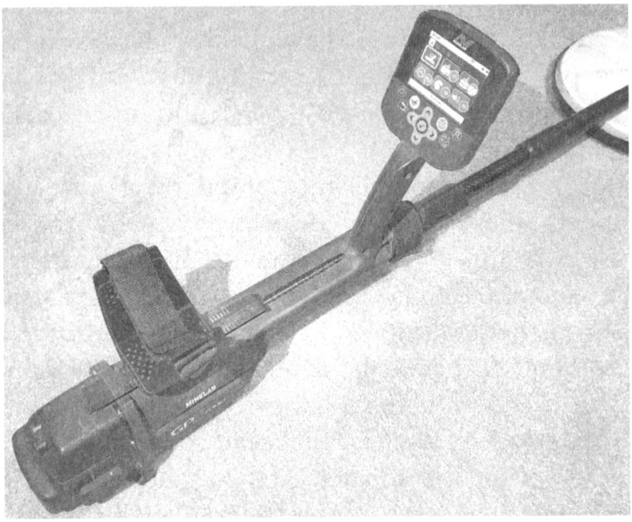

Purchasing the latest top of the line detector like this GPZ 7000 from Minelab is a fairly expensive proposition, and with accessories it's even more.

If you really are going to be seriously detecting for gold on a regular basis, in most cases you probably would be better off with a PI machine. There are a lot of choices in this class of detectors, and I will talk about my recommendations in a few pages when we talk in more detail about the detectors on the market. I recommend that you take your time and think about what you want to do, research the different detectors on the market and ask other prospectors for their opinions. Don't rush too fast to make a purchase, and as I said, get the very finest detector you can possibly afford. If you are serious about getting into detecting for gold, the price of your detector is one area in which you don't want to cut corners.

2. How Deep Will My Detector Go?

This is another common beginner's question and many new prospectors expect to get a single depth number. They expect an answer like this detector will go 8 inches but this other one will go 10. Instead, it's a complex and difficult question to answer to answer, as it depends on many factors. There are 4 main things that affect the depth of detection of a target. They are: 1) size and nature of the target, 2) mineralization of the soil; 3) Size of the coil; 4) the detector itself. Let's look at each of these depth of detection factors individually.

First: The larger the nugget, the deeper it can be detected, so what size nugget are we talking about? Also, wiry or mossy gold is not detected as deep as solid nuggets. Also, the shape and orientation of nuggets in the ground changes depth. A flat gold flake will be detected more deeply if it is oriented flat than if it were vertical. The alloy of the composition of the nugget also matters.

Second: The mineralization of the soil also greatly affects the depth of detection. Mineralization greatly affects the operation of VLF metal detectors and to a lesser extent, PI detectors also. Strong mineralization dissipates the transmitted field and decreases the depth a target can be seen.

Third: The larger the coil the deeper your machine can detect big nuggets, but larger coils have a hard time seeing the smaller nuggets and may miss those smaller targets completely. Not all coils are the same design, and the layout of the wire in the coil also affects detection depth of detection.

Fourth: They type of detector technology matters. PI detectors are sensitive to electromagnetic interference from certain radio transmissions and power lines as well as lightning storms. In some places these are a problem, but in other places they are not. However, in places where it is a problem, it can significantly reduce the depth of detection for a PI detector. However, because PI detectors are more resistant to mineralization, one can use those bigger coils that see the gold deeper. So, the answer depends greatly on many factors and it's not a simple question to answer. In general, VLFs are best with small to medium nuggets up to about 6 inches deep in low to mildly mineralized soil. PI detectors may miss some of the smallest gold, but will find a bit larger nuggets slightly deeper in mild soils but can penetrate much deeper in heavily mineralized ground, and with larger coils sometimes up to a foot or more. Both technologies can go deeper on very large multi-ounce nuggets, but if the gold is more than about 3 feet deep, it's beyond the reach of nearly all the available metal detectors on the market.

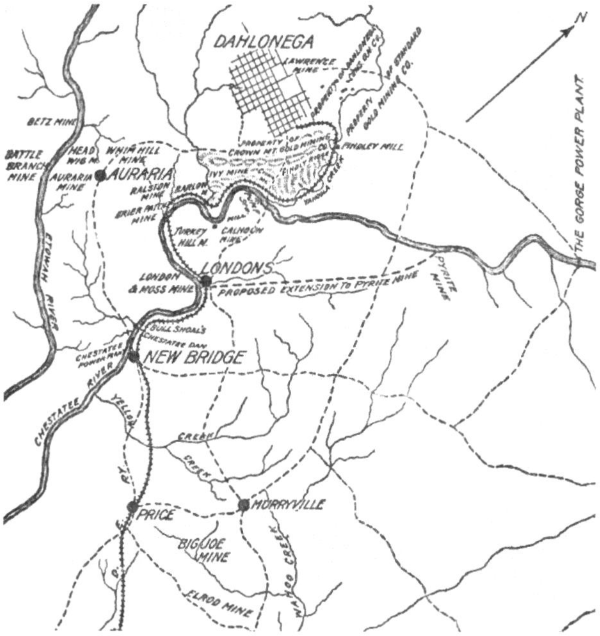

There are some locations in the eastern USA which have yielded coarse gold, like Dahlonega, Georgia. Because most of the mine property is private land, prospecting access is often difficult.

3. I Don't Live in the Western USA, Where Can I Detect Nuggets in My State?

Unfortunately, the US states are not created equal when it comes to gold, and just because gold has been found in the past in a certain area, it does not necessarily follow that those places are always good places to detect nuggets. The same fact is also true in other parts of the world, including Australia. It is no accident that the vast majority (like 99%) of all electronically detected nuggets in the US are found in the 12 western states (not including Hawaii), and more than half of that 95% are found in just 4 states, AK, AZ, NV and CA. Nuggets which are large enough to detect, and close enough to the surface to be detected are a lot less common than fine gold and gold buried deeply. There is plenty of detectable gold out there, but it's just not spread out evenly everywhere, and it's not common or easy to find. Gold veins and nuggets occur in certain regions and districts with the right rock types and geology. Where the rock types are wrong, and the geology is wrong, there is simply no gold. Even in states like Arizona, where lots of gold has been found with metal detectors, some portions of the state are favorable and have gold deposits, and some other places are so unfavorable that they are just plain barren and have no gold. This concept is true all across the globe - gold is found in only a limited number of locations.

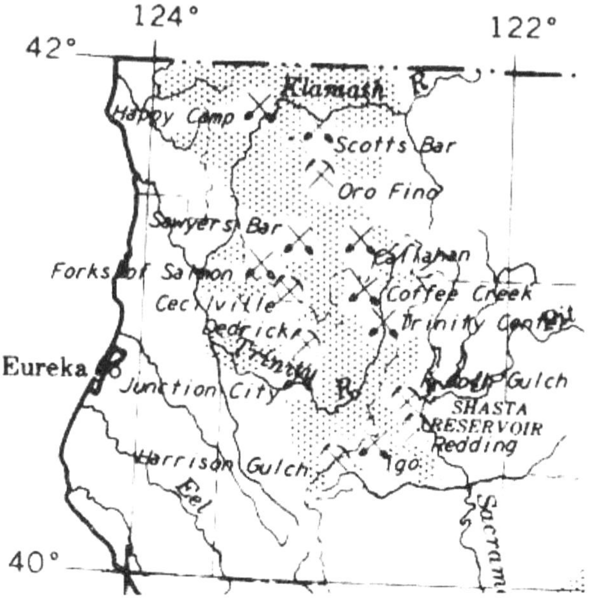

Gold location maps, such as this map for the Trinity / Klamath area of Northern California, can be very helpful to the prospector researching for places to hunt for gold nuggets and specimens.

In addition, many gold districts have only fine dust sized gold and in many other places, what big nuggets are present are buried far too deeply to be detected with any detector. In many eastern locations, the gold tends to be smaller in size and for those few locations with some larger gold nuggets near the surface VLFs may be the preferred type of detector because of the large amounts of trash found at these sites.

On further research, you will probably find that there is only a limited amount of detectable gold in the US outside the western states. I'm not saying there is none, just that it would be a fairly rare occurrence and much less common than in some of the western states. There are some places in the eastern US where gold was mined, and historically both of the Carolinas as well as Georgia each produced some good-sized nuggets, but nearly all the land is privately held and not open to public prospecting. In some of those locations you might get permission of the land owner, especially if you already know them - but that is a special case. There are also a few eastern areas which are generally open to use by prospectors, but most are "pay to prospect" fee locations. Access is a real problem in the east, where nearly everything is private property. In the west there is a considerable amount of open federal land. That's why I recommend new prospectors joining a club to find out more about their area from guys with experience. There are some eastern prospecting clubs with access - you might want to look them up. For those who live out west, there are many different clubs, but GPAA is the only one that is in any way close to nationwide in its scope.

4. I Do Live Near the Gold Country in the Western US (or Elsewhere). Where Should I go to Find Some Nice Nuggets for Myself?

This is the question most commonly asked by new prospectors after they have made the plunge and bought their first detector. It's perfectly good question, and in fact, this is the great and unending question most prospectors find they are asking themselves over and over again. It's a question that a good prospector will never outgrow. You will find that in general, finding those good spots is a difficult task and prospectors jealously withhold what they know about locations that have produced good detectable nugget gold. Giving away gold is very much like giving away money. Think of it like approaching someone that you do not know very well and asking for him to give you a thousand bucks - not a loan, but a gift. It may not seem that way to new prospectors, but if I tell you of a secret spot I know and you go out and find a few ounces, you will have a ten thousand dollars' worth of gold and I'll have nothing - nothing except the knowledge that if I had kept my big mouth shut, I'd have probably found that gold for myself. From my perspective, it's like I gave you a gift of thousands of dollars' worth of gold nuggets. Good gold hunting spots are not easy to find and are considered as precious to the folks who have put in the time and effort to find them. Generally, in books and on the internet, you will find folks very willing to talk about techniques, tools and equipment, but mums the word on exact locations to find detectable gold. Giving away those spots really is like giving away free gold nuggets. It is a very good friend indeed that will share with you the location of good spots that at one time were very productive, even though now they have pretty well been worked out. The bottom line is that to be successful in the long run - you need to learn how to study and do your own research, hit the field and explore to find your own spots. That's what the most successful prospectors do, and it's a really big part of just why they are successful.

Many friendships have been lost over gold patches that have been found. Gold brings out the worst in many people, as fair concepts of share and share alike go out the window when gold becomes the chief and only focus. Pacts and agreements to split the output are forgotten as gold mesmerizes one or more of the partners. Many long-time prospectors have been badly burned at one time or another by former friends or associates, and so are very wary of sharing too much information with anyone. If a more experienced prospector takes you under his wing and does show you some good spots, you should be very thankful. That is very much the exception and certainly not the norm. So don't expect that your fellow prospectors will reach out and tell you about locations that are still currently yielding significant amounts of gold nuggets. Old patches still have some potential to yield additional gold in the future when new technology is implemented into future detectors. We don't know when that might be, so we keep these known patches as reserve knowledge. In addition, old patches often continue to yield small amounts of gold even with the existing technology. So it's easy to see why people are very hesitant about giving information of exact spots where gold has been found. I doubt that this is something that's likely to change in future.

As a new prospector, I'd suggest you join a local prospecting club. They often have claims you can legally prospect on that are within a reasonable drive. You can meet other guys in your area interested in prospecting and learn a lot from the other club members. It really is a good place to gain useful prospecting experience. Finding gold isn't easy and the knowledge and experience you will gain are what make the difference in finding gold or getting skunked.

Read "location books", then get out in the field and explore around for yourself. Most location type books have references to other books - research them as well. Some of these obscure old

Places like Moore Creek (shown above) offer the prospector a special environment where he has an excellent chance to find gold with a detector.

references can be awful hard to find - though it's amazing what can be found on the internet. The libraries of universities with decent geology programs can be a great resource. Always be on the lookout during your research for information describing the source the gold is coming from. Use topographic maps, use aerial photos like Google Earth, and also use geologic maps - put them all together with what you learned in the books, and select some field targets to explore. Look for areas with geology the same as nearby productive areas, but located on the other side of a mountain range or a few miles away. Many times, these fringe areas have been overlooked and will still be productive to the prospector who goes out and explores them. Use your GPS to put yourself at those target points and keep on testing in the field until you find your own patches. Learn a bit about mining law and stay off claims unless you have permission. It doesn't hurt to spend some time learning the basics of minerals and geology. Later chapters in this book will cover in much more detail the elements of the topics I have briefly touched here. So don't worry, we'll talk lots about how to do your own research, use geologic maps and related topics on finding gold later on in this book. Learn what you can from whoever you can - just don't push people when they hesitate. Get whatever good information you can, but don't pump them, as most people will quickly figure out what you are doing. Take the time to make some genuine friends, because in the end some friendships will be even more valuable than the trophy nuggets you hope to find.

Good prospecting books and magazines like the GPAA Gold Prospector can be very helpful in adding to your knowledge base.

5. I've Never Prospected for Gold Before, but I Want to Learn. What Should I Do?

I've mentioned this before, but as a new prospector, I'd suggest you join a prospecting club like GPAA or one of the well-known regional/local ones. Nugget detecting isn't easy and a club is a good place to start. Go out on the club prospecting trips and watch with a careful eye, especially those guys that are successful. There is a whole art of prospecting and it's not something that can be picked up and learned in a few minutes. Some prospecting equipment dealers will offer lessons - those may be helpful for you as well. For those of you who have never prospected before, never found any of your own gold in the wild, I'd suggest first buying a pan and going out and learning some of the basic techniques of prospecting for gold. Then consider a detector at a later date. The experience you gain learning where to find gold with your pan will help you later when you buy that expensive detector. You don't need to wait forever, but don't overly rush things either. It makes sense to let your equipment inventory grow in tandem with your experience. If you really have the fever, you'll be genuinely thrilled even by those first few colors in your pan skill (trust me, I know). If you already have a detector, take it out to a nearby gold field and practice burying small targets and learn what sounds your detector makes in response. Lead makes a good stand in for gold (they sound just the same on your detector), and if you lose a small bit of lead, it's not a terrible loss.

Buy some books on prospecting and read them carefully - you can learn from books, too, although just like old timers, some books are more worthwhile to listen to than others! Just by coincidence, I happen to think this is one of the better books, but there are others. You can even ask the more experienced guys for their recommendations on books - maybe they will even loan you a few. YouTube also has lots of information, and I think my channel (**Chris Ralph, Professional Prospector**) is among the best!

6. I Bought That High Quality Nugget Detector, I Learned How to Use it and I've Spent Lots of Time Trying to Find my First Nugget. How Come All I Find is Trash?

The truth is that as an electronic prospector, you will dig a lot of trash - this is true in most locations whether you are a new guy or an old hand. In some places the amount of trash left by the old timers is simply overwhelming. Now some places have far more trash than others, but even in remote places it always seems like there is at least a stray bullet or two. It's a rare spot indeed where there is more gold than trash present - they are out there, but very uncommon. On the other hand, if you are successful, you will be finding some gold along with the loads of trash. There are a number of common problems the new guys often get caught up in. Here are a few things to keep in mind while you're out there that will help make you more successful in the field:

1) Detectable gold is far harder to find than most new guys imagine (how many times have I repeated this now?). No matter how powerful your detector, or how well you know how to run it, the detector can't find gold if the gold isn't there in the ground to begin with. To be detected electronically, gold must be both large enough and close enough to the surface for your detector to see it. In my opinion, the No. 1 reason new (and experienced) prospectors fail to find gold is a failure to put the coil over detectable gold. You have to put the detector in the right place - a location where detectable nuggets are present close to the surface. How certain are you that your spot has gold large enough for your detector to see at a depth shallow enough for the detector to see it? Do you know other guys who are detecting gold in the same area and are they also using similar detectors? GPAA and many other club claims are commonly very well pounded, and most of the easy gold has already been collected.

This seven-pennyweight nugget is quite a find. It was found at depth, and the target response was faint before it was dug out of the ground.

2) Gold is a far fainter signal than trash the vast majority of the time. You might be ignoring little soft blips and beeps that are far more likely to be gold, continuing to detect until you get a good strong powerful signal. Then you dig out a nice bullet or brass casing. Many signals that are gold are too small for the many VLFs to get a good discriminator reading, so you might ignore it because you are not sure if its trash or not. It is important to dig all of those repeatable signals your detector cannot fully identify.

With Minelab's PI units, you'll need to be able to recognize faint target sounds as well. Because of size or depth, many nuggets sound much fainter than trash targets - those who learned to metal detect as coin shooters are used to strong signals from good sized targets like coins, rings, jewelry or even pull tabs. No matter the type of detector you have, new operators need to learn to hear and pay attention to those fainter signals, or you will walk right past the gold to dig only the trash.

3) Going too fast is another common problem. Your detector will make a far more recognizable signal if your coil is moving slowly over a weak target than it will if the coil is moving along fast. Again, coin shooters are used to moving quickly through parks and schools whipping their coils and vacuuming up shallow, loud targets. Gold is different! It's important to keep the coil low and close to the ground and move the coil slowly in order to get the best results.

7. What Kind of Rocks Should I Look For in Order to Find Gold?

This is a lot more complex question that it seems like. Yes, it's very worthwhile to learn a bit about rocks, minerals and basic geology - but you don't have to be a geology expert to find gold. Many new prospectors would like to have a simple rule like "wherever you see vein quartz, there will be placer gold there" or perhaps "if you find red stained schist, there will be gold in it". I wish it were that simple, but it's just not. When I visited Moore Creek in Alaska - I found out that this historic and rich placer area was virtually devoid of vein quartz, and yet this site has yielded many huge gold nuggets. The problem is that rocks which are key and great indicators in one mining district are no good at all elsewhere. The question needs to be phrased as "What kind of rocks should I look for in order to find gold in my area?" and that answer depends on the area where you are prospecting. Yes, in general, vein quartz is a good indicator as are rocks like schist and slate. Iron-stained areas like gossans, can be excellent indicators as well. Later on in this book I will go into more detail on good indicators to look for. However, only certain indicators work in certain areas - none are universally true.

To learn what kinds of rocks and deposits produce gold in certain districts, one has to study mining reports and geologic bulletins. Usually these will tell you where the gold was coming from in a particular district, and once you know that, you can key in on those kinds of rocks or deposits and that will help you be more productive in your searching. Knowing a little more about geology will help you interpret and understand these reports. This book has a practical section on geology that can help you do that. There is also a section a bit later on in this book on favorable locations to search for hillside (eluvial) gold that may be helpful.

8. Can I Make a Living Prospecting for Gold With a Metal Detector?

Even at over $3,000 an ounce it's really tough to score an average over the long haul that pays for bean and gas money, and all the other things we need to survive. A lot of the guys who prospect regularly have other jobs that allow this or they are retired and have enough savings and retirement pension to cover their needs so that the gold is just extra. Most who live only on prospecting live in the field with a very low-cost hand to mouth kind of existence. Although they may have some very productive days, their take is normally less than minimum wage. A very few are able to make a reasonable living, but this takes lots of skill (certainly not the skill level of an inexperienced new prospector). Although some days are more productive and some are less, a typical average take for a very experienced prospector in the US might be around 2 grams per day over the long term. At $3,000 per ounce, this is about $200. On the other hand, when I was working before retirement, my job was a whole lot more reliable than my prospecting.

Yes, by camping outdoors and watching every dollar one can eek out a survival on that kind of money - but it is certainly not a life of luxury, and requires careful saving for times when things go poorly. Inexperienced prospectors should expect far less, perhaps less than $10 per day average over the long haul. The professional living off his finds must constantly go out and find new patches - well worked patches eventually yield so poorly that one is forced move on. What began as an exciting odyssey questing for gold becomes in time a more mundane grind - a hunt for gold to survive. The gold you find cannot be saved as a trophy, it must be sold quickly to pay the bills - sometimes it must be sold so quickly that you cannot wait to get top dollar for your finds. Learn to walk before you try to run - try prospecting on weekends and see how you do. Don't count your chickens before they hatch. I cannot tell you how many newbies I have seen go out to try prospecting for a living and have failed - the number is large. You need a big nest egg to live off for the months you need to learn the prospecting skills of finding gold.

9. How Much Gold Can I Expect to Find With My Detector?

When I average a pennyweight per day, I feel like I am doing OK on a minimal level. A few dwt per day is much better, and I am real pleased if I get a quarter ounce or better in a day - but that does not happen nearly as often as I'd like. I have averaged between 1 and 2 dwt or about 2 grams per "full" 8 hour day of detecting over the last 2 years.

Ownership of a high-quality detector in no ways guarantees that you will find gold. Nugget hunting with a metal detector is hard work.

All that said, new and inexperienced prospectors heading out for the first time with a metal detector may take months to find their first nugget - a year to find your first nugget is not unusual. A new guy might well expect to "average" zero gold per day for the first six months. You cannot expect to just drive out to the desert and do just as well as someone who has been prospecting for 30 years - that's just not reality. Finding gold is not simply about random searching and luck. This is a hobby / passion/ sport where knowledge, skills and experience count for a lot and over the long run, they make a huge difference in what you will recover. That's also why I advise new fellows interested in prospecting that if you're in it primarily to make a buck, you will eventually be sadly disappointed. It takes a long time and some significant effort reading and researching and trying stuff in the field to climb up that steep learning curve ladder. Persistence and hard work are the keys - that and a big bankroll to tide you over while you learn. As I mentioned right at the start - dogged determination is a necessary asset for the new electronic prospector who wants to be successful in finding gold nuggets.

10. What Accessories Do I Really Need for My Detector?

This kind of depends on where you are and what you are doing in your prospecting. Many VLF detectors that are designed for prospecting come with coils well-suited for nugget shooting. Many "all purpose" detectors come with coils that can be used for nugget shooting but are really better suited for hunting coins and jewelry. Smaller double D nugget shooting accessory coils are available and are a worthwhile purchase for these detectors. More detail about coil selection will be discussed later in this chapter.

For Minelab PI type detectors, many operators in the US use Mono coils almost exclusively or at least regularly. Many US detector operators choose to get at least one small Mono coil. Having a few aftermarket coils of various sizes to supplement those which came with your detector can be very helpful, but is not necessary to buy a full complement of dozens of specialized aftermarket coils. I have several aftermarket mono coils I use regularly, and a larger diameter coil I use only infrequently. There are lots of accessories out there in addition to coils, from picks to lightweight batteries to special support systems which take the weight of the detector off your shoulders. One can spend a large amount of money on these accessories - even thousands of dollars in total.

Not everyone uses these scoop spoons, but I find they are handy and help speed up the process.

Some guys really do spend that much, but for the most part that's really not necessary. There is a more detailed discussion of these accessories further on in this chapter.

11. There Are Claims Everywhere in the Gold Fields. How Can I Find Places Where There are no Claims?

Although it would be great if they did, counties don't distribute maps showing the location of all valid claims - and neither does the BLM. Unfortunately, there is just no such thing - mostly because there are new claims being staked and valid claims becoming invalid claims all the time so it changes all the time.

It's important to stay off claims that appear unmarked as well. Any VALID claim is one you should stay off. People steal or remove claim markers all the time (even though it's illegal to do so) - so even if no markers are visible, the claim is still legally valid. In the Nevada desert, campers often take wooden claim posts and use them for campfires. So how does one determine which claim is a valid claim and what land does it cover? Well, that's a fairly complex subject and there are whole books written on that topic! (The chapter on Mining claims in the appendix has much more information and detail). That's why I recommend that new prospectors consider joining a club. It will give you a chance to learn to walk before you try to run or fly a jet airplane solo.

If you have a certain area you are interested in, you can look it up on the internet using the BLM's

MLRS (Mineral and Land Records System). The website portal can be found at https://mlrs.blm.gov/s/ or you can just Google MLRS. The new system has a map-based research tool. If you are logged in or not, you can click on the research tab on the MLRS home page to access the map of claims. You then zoom in to the areas you are interested in. The map is a general one showing that there are claims in certain areas, but it is not very exact. Claims are not put onto the MLRS system for a month or more after they are submitted; so, loads of new but perfectly valid claims just don't show - at least for a while. So, if an area seems open and unclaimed on the MLRS, there is still a chance it is claimed. The claims viewer research tool can also be set up to display the grid of the Public Land Survey, and the underlying layers can be changed to topographic maps, satellite images or many other useful map bases. This map-based viewer is great for rough checking of an area for claims. It's a high-speed way to check over an area for any current or closed claims might be located. I highly recommend it. I also recommend generating a written report on the township you are interested in. They have additional detail that is easier to review in report format than on the maps. The map will show if there is at least one claim in a quarter section, but there may be one or many.

Trying to figure out the claim status of an area can be a big problem for a new guy coming in from out of state. If you live in the area, then you have the time that you can go round and try different areas until you find someplace good. It may take a while - like months of weekends. On the other hand, if you are from out of state, and only have a week or ten days, then you can't afford to just bum around and explore until you find a good spot. Doing your own research to find places with gold is a whole art in and of itself. So, I am repeating again that I honestly think the answer is to join one or two of the prospecting clubs. The cost is comparatively inexpensive, and then you'll have ground to prospect that does have gold. You still will find yourself hunting around on the claims. One of the things I do not recommend for beginners (though many are interested) is to buy your own mining claim. There are a lot of mining claim wolves looking to pounce on any greenhorn prospector with a dream of owning his own claim. Until you have the skills to fully evaluate a claim and decide if the purchase is worthwhile, stay away from buying one.

12. Do I Really Need the Latest and Greatest Model of Nugget Detector?

This is a special concern of those who operate pulse induction detectors marketed by Minelab, but really there is new technology and various new detectors coming onto the market every year. Ever since the first SD 2000 hit the market Minelab has continued to produce upgraded detectors marked by small but significant improvements. The biggest technology jump so far produced in this series was the step from the GP series to the GPZ 7000. Improvements in this line of products has produced a somewhat smoother threshold, a greater sensitivity to small nuggets, better ability to manage or block the effects of electromagnetic interference and a rudimentary, if unreliable, discrimination for iron targets.

This 75 ounce gold nugget was found with a metal detector in northern California. It is a rare find but large nuggets are out there to be found by skilled prospectors out searching for gold.

There have been some small increases in overall depth sensitivity as well. There are also been other small but notable improvements. Each of these improvements has allowed better operations in certain types of detecting environments, and for prospectors working those types of environments the improvements can make a significant difference in the amount of gold that can be detected. If those kinds of conditions are not found where you do your prospecting, then you may see little if any improvement by upgrading. Minelab's current top-of-the-line model is the GPZ 7000, which is designed to see mossy and wiry types of gold at depths greater than the older models. Minelab releases new models regularly. Previous SDs and GPs were generally designed for optimum operations in prospecting conditions with big nuggets and deep very mineralized soils as seem to be fairly common in many parts of Australia. Here in the US, big nuggets and deep very mineralized lateritic soils are very much the exception (in fact, gold bearing lateritic soils common in Australia are nearly unheard of here). As a result, many US operators may be able to use increased gain or other new settings to see some significant benefits.

So does one need the latest and greatest detector in order to find gold? The latest GP 6000 detector is reputed to generally have more problems with hot rocks compared to the earlier GP series. However, hitting more hot rocks with GPs are a tradeoff for increased sensitivity to small targets. Although folks have been arguing about SD vs. GP since the first GPs came out, there is no doubt all are powerful detectors capable of finding gold. The metal detector is a tool just like a pan or a pick. You can have the greatest machine and if you can't put the coil over detectable sized gold near enough to the surface to be detected, then you won't be finding anything. Skill in finding nuggets is the most important factor in determining the success or failure of the operator. If that were not so, very experienced operators would only have a moderate advantage over a totally inexperienced newbie armed with a new top-of-the-line detector. Instead, the truth is that very experienced guys rarely get skunked, while a newbie might scrape around detecting for a year to find his first nugget - in spite of the fact both are using the exact same machine.

On the other hand, if you have inherited grandpa's old 1970 BFO or other decades old VLF/TR type metal detector designed for coin shooting, I'd say it's a fine museum piece, or maybe you could use it to find coins at the park, but don't waste your time prospecting with it. The major manufacturers have not made new BFO detectors for almost 30

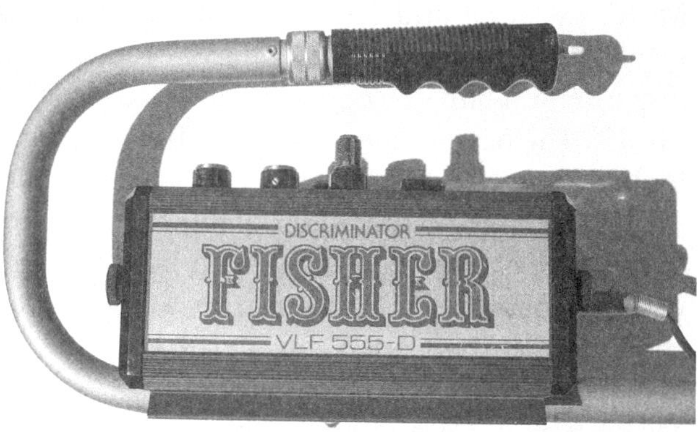

If you inherit dad's ancient VLF/TR unit like my old Fisher detector here, you might be able to find shallow coins, but you are not ready for nuggets.

years – because it's an obsolete technology. While you don't need to own the very latest and greatest detector to find gold, don't handicap yourself with some ancient detector that just can't do the job - get a good, modern machine that's designed with nugget detecting in mind, and get the very best that you can afford.

Research and exploration are important but there is still more to it than just research. Most of the inexperienced prospectors armed with new GPX 6000s are not out prospecting the sandstone hills north of the Grand Canyon or the limestone rocks of central Nevada (places with no detectable gold). They are out on club claims and other areas in known goldfields. They are out detecting around well-known gold fields armed with the same equipment that the most successful guys are using - so why are they not finding anywhere near as much gold? It's because there is more to successful electronic prospecting than just owning a great machine and driving to a location where gold was found in the past. The successful guys know that there is a lot to be learned from experience and past success – they have learned how to read the ground. It's not something that can be explained in a paragraph or two, or even a few dozen pages, but knowledge and experience do count for a lot, and there is no doubt that the best kind of experience you can gain is the experience gained by actually finding gold!

So again I ask: Do you really need the latest and greatest detector? A part of that depends on how often you'll be detecting and how much money you actually have to spend. If you have the finances and want only the best then go ahead –

there is certainly nothing wrong with having the very best equipment. Getting the best detector you can possibly afford is a big first step but no more than that - it is just a first step. One does not really need the latest and greatest to find gold. In my honest opinion, experience in finding gold and knowledge of prospecting methods are far more important than having the very newest top-of-the-line detector.

My Current Opinions on the Best Metal Detectors for Hunting Gold.

I've decided to offer a listing of what in my opinion are the 5 best gold-oriented metal detectors on the market at this time. You are fully welcome to have your own opinion, but some readers will be glad to read this list and find it useful. The manufacturers are always working on new models, and some of these make it to market, while some never do. So if you are reading this later, there may be other great detectors on the market that I am not mentioning.

There are also other manufacturers that are making good detectors now which I am not very well familiar with like Algoforce and Notka. I am not saying that their products are not good – I just do not know as much about them. I am also not including any list of the more general-purpose detectors that may be good for coins and jewelry and ok for hunting gold too. I am just making a list of top-quality detectors I am familiar with which in my opinion are the best 5 gold-oriented metal detectors currently on the market. I am also including comment on each of them concerning what I like about each of them and talk about their strengths and weaknesses. I think just the discussion of what is good and what is important may be helpful to prospective buyers thinking about what features are important and which ones are not. Metal detectors are tools – and each has its own best use purposes. There is no one detector that is best at all situations. So, these are - in my opinion – the best gold-oriented metal detectors on the market today. They are not listed in any particular order.

Minelab GPZ 7000; This Pulse Induction (PI) detector is in my opinion the most powerful for detecting deep targets, especially where the gold is wiry or mossy in nature. All the coils for this detector are fairly large and it comes standard with a 14 inch coil. It's also fairly heavy, weighing in at 7.3 pounds. I have found some really deep, good sized nuggets with this detector. The biggest downside is the price which is roughly about $9,500 at this time.

Minelab GPX 6000; This PI detector is the newest of these five detectors. It is very sensitive to the smallest gold and I would rate it as second to the GPZ 7000 as the most powerful for detecting targets where the gold is wiry or mossy in nature. It has become a favorite for many prospectors. The downside is that it seems to be more sensitive to hot rocks and mineralization. Certainly, nowhere near as likely to see hot rocks and mineralization as a VLF type detector, but more sensitive to them than other PI type detectors. The price is roughly about $7,000 at this time.

Garrett Axiom; This is the only detector in this group that is not made by Minelab, and is a very sensitive and versatile detector with the best discrimination for any PI type detector. It also has a very long-lived, built-in battery that is charged with a USB-C connector, same as my phone. It also comes with wireless headphones. I have one and really like mine. The price of the full version is roughly about $3,000 at this time.

Minelab SDC 2300; This PI detector is very sensitive to the smallest gold for a PI type detector and folds into a very portable small shape. It's been a favorite of mine since it first came out and I've found a lot of gold with mine. It is very simple to use and resistant to hot rocks and mineralization even though it sees small gold. It also is the only one of the detectors listed here which is waterproof. The others on this list have water proof coils and shafts, but the controls and electronics are not water proof. The downside it that it is not very sensitive to large, deep gold. The price is roughly about $4,000 at this time.

Minelab Gold Monster 1000; The Gold Monster is the only detector employing VLF technology on this top five listing. But the VLF detectors certainly have their place. With the small coil, this detector is amazingly sensitive to the tiniest gold. It is lightweight and very portable with fully automatic ground balance. It also has a very efficient discrimination that when properly used does a great job of identifying iron-based trash in the ground. I have found a lot of gold with my Gold Monster and use it frequently when conditions are right for a VLF. I recommend that prospectors considering the purchase of their first metal detector to find gold give a very serious look at the Gold Monster. The price is about $850 at this time.

Fisher Gold Bug Tracker; I am adding this 6th detector not because it belongs in consideration as the best, but because it is the cheapest. The best all cost a lot of money, and the tracker is much cheaper. It is nowhere near as good as the best, but for new prospectors on a tight budget, it offers a possibility. It has automatic ground tracking and is simple and easy to use, and is sensitive enough to find small gold. The price is about $350.

Practical Considerations of Successful Nugget Detecting:

OK, so we have spent some time explaining the basics of metal detecting and considering the typical questions asked by beginners. I hope you got the message that metal detecting for nuggets is not easy and requires knowledge and skill. So now let's now take a look at the full range of skills and knowledge that go with successful nugget hunting in the field. The truth is, that in metal detecting for gold, a comparative small percentage of prospectors make a very large percentage of the finds. I don't think that any one has ever done a real survey, but maybe 15% of the guys who own a detector and actively prospect for nuggets find 75% or more of the gold. The vast majority of electronic prospectors find little or no gold. So what makes these few guys so successful? Well, its more than just one little thing, but these guys know how to get the most out of their detectors and where to look to find gold. They do research to find new locations and work real hard at finding gold and put in serious hours operating their detectors. For them, their knowledge and experience are their most valuable tools. I can't somehow magically infuse you with 30 years of experience, but let's take a look at some of the things you need to think about to become one of those successful few.

Consider getting training – I don't offer that but there are others who do. Don't assume it's easy and just assume that you will be picking up handfuls of gold all the time. It just doesn't work like that, and persistence is one of the keys to finding gold. It takes time to learn to successfully find gold - accept that as true. If you don't get out that often, you will find it takes longer to learn. If your expectation is to go out and strike it rich immediately, you will be disappointed. You need an attitude that you decide in advance that you will - Never give up, never surrender - plan on that. You need time to be able to get out and use it – consider the commitments and time available that you have. Know that it takes time to learn. A detector parked in a closet, gathers only dust. Most guys take months to find their first nugget with a detector and some take a year or more if they don't get out much. Yet other folks have had great success with theirs and if you work on it, you can too. Just accept that it is a learning process.

Getting the Most Out of Your Detector

Every metal detector has its own unique features and adjustments, and of course the available features and adjustments change over time as the technology changes with new models of detectors. To operate your detector at the peak of its efficiency, I urge you to read and study the operating manual of your own detector. It will give you the specifics of adjusting your own detector. I cannot produce a chapter that will spell out the owner's manual information for all available prospecting related metal detectors, but I can give some general thoughts on the basic concepts of getting the most out of your detector – that is what I will do in this next section.

One thing I want to emphasize is that you should always check the settings of your detector before you begin detecting or if you suddenly experience odd behavior from your detector while you are out prospecting. In the field and during transport, it is not all that unusual for knobs and switches to get bumped and moved to undesirable settings that cause your detector to behave oddly, and fail to achieve optimal performance. That's why I always re-check things when I start out and at least once or twice during the day just to make sure I am properly adjusted to the settings I want to be operating on.

Setting a Threshold

The threshold is the background noise that your detector makes when it isn't finding any target at all. It's that small hum of a sound which the detector makes all on its own. Nearly every metal detector has some sort of adjustment that allows you to set the threshold at the point you want it to be. Because we all have different levels of hearing, with some able to hear fainter sounds than others, an adjustable threshold is pretty much mandatory on any metal detector.

Getting the right settings for your metal detector can make a big difference in how successful you are. The conditions of different areas can require different settings to produce the best results.

You should adjust the threshold so that it is a small sound - one that is just within the range of your hearing. You do not want to adjust it to a point where you cannot hear it. Some operators who are used to hunting coins, jewelry and other larger targets adjust their metal detectors to a silent threshold. This means that you will miss faint targets. When hunting coins or jewelry, missing little bits of trash or faint mineralization sounds may be no big deal, but when detecting for gold nuggets, being able to hear those faint sounds is extremely critical. You must adjust your threshold to a point that you can hear it or you will definitely miss faint targets that are gold. Exactly how much you will miss does vary somewhat, but if you make your setting so that you cannot plainly hear a soft threshold hum, it is likely that you will miss between 10 and 30 % of the gold you would have found had your threshold been properly adjusted. In pounded spots that have already been hit hard by other guys, the loss could be 50% to 100% of the gold because all the loud and obvious targets have already been removed.

On the other hand, it's important not to make the threshold too loud because it will lead to fatigue and you'll miss some targets simply for the fact that your ears will be tired of listening to it and a faint target may not stand out against the variations of a loud threshold. A threshold set too loud can also be very aggravating over the period of an entire day's metal detecting. Remember that most gold targets are faint, and most booming targets are trash, so setting the proper level of threshold so that you can hear it, without it being too loud is quite critical for successful nugget hunting.

The Minelab SDC 2300 PI detector is made to fold up into a small package like this. It is fully waterproof and sensitive to very small gold. With automatic ground balance, it's easy to operate.

Ground Balancing - Manual and Automatic

When the magnetic field generated by your metal detector's coil passes over the ground, it generates a return signal from the iron bearing minerals in the soil. This is in addition to any return signal that is generated from a target. In some cases, the return signal from the iron in the soil is actually thousands of times greater than the return signal from a small target. Your metal detector has to be able to differentiate between these two, otherwise your detector would never be able to find the target. All metal detectors designed with finding gold nuggets in mind possess some sort of ground balance system that is used to eliminate the natural response of iron minerals in the soil, so that the detector can differentiate between the magnetic response of the ground and the magnetic response of the target. The electronic systems that actually accomplish the ground balancing adjustments can be fairly complex, and an explanation of their exact methods is beyond the scope of this book. However, an accurate ground balance to remove the effect of iron minerals is critical, and keeping your detector properly adjusted and balanced is an important skill that all successful electronic prospectors must learn and master.

Because iron is the fourth most common element in the earth's crust, it is very common to find iron minerals in the soils and gravels of many goldfields. In addition, many of the geologic systems that deposit gold also leave behind deposits of iron as well. That is why many goldfields have a tendency to be much richer in iron minerals than the average location that is unmineralized. However, there are some exceptions and in some gold districts, the amount of iron in the soil is fairly low, while in other locations it is extremely large. In spite of those exceptions, medium to large amounts of iron are far more common in productive gold fields than locations where the amount of iron is low. As a result, our detectors have to be able to deal with that iron, and we need to know how to keep our detectors properly adjusted and balanced so that the iron does not cause false signals. There are also other types of mineralization in the ground, including the presence of salt, which when damp, is electrically conductive.

The quick way to determine if your metal detector is properly balanced to the ground mineralization is to simply move or pump the coil up and down from the surface to a height of about 3 inches. If the ground balance settings are correct you should hear little or nothing while moving the coil up and down. If you hear some significant signals, noise or other target sounds, your detector is not properly balanced. Many modern detectors have an automatic ground balance system that can be turned on and the detector will adjust itself to a proper ground balance (usually you will need to keep the coil moving during this automatic process). Other detectors require manual adjustment from the operator by turning knobs. Be sure to check the operator's manual of your detector to find out what you need to do to properly adjust your machine. When you select a spot upon which to balance your detector, make sure you have selected a representative spot for the area, because balancing on the "wrong ground" (read: misrepresentative ground) can result in a significant loss of sensitivity. As you get fully used to operating your machine, you'll know detector is out of ground balance. When the threshold has gotten a bit noisy, you'll learn to check and adjust the ground balance as necessary. One of the characteristics of the iron present in mineralized ground is that the signal response from the ground decays very quickly, usually much more quickly than the eddy currents which are generated in real targets. Because pulse induction metal detectors switch off for a brief time before they look for response signal, a significant portion of the ground response decays and disappears before the detector switches back on and starts to look for a target.

Salt flats in the Nevada deserts. Even a small amount of damp salt in the soil can play havoc on the operations of your metal detector.

This is a part of why pulse induction detectors generally perform better than VLF detectors in highly mineralized ground. High frequency VLFs which are designed to find gold are very sensitive to ground mineralization, and keeping your detector in proper ground balance is critical. Because VLF metal detectors are constantly transmitting and receiving at all times, areas with high ground mineralization significantly reduce the depth of penetration to which a target can be detected. The same sensitivity which allows the VLF detectors to find very tiny pieces of gold also makes them very sensitive to the effects of ground mineralization. This is why many of the newer VLF detectors are made with automatic ground balance to keep your detector in tune.

There are a number of problems that are caused by operating your detector out of proper ground balance. The most common is lots of false signals and excessive noise from your detector that is caused by being under balanced - or not compensating enough for the mineralization. In the worst cases there are so many sounds caused by ground noises that you simply will be unable to distinguish many of the real targets caused by the presence of metal items. Being out of balance in the other direction (too much compensation) will make your detector miss targets by ignoring them. That's why it's critical to keep your detector properly balanced at all times. It's true for both types, even though PI detectors generally handle mineralized ground much better than VLF units. In bad ground, a VLF operator may need to adjust his machine every step or two. In the same ground, a pulse machine operator might only need to adjust his unit every half hour to an hour.

Damp, salty ground is especially difficult for VLF units because the ground balance for salt is completely different than the balance required for iron minerals. In many cases where iron is present in salty ground, it is almost impossible for VLF to achieve a decent ground balance that cancels the two different issues. Some extreme cases of damp salty ground, also known as alkali soil in the Western US, can even present problems for pulse induction units.

There are actually a number of issues related to the ground - one is the mineralization in the ground itself - which is what we have been discussing, but there is also a problem when the ground contains rocks or pebbles whose iron and magnetic characteristics are totally different from the soil around it. The rocks which are significantly different from the surrounding soil are termed by detector operators as hot rocks. Depending on their size and actual makeup, as well as the surrounding soil in which they are enclosed, these hot rocks can make a sound that is very much like a good target. Sometimes however the sound they make is different enough from the sound of a good target that the two can be differentiated by ear.

Often in a certain area the hot rocks have a particular appearance that is consistent for that area (Its common that they are dark with a high iron content, but really they can have almost any color or appearance). The successful operator will learn to recognize the characteristic appearance of the hot rocks in a certain area and when they are exposed on the surface so that they can be seen, they can be pushed aside with your foot, or otherwise ignored if they are too large to push aside. Always be sure to re-check the spot once

With or without the vesicle air holes, dark chunks of basalt like this rock are often hot rocks. They contain iron and many are magnetic enough that they are slightly attracted to a strong magnet.

you've pushed aside a hot rock, because it is possible that the sound is coming not from the hot rock but from a real target.

Sometimes you'll find a situation where hot rocks literally pave the surface of the ground. There are some timing settings on certain PI detectors which are made to compensate for this, and this is the exact situation for which those timings were intended. There is another technique that is well known for balancing out excessive amounts of hot rocks, and that is by ground balancing on them. You'll basically perform a ground balance adjustment right over the top of a hot rock. In some situations, with certain detectors, that is the only way to deal with very heavy amounts of hot rocks. However, this is not a desirable solution and should not be used unless there is no other option because it does result in a serious loss of sensitivity. However, in certain good gold patches it may be worthwhile to make this sacrifice in order to find some gold.

As I've noted, many modern gold detectors feature an automatic ground balancing system. Most of these allow the selection of fixed ground balance or automatic (tracking). This is basically to say that while the detector is set on tracking, the detector itself is automatically adjusting the ground balance setting for you every few seconds. When set in a fixed mode, the detector keeps the ground balance setting fixed and is not making any changes for you. While it may sound like the automatic adjustment of setting your detector on tracking is the way to go, in most situations a properly balanced detector set in fixed mode will allow detection of smaller and deeper targets. This is because the tracking system has a tendency to track out faint but real targets - and normally it's those faint targets that you are really interested in. The tracking system basically is not able to differentiate very faint target signals from very minor changes in ground mineralization. After you swing your coil over the target a couple times, the automatic ground balance adjustment simply makes that faint target disappear. You can rebalance away from the target and normally you will get your target back at the spot you were previously looking, but if you were not paying enough attention, you might not have noticed that weak signal to begin with. In my opinion, with both VLF and pulse induction detectors, unless the ground is screamingly variable (which does not happen that often), you are better off to run your detector in fixed mode under most circumstances. Just be sure to keep adjusting the ground balance as needed.

Sensitivity (Gain) Settings

The gain setting adjusts the amount of signal amplification applied to the target response received back from the coil. While increasing the gain increases the response of weak target signals, it also increases any response that might be received from ground mineralization and hot rocks. So, achieving greater depth of detection by turning up the gain is really a two-edged sword - it can help the target response, but also create noise problems with significant ground mineralization. Because of this, the goal of adjusting your gain is to achieve the most distinct target sound - the situation where the target stands out best from any background threshold noise.

With PI type gold prospecting detectors, , the gain setting is critical when detecting for gold. Some prospectors think the main key to detecting small gold with both VLF and PI detectors is cranking up the gain to the highest level where a fairly stable threshold can be maintained. Be careful with this because many people do set the gain too high - don't forget that it is the contrast of the target sound vs. the background noise that is the most important. If you are getting too much ground noise, it is important to reduce the gain back to the point where your detector is not producing excessive noise. Unfortunately, when strong and variable mineralization makes it necessary to reduce the gain, some of the sensitivity to small gold and other weak targets will be reduced. However, where ground conditions are mild, the gain may be cranked up to the maximum settings without creating excessive noise. Perhaps the best way to make these adjustments is by doing some tests with a practice target in the field. It is important to use a small target - I recommend a small piece of lead, as gold can be lost, but some experts say you ought to use a real gold nugget as a target. The two are similar enough that no one can reliably differentiate a small gold target from a small lead target of the same size and shape. Glue your test target to a plastic poker chip or some other plastic base. When you are setting up at a new location adjust your detector based on the responses you get to that test target and the ground around it. Try a variety of combinations, remembering that your goal is to get the most distinct target response possible on your small test target.

The iron probability is shown on the screen of my MXT. At 80%, the machine "grunts" and you can be pretty sure that it's an iron target.

Discrimination, Both VLF and PI Detectors

Unfortunately for us, in many areas the old timers left behind a lot of metallic trash. In fact, in some places, modern folks have come in and left an additional share of metallic trash as well. Our detectors find metal, but we don't really want or need to dig metallic trash left behind by our predecessors. There is no question that every prospector would love to have a detector that sounds off on gold only, but unfortunately, that's not what metal detectors do. Because of the obvious need, the electrical engineers who design our detectors have made an attempt to provide us with the ability to discriminate - to differentiate and tell the difference between various types of trash and the golden targets we hope to find. Their efforts have had some success, but discrimination on modern metal detectors is by no means perfect. The only 100 percent perfectly reliable discrimination system is to dig it and see what it is. This is in part why many prospectors say that one must dig all targets. Among various shortcomings, none of the discrimination systems currently on the market discriminate clear to the maximum detection depth.

Old cabins, mines and campsites are often the source of lots of trash and other junk targets, but can have valuable coins and other items.

Accurate discrimination requires a strong and solid signal for the detector to evaluate and read. They work best with near surface trash, and luckily, at most locations the bulk of the trash is near the surface. Sometimes, especially on areas littered with surface trash but known to contain gold, using a magnet to collect surface iron trash can be helpful. A strong magnet is passed over the surface (the one I use is mounted on a handle) and it collects all sorts of small bits of iron trash - some high iron hot rocks will often be collected at the same time. Magnets are cheap and effective, but when the trash is thick, multiple passes with the magnet may be necessary. This process of "magnet sweeping" can go a long way to quickly eliminate junk iron sitting on the surface from a select spot and allow you to detect the gold in these locations. Of course, if much of the iron is buried, the magnet will not collect any buried material. In those cases, a light surface raking that stirs up the top inch or two may allow the magnet to collect additional junk.

The discrimination systems of the Pulse (PI) detectors that are currently on the market leave much to be desired. My current favorite is the system on the Garrett Axiom. Many PI detectors can be fooled by small iron targets like old rusty boot tacks, which show as good, and large gold-quartz specimen nuggets that show as junk. While I am not upset to dig a few small pieces of rusty iron, leaving behind a big multi ounce specimen nugget of gold in quartz is something I really have a strong desire to avoid. So if one is going to use discrimination, one must be careful and use it sparingly and only decide not to dig when the detector has a strong signal and is sure and that it is trash.

Some prospectors do use it, so if you decide to try it out, just be aware of its limitations. The Axiom pulse induction detector from Garrett has a much better discrimination capability than most PIs. The discrimination capabilities of the VLF detectors are far better than those of the PI technology, and are very unlikely to tell you that a large piece of gold is only junk. They are however, sometimes fooled into showing some junk as good. When you are testing a target with your VLF discriminator, and deciding whether or not to dig, be sure the coil is centered right over it. You may get inaccurate iron discrimination readings if the coil is not properly centered. This could cause you to loose nuggets. Many small nuggets below about 2 grains weight do not give a reliable or stable discrimination indication. Repeatable, but weak targets should always be investigated and dug. The key is that if the target is repeatable, the magnet does not remove it, and does not seem to move around as the coil is passed over it, it probably is not iron (it is most likely to be lead or a gold nugget).

Adjusting Discrimination Settings

Essentially all detectors with a discrimination system have some sort of setting adjustment to provide more or less discrimination. Carefully review the manual and test out your detector as often times setting the discrimination too high will cause some gold targets to be rejected. This is especially true of multipurpose VLF machines that are capable of discriminating out all low conductivity targets like pull tabs and foil. Strong discrimination must be applied with great care when prospecting for gold nuggets, as impurities in the gold, hot rocks or mineral concentrations in the soil around the nugget may confuse the discriminator, resulting in a false reading telling you that a gold target is trash. Be sure you fully understand your VLF detector's discrimination system and how it performs.

Some VLF detectors have an iron probability reading. When in prospecting mode, what does that really mean? When using the prospecting mode, probably anything that does not make the "grunt" sound (80% iron probability) should be dug. While the target ID on the Gold Monster is excellent, I've found it is not perfect. The ID does not work as deep as the detector will detect. In high iron soil areas, it is possible for small nuggets near the edge of their detection limit to score as high as 70% iron probability. In my experience, most junk iron gives a reliable signal with the discrimination. However, if you are looking for meteorites or in an area where they have been found, a strong iron target may indicate a meteor and these can be valuable. Another problem is gold in quartz. Ribbons of gold in quartz may respond oddly - giving readings indicating foil to the detector. It is necessary to all dig targets that are not clearly in the trash range, as they could be gold of one form or another.

It also makes sense once you have adjusted your discrimination to use a test nugget or a small lead target to see how your detector responds in a particular area. Bury a test target (use one of about 2 grains size for a VLF, use about a 4 grain piece for a PI) about an inch deep at the site you are testing, and see what the response is. In order to be sure you do not loose the nugget, most people will glue the test nugget to some small plastic piece like a plastic poker chip. Some detector owners have a full set of test nuggets of various sizes that they use. I use bits of lead, as I don't like to take gold back out into the field because of its potential for loss.

Volume and Amplitude Amplification

This is an adjustment of the loudness of response - not the signal gain discussed above, but the machine's output volume and amplitude. Its not unusual that folks loose some of their hearing as they get older - whether its from working in a loud industrial environment, shooting guns, old age or just too much rock n' roll, there are a lot of factors that can lead to hearing loss. This is something the user needs to adjust and set for his own unique hearing needs. This adjustment is also necessary when using a different output system, such as when one goes from headphones to an external speaker, the amplification needed for one system will be different that the amplification needed for the other.

Properly adjusted earphones are important. Tone adjustments can help those with hearing loss.

Tone Adjustments

Some folks have hearing loss issues that are much worse in certain frequency ranges. Tone adjustments allow the operator to change the pitch of the detector output to a range that works best for your hearing range. If you have suffered some sort of hearing loss, you might want to experiment around and see if you can't improve your ability to hear things using this type of adjustment, as many times hearing loss is more acute in certain tone ranges than others.

Timing Adjustments on PI Detectors:

Pulse induction (PI) detectors put a pulse of electromagnetic energy into the ground, wait and then take a reading of the response. The duration of the pulse and delay waiting have a great effect on the size and depth of nuggets detected. Some PI detectors allow direct manipulation of the sample delay, while on some other detectors timing changes are selected by switches and programmed settings. There is no one setting that is best for all conditions, and you should adjust your detector to match the detecting conditions of the location you are working. In general, short delays favor small gold that is shallow, but other timing settings may favor bigger gold down deep. It is important that you read your manual and understand the pros and cons of the various timing settings and set them to what is appropriate for the particular area in which you are prospecting.

KHz settings on a VLF and what that means:

In a VLF detector, the current flowing through the transmit coil changes polarity quickly - it alternates back and forth, with the current flowing one direction, then reversing and flowing the other. Each time the current changes direction, the polarity of the magnetic field created by the coil changes. The VLF detector derives its name, "VLF" (which stands for Very Low Frequency), from the frequency of those reversing cycles. Some early metals detectors used much higher frequencies - reversing in the millions of times per second. All modern detectors reverse much more slowly. For the modern crop of VLF detectors, these voltage reversals are measured in thousands of cycles per second, also called KHz. In general, gold detectors operate at higher frequencies, which allow them to find small nuggets, while coin and jewelry detectors work at lower frequencies for better discrimination and deeper penetration. The gold-oriented VLF detectors will be in the 40 KHz and above, the general-purpose machines will be around 19 KHz, and the coin and jewelry detectors will be less than 10 KHz. A high KHz allows the gold-oriented VLF detectors to see some very small bits and flakes of gold, but also makes them more sensitive to the effects of soil mineralization and hot rocks. In general, the multipurpose detectors will have a lower frequency than the pure gold units, with their frequency set at a compromise in between that of pure gold VLFs and the lower frequency numbers of the typical coin and jewelry machines. The lower frequency of the multipurpose units allows for better resistance to soil mineralization and better discrimination capabilities, but also limits the multipurpose detector's capability to see the smallest pieces of gold. Also, as I noted above, the lower the frequency used by the detector, the deeper it can normally penetrate.

Complexity and Programmability:

In considering all the complexity and potential adjustments possible on modern gold detectors, the many choices can seem formidable to the new user. This is especially true of the GPX-5000 and GPZ 7000 detectors by Minelab. On the one hand, the greater the adjustability and programmability, the greater the increased flexibility that is available for the user to compensate for different conditions out in the field. Years ago, the programmability trend started in coin and relic machines like Whites DFX, and there is little doubt that it will surely continue on in the future with prospecting detectors. While it may at first seem difficult to the user, once you have mastered your machine's potential operating ranges, you will have a considerable power and flexibility to compensate for those unusual conditions you may find in the field. Take time to study the operator's manual and learn about the capabilities of your machine - it is well worth the time invested.

Metal Detector Coils:

On one level, your coil really is just a loop of copper wires wound into a certain shape and connected to a plug, which allows the coil to be connected to your detector. Although that may be true, there is a whole lot more than that to coil design and construction. The wire loops need to be held firmly in place, insulated from static electricity and protected from dirt and damage. While a few detector operators with advanced knowledge may construct their own coils, most will not. Luckily, most gold detectors, including both VLF and Pulse machines, offer a wide range of commercially available accessory coils. I strongly recommend that you purchase several coils of differing sizes and types, as they can help greatly in finding gold. The available array of choices is such that you will need to think carefully about the places you detect so that you can select the right coils for the situations you will be working under. In recent years, many of the manufacturers have recognized that having a full line of accessory coils is good for the buyer and profitable for their business as well. These accessory coils are important and allow the operator to alter the behavior of the detector to meet certain conditions in the field, better helping him to find those golden nuggets.

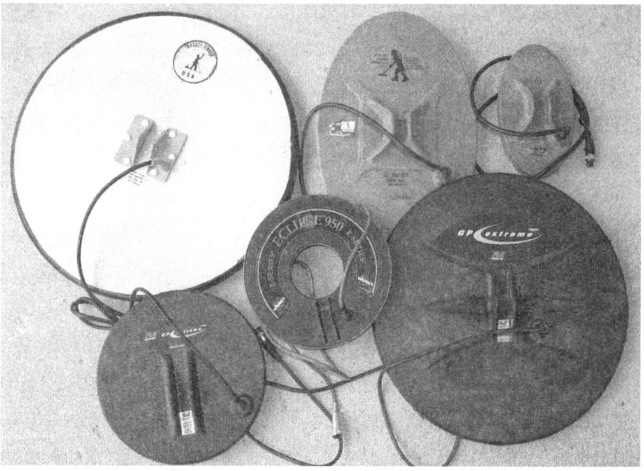
Accessory coils are an important part of getting the greatest performance out of your detector. Small coils see smaller targets, larger coils see deeper.

Coils are designed to be used with specific models of detectors. A coil designed for a VLF detector will not work properly on a PI, and in fact a coil designed for one VLF may not work even on another VLF of the same manufacturer. When considering the interchanging of various coils, be sure to consult with the manufacturer. Often this data is available on their website. In general, large coils perform their best detecting large objects deep and small coils are best at detecting small objects that are shallow. This is true for both pulse induction (PI) and VLF detectors. Large coils create a bigger, but less dense magnetic field - therefore they can see farther down into the earth, but because the field is not as dense, it takes a larger object to set it off.

They can miss small things that just do not create a large eddy current in the ground. Small coils create a smaller and denser field. This is why they don't reach down as far, but can see smaller objects if the target is close enough to the surface. So, in making a coil choice, you need to think about what types of locations you will prospect. Is the gold small and the bedrock shallow? That would be a good location for a small coil. If the gold is big and the bedrock deep, then a larger coil might be in order. Often you may use a medium-sized coil as a compromise between the two. Practically speaking, it is best to have a set of accessory coils in order to allow you to get the most from your machine in a variety of situations. Another general coil characteristic is its resistance to water. Coils are made in varieties that are completely waterproof, and some that are only partly water resistant. Waterproof coils can be completely immersed up to the top of the cable (a few detectors are fully waterproof, but in most the electronics box is not waterproof and if water gets in, major damage will occur). Water resistant coils can be used over dew-covered ground or in a light drizzle, but should not be immersed underwater. For some reason I have yet to determine, a large percentage of the VLF coils are water proof, but only a very few of the PI coils are - be sure to check with your dealer or manufacturer to find out about your coils - you don't want to submerge a coil that isn't truly water proof.

Detector Coil Selection for VLFs:

For VLF detectors, I recommend the use of DD type coils when prospecting when they are available. Coils of the DD type have two loops of wire in a "D" shape, with the flat parts of the "D" being back to back (one "D" is reversed). While many coin and relic oriented as well as multipurpose detectors come standard with concentric coils, I believe these are inferior to DD coils for nugget prospecting applications under most circumstances. Most VLF detectors that are specifically designed for prospecting come standard with DD type coils, and that is done for good reason (the Gold Bug 2 by Fisher is an exception). Concentric coils are often a poor choice for most nugget shooting applications because they are very sensitive to hot rocks and iron mineralization in the soil. Most gold bearing areas have at least moderate iron mineralization, and the DD type coil will do a better job in those locations. In my opinion, a small DD coil in the range of about 6x3 inches is the best nugget shooting coil for a VLF. Some detectors have round coils about 5 inches and these are also good. Coils like this are good for smaller sized gold, and able to pick through trashy areas with high mineralization. The small coils are also good for uneven terrain type nugget shooting - they fit in tight crevices where nuggets might hide. In trashy areas there are problems created by large coils, because when you get multiple targets under the coil the detector will combine them together. Small coils allow good separation of junk from valuable targets like nuggets. A good application for coils like this are old mine dumps where the gold is small and there is often a significant amount of trash. If you intend to detect for nuggets with your VLF, I recommend getting either this coil or a size about 6x10 in the DD format (or both). I own both little coils and medium size coils (the 6x10) for my VLF detectors and they are handy.

The medium DD coil in a size of roughly 6x10 is standard on most VLF detectors that are designed for prospecting. This is a very good DD coil for general purpose prospecting with a VLF. If you don't get one of the coils larger than 6x10, this one makes sense, but if you have one of the larger DD coils, you probably don't need a medium sized 6x10 inch coil in addition to the larger one. Personally, I have a medium sized coil for prospecting with my VLF, but do not own a larger size VLF coil because there are negative effects of large sized coils in typical gold prospecting environments.

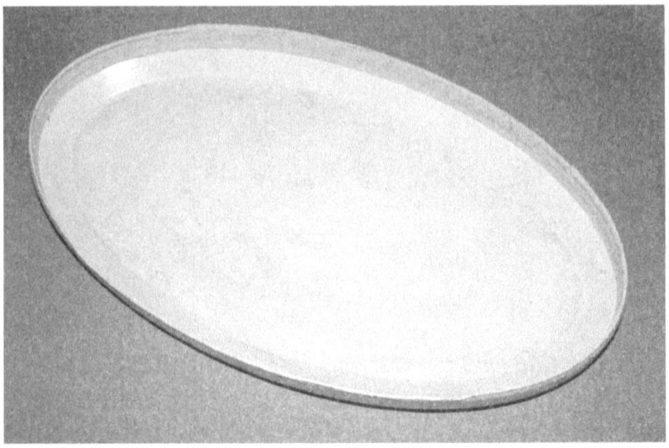

One of the most important accessories is the scuff plate that protects the bottom of your coil.

The larger DD coils for VLF detectors cover a large sweep pattern. These come in various sizes including 12 and 14-inch lengths. These larger coils will give your VLF a bit more depth, especially over moderately mineralized ground, but will sacrifice some sensitivity to smaller targets, and they are heavier. If you get one of these, it will probably be best for sweeping larger areas and patch hunting open ground for nuggets. The big problem with using larger coils on VLF detectors is ground mineralization sensitivity. In many gold fields, the ground is mineralized with iron and hot rocks. Where the ground is significantly mineralized, you will need to turn down the gain on your VLF detector to achieve a stable threshold and the problem is made worse with larger coils as they "see" more iron. Often you will find that you have to turn the gain down so much that you lose the depth advantage of the large coil. In heavily mineralized areas, you will need to do the same with medium coils as well, but the larger the coil, the worse the problem with mineralization for VLF detectors.

The inside of a DD coil has two overlapping D shaped copper coils arranged facing back to back. These coils resist the effects of mineralization better than other designs.

PI Coils: Mono and DD Types

There are actually many more possibilities concerning accessory coils for the PI type detectors. Part of this is because the PI detectors have both a DD and mono coil options for nearly all sizes. Coils of the DD design for the PI detector are similar in configuration to those of the DD coil for VLFs, with two D shaped loops of wire back to back. As a general thing, mono coils punch deeper, while the DD coils are more resistant to mineralization and hot rocks. Additionally, because the PI is less affected by problems with heavily mineralized soils, very large coils can be created for operation with a PI. I own a 19 coil for my GPZ that I purchased to search for big nuggets down deep with my Minelab. The various coils designed for the SD and GPX series detectors produced by Minelab are interchangeable but the Minelab GPX series coils from older models like the GPX 4500 and 5000 are not compatible with the GPX 6000. Many different coil sizes and designs have been produced for PI detectors, and more are coming out regularly. The many accessory coils for the Minelab detectors have partly overlapping capabilities, so carefully evaluate what you really need to hunt the conditions you are likely to encounter. It's normally worthwhile to have a small and a medium mono coil, as well as a medium to larger DD coil. It's easy to end up with too many coils - it's tempting to buy different sizes for their different capabilities.

The considerations for selecting a mono coil vs a DD on your PI detector have mostly to do with the amount of ground mineralization or interference that is present. Mono coils give a sharper, faster signal. They generally penetrate deeper and are more sensitive to weak signals. DD coils are more resistant to the effects of ground mineralization as well as electrical interference (EMI), but give broader, less intense signals. Generally, where the ground or interference will allow it, it is better to use a mono coil. The electrical interference cancel mode on the Minelab detectors can only be used with a DD coil, and the same is true of the discrimination mode - it only works with a DD coil.

On some Minelab PI detectors, DD type coils can be run as a mono, simply by flipping a switch which then uses only the left side of the coil to both transmit the pulse as well as receive the signal, just like a mono coil. There are some advantages to using this setting - in that it can help make the signal sharper and it can be helpful in pinpointing, but maximum depth will be obtained in the normal DD mode using both sides of the coil.

Most of the PI coils come in oval shapes or round. The simple trade off on this choice is that the rounds go slightly deeper, while the ovals cover a bit more ground with each sweep. The oval is also considered by some as very slightly more sensitive at the coil tip, but the difference between the two is minimal. I have both, and in the medium size coil I use for patch hunting when I want to cover the most ground, I own an oval. As a sensitive coil for patch pounding I have an 11 inch round mono. To compare the two, one needs to add the two sides and divide by 2; thus a 14 inch diameter round would compare with a 17 x 11 oval (17 + 11 = 28, and 28 divided by 2 equals 14). Another slight advantage to the oval coils is their ability to stick the narrow nose of the coil into tight spots where an equivalent round coil might not be able to reach. The scale of larger and smaller sized coils is different when considering VLF vs. PI detectors. Because of sensitivity to ground mineralization, VLF coils intended for gold prospecting are generally much smaller. A medium to larger VLF nugget coil may be sized 9 inches by 6 inches, yet the same sized PI coil would be considered among the smallest available. The fact that PI coils are larger, explains part of the depth advantage PI units have over VLF detectors with larger gold.

Oval coils have certain advantages: the tip is sensitive and they can be poked into the brush.

As with VLFs, small coils on the PI detectors are used principally for small gold that is located near the surface (of course it will also find larger pieces near the surface as well). Some of the likely spots that work best with small coils include shallow bedrock along streams and other drainages, old hydraulic mines and dry wash tailings. Because small coils are more sensitive, they generally have greater issues with hot rocks and strong ground mineralization. PI detectors are in general much less sensitive to hot rocks and ground mineralization, but they are certainly not completely immune to those effects. Where the ground mineralization is really intense, these problems overcome the detector, and you will likely see the worst of these situations when using small coils.

Medium coils are a compromise between the strengths of and weaknesses of both large and small coils - to be honest its medium sized coils that I use most often, just for that very reason. They are heavy enough that I need a support cord, but small enough to get into many bushes and some of the other tight spots where the gold may hide - perhaps not as well as the little coils, but lots better than large ones. They also cover some good round area and still provide some sensitivity to smaller gold.

Because of their ability to deal with ground mineralization, it is possible to have PI coils that come in really large sizes. The "large" coil in my arsenal is 19 inches in diameter. It is an accessory coil made for the Minelab GPZ 7000 detector. While that is a large coil, there are some available for purchase, which are even much larger! Larger coils have a great advantage in that they can see big nuggets deeper, so they lead to some big, spectacular finds. They also cover a larger amount of ground quickly as they sample more ground with each pass. The downside of this is that they do not see the smallest gold that a smaller coil could detect. While less sensitive to small gold, they are also less sensitive to ground mineralization as well. However, they have their own set of disadvantages. Larger coils also act an antenna, and are therefore more sensitive to electromagnetic interference. If there are sources of interference around, your large coil will show it to you. They also are heavier and at the end of a full day of swinging them you will feel it, even with a bungee support or other support system to relieve the weight. Because of the deeper targets, your holes will be generally larger as well - so you will work to dig both deeper and wider dig holes. They also can be more sensitive to false noises just caused by the shape of your dig hole. Because a big mono can generate noise off the sharper edges of the hole as you dig, it is necessary to dig a broad dish shaped hole as you search for your target. The problem with false noise is that it can become more pronounced the deeper you go. Finally, it's hard to pinpoint with these larger coils. I often find myself switching back over to a smaller coil to pinpoint and recover the target itself. Some makers also produce what is called a "Salt", or Anti-Interference Coil. These are a specialty type of coil, which is made for working in areas of extreme electromagnetic interference, or along salt marshes and alkali lakes. If your area includes these, you might want one but most guys don't find they need to own one.

I've covered some of the basic advantages and disadvantages concerning a large number of different types of coils. The reason it is good to have a handful of accessory coils is that no one coil does everything. That's why it's good to have an arsenal of accessory coils - at least a few. You don't need 15 different accessory coils, but having several different ones will allow you to use your detector in different ways under different conditions. I use mono coils on my Minelab nearly all the time, both in California and in Nevada - but it is because those coils are the best for the conditions I generally work most often. The ground conditions, the gold that is present and what you are doing determine what coil is best for your specific situations. If you are out exploring for nugget patches, and hoping to cover a lot of ground, you want a good-sized coil - I use a 17x11 oval in those situations. On the other hand, if you have found a good spot and want to pound it for every nugget you can get, then you really want to hit that spot with a variety of coils to get all the gold you can. One coil will hear gold that another will simply miss, so that's why you should go over a hot spot like that with every coil you have.

Power lines like these run through the gold country and may interfere with PI detectors.

For those of you swinging a Minelab metal detector, I have my special word of warning about changing coils. Detach your battery first, and reattach it to your detector only once you are sure the coil is firmly attached. If somehow in your excitement while out in the goldfields, you somehow turn on the power to your Minelab detector with no coil attached, it will fry itself in a few seconds. There is only a small diode that needs to be replaced on the detector's electronics board, but your prospecting trip will be over. Since repair requires shipment back to the factory, this can be a pain and take a while. So, avoid the problem - I learned this the hard way!

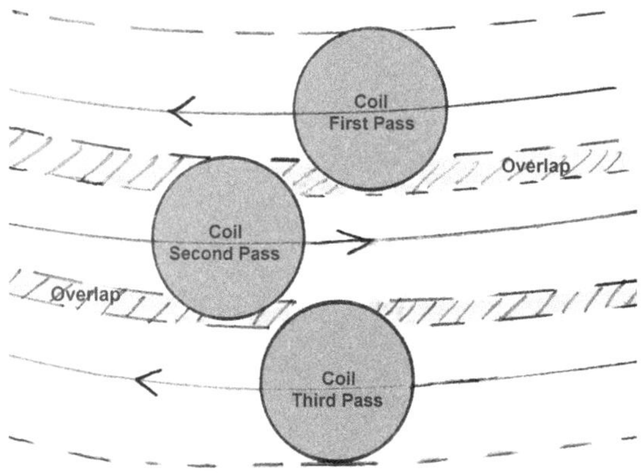

Always make sure you overlap your coil swings, especially in areas known to produce nuggets.

General Coil Considerations:

One of the most important considerations in the operation of your metal detector is how you move and operate your coil. One of the things I have observed about the guys who are consistently finding gold with their detectors is that they have good technique in moving their coil across the ground. "Keep it low and slow" is the motto of many successful detector operators. Swinging your coil wildly, moving fast and lifting it off the ground at the end of each swing will lead to missed targets, guaranteed. The whole concept of coil control and proper technique with your detector is to 1.) Keep the coil just above the surface of the ground and parallel to it; 2.) Move it slowly; 3.) Carefully overlap the area covered by the coil on each swing. All three of these are important, so let's consider each concept individually.

First up is the concept of keeping your coil flat to the ground and close to it. Every inch that you hold your coil above the ground is an inch you won't be able to penetrate into the ground to see a nugget. For practical purposes, under most conditions you have to keep the coil a bit above the ground to keep from banging it on every little stone, but you should keep it as close to the ground as reasonably possible. Additionally, the deepest penetration of the coil extends outward in the direction perpendicular to the coil, so you need to keep your coil parallel to the ground surface to see as far down as you can. When you find a patch of gold nuggets, scrub the coil slowly right on the surface - the coil cover will protect it. Scrubbing your coil in patch areas and actually touching the ground will help you get all the gold you can out of these areas.

Second is the idea of moving your coils along slowly. So, what is the best coil sweep speed when hunting gold? While many detectors may not seem to be overly sensitive to the coil sweep speed - it is important. There is no question that when hunting very faint targets such as small or deep gold nuggets slow coil movement helps make the target response stand out against the background threshold sounds. Because of the quick target response of most detectors, fast coil movement works just fine for coin shooting or relic hunting. As a result, many folks who started out hunting coins learn to work their detectors quickly through parks and schools, as it tends to increase their finds. However, this is a terrible idea when prospecting for gold - this is one case where one needs to go slow or nuggets will be missed. Slow coil speeds will improve the response of most detectors - especially on the smaller and deeper gold that is near the limit of the detector's ability. A faster sweep speed can also increase the signal response on hot rocks and mineralized ground, creating "noise" that will make you miss faint targets. You can do a lot to eliminate or greatly reduce unwanted ground "targets" by simply slowing way down on your coil swing speed. On the other hand, with a slow sweep speed, a good metal target will actually increase in its signal response! When exploring a larger area and the goal is determining if detectable gold is present, one can go at a slow to moderate coil sweep speed, but when you find some gold and want to pound the immediate area to locate any additional nuggets, its best to slow way down. At a very slow speed, it should take perhaps 10 seconds to cover the roughly 8-foot arc length of a typical sweep.

Third, carefully overlapping the path of your coil swings is needed to get maximum depth of detection. No matter the coil design, the deepest detection point will be located underneath the center part of the coil. Under the edges of the coil, penetration is just not as deep. That is why having some coil sweep overlap is important. While generally exploring or patch hunting, I overlap about a quarter of the length of the coil. However, in patches where gold has been found, and the gold bearing material layer is deep I may alter my coil sweep to overlap about 1/2 the length of the coil.

When you add up all these factors, and add in a couple other things, you get my special prescription for extracting gold from pounded patches that others have already discovered and worked. Remember that the keys are to scrub the spot with your coil right on the ground, move your coil very slowly, and overlap the coil paths closely. As applicable, you can then add the following ideas for pounding old patches: 1) work the perimeter of the old patch area; 2) put your coil into areas where others generally won't, such as rock piles, into clumps of bushes and cactus, or other hard to reach spots, etc.; 3) If there are smaller rocks lying on the surface, rake them aside to allow your coil to ride right on the surface, improving your ground scrub; 4) Use multiple coils (every one you have) on a patch will help to get everything your detector can find.

The Targets You Will Find:

If you follow the proper coil techniques, you will be finding targets in the ground - and sometimes a lot of them (especially where old prospectors left their trash behind). Finding something with a metal detector is not all that hard in many areas, although there is a catch. I hate to be the bearer of bad news, but unfortunately, the vast majority of the targets in most goldfields will be trash or some other thing that is not too valuable. If you are persistent and look in the right locations, some of your targets will be gold. If you find yourself digging a lot of trash, don't get discouraged; just remember that finding gold is not an easy task - it's a challenge. If finding gold were easy to do, gold wouldn't be so valuable - one little ounce is now more than US$3,000! You will need to learn to interpret and recognize all the sounds (and meter readings) that your detector produces. Let's consider the things you might find as a target with your metal detector. The target signal sounds of a VLF and a Pulse detector are significantly different. VLF detectors tend to give signals that

Unfortunately, the bulk of the finds you will make with your detector will be nothing more than trash.

are short and sharp on targets that are not made of iron. Nuggets are sometimes described as making a zip-zip sound. The target signals on a PI are broader and move up or down in pitch much more slowly - they are often described as a wee-ooo or an ooo-wee sound.

On both detectors, iron tends to make a broader type of sound. The operator can often determine size of very large targets, because the size of the target is generally indicated by the area covered while the metal detector is sounding off and the duration of the signal. You will get a feel for this once you have detected some pipes or other large objects. While it is generally difficult to tell a half-inch target from a one-inch target, you can often tell a small (like one inch) target from a six-inch target. The deeper the target, the more difficult this type of distinction is. This can roughly be used for discrimination, because targets that are several feet long or larger are very unlikely to be gold. Heaven help you however, if you walk away from a 2-foot target and it turns out to be a gigantic nugget! Rich pockets of gold quartz with lots of gold can give really broad signals, so you need to be aware of the possibilities where you are detecting. When in country known for rich pockets or gigantic nuggets, you have to dig it all, even if the target seems too large.

One of the interesting things about PI detectors is that the sounds of the targets detected are inverted depending on the size and nature of the target. The target signals on smaller nonferrous targets like

This gold is distributed through quartz rock is not solid - one of the types that respond badly on a PI though there is a good bit of gold in this specimen.

gold nuggets or lead bullets give an up-down tone where the tone rises in pitch and then goes down (often called a wee-ooo sound). With iron objects or larger non-ferrous targets, the tone sound goes down in pitch first and then rises back upward (often called an ooo-wee sound). This situation is caused by the nature of the ground balancing systems of the detector. It can sometimes be used as a simple discrimination system, but on the PI detectors, it is crude and easily fooled, so therefore it is very unreliable because there are a fair number of exceptions to the rule (small very rusty iron items also sound like non-ferrous targets, and larger nuggets will give an ooo-wee sound like a piece of rusty sheet metal). One of the most important things for guys who have not found a lot of gold with a detector and wish to be much more successful, is to focus on listening to small and faint but repeatable sounds. Gold is often small and its target sound does not BOOM through your detector like a near surface coin or a big rusty bolt (yes, I've repeated this point a number of times - I want to emphasize it). The successful electronic prospector has trained his ear to hear faint sounds that may indicate gold. Often this is just the opposite of the coin hunter who trains his ear to ignore the slight sounds of mineralization and bits of foil or other small trash. Because of this, experienced coin shooters may need to make an effort to "re-learn" what to listen for. Trash targets often scream real loud because they are normally near the surface and can commonly be quite large. Yes, there are exceptions when you find a spot no one has ever gone over with a metal detector before - then you might find a large nugget near the surface. However, it remains a fact that large near surface nuggets are rare and near surface very loud targets are commonly trash. In spite of the fact that larger near surface targets are normally trash, many times you will need to dig it all to find out what it is, and no one wants to miss a large nugget. Discrimination systems on most currently available PI detectors are unreliable, although the Garrett Axiom is actually pretty decent. Some new technology on the horizon may change this in the future. Discrimination systems on VLF detectors are much better, but still not 100 percent perfect. For more discussion about discrimination, see the portion of this chapter on that topic.

Part of deciding what to dig and what not to dig depends on your location. In some desert areas of the west, with new nugget patches that may have never been found by the old-time miners there is very little trash present. Digging all targets is not too difficult under those circumstances. On the other hand, I have been in places that are just littered with trash everywhere. Some trashy places are candidates for a good discriminating VLF, and in addition, some may require magnet sweeping to remove the bulk of the surface iron. Don't forget about the potential of the magnet, as there are some circumstances where it can really make a difference in clearing off iron trash. Big, powerful magnets are easily available for reasonable prices, and mounted on a handle they can make a worthwhile tool to carry along with you to the hills when you head out. There are also other techniques for dealing with trashy locations, and these focus on working the periphery of the placer area: benches and small tributaries in stream areas, or around the outside parts of hillside (residual) deposits. Often locations right in the gut of a stream or the center of the diggings or near old cabins or camps are just over loaded with trash. I have found that sometimes all you need to do is spread out a bit, and get away from the center of the old timer's work and sometimes you will see a great reduction in the amount of trash you will find. As I have noted, some detectors (notably VLFs) have an ability to discriminate, but no discriminator is 100% accurate and only digging all targets will allow you to see what your detector has found. On the other hand, digging it all in a trashy area can be rough - again, you'll have to make the decision on what levels of trash you can tolerate. Be careful with discrimination and only walk away from signals that the detector is 100% certain is trash. The old timers often camped right on their diggings and dumped trash in places with decent gold. Many prospectors will walk away from trashy spots, so if you have the patience to sort through the trash with a good VLF, you may find some good gold right in the middle of the old diggings.

Interpreting the sounds and meter readings that your detector produces becomes easier as you get used to your detector. You will learn in time to recognize those subtle sounds and responses. Every design of detector has its own set of responses and it just takes time using it to learn to master it. Don't worry too much about this because you can still find some nice gold even without mastering all the slightest responses of your detector. As you are swinging your coil in the field, you may get some false sounds, especially with Pulse Induction (PI) units. Falsing or false signals are commonly caused by the coil cable touching a rock, plant or branch, but can also be caused by sand and other material that has gotten into the space between the skid plate on the bottom of the coil and the coil itself. You may find that you need to pull off that skid plate once in a while and clean out the space under the coil. Keeping the skid plate well sealed off to prevent the intrusion of sand is also a worthwhile effort.

For PI units, false signals can also come from electro-magnetic interference caused by such things as power lines, lightning, airplanes, long unpowered metal fence lines, etc. Depending on your location and the time of year, these sources of false readings can range from having no noticeable effect on the detector operator, to being so extreme as to completely prevent you from using your detector. Sometimes simply waiting will allow these things to pass (like lightning storms). Sometimes adjustments or other settings such as "cancel mode" and DD coils can compensate for the problem. Check your detector's manual to determine the available adjustments to compensate for the problem.

Some of your non-gold finds may be interesting, including coins and other valuables. This 1800s pocket watch was found by the author while prospecting in an old gold mine dump.

When you find a target, pass your coil back and forth to test it out with your detector and see what it tells you. Repeatability of the response is an important key - it should be hearable every time you pass your coil over it, even if the repeated response is very weak. Intermittent and extraneous signals are heard once, and then gone. If you are wondering about a signal, re-ground balance a foot or two away from the target spot and pass your coil over it again. If there really is a target present, you should be able to hear it. Now if you are operating in a tracking mode with automatic ground balance be careful not to linger with your coil over the target or pass your coil only over the top of the target. In many cases keeping your coil too long over the target when in automatic ground balance mode will cause the target to be ground balanced out - that is the detector will compensate, and include the target response in the ground mineralization reading that it is trying to cancel out. Then once it is cancelled out you will no longer be able to hear it. If the target seems to simply disappear, re-ground balance away from the target, then retest the target spot, and in most cases it will re-appear as the detector has gone back to the original (and correct) ground balance setting. Sometimes in evaluating a target it will be better to go into fixed ground balance mode to keep everything stable.

Step one *in recovering a target is to accurately pinpoint its location in the ground.*

Most detectors are a bit more sensitive when they are in fixed balance mode anyway (which is why I try to operate in that mode whenever it's possible to do so). One of the keys to being productive in a goldfield is to be able to quickly evaluate, pinpoint and dig a target once your metal detector has sounded off. If you spend too much time trying to identify, locate and get the target out of the ground, you will have less time actually looking for new targets. If you can get the target out of the ground quickly, you will be able to move on and dig more targets. In most situations, that means in the long run you will dig more gold. Good skills at pinpointing means that you are digging targets quickly and efficiently. I will normally spend an extra few seconds before I dig to accurately pinpoint the target - accurately located targets mean you can often dig smaller holes and have the target out more quickly. I do this by swinging one direction and identifying an imaginary line on which the target is located, then turning 90 degrees to face the other direction, I line up another imaginary line on which the target is located, and the center of the X where they cross is the location of the target. Pinpointing with DDs is done with the idea that the strongest target response is from the section in the middle where the two coils overlap. On a mono coil designed for a pulse unit, the best spot for pinpointing is the front tip of the coil.

Step two *is to dig the hole, testing until you have dug the target out and onto the dig pile.*

All About Hot Rocks

One of the things that cause many detector operators to go crazy in certain goldfields is the presence of hot rocks. These are stones that have a mineralization and response significantly different than the surrounding soil and gravel. They contain mineral levels either much more or much less than the soil the detector is ground balanced for, so the detector often sees them as faint targets (sometimes they sound off as very loud, strong targets as well, but more often as faint). Like a metallic target, the larger and closer to the surface the hot rock is, the louder it will sound off on your detector. These hot rocks can be found in a variety of situations, but perhaps one of the most common is in stream or gravel deposits where the forces of erosion can bring rocks of different make up together, so that a few may be much more mineralized than the mass as a whole. Many types of rock can, under the right circumstances become a hot rock. Common examples include chunks of magnetite and dark colored basic volcanic rocks like basalt or andesite. They can actually be a variety of differing colors, but it is typical that hot rocks of a given area will typically have a similar appearance, so it is helpful to learn to recognize them as you detect. VLF detectors may also experience problems with cold rocks - rocks that are less mineralized than the surrounding soil.

At times the hot rocks will make a broader, less sharp sound that experienced operators can

***Step 3** is to sort through the dig cup, separating the target from the soil a handful at a time until you have found what you are looking for.*

distinguish from normal targets, but that is not always true, and sometimes the sound made by a hot rock is not easily distinguishable from that of a good target. Overall, VLF detectors are much more sensitive to the effects of hot rocks, but even PI units are not fully immune from their effects. Where they cannot be easily distinguished from good targets, they must be dug and treated like a target. If there are only a few of these hot rocks, it may not be any big deal. However, the bad part of this is that under some circumstances, a gold bearing area may contain tens of thousands of hot rocks. The result is something like an extremely trashy spot - it is an area that cannot be effectively worked with a metal detector, unless some adjustments and compensations are made.

Luckily there are some things that can be done to deal with the effects of hot rocks. Perhaps is the first you should try is to re-ground balance your detector, as a detector which is out of tune can be overly sensitive to ground mineralization, and as you move around the area you are working, the mineralization does tend to vary somewhat. The next step might be slowing down your coil movement some - fast swings over hot rocks tend to make them sound better - more like a good target. Another characteristic of hot rocks is that they quickly die out as you raise your coil - they die out much faster than a metallic target. As a result, hot rocks are normally shallow, at or near the surface (there is one terrible exception, which we will discuss shortly). The deepest hot rocks I can remember digging were at about 5 inches, and

***Step 4** is to carefully fill in your dig hole.*

these were roughly the size of a bowling ball. A tin can of similar size would respond loudly at depths well over a foot. So, one method of dealing with hot rocks is to slowly raise your coil as you swing over it, listening to see how fast the target disappears. If the target quickly disappears, it's probably a hot rock. Some experience and practice are required in getting the hang of this method of dealing with hot rocks. The problem with this method is in evaluating faint targets - ones that are already near the edge of detection. For very weak targets, both hot rocks and valid metal types, they disappear fairly fast. Still another good method of dealing with hot rocks is by switching to a DD type coil. DD coils handle hot rocks better than most other types of coils. In some unusual conditions where the hot rocks are only on the surface, it may be possible to put on a smaller coil and hunt around them.

Discrimination on some VLF units can be used to identify hot rocks with reasonable reliability. VLF detectors with good discrimination that can discriminate based on relative conductivity can normally see that most hot rocks are not in a discrimination range that could possibly be mistaken for gold. For VLF detectors that have this feature, this is perhaps the best answer to deal with areas containing loads of hot rocks.

Another technique, one that is undesirable under most situations, is ground balancing to the hot rocks themselves. One locates a hot rock and then ground balances over it. While doing this will allow you to ignore most if not all of the hot rocks, this is an extreme technique that has some notable negative effects. The most critical is that you will be reducing your sensitivity to gold and other valid targets in a very significant way. This method is something of a last resort, but there are places where this is pretty much all that can be done (especially if you are using a VLF). With a VLF and some PI type units, many weak hot rocks can be eliminated by turning down the gain setting. While this allows weak hot rocks and hot ground spots to be ignored, it is a problem if you turn down the gain too much as this also reduces your sensitivity to those golden targets. There is an artful balance in turning down the gain sufficiently to achieve a stable threshold and the elimination of certain weak hot rocks without sacrificing too much sensitivity. As noted, PI detectors are for more resistant to hot rocks and can ignore many hot rocks that would respond on a VLF detector. Many successful detector operators use PI detectors primarily because of their ability to ignore hot rocks and ground mineralization. As noted above, they are not completely immune to hot rocks, and some PI units have timing adjustments that also assist further in enabling those PI detectors to ignore hot rocks.

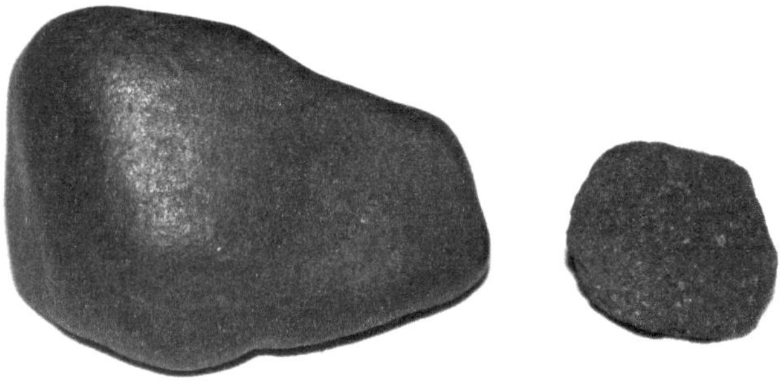

Left – a smooth, river rolled chunk of magnetite, right – a rough, natural piece, both are strongly magnetic, and make for a fairly strong hot rock.

Finally, we come to the king of the hot rocks, a mineral known as cuprite. It is to metal detectors like Kryptonite was to superman. It sounds off just like a metallic target on both PI and VLF detectors, and unlike most hot rocks, it can be detected at great depth - the same as metal targets. It is the exception to the rule. A bowling ball sized fragment might sound off on a PI detector as deep as 3 feet. About the only thing good I can say regarding this mineral for detector operators is that it is rare - most operators will never encounter it. I know of only a couple placer deposits in the US where it is a problem, although there are probably more than that out there. It is a secondary copper mineral, reddish with a grey metallic sheen, but sometimes the sheen dominates and it appears grey with a slight reddish color. Often there are green to blue-green copper minerals associated with it. It is found near copper mines and other copper deposits. If you try to detect old copper mine dumps which contain cuprite, it will give you fits. Because it responds like metal, there are no easy solutions to deal with it. Another, even rarer copper mineral, covellite, is reputed to have the same effect, but most copper minerals are no problem.

This photo shows a piece of Cuprite copper ore. It is green on the outside, but the inside is red.

We should take a little space to talk about hot spots and ground noises as well. Ground noises, also known as hot spots are a close cousin to the hot rocks we have been discussing. In the same way hot rocks are individual stones that have a mineralization and response significantly different than the surrounding soil and gravel, ground noises are patches of soil that are much more mineralized than the surrounding soil and gravel. When you dig up a hot rock, you end up with a stone that was your target, but when you dig a ground noise spot, as you pull the mineralization out of the ground the soil gets spread around on your dig pile and the noise often just disappears. That's the hallmark of ground noises - they get weaker as you dig them. Like hot rocks, problems with ground noises can vary from one extreme to the other - you may have little or no problem with them in some areas and extreme problems and difficulties in others.

Like hot rocks the signal response tone of ground mineralization spots often sounds broad and not sharp. However, you cannot always be sure about ground noises and hot rocks - it is important to be very careful when you dismiss a response from your detector as a worthless ground noise or hot rock and simply walk away. Sometimes gold will sound exactly the same as these spots of mineralization. A while back, I found a 7 dwt nugget, and only about a foot away, a 19 grainer (about ¾ dwt). This is a hillside location, and there is no stream or river very close by. I found them in an old dozer scrape. Of course, the first thing I did after I found those two nuggets was search the nearby area in that dozer scrape for more gold. I found a little spot where there seemed to be a target about 5 feet away from where I found that 7 dwt piece. It was very weak. I dug down about 3 inches and the target sound didn't get any stronger, in fact, it seemed to get weaker. I walked away, thinking it was a bit of ground noise. It was very weak and sounded just like a normal bit of indistinct warbling - a little bit of ground noise of just the kind we all walk past regularly, and for good reason. On the rare occasion when a mineralized spot gets dug, the apparent target just disappears when it is dug and spread out on the pile. Ground noise is not common at this location, but it does occur. Besides, it was the end of the day and time to go anyway - and I knew I'd be back later to check the whole area out further.

Two weeks later, I went back to the same area to cut some firewood and do some more detecting. A friend and I cut about a cord and a half of firewood, stacked it in the pick-up and trailer we brought with us, and then in the time we had left we did a bit of metal detecting. I tried a few spots, but came back to the spot where I found the two nuggets. I had thought about it in the couple of weeks since I was there, and decided I really wanted to check out that spot of ground noise I had walked away from. I carefully ground balanced my detector, and sure enough, that ground noise was still there in my old dig hole (a good sign it might not actually be a mineralized spot). In looking at it closer, I found that from one angle, it sounded just like a very weak and warbly ground noise, but if I turned 90 degrees and swung the coil from the other direction, it sounded like a weak (but good) target (I was using a Coiltec wallaby mono). This is still not outside the pale for mineralized spots, but I decided to dig out another 2 inches just to see. When I re-tested the now deeper hole, the target had definitely gotten a bit stronger - it now sounded real good in one direction and like a weak target in the other.

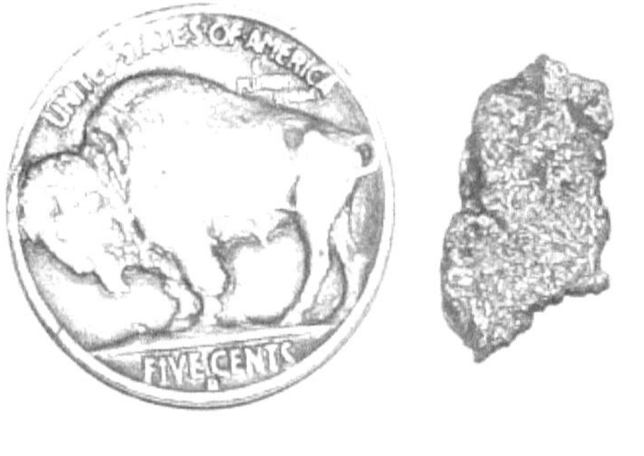

The nugget on the left is the 1.1 dwt chunk of gold that sounded off like a bit of ground noise.

I told my friend who had come over to watch that I was for sure going to dig this now and it was either going to be a bit of mineralized ground or gold - and the odds were in favor of gold. Nearly all the trash here is shallow and I was now down about 5 inches from the original surface in hard but weathered rock in place. I dug another couple inches and it was finally a strong and very good sounding target, but it was still in the hole. No doubt it was gold. Another inch or so of digging and it was out of the hole - the ground was getting really tight and hard - it would have been hard to dig much deeper. I sorted through the pile and found my target - it was a gold nugget, and weighed in at 1.1 dwt. It is fairly thin, and I think the poor response until it was very close to the coil was due to it being in the ground on edge. I rechecked the hole afterward - and that funny "ground noise" was gone. The moral of this story is that you need to be careful about walking away from repeatable target sounds in the ground. Yep, hot spots in the ground exist, and can be fairly common in some areas. Just be sure to investigate sufficiently enough to be sure you are not actually walking away from gold. The finder of the largest nugget ever discovered with a metal detector, the 876 ounce "Hand of Faith" nugget from Australia, reported that he almost walked away from it, thinking it was simply a ground noise. Imagine walking away from an 876 ounce nugget! Sure, my nugget in this example was an awful lot smaller, but I found that nugget because I thought about things after I left the first time. I decided I ought to dig any repeatable sound on my detector even if it didn't sound like a normal target - and especially because I was only 5 or 6 feet from the spot where I'd already detected two nice nuggets. Now that may sound like a no brainer decision, but it took a while to sink in through my thick skull. That little nugget told me a lot about my detector and the true meaning of the responses I was hearing.

As noted, the main sign of a ground noise is that when you dig out the apparent target, it seems to disappear. It is also fairly common that you will find that your dirt pile starts to give a faint response, while the dig hole also still gives a faint response. If, however, you swing your coil at the original level of the ground over the hole, the original target you were hearing at that level has disappeared. Ground balance your detector over the pile of dirt that came out of the hole, and normally the target will have completely disappeared. This all happens as you dig up and spread out the soil that made up the hot spot on to your dig hole pile.

The methods of dealing with ground noises are basically the same steps you would take in dealing with hot rocks. Normally mineralized ground spot sounds will be weak, but as I noted in my example above, sometimes deep nuggets will sound weak and broken up as well. Re-ground balancing your detector, using DD coils, raising your coil up off the ground, slowing down your coil speed - all the techniques you will use to deal with hot rocks will also help you to recognize and deal with mineralized ground spots as well.

Through all that trash and mineralization, if you are hunting in the right locations, you will of course be finding some gold as well. Gold can be almost any size or shape, alloyed with various other metals, be mixed in with a wide variety of other minerals, and be found at any depth from the surface on down. Certainly, that makes for a lot of possibilities, and there is no question that detectors respond differently to gold depending on the various forms it assumes. Certainly, the larger the gold is, in general the larger and more lasting the response eddy currents it will create and the better your detector will respond to it.

Flaky, thin pieces of gold like these shown here can be harder for your detector to find.

However, given all the shape, alloy and mineral mixture possibilities, the issue is much more complex than a simple size-response relationship. It is also important to note that the more solid a nugget is the better your metal detector will respond to it. Gold that is spread out in small pieces through quartz or iron oxide minerals just won't sound off as well as a solid chunky nugget of fairly pure gold - sometimes even if the total weight is significant. Wiry and mesh like pieces respond to a detector in a manner similar to a solid nugget of much, much smaller size. On top of this, not all metal detectors are created alike in this regard. VLF detectors are far more sensitive to small wiry, loose, mesh-like crystalline gold in comparison to PI types. The PIs respond poorly to this kind of gold and do much better with solid pieces. The gold in hard rock mine tailings often has a tendency to be more wiry and less solid, therefore that's a place where PIs commonly do poorly (they often tend to be trashy as well).

The response of PI detectors on loose, wiry and mesh like gold has led to discussions of "invisible nuggets" - gold that is significant in its total weight, but so spread out as to be difficult or even impossible to detect. This is especially true on the PI type detectors; although the most recent models of the Minelab GP series (the 4500 and the 5000) do better than previous models with these spongy, porous nuggets where the gold is dispersed through various other minerals. It is important to note that if the gold particles are small enough, even a VLF will have difficulty detecting some of these pieces. Gold is found in every step of the full spectrum from very solid through somewhat solid to wiry to delicate crystalline mesh to tiny specks disbursed through the vein rock. Certain districts have a tendency to produce nuggets that are more solid and others more porous. Olinghouse, a district located about 30 miles east of Reno is well known for spongy, wire mesh placer nuggets that look like a small piece of coarse steel wool made of gold.

Pure and solid nuggets like this one respond much better than nuggets which are wiry, mesh-like or mixed in with lots of rock.

There is no question that the nature of the gold being detected greatly affects the strength of its response to magnetic currents, and the total weight of the gold is not the only factor. The shape and solidity are important, certainly delicate wire mesh and quartz gold specimens do not respond as well as solid pieces of the same weight. It's a complex topic and while the prospector cannot choose what type of gold he will find, knowing the types of gold you will be searching for and its nature can help the knowledgeable prospector in choosing the type of detector, coils and settings that he will operate while he is out searching for the precious golden metal. Be aware of the nature of your target, and adjust your strategies accordingly.

When you have retrieved a target out of a hole, always re-check your hole once you've taken a nugget out of it - multiple nugget holes are not that rare, and you don't want to leave the rest of the gold for another operator. Whether its bedrock crevices or hillside patches, nuggets do sometimes tend to occur in groups. More than once a prospector has ignored this rule and left me some nice gold. I have also dug gold making a second pass over my own holes before I walked away – so always be sure to recheck before you move on to make sure you don't have multiple golden targets in the same spot.

Even if you don't find a second nugget in the same hole, it is normal that nuggets will occur in groups. Single isolated nuggets are found, but my experience is that groups are more common. If you find one nugget it is normal that the geologic environments that created that first nugget also existed to lead to the formation of more gold. As other nuggets may be found nearby, careful concentration and detailed exploration of that spot is well justified. An area that yields multiple nuggets is known by detector operators as a nugget patch or just a patch. Patches can be found in a number of different types of environments including hillside slopes, stream drainages, areas of old dry wash tailings, old hydraulic mines and other locations. Always do your best to turn a single nugget find into a patch of multiple nuggets. Because nuggets commonly occur in groups, when you find one nugget, you should slow down and concentrate very closely on that immediate area because you have a good chance to find more.

Locating Fine Gold Pay Streaks

Another topic I'd like to discuss is searching for fine gold concentrations using the "mineral" settings of a metal detector. Lots of fine gold is too small for any detector, and nearly every detector maker suggests in their marketing and other written materials that good deposits of fine gold can be found by hunting for concentrations of black sand with their metal detector. That type of thing has been claimed since as far back as the 1970s and perhaps even before. For those who had someday planned to try this technique, I hate to be the bearer of bad news, but in my experience, that's nothing but a bunch of hogwash. Experienced operators virtually never try to find gold this way, and for good reason. As there are no metal detectors that will find only gold, there are no detectors that will find only black sand. If you try to set up your detector to find mineralization, you most likely will quickly find some - for the simple reason that mineralization is present all over the place in most goldfields. The unfortunate thing is that what "mineralization" you find is unlikely to be black sand associated with gold. There are many reasons why your detector might respond to ground mineralization, and only in the rarest, most unlikely of occasions could it possibly be a concentration of black sand containing significant fine gold.

Here is the reason why - When the magnetic field generated by your metal detector's coil passes over the ground, it generates a return signal from the iron and other conductive minerals in the soil. This is in addition to any return signal that is generated from any target that might be present. In some cases, the return signal from the iron in the soil is actually thousands of times greater than the return signal from a small target. All metal detectors designed for the purpose of finding gold nuggets possess some sort of ground balance system that is used to eliminate the natural response of the iron minerals in the soil, so that the detector can differentiate between the magnetic response of the ground and the magnetic response of the target. When the dampness of the soil changes, when the depth to bedrock changes, if there is a slight bit of salt in that spot, if the area has been burnt, all of these and a hundred more time wasting possibilities that are not gold can cause a "mineral" response on your metal detector. Hunting for mineralized spots with your detector can be a frustrating experience as you are likely to find many, but virtually none of them will produce any gold!

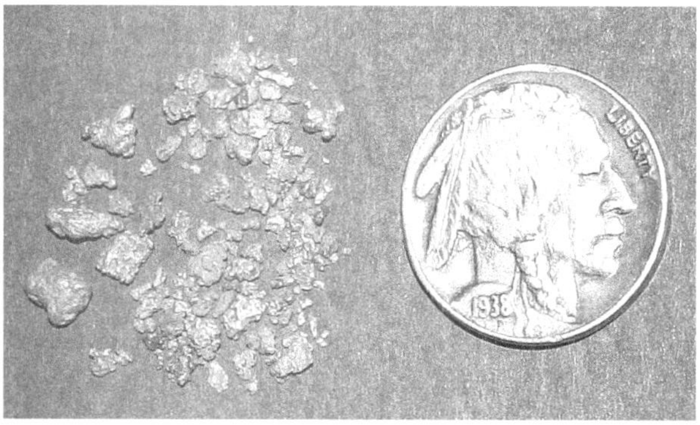

Here is a couple pennyweight of some smaller gold dry washed by the author from a pay streak located using a metal detector.

So, if hunting for mineralized spots with your detector is simply a waste of time, then what are the better methods of finding a good pay streak of fine gold with your detector? It may sound so obvious as to be ridiculous, but the best indicator of the presence of gold is the presence of gold. So if you find a spot with your detector that has a number of nuggets concentrated in a small area, there is a fair chance that it might also have some smaller fine gold associated with it. To get that smaller gold, you may want to dig the area up and process it with a sluice or dry washer. The sluice is preferred where water is available because of its higher recovery rate, but in dry desert areas, the dry washer may be just the equipment to pick up those fines. The old standard locations of inside bends, exposed bedrock and behind boulders are great places to look but the question where to search is a whole subject all to itself and will be covered later on in this book. Now not every patch of nuggets found with a metal detector is going to have a worthwhile amount of fine gold associated with it. Sometimes, the gold all falls into a certain size range and there just are no fines. However, there certainly are places where coarser gold indicates the presence concentrations of fine gold, and at the current prices, the chance to pick up even a few pennyweight or a quarter ounce of fines may well be worth going after for most prospectors.

In general, the best areas to find concentrations of fines are those places where the concentration has been created by the effects of water flows, ie., stream born paystreaks. These deposits almost always contain at least some portion of fine gold. Often these types lay down pay streak concentrations of various mixed sized gold depending on the speed of the water flow. Coarse gold lays down in faster water, while fine sized gold requires slower water conditions to settle out. In many situations, pay streaks are formed with zones of gold with the coarsest toward the head and outside and finest to the inside and tail of the pay streak. This will be discussed in great detail in the placer geology chapter.

Nugget patches which are residual in nature are often more spotty and commonly don't make for great spots to work with dry washers. Often there are only a few coarse nuggets spread thinly here or there. Certainly, there are exceptions, but generally speaking pay streaks laid down by the flow of water make for a better target to extract additional gold.

Rounded gravels in a river bank, indicating river deposited gold – look for any exposed bedrock.

So how do you figure out what type of deposit you have? How do you know if your nugget patch was formed by the residual actions of erosion or by flowing waters? Well, the first thing to consider is the location and nature of the material in the patch you are working. Are the rocks rounded, and are you working in or near a stream? Or are the rocks angular and instead you are working on a hillside? Is the gold all rounded and well-worn or is the bulk of it angular and sharp? By considering all these factors, you should pretty well be able to decide what type of deposit you have found, and then take the appropriate actions.

In the end, the best way to evaluate the situation is to test the ground in question. You can analyze and guess forever, but the only way to know with certainty if there is fine gold present in a patch of gravel is to sample it. Plus no one wants to walk away from a nugget patch wondering just how much gold he left behind. Take a few buckets of gravel from your patch to make up a representative sample, perhaps from more than one location and then pan them out. Weigh up the gold you recover and see what it totals out to be. If you are getting good fines, enough to make it worth your while, then you may need to see about treating the gravel in your patch to recover the fine gold.

It can be well worth your time, because sometimes there is actually more of the fine sized gold present in a pay streak than the gold contained in the coarse nuggets which are large enough to find with a metal detector.

Digging also exposes more bedrock for metal detecting - this is a byproduct of digging out the material for dry washing or sluicing. When you dig down to process the gravels, always carefully check the bedrock you expose with your metal detector, as nuggets can hide down in fractures and crevices that are not easily visible to the eye. Your metal detector can tell you if there is gold present that you have not seen. Sometimes you may even dig out the overburden to test an area for dry washing - in the same way, this type of digging should always be checked with a metal detector.

Other, Less Common Detector Finds

Finally, on the topic of targets, it's worthwhile to note that there are some other valuable targets that you might find with your detector while searching for gold nuggets. Most obvious is the man-made metallic stuff left behind by old time miners and residents of the area such as coins, etc. I once found a pocket watch while out hunting for gold. A friend found a gold wedding ring in some old diggings. Another favorite find of mine is an old blasting cap with teeth marks - the old miner crimped that blasting cap down over the fuse with his teeth, risking blowing his face off rather than using a set of pliers - which is the way it's supposed to be done (the cap is over a century old, the fact that it is still in one piece shows that this particular cap was only a dud). Other targets you might find include metallic copper, platinum and silver nuggets. All are fairly rare, but at some locations in Arizona, copper nuggets are not all that unusual. They are formed by secondary geologic processes near the surface. I know of no locality where native silver is common, but it does occur in some places, and it is also formed by secondary processes.

Another worthwhile find made by prospectors are meteorites. Years ago, prospectors working in the Gold Basin area of Arizona found a good number of very strong hot rocks. They were common and gave the prospectors fits. Eventually one of the guys came up with the idea that these rocks looked like meteorites, and so he had one of them tested. Sure enough, they were identified as the genuine article. Many thousands of them were gathered

Many meteorites contain iron which sounds off loudly on nearly any metal detector. They are normally strongly magnetic.

up from old nugget dig areas, as the other prospectors simply tossed them aside, commonly leaving them right next to their old dig holes. Many meteorites contain iron and nickel metal, and these are the metals that your detector responds to. Since then, quite a number of meteorite finds have been made by prospectors looking for gold. Some even end up specializing in searching for meteorites rather than gold! Be sure to keep you eyes open for magnetic rocks that when cut or broken open show visible metallic iron. If you find such a rock, contact the nearest university, and you just might have found a new and valuable meteorite strewn field.

Detector Accessories:

In addition to coils, there are a wide variety of other types of accessories that are available for the prospector to use with his detector in the field. You may not need all of these but some of them are pretty handy. I have included my thoughts and recommendations on the various categories.

Metal-Free Shoes and Boots

This is not so serious a problem with VLF detectors, but with PI detectors, and especially with PI detectors when using larger sized coils, you need to use metal free shoes. The larger coils will pick up a response from the steel eyelets, shank and other metallic parts of the shoe. This will cause apparent false targets, lost targets and other frustrating situations. There are several types of special non-metallic boots on the market and detector dealers commonly sell them. As far as I am concerned comfort in hiking around is the most important consideration. Several of the big-name tennis shoe makers also produce non-metallic tennis / basketball type shoes as well. I personally prefer the lighter tennis type shoes myself, but I also own and use the boot type also.

Headphones / External Speakers:

Nearly all prospectors will want to use some sort of headphone with their metal detector. Although essentially all VLFs do, some PI gold-oriented detectors do not have a built-in external speaker. There are many reasons for using headphones. One is that you will be far better able to hear any faint target sounds through headphones, which can make a big difference in detecting those marginal target sounds, which gold nuggets often are. Also, most of the better headphones have some sort of volume control. There is no direct volume control adjustment on the MXT and some other prospecting oriented detectors, so the volume adjustment on the headphones can help to control the loudness. Additionally, when the operator has much different hearing in one ear as opposed to the other, independent volume controls can do a lot to help compensate for that. Finally, the beeps, chirps and other noises made by any external speakers can be annoying to others when working in public areas. So while headphones can be rather hot in warm weather, they are pretty much a necessity in my opinion.

Some folks swear by external speakers that can be chest mounted. I have tried them, and would agree that they can be helpful in certain circumstances. They are much higher quality than the typical internal detector speaker, and they allow you to hear the sounds that are around you. When there is an extreme danger of rattlesnakes, being able to hear the snake rattle and other danger noises around you may be critical. They are also much cooler to wear when working in hot weather. Some of the headphones that come with the detector from the manufacturer are reasonably functional, but not the best. It is not absolutely necessary to get custom headphones but there are better quality types on the market and it may be worthwhile to get a pair.

Picks and Digging Tools, Rare Earth Magnets, Plastic Spoons

You will be digging targets, so you will need some sort of digging tool. I dislike carrying heavy full-size pick, so I used carry only a standard Estwing rock pick (it worked find with a VLF). However, especially with PI detectors that you will need a bigger pick for digging deep holes. Because of that I have gone to carrying a medium size pick that is bigger than my old rock pick, but smaller than the full-sized picks. A wide variety of choices are available. I always keep my big pick in the car, and if I have to hike back to get it, I will. I think that for the VLF detectors you can get by with a standard rock pick, but with the PI detectors, you should own one of these bigger picks and keep it handy. They work well for those really deep holes or those tough holes in hard caliche cemented gravel.

I have strong magnets attached to the head of my pick. These do a great job of sucking iron trash out of a dig hole. This saves me sorting through the dirt in my dig pile for small bits of iron trash, speeding up the process of target recovery. High-power rare-earth magnets are strongly recommended for your pick to help you quickly capture trash iron targets. I also use a spoon or cup when sorting through my dig pile for a target. Some guys just use their hands, but the cup speeds things up and prevents things like insect bites and stings, or accidentally grabbing some cactus spines should some of these be mixed in your dig hole with the target.

Batteries and Power Systems

All detectors have some sort of power system. Most will run on batteries for a reasonable time, but having a rechargeable set of batteries can save money for those who use their detectors often. High power systems like the Minelab PI detectors come with rechargeable batteries, but the older lead cells are quite heavy. There are a number of lightweight battery systems available to replace these older, heavier battery units. They are not cheap but do make a big difference in the amount of weight you need to carry around with you in the field. Eliminating the big weight can make a real difference at the end of the day. The newer Minelab PI detectors come from the factory equipped with lightweight batteries.

Here I took an inexpensive pick and welded on a flat section of mild steel for better digging.

Bungee Cord Weight Support Systems

I nearly always use a bungee cord support system with my PI detector and find its pretty much essential with the larger coils. Deciding as to whether or not you need a support system with your detector depends on the weight of your detector and any issues you might have with back and shoulder problems. Any accessory that prevents injury, keeps you comfortable and able to spend more hours in the field searching for gold is a good idea in my book. Additionally, the cost on these things is usually quite minimal.

External Amplifiers, Signal Amplifiers, etc.

Some detectors are significantly helped by using an amplifier to increase the signal. For those with hearing loss, amplifiers can be very helpful even on detectors that have sufficient signal amplification on their own.

Backpacks, Water Bladders

Carrying your gear into your detecting spot may take some effort and a backpack is often just what is needed. It's always important to take plenty of water with you. Personally I like to take water in the plastic bottles such as one can purchase at the grocery store. However, some guys swear by the soft plastic water bladders which are sold by some of the sporting equipment outfits.

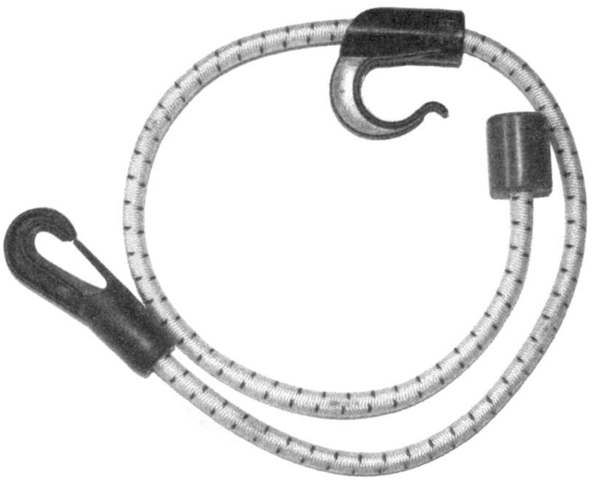

A simple Bungee cord support like this one can make a huge difference by reducing back and arm pains from using a heavy detector and coil.

Pinpoint Locators for Nugget Detectors: I've never been a real big fan of pinpointers. Some people like them - I really don't see much need for them. With good training and experience, you should be able to center the coil on the target and pinpoint it accurately. I have already described good pinpointing procedures in this chapter. Pinpointers for nugget detectors come in two types, the separate standalone unit kind which is good for any metal detector, and the attachable small probe type, that is like a tiny coil and is specific to the type of detector you own. The standard separate unit type is often not very sensitive, and is easily lost - they just can't see small nuggets from any distance. I find they slow me down in target recovery more than speed me up. The probe type I just don't see much use for, but I will note that when using very large coils like my 20 inch round (a coil for my Minelab PI), I will often switch out to a smaller coil for final pinpointing of the target, because even with good procedures, its just pretty difficult to pinpoint down in a hole with a real large coil like that. Pinpointers are good for coin hunting, and also in digging with big coils.

Spare Parts and Supplies

It's normal that when you go detecting for nuggets you will be out in some remote locations. You may be many hours drive from the nearest prospecting shop where you could purchase spare parts and supplies. I always carry extra parts and accessories to make field repairs (both for my car as well as for my detector, but we are talking about detectors here). I carry with me a spare power cord, spare plastic nuts and a spare bolt that holds on the coil. In the car I will have extra batteries, lower shafts, and extra coils. One important thing is to make sure everything on your detector is working properly before heading away from your vehicle. Take a few seconds before leaving the vehicle to inspect the condition of your detector and gear. Turn it on, and make sure its operating properly before you take off. It's a real pain to hike in several miles only to realize the detector is not working right because a battery you thought was charged is actually dead, etc.

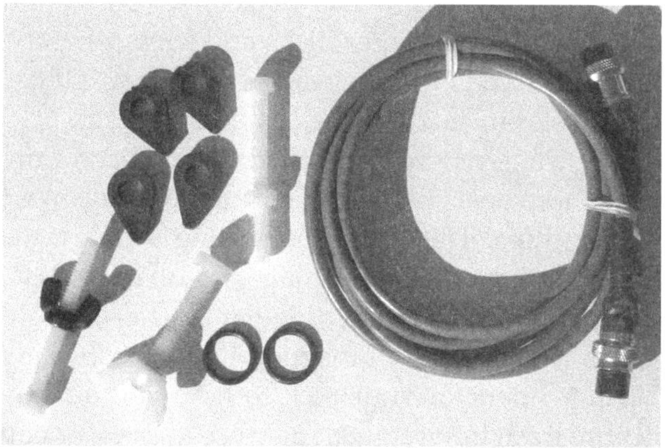

Some parts are prone to break, and if they do you are done detecting. Here are some of the spare parts that I take along with me into the field.

Detecting Safety: Snake Chaps, Walkie-Talkies, Handguns, First Aid, Satellite Transponders, etc.

Finally, we have safety accessory equipment. Detecting is often done in remote locations, and there are potential dangers - falls, wild animals, equipment breakdowns, you name it. It's important to protect yourself, be prepared for unexpected problems and keep safe. The need for safety related accessories varies depending on the areas you will detect. Typical accessory equipment in the category include: snake chaps (to fend off rattlesnake bites); gloves (to protect your hands); plastic spoons (also to protect your hands); eye glasses (to protect your eyes from flying chips of rock); walkie-talkies (to call for help and other wise communicate with your prospecting partners); air horns (to scare off bears); handguns (to shoot snakes, signal for help and scare off bears). All of these are useful and may be needed at one point or another, depending on the area you are prospecting in and how you want to handle things. In recent years I have begun to carry a satellite communications transponder. I often go to places with no cell phone signal whatsoever. It allows me to send texts over the phone satellite system at a cost only a tiny fraction of the cost of a satellite phone. My unit is made by Zoleo, but there are a number of different service providers.

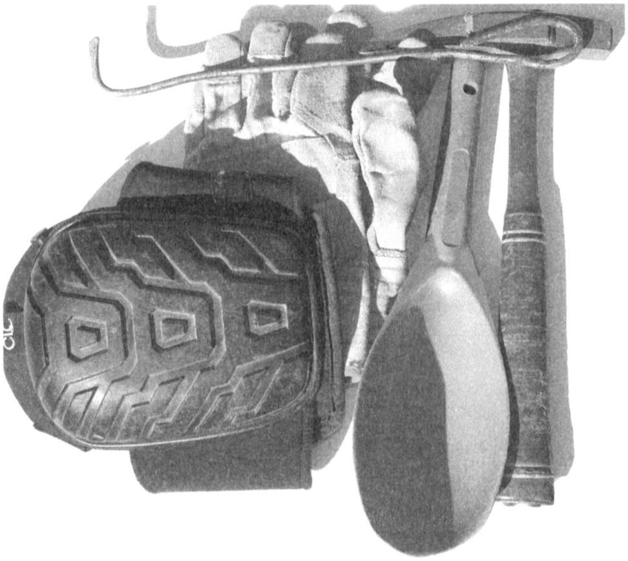

The author uses a number of accessories – gloves, knee pads, pick, spoon and crevice tool.

Finding A Location to Prospect:

Ultimately, the location you will prospect is the most important choice you will make, this decision will be the one that determines if you are successful in finding gold with your detector or not. This is because in the end, it is not the metal detector that finds gold - it is you, the prospector. You have to find that place where the detectable gold is present and put your coil on top of it. Yes, the most difficult thing about finding gold is not learning how to operate your detector or purchasing the right accessories, instead, its learning how to find productive spots where there is gold large enough and close enough to the surface that your metal detector can find it. Not every gold location has this - in fact most locations worldwide do not have much gold that would fit into those categories.

In many locations the gold is too small or too deeply buried. The idea of large gold close to the surface is something you want to keep in mind while you do your research and when you are in the field, as this is the type of place you want to find. It will take time and effort out in the field exploring to find a productive area that yields good nuggets.

Your detector is merely a sampling device - it samples the soil underneath the coil for detectable gold. The prospector or operator decides where to place the coil and take a sample. That in my opinion, is the biggest difference between successful nugget hunters and the unsuccessful ones - the successful guys know how to consistently find the places where they can put their coil over a nugget. Knowing the right landowner contact persons, being able to read the ground and see where the gold is, researching the regions where the old timers were successful, having the experience and skill to find your own patches - those are the skills that enable the successful prospector to put his coil over the gold on a regular basis. Let's take a look at some of the things you should think about in selecting a place to prospect with your detector.

Exploring for New Patches vs. Re-Working Old Ones:

There is no doubt that many of the well-known patches here in the US as well as Australia have been pounded (flogged) pretty hard in the last decade plus, often with Minelab PI detectors. The first operators on the scene normally got the best and largest gold. Still, many times if you work these well-detected areas very hard, you can pick up a nugget or two that has been missed by the previous operators. Sometimes you might find a previously unknown extension of a known patch nearby. However, it is rare that old flogged patches will yield well and produce a good number of finds. Normally this happens only where flooding moves material around in stream drainages either exposing or depositing new gold that was not previously detectable. Some of these patches can be effectively "renewed" - but that is the exception, not the rule. Generally, to make big finds, you will need to find your own new patches that have not been previously worked hard by other folks. If you want success, there is no question that you need to get off the well-trod paths and out into the bush. While the old saw "gold is where you find it" is true, it's important to remember that there are also real causes and effects - forces that originally put the gold in the places where it is found. Gold is not just randomly flung around without any rhyme or reason – even when that's how it sometimes appears. All gold is there for a particular set of reasons: the geology was right; weathering and erosion moved the nugget into position; etc. If the gold is emplaced as a result of various forces of nature, you can learn to see those results in the field and then go on to recognize those patterns and read the ground. This will in turn allow you to solve the puzzle and find the gold.

How to be Successful in Nugget Shooting:
The guys that are really successful are out hunting their own locations, and the sooner you are able to become proficient at finding your own patches, the better off you will be. Doing a lot of research homework to come up with new spots to explore may seem like an unpleasant task; however, this is what needs to be done to be successful. I recommend you get used to the concept of reading old reports, studying maps and talking with anyone who will give you information - research is a necessary part of the golden quest.

There is a lot involved in coming up with ideas for exploring. To research for potential good nugget patch areas, you will need to understand the right kinds of places to look. Develop a plan for yourself of places to explore and things you want to look for when you are out in the field exploring. Stick with your plan, apply what you have learned, and keep on looking even when you are coming home empty handed. In time your perseverance will pay off.

This book has an extensive section on placer geology, hard rock geology and information on how to research new locations to prospect. Carefully go over the placer and hard rock geology sections of this book, as I don't want to duplicate all that material here. When you do your research, you will want to be sure to understand the placer geology of the location you are searching. Read specifically to see if the reports you are searching have information about the source of where the placer gold is coming from. Is the gold coming from quartz veins in a certain type of rock? Is it coming from the erosion of old Tertiary River channels? It's good to know that.

Old mine records can contain a considerable amount of information about places that were once mined for gold. They are a great resource.

Knowledge is an important tool, and knowing where the gold is coming from will tell you a lot about where you should be searching with your detector. When you are out prospecting for new patches where the terrain is gentle, work along a path that is at right angles to the slope of the hillside. Gold washes downhill and a path that cuts across the direction of the tiny drainage ravines will maximize your chances of cutting across any nugget distribution lines that are present in the area.

As I mentioned already in this chapter, nuggets often occur in groups – where there is one nugget, there is often more. These areas that yield multiple nuggets are known by prospectors as patches. When you find a nugget, you should slow down and go over that area carefully. If you find more nuggets, it will be worthwhile to go over the area systematically, so that you swing your coil carefully over every inch of ground in that area. This process is known as gridding. The process of gridding patches is not too terribly complex. One has to identify the width of your normal coil swing

The author dry washing the site of a nugget patch he found with a metal detector to capture fine gold.

and set up imaginary lines on the ground and follow them. As an example, if your normal swing is 8 feet wide, you might set up a series of lines 7 feet apart, which allows some small amount of overlap at the end of each swing. As you complete one line, you will step over 7 feet and do the next line. In this way, you can be sure you are checking every square inch of ground with your coil. I have seen rocks used to mark these grid lines, some prospectors use small plastic cones, some prospectors drag a length of heavy chain behind them so they can see the lines they have traveled (because of this, gridding is also sometimes called chaining). Depending on the ground, sometimes this level of detail is not necessary - in small narrow streams you may be able to cover the entire width in one pass, others you may cover one side on the way up, and the other on the way down. The main point is to carefully and systematically cover every inch of ground in areas where you have found gold. Gridding is best adapted to open areas with flat to moderately sloping ground.

Another well-known technique of working patch areas that have yielded gold is by raking the surface to remove pebbles and rocks allowing your coil to get closer to the ground. If you are out detecting and discover an area on a flat to moderate slope where all the little rocks have been raked into windrows of material, you have likely stumbled onto someone's nugget patch. It may pay to go carefully over the area in case your machine is better than the one used by the original finder. In fact, in a patch area, you should have your coil rubbing right on the ground, getting as close to the surface as possible. Where there are larger rocks sticking up like icebergs in the middle of a good gold patch, move them and detect the space underneath them. Search the surrounding area well as there may also be other nugget patches nearby.

If you locate an exceptional nugget patch area, there are ways of working it so that it yields the maximum number of nuggets. The problem with a metal detector is its limited depth capability. Even the best and most powerful ones can only see so far into the ground. There are various ways of working these deposits: for those that occupy only a small area, hand digging out the soil is a possibility; for ones with lots of small rocks on the surface I have mentioned raking off the stones; and the most extensive is to move the surface soil with a bull dozer. Using heavy equipment like a dozer, front end loader or backhoe on federal land requires permits, so be sure to look into all the legal requirements before you do anything along those lines. Use of hand tools is usually not a problem.

In gold areas be sure to hunt the roadways as they are often over looked. In northern California I found a half ounce nugget and a one pennyweight piece a few feet away in the middle of a well-traveled dirt roadway, next to a forest camping spot that clearly had been used a number of times. Some of the later chapters of this book have significant information on signs of gold mineralization that you may want to look for. Read those chapters carefully as they are full of good information. With experience, you will recognize areas with good potential based on your experience. This includes seeing geologically favorable locations as well as recognizing the signs of old timer workings.

Location Selection Ravines vs. Hillsides

Many successful detector operators seem to specialize in one form of detecting or another. For one reason or another, a number of guys seem to specialize in detecting ravines, while others specialize in detecting hillsides. Most long-time operators have done at least a little of both, but guys tend to gravitate back to the types of locations in which they have had their best success in the past. Each of these types of locations have their good and bad points. The area in the bottom of ravines is limited in extent, and not hard to find with a topographic map, so the majority of ravines in gold bearing regions have already been prospected with a metal detector. Where previously worked by the old-time prospectors, they tend to be trashy. However, with flooding and heavy rains, these drainages are "renewed" to a certain extent as gold gets moved around and previously buried sections become exposed. The renewal is certainly not a complete renewal, but these sites continue to yield at least some gold year after year.

Hillside areas are much larger and widely dispersed – I am defining hillside locations as basically all the locations that are not in the bottom of any drainage. There is just a lot more hillside area out there to explore. I would say that the majority of the undiscovered patches that are still out there undiscovered are in hillside locations. Many of the really spectacular finds that are made by detector operators are made in these hillside types of locations. The downside is that hillside patches can be more difficult to find because they are so spread out. Your choice of target area will depend on a number of factors, part of which is the style of gold mineralization in your area (see the chapter on doing research).

A patch of nice sized nuggets like these are something you would want to keep quiet about until you got the new patch all cleared out. You want the chance to collect them for yourself.

Keeping Quiet About Your Finds:

When you do make a nice nugget patch find, I'd strongly advise you to keep quiet about it. Big nuggets and large takes attract other detector operators just as surely as can be. Many long friendships have been broken up over gold finds. If you have partners be clear and agree in advance on how the proceeds from a nugget patch will be divided. Even if you have no partner, don't brag or speak loudly about your gold finds, in most cases you will find keeping quiet is the way to go. I was recently shown a 6+ ouncer found recently in southern California at a location not known for its larger gold. I can say that if I got into a patch with larger nuggets I would not soon be publishing it abroad on the internet forums - I'd not say a word until I thought the entire area was fully cleaned out. I guess that this was the case with the nugget that was shown to me, as it has not been spoken of. When in doubt, it's always best to keep quiet.

Examples Of Places To Prospect:

The following is a general listing of the kinds of places you might want to investigate potential nugget detecting sites, and the basic characteristics about them which you are looking for. There may be some locations that do not fit into any of these categories well, but this listing is sufficient to get the prospector going in research and exploration out in the field.

Old Placer Mines:

There are many old placer mines in the US, Australia and other parts of the world. Many of these have good potential to yield nuggets to a metal detector. Placer mines come in many varieties. Some have been dug with earth moving equipment, some were worked underground (deep leads and drift mines) and others were processed by the flow of water (See specific discussion of hydraulic mines, below). The exposed bedrock, old working faces, tailings, open cuts and access roadways in and around such mines may represent places where the modern prospector may find gold with his metal detector. Even underground placer mines may yield tailings piles worth checking out with your detector.

Streams, Ravines and Drainages in Gold Bearing Areas;

This is a very general category that encompasses a number of possibilities, both in the desert and locations where water flows year 'round or only intermittently. This category also includes benches of streams where the water no longer flows. In stream environments where the water flows at least sometimes, the water will stratify the materials and the densest materials, like gold, will be near bedrock, or sitting on a false bedrock. These types of locations are very popular with many prospectors. Tributary ravines with exposed bedrock have been the source of many nuggets - in areas with sparse gold, the nuggets will get concentrated in these ravines and small tributaries through the processes of erosion. The names for these little drainages vary quite a bit, and depending on your location, they might be called an arroyo, a gully, a ravine, a piddler, a pup, a wash, a hollow, or a gulch. In the Middle East they are called waddi. There are probably other names in addition to those I have named. Whatever the name, they are small drainages that collect materials eroded off hillsides and mountains. If that material contains gold, the drainages will also contain gold. The larger drainages normally have their bedrock buried deeply under loads of gravel, however, the small tributaries high up in the drainage system will often have sections of exposed or very shallow bedrock. These areas where the bedrock is not deeply buried are good locations to detect gold nuggets that have become caught in the cracks and crevices of the bedrock. A considerable amount of gold has been detected in these types of locations. For more information on these types of placer deposits and where one might search them to find gold, see the chapter in this book on placer geology.

Hillside Slopes in Gold Bearing Areas:

This also is a very general category that encompasses a large number of possibilities. For this discussion, I am defining hillside locations as basically all the locations that are not in the bottom of any drainage. The incline of the "hillside" may be steep, moderately sloped or flat land - or any grade of slope in between. Many of these locations are residual placers, sometimes enhanced by wind driven erosion in desert areas.

Old hydraulic gold mines like this one have yielded considerable gold to metal detectors.

They are among the hardest to find, but also among the least worked and most productive when they are found. They tend to produce the majority of truly spectacular finds made by prospectors armed with detectors. Although they are a bit harder to find for modern prospectors, they were also hard for the old timers to find - but the old miners did find some of them. An excellent sign of this type of deposit are hillside locations where the old timers worked hillside locations by dry washing. Similarly, in areas with significant water, the old timers worked these locations by ground sluicing. Unlike streams, the material is not sorted by water flow and therefore gold may be found from the surface on down. In fact, because of the wind and minimal effects of rainwater, the very best gold may be near the surface and the deep deposits may not be nearly as good. They also tend to be low in trash, except where the old timers worked them.

Finding hillside patches is a bit more difficult and requires the prospector to learn signs of mineralization and other indicators of gold deposits. It also requires that the operator be willing to endure long periods of drought. This type of prospecting is more hit or miss - and while the finds may be good if you hit a patch that has never been worked before, you may spend long periods searching and exploring without finding anything before such a patch is found. A detailed description of residual placers and the signs of mineralization that a prospector should look for in hunting these types of deposits are discussed in the later chapters of this book.

Drainages Below Placer Workings;
Many of the old-time placer mines did not get good recovery with their sluice box systems. If all of the tailings from a certain placer mine went into a ravine, the spot below the dump point may be a good location to prospect.

Rows of hand stacked rocks like these in a creek bed or on a hillside often indicate the site of an old placer gold mine.

Bucket Line Dredge Tailings;
Bucket line dredges are normally pretty efficient at capturing gold, but there are exceptions. Most dredges had a small screen size and nuggets over this size went to a nugget trap that was not always very efficient. So, where these dredges worked placers with large nuggets, the oversize piles may contain a few large nuggets. The other possibility here is gold laden quartz. Such pieces are not normally captured in the nugget trap or other equipment unless the specimen is mostly gold. Specimens with significant quartz will show up in the tailings and be visible to your metal detector. Unfortunately, these tailings normally contain far more deeply buried trash than gold laden quartz.

Old Hydraulic Mines;
Much of the placer gold recovered in California came from hydraulic mines. Such mines are also found in Oregon, Alaska and other states where placer gold was found and water was available. There are even some of these mines in the southeastern part of the US Appalachia ranges. These operations used pressurized water to wash away great depths of gravel, and the gold was recovered in sluice boxes. For environmental reasons, most of these types of operations were ended in the 1880s. Most of the hydraulic mines were located along river benches, but in California, many were located along the pathways of ancient channels, called Tertiary channels, the remains of rivers that flowed long ago during Tertiary times. These hydraulic mines cover many tens of thousands of acres in the western US gold country. Although the gold varies from site to site, there were some coarse nuggets recovered from these mines. Today, these same hydraulic mining areas are usually good for prospecting with a metal detector. My favorite type of hydraulic mine site is what the old timers called a ground sluicing operation. These were smaller scale operations and sometimes did not have the careful clean up of bedrock that the larger mines did. I have generally done well at these types of sites and have picked up some nice nuggets.

Look for areas of exposed bedrock or at least the areas which have only shallow gravel cover at these sites - these are the places where the old miners may have left gold behind. It is normally best to set up your detector to look for small gold, including smaller coils. The amount of trash can vary quite a bit, even within a single mine site, old nails and broken bits of rusty can will cause problems. Discrimination may be needed in places with a significant amount of iron trash. Hot rocks can also be an issue at some of these sites as well.

Desert Dry Placer Areas:
Desert placer areas are very popular sites for prospecting with a metal detector, and my discussion here includes dry wash areas. This discussion would include both drainage ravines and hillside locations. In forests and other heavy growth areas, trees and plants leave a barren mulch layer made from leaves, twigs, bark, etc that impedes a metal detector. Often these layers are a foot or more in thickness and any gold present lies beneath them, making it difficult to detect any gold there. In the desert regions, because of sparse growth, no such mulch layer exists, so these spots are easier to detect for nuggets. This is not to say gold cannot be found in forests, but the desert areas do have their advantages. The Rye Patch region in Humboldt and Pershing counties in Northern Nevada is probably the best-known nugget detecting area in Nevada. The main area

south of Majuba Mountain has been very well prospected over the past 20 years, but there are still nuggets being found. Good nugget detecting areas are places where the gold is not too deep - this place fits the bill. The bedrock here is a shale formation. There are a lot of areas here where there are dozer scrapings, but the soil layer is not very deep and many places the dozer cut down a foot or so and hit the country rock. Geologically, most of the dozer scrapings were in areas where there were concentrations of iron-stained quartz and brown cubes of oxidized pyrite on the surface. These quartz areas and the ravines draining between them are the places where one would look for gold. The productive region covers a large area and out away from the well-known patch areas, so there are other locations in this region where new patches sit, waiting to be found.

There are actually a number of desert placer sites around Arizona, Nevada and California. The old timers, who had nothing like a metal detector available to them, mined the desert placers with dry washers. Dry washers are not the most efficient gold recovery devices, and in most instances the placers had to be pretty rich to make their mining worthwhile. The old tailings of the dry washers can actually be good locations to search with a metal detector, as large nuggets went over the coarse screen and sometimes smaller nuggets did not get caught in the dry washer itself.

Rich Hill, Arizona – Desert dry placer locations were tough for the old timers to work and have yielded thousands of ounces of gold to prospectors armed with metal detectors.

Some of these dry wash piles can be large and a rake can be helpful in spreading them out to allow your detector to prospect them fully. There are many well-known patches in the desert regions that have been worked with metal detectors for years, and sometimes you can still find a nugget or two in these places. If you want to make a more spectacular find however, you will need to get out to locations where 10,000 other folks have not already run their detector. This takes research and time spent exploring. Desert sites can be one of the locations where a Pulse type detector may do significantly better than a VLF, but gold can still be found in these spots with a VLF detector.

One problem you may encounter in desert areas is caliche (known as calcrete in Australia). These gravels are cemented by calcium carbonate and can be almost as hard as concrete. Another desert oddity you may encounter are locations with wet, alkaline soils (salt in the soil) which sounds off on for your detector. These may be difficult to work with a VLF. Luckily, these alkali spots are small and far between in the US, so they were not really a factor for many detector operators. In some locations in Australia, they are far more common. Desert areas can vary quite a bit in their trash content. Locations not known to the old timers and little worked in the old days can be fairly low in trash, but those locations well known to the old timers or too close to modern cities may be full of junk.

Hunting Around Old Hard Rock Mines:

Old hard rock mines that yielded coarse free gold can be good sites to prospect with a metal detector. For various reasons, many prospectors never think about metal detecting the old hard rock mine dumps. In many old mines, the old miners had to sort out the ore down to only the highest-grade material as they had to haul it miles to the nearest mill for processing. Sometimes they were not so good at sorting and a lot of very high-grade ore was accidently tossed out on the dumps (many times the sorting was done underground). Sometimes small mines and prospected produced some higher grade, but never got enough to make a worthwhile shipment to a mill. Often these piles of ore are set aside and not in the old dump itself, but li located along side. A VLF detector made for gold is just the tool for prospecting old gold mine dumps. Watch out for old mine workings - they can be very dangerous, and I strongly urge you to stay out of them. Stick to the mine dumps on the surface. Scan the old dumps with your detector and save and carefully scrutinize any rock that gives a good metallic signal. I've found pieces with no visible gold on the outside that the metal detector indicated it had gold. When I broke them, there was nice gold within. Take a bucket for your finds, as you may get a number of rocks that respond. If you get any high-grade samples, ones with lots of visible gold, these pieces can be most valuable as specimens for collectors. Some may show wire, leaf or other forms of valuable crystalline gold. Ore specimens with only a little visible gold can be crushed and the resulting sand processed with a gold pan - and you still may find some nice gold because the smaller gold adds up. I've found some good gold in old hard rock districts - even ones with no recorded production and only smaller sized workings. Be sure also that you have the legal right to prospect for gold around the old mine. While there are many old abandoned mines out there, there are also many that are currently under claim.

The dumps of old hard rock gold mines like these shown here can produce some good gold specimens. Stay safely out of the old workings as they often hold many unseen dangers.

The soil conditions of these dumps will vary much depending on location. These are normally not heavily hunted sites, in general. You'll need good shoes as dumps can be steep and they can run as well. The dumps commonly contain a significant amount of trash and discrimination may be needed. Be sure to set your discriminations low as you do not want to discriminate out any fine sized gold. The DD coils are preferred because of their ability to sort through the trash and better handle the mineralization. I try to set the gain as on my VLF high as it can go without excessive chatter and noise.

Legal Issues Of Electronic Prospecting:
I'd like to spend a bit of time looking at the legal issues related to prospecting with a metal detector. As a prospector you have the right to explore most federal lands under the management of the Forest Service or the Bureau of Land Management which have not been set aside for special purposes. Because gold nuggets are natural and not man-made objects, they are not protected under the Archeological Resources Act of 1979. The use of hand tools is normally considered acceptable casual use, but heavy equipment requires special permits. Hunting on US government property at old ghost town sites can be dicey - while coins are specifically exempted as not covered under the Archeological Resources Act, other man made items may be off limits. Locations designated as historic sites can be off limits. Detecting at US National parks and designated historical sites is strictly illegal.

State parks and historic areas are also likewise illegal to detect. Detecting on otherwise unallocated BLM or Forrest service land is legal, but the 1979 Archeological Resources Act may protect certain man-made items. While that act covers nearly anything that has been made or handled by humans, it does specifically exempt coins and all items less than 100 years old are also exempted. It is important to do your research, know what and where you are hunting and your legal rights to work there. In recent years, US government agents have become very zealous about protecting publicly owned relics - so do your homework first. It may even be worthwhile to carry a copy of the 1979 Act with you in the field as many agents are not really aware of what is protected and what is not. It could save you quite a headache, because there is no question that naturally occurring gold nuggets are not covered. However, your detector will respond to metal relics just as it will toward gold, and some operators will be tempted if they should find a valuable metal relic.

Your metal detector does not give you any rights to trespass on private property or prospect on mining claims belonging to others. The gold on mining claims owned by others belongs to those claimants and removal of their gold is simply theft, and punishable as such.

This pile of old boards represents the remains of a collapsed miner's cabin. Some relics from here may be protected under the law.

The same holds true of gold on private property and you would likely be guilty of trespass as well. Know before you hit the field where the existing claims are in the area you want to work. Respect the rights of mining claimants and private property owners; if you can't get permission from them, you simply can't go on their property. For more information on mining claims and researching their locations, see the chapter of this book about research, and the appendix on mining laws and claims.

Alternative Detector Technologies:

Great strides have been made in metal detector technology, but rumors of new technologies are always being heard. Before I finished off this chapter, I thought I would take a brief look at some technologies that are "out there" but not really practical at this time. Perhaps one of these might be available to future prospectors.

Ground Penetrating Radar Technology

From time to time I hear rumors that someone is doing research to make ground penetrating radar (GPR) equipment that is suitable for use by the individual prospector to locate nuggets. While that may happen someday, and I have heard that there is some GPR research being done for personnel land mine detection, as of this time, there really is no practical GPR device available for nugget detecting. Radar systems make use of electromagnetic reflection, as opposed to metal detectors, which use electromagnetic induction. Ground penetrating radar (GPR) works in a manner similar to regular radar, using pulses of electromagnetic radiation which are in the microwave band of radio frequencies. The reflected signal is read and interpreted by a computer to detect subsurface structures and objects. Often, a considerable amount of expert "interpretation" of the output data is required. Typical uses include locating large buried voids/cavities, underground storage tanks, sewers, buried foundations, or ancient landfills. It can also be used to characterize the depth to bedrock in some placer deposits. At the present, this technology is used to identify buried metal fuel storage tanks and similar larger items like buried 55 gallon drums. Small objects are difficult to identify using GPR. Since virtually no gold nuggets are as large as a 55 gallon drum, other than bedrock testing, this technology as yet still has no real applications for the individual prospector seeking to find placer gold, but perhaps someday it might. There are some other problems with GPR in that it does poorly in wet and salty ground. If GPR does become available in a version useful for detecting gold in the 1 gram and larger size range, such a unit will likely cost 5 to 10 times the price of today's most expensive detectors, and require it least two people to operate it properly.

Since I originally published this book in 2010, no real advancement has been made in the technology of GPR. Nugget detecting is still done with VLF and PI technology with metal detectors of the standard technology. Two of the major metal detector makers have gone out of business since 2010 - Whites and Tesoro. On the other hand, the remaining makers are continually developing new detectors and coming out with new products with small but significant improvements to their products. Potential new products are always being tested and some will results in new products, while other prototypes will not result in any new products. New detectors will come out over time and be used to find gold nuggets around the world.

Hand Held XRF Sensors

While there may not be any such thing as a "gold only" metal detector, there is a hand held device reliably capable of distinguishing gold from all other metals. It is a hand held X-Ray Fluorescence

(XRF) sensor. These units expose the material

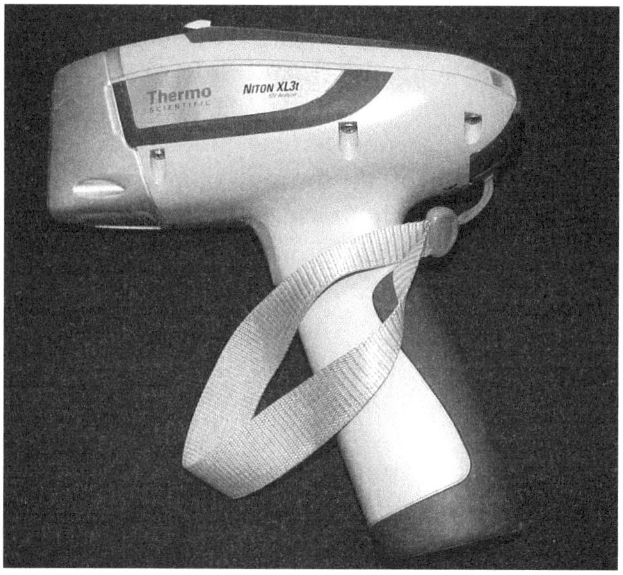

This Hand held XRF sensor can tell gold from other metals, and make a quick assay of ores.

being tested to high energy X-Ray light, and the device measures the type and quantity of the X-Rays which are reflected back. Because of the nature of quantum physics, the wavelengths of the returned light indicate the elements which are doing the reflecting. Based on the amount of returned light at the specific wavelengths, the percentage of the element in the sample can be estimated. Unfortunately, the scan window is only about ½ inch square and the units penetrate less than an inch through rock materials. Additionally, they typically cost more than $40,000 apiece. This makes them impractical for individual prospectors. Because some hand held XRF units have a radioactive source, many of them require some sort of license from the Nuclear Regulatory commission in the US if you would like to own one. Their most practical use to the mining industry is to obtain rough assay numbers. A bit of ore can be scanned with one of these devices and in seconds it will reveal the nature and percentages of the ore's metallic contents, such as gold, silver, copper, etc. The problem is that while they are fairly sensitive to silver and copper, they require the ore contain about 1/3rd ounce per ton before they will register any gold content.

Long Range Locators - LRL / LRD

I am being honest in telling my experiences and opinions in this book, so I am going to tell you the truth concerning long range detectors (LRLs or LRDs). They are essentially sophisticated looking pieces of equipment intended to be used for conducting electronic dowsing. There has never been a single shred of scientific evidence that dowsing actually works, and while claims are occasionally made, further investigation of these announcements always show such claims as bogus and false. My knowledgeable and experienced opinion is that "Long range locators" / "Long-range detectors" are a complete and total waste of time and money; they do not work in any way whatsoever. The folks who sell them are less than frank concerning the science involved in them - they will often imply that much real science is involved in their operations, but that is simply untrue. Prospective buyers are told proper operation requires that they "believe" in them. Electronics experts who have taken their circuits apart have shown that these strange looking devices are simply electronic systems which make a series of blinking lights go on and off at random times when the unit is turned on. They are nothing more than that. The electronic LRL/ LRD salesmen charge huge sums of money for these random blinker devices, and they simply do not work for finding gold or other valuable materials. While I do not believe in dowsing, if you decide that you want to mess around with it, just use one of the inexpensive manual systems made out of metal rods. A pair of bent coat hangers is just as accurate as the most expensive electronic unit, but the LRL/ LRD salesman could not take you to the cleaners for $10,000 by selling a couple pieces of coat hanger metal - for that kind of money it has to at least look like its high tech. Don't waste your money on one of these high priced LRL/ LRD electronic dowsing rod systems because you will surely regret it afterward. There are folks who will doubtlessly disagree with some of my comments above, but I will stand firmly by my opinions that LRL/ LRD electronic dowsing systems are a total waste of money.

Some Thoughts on Metal Detecting and Keeping a Positive Attitude:

New prospectors who decide to take up metal detecting for gold face a double learning curve when they are starting out: they need to learn about both the operations of their metal detector as well as the skills needed for finding gold. I can tell you that the skill of learning to find gold is the more difficult of the two. Because of this, it can take months to a year or even more to find your first nugget, and even after you find some gold you may go through an extended dry spell. Failure is tough to take for all of us and it is not all that uncommon that an electronic prospector will get discouraged and toss his detector into the closet for an extended period. If finding gold were easy, gold would not be so valuable!

Being successful in the long run has much to do with keeping a positive attitude, being dedicated to the quest for gold and sticking to it. Know before you start that it is not easy, and make the commitment in your own mind to persevere through the difficult times. Successful detecting and finding nuggets truly requires keeping a positive focus and putting in the necessary time to explore. This is actually equally true for all types of prospecting (which is why there is a chapter on this topic later on in the book), but it is especially important for electronic prospectors because of the higher-than-average likelihood of getting skunked regularly when you are out with your detector. If you allow yourself to get discouraged, it is very easy to just give up. More than a few new prospectors do exactly that. Sometimes good deals on buying metal detectors can be had from other prospectors who decided to give up.

Learn what you need to know by reading books like this one. Maybe get some training - there are detector dealers and other who offer basic training services. I do not do this but there are a number who do. Join a prospecting club ad maybe meet some folks who have been successful and talk to them. In fact, the best knowledge you can gain about finding gold is in actually finding gold, and that can only be done out in the field detecting. Long term success in detecting for gold is never easy. You need confidence and determination to hang in there during times of drought. Learn to enjoy being out in the field for its own sake, even if you don't find any gold that day. Whether it's the mountains or the deserts or where ever, nature has a special beauty all its own. Don't doubt yourself! Do some serious research, spend the necessary time in the field and work hard at it and you'll find the gold. It isn't easy but the gold is out there and if you are willing to put in the necessary work, you can find the gold just as well as anyone else. If there is some magic secret to success, simple as it sounds, that is it. Other guys really are finding gold, and you can too, but you have to stick to the tried-and-true methods and procedures that lead to success. Make a commitment to yourself to do what it takes and then stick to it.

Making That "Big Discovery": Put in the work and the next big find may be yours!

Commercial Scale Small Gold Mining Operations

Many prospectors who are successful and enthusiastic about the world of gold mining will eventually consider the possibility of going to a larger scale. A larger scale means more gold, right? Not everyone will reach the point where you consider going commercial, but some will, and that's why I am including a chapter here on commercial operations. So if you think you want to go in that direction, I am going to offer some thoughts you ought to seriously consider. Perhaps the first and most important thing I want to say is that commercial mining is very different from recreational prospecting - in commercial operations your focus is making money not having fun. Sure you can enjoy running your mining operations, but when you start investing significant sums of money, taking big risks and putting yourself in a position that you may be cited with very significant fines if you don't obey all the laws, then its time to get serious. When mining is an avocation or hobby, the money is still important, but you can afford to just blow off those occasions when you get skunked or are not producing any significant gold. You can remember the fun you had, and just not worry about it. However, with commercial operations it's just not that way - you need to always make enough gold to cover your costs, or face financial disaster. Far too many guys who go this route still treat their mining like it's a hobby and don't use any business sense. That I guess, is the chief point I want to make in this chapter - that commercial mining is a serious business and absolutely must be approached that way. If you go commercial, mining is now your job; mining is your full-time business.

The first thing you need to think about is what you have and what you want to do with it. What kind of mine or other property do you have? It is a placer or lode deposit? We will talk mostly about placers, but I'll cover some options for processing lode deposits as well. There is simply no way that I can cover all the possibilities and make detailed recommendations on all of them, but I hope that my comments here will be sufficient to get you pointed in the right direction so that your commercial mining venture will be a profitable success. There are a number of various small-scale commercial mining options and possibilities. These will include things like placer, heap leach, gem mining, etc. All of these hold potential opportunities for the small miner. Placer is by far the easiest to get into, but is the most difficult to sample, and the materials you mine will be spotty with some parts rich and other parts nearly barren. Gem mining could present an interesting opportunity, but you have to find a market for the rough or gems you produce. Hard rock mining, including heap leaching has a lot going for it, but it is by far the most difficult to obtain permits for. Good planning for your commercial operation is essential, and it's important to know and consider all the possibilities.

Small scale commercial operations can be as simple as metal detecting behind a bulldozer.

So What Constitutes a Successful Commercial Mining Operation?

It's not about size - small mines can actually be very profitable. In the broadest sense a mine may be said to consist of a body of ore sufficiently large and rich to pay for the original purchase price of the property, all costs of exploration, mining, equipment and facilities together with a large percentage of return on the original investment. When I was attending the Mackey School of mines, they drilled into us in my engineering classes that it's not ore if it can't be mined at a profit! In order to be profitable, you have to know and control your costs of production. When making estimates for the operating costs of a mine, it is necessary to consider every item of possible expense chargeable against the property, all the physical and geological conditions, environmental issues, access and any other problems you might face in the operation. The location of property is exceedingly important, both with respects to permitting issues as well as access for equipment and employees. In this connection it becomes necessary to consider availability of water, electrical power, shipment of fuels to the property, space for the treatment of the ore, and for storage of the tailings. Good planning is essential when getting ready to start a commercial mining operation. When you step into the realm of commercial operations you'll be dealing with a whole other world of legal compliance. The complexity of the rules that you will need to know and be responsible for will require your full study and attention. The truth is that in many small scale commercial mining ventures you will be better off hiring a consultant to assist you in determining compliance. You can often save a lot of money by planning and considering your efforts for reclamation and coordinating them with the progress of your operations to minimize the amount of reclamation work you have to do. As far as federal permitting, it is usually a lot easier to obtain the permits you will need on BLM lands as opposed to those administered by the Forrest Service. Federal permit requirements are the same in all the states, and each state will be subject to the same federal laws, but the state and local requirements vary considerably based on what location you will be operating in. It's impossible to specify all possibilities for all the state and local authorities, but it suffices to say that you will need to deal with them and some will be easy to deal with and some very difficult. Again, with anything complex I recommend that you seek a consultant to assist you. Be sure you know what your permits say and exactly what they require of you so that you can remain in compliance with any restrictions or limitations that the permits establish. The regulatory authorities will expect you to comply and if you don't know what you're required to do in your permits you're simply begging for problems.

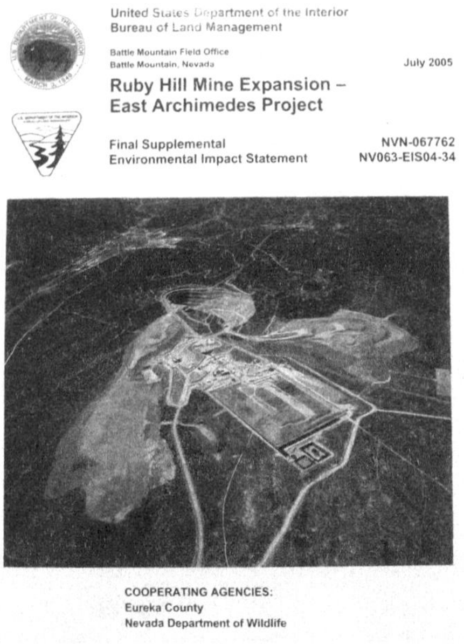

Commercial mining will require a whole new level of governmental permits and compliance, including Environmental Impact Statements.

You will also need to consider and deal with issues related to employees and MSHA. The mining industry actually has its own special version of OSHA, and it's called the Mine Safety and Health Administration (MSHA). They have their own strict set of rules that you will need to deal with for any employees that you might have. Again, it's important to know the rules that will apply to your operations. Most small operators I have known have studied books and attended seminars or classes to learn all the regulatory things they need to know. It does take time and effort to learn all that stuff.

One of the most important things is to know what you have as far as the ore on your claim. Perhaps the biggest difference between large commercial operators and small-scale miners is the amount a sampling that they do. The large operators invest millions in sampling and deposit analysis before they are willing to commit a dollar toward equipment for mining.

When putting a lot of money into a placer operation be sure to take care of all your sampling. Years ago, I knew a guy who had a placer property and decided to go commercial so he could produce more gold. He borrowed $300,000 against his house and went out and bought a nice little Gardener Denver placer plant. He also purchased a small bulldozer, a front end loader and a dump truck to dig and haul his gold placer gravels. I asked him before he got started if he had done enough sampling. He told me he didn't need to do any sampling because he knew the deposit was rich. He and his two sons went out, got their placer plant all set up and started digging. They spent three months digging on the site always hoping that when they got down to bedrock the gold would get better. During the three months they processed many hundreds of yards of gravel, but recovered only a little more than 3 ounces of gold. The gold was not enough to even cover the cost of their diesel fuel, let alone pay the loan he had borrowed against his house, or pay any wages to himself and his sons. In the end, he lost his house and almost everything else. Don't just assume that your project will be profitable - make sure you've sampled it sufficiently to prove that working it will be economically profitable.

Commercial operations are expensive – and unless you are supremely wealthy, you'll need to think about how you will be funding your commercial operation. This is perhaps the most difficult part of small-scale mining. One of the most common problems I see with small operators is that they are normally badly undercapitalized. The only mining equipment they can afford is ancient junk, and when things break down or the ore just doesn't contain as much gold as originally hoped, they are quickly forced out of business. This is in spite of the fact that they have problems which could have been solved. If they just could have had more time to deal with things, or could have started off with better equipment, their problems could have been overcome. There are various options for capitalizing a mining venture. It is not often that an individual could save enough to get started from their regular income sources. Normally, capitalization involves some sort of borrowing. I know guys who have borrowed against their house and lost their shirts (and later their houses as well). Be careful in the commitments that you make, and the money you borrow. One possibility is partnerships with others having the same interest. This allows the risk and investment to be spread over a number of folks. However, be careful with whom you partner, as that type of agreement carries its own set of risks. Be sure your partners are honest and reliable. Another creative option that applies in some situations is the possibility of raising part of your cash with pay to mine tourist operations. If you have possibilities for people to dig for gold or gems on your property, you may be able to start a business based on those possibilities - there have been some operators who have done this quite successfully.

A trommel is a cylindrical rotating screen and they are used on many placer plants to wash and sort the gravels. This screen has ½ inch round holes.

Part I: Placer Mining Operation Options:

Let's consider first the possibilities of a small-scale placer operation. This is the route that a lot of small scale miners choose, partly because it's easy and partly because it's quite a bit lower in it's initial cost. This type of operation would include all types of placer mining efforts, both wet and dry. Let's take a look at a few possibilities:

Large Dredges and Dredge Teams

Perhaps the simplest would be to buy a couple of large dredges and work them with teams of guys. This can be fairly low cost, but somebody will need to be able to keep these operators constantly working in a series of paystreaks where they can make money. Someone will need to be doing exploration for the team to find new paystreaks before the existing ones run out.

Dozer Scrape and Detect

Another very small-scale commercial placer possibility is to simply scrape off the surface of nugget patches and then detect the surface and re-scrape and repeat until the yield is too small to make it worthwhile. Again, to make this type of operation worthwhile someone will have to be regularly supplying a series of suitable nugget patches to be scraped.

Small Scale Ancient Placer (drift)

Both old Tertiary placers as well as high bench gravels can be processed by underground mining along the bedrock. This allows the operator to mine only the richest gravel at the bottom of the deposit and leave the lower grade materials behind. Additionally, these are larger scale deposits that can have significant gold reserves. Underground mining however, can be expensive.

Larger Portable Dry Systems

Normally dry processing of gravels in the US is done only on a small-scale. While large scale dry placer operations are quite unusual in the US, in Australia there are a number of small operators who actually do dry processing of dry desert gravels on a scale large enough that the equipment must be fed by bulldozer or front end loader. It is certainly possible that similar large-scale dry blowing equipment could also be used on desert placers in America, Mexico and other arid locations.

Wet Placer Systems Fed by Loader or Dozer

The most common operation for small scale commercial miners is the wet placer system fed with a dozer, truck or front end loader. These systems can be purchased on a turnkey basis ready for operation or even built by the operator if he has the fabrication skills and necessary equipment. Larger gravel deposits can be processed, such that one plant may operate for years working a single deposit. Where the property being worked is privately owned, and near enough to markets, it may even be possible to profitably sell rock and aggregate materials from the operation as a profitable byproduct. A discussion of the placer plants use by these types of operations follows:

Placer Plants and Gold Recovery

In the various forms a larger scale operations be it a drift mine or a wet placer system fed by a loader or dozer, the operation will need some sort of gold processing plant to separate the gravels from the precious metal. There are a number of different types of equipment that are used in these plants including trommel screens, sluice boxes, jiggs and vibrating tables. The exact design of the system varies depending on the operation and the types of gravels being processed.

Shaker tables, like these shown here, do a great job of catching fine gold particles. They are used at both placer and hard rock operations. To collect small gold particles.

In the larger commercial operations, recovery techniques for fine gold capture are often very site specific. A recovery system that collects a high percentage of fine gold from one deposit may not perform as effectively with materials from a different deposit. Many factors, such as particle size, clay content, gold size distribution, mining methods, and character of the wash water, affect the amount of gold recovered in gravity based processing of fine gold deposits. One of the keys is properly screening your feed material and getting your sluice box optimized with the correct flows, liners, etc. See the section of this book on sluice box operations for further details on proper sluice operations. Depending on the nature of the process, the commercial operator may use only sluice boxes, or may also use jiggs, tables, bowls or a combination of these. Some experimentation and testing is usually required to design a system for optimum gold recovery in a commercial mining system.

In general, the commercial concentration of placer gold gravels involves a series of steps. The first is roughing in which the bulk of the waste is removed in equipment like sluice boxes. The object of this initial concentration is to separate the raw feed gravel into two products - concentrate and tailings. The goal is to separate all the gold will into the rough concentrate. The next step, cleaning, involves the separation of the gold from the other heavy materials in the concentrate. This process may be as simple as washing black sands in a gold pan, but for commercial operators with large amounts of concentrates, there are normally several stages of cleaning before a final concentrate is produced. Various types of equipment are used commercially for this process, including screens, bowls and vibrating tables. Amalgamation is still used by a few operators, but many choose to avoid the risks of mercury (it also makes permitting a lot easier). Concentrates are cleaned until the desired grade of purity is obtained. For the commercial miner, this usually means a gold product clean enough that it can either be sent to a refinery or sold to jewelers and mineral collectors (for more information see the chapter on final black sand clean up operations).

For some operators that are working on private property and patented claims, the rock and gravel you handle may make a saleable product and you might consider adding some additional screens so that you can produce various rock aggregate products from your operation. From Rounded boulders that are used in landscaping to sand suitable for concrete, rock products can be very profitable. If you have the markets close enough to you to justify it financially, producing aggregate by-product can make a huge difference in the profitability of your operations. On standard mineral claims on government land, you do not own the rock and gravel (it is not subject to claim laws) and you cannot sell it. In some cases however, you can buy it from the government to re-sell.

Part II: Hard Rock Options:

Many years ago, there was a very robust small-scale hard rock mining industry in the USA. Small operators could ship their ores directly to one of a number of custom mills or smelters. There was a time 60 or 80 years ago, smelters would buy small lots of high-grade ores, there were also small custom mills across the west that would process the ores, and pay the miner for the metals recovered. That was long ago, and things have changed. The nearest smelter that takes ores from miners is in Mexico as our US government has forced the closure of most of them. Yes, there still exists a small-scale hard rock mining industry, and it can still be done, but it's a lot harder to make it work than it once was.

Heap Leaching

Hard rock mining is usually done on a large scale, and although a few small-scale hard rock operations still exist, it's not an easy operation to start up and run profitably. The main problem for individuals or small groups interested in hard rock gold deposits is the high capital costs for the equipment to crush and process hard rock ore in order to extract the gold from its solid rock enclosure. Because of this, many prospectors who look for hard rock gold seek instead to sell their finds to large mining companies that possess the resources to develop them.

The author's small scale heap leach operation in Nevada. Settling ponds right and left foreground, piles of material being leached upper right.

In this day it is a lot harder for the individual operator to get involved in hard rock mining. This type of metal mining is a lot tougher than it used to be for several reasons, including modern environmental and other restrictions and perhaps most importantly, far fewer opportunities to sell or process your rock. The issue for a would-be hard rock metal miner then becomes what would you do with your ores? Small scale heap leaching is a possibility - and it is a process that is worth considering.

Heap leaching was developed as a low cost technique for extracting gold and silver from lean and low grade ores. The technique was developed in the late 1960's at the now disbanded US Bureau of Mines. Heap leaching is very inexpensive on a cost per ton basis. Most mines will put anything that grades .01 oz/ton or better on the heap. All a heap consists of is a large area covered with a heavy duty, thick plastic sheet, graded to drain to a central point. Gold bearing rock is stacked in a big heap on the sheet up to 50 or even 100 feet deep and then sprinkled with a weak solution of cyanide. Millions of ounces of gold are recovered annually from this type of operation all over the world. It is very common in Nevada. The cyanide solution dissolves gold and silver and the liquid drains down through the pile to a central point where it is pumped through a recovery plant and the gold and silver are extracted. Additional cyanide is added to the solution, and it is then recycled and put back onto the heap to dissolve more precious metals.

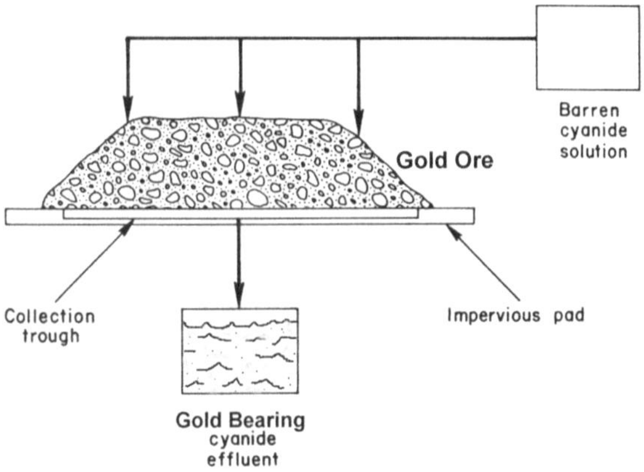

A basic diagram of how the heap leach process works – it can be simple and easy to operate.

In recent years, many issues with the permitting of projects using cyanide have arisen, making the construction of these types of projects much more difficult. Activist "groups", many based mostly on seeking donations and gaining notoriety, have made the use of cyanide a political football. They work to inflame local political issues and promote NIMBY ("not in my backyard") objections. In spite of the fact that we use thousands of toxic chemicals in our everyday lives, the name "cyanide" still strikes fear into the minds of many people who are not familiar with it, and when many folks are upset, politicians react and vote against permits. Often one of the most important considerations in permitting is containment of cyanide bearing fluids, including leak prevention, leak monitoring or detection and methods for controlling situations where too much liquid gets into the system due to rain or snow events. Some years ago, I actually helped to set up and run a small heap leach project in central Nevada in the early 1980s. A significant portion of what we were leaching was old mine dumps. We mined some new ore, but the bulk of the production was from the dumps. Many old mine dumps contain significant gold and silver that could potentially be recoverable by heap leaching. In fact, there are still a number of old mine dumps and mill tailings piles around the world that might be profitably worked as a small-scale project by heap leaching. One concept would be to have a central treatment site for the leaching operations, and bring in ore rock material from various locations nearby.

Before moving forward with a small-scale heap leaching project it is important to determine if the ore is suitable for cyanide treatment. There are a number of potential problems, so testing must be done to characterize the ore and see what the recovery percentage might be. Ore materials containing antimony sulfides, graphite, carbonized wood, oxidized copper minerals or pyrrhotite may not yield gold to the cyanide process as might otherwise be expected. The minerals listed above can consume or destroy the cyanide in solution. In addition, it is always necessary to keep a level of protective alkali to prevent the solution from becoming acidic. Sulfides naturally react with oxygen to form sulfuric acid, so cyanide treatment of some high sulfide ores may also prove difficult. Cyanide treatment is best and most easily done with oxidized ores. Many old mine dumps which originally contained some sulfides have now been largely oxidized. For those who are serious about the possibility of small-scale heap leaching, I strongly recommend seeking a consultant to assist you in the details of project evaluation, planning and permitting.

Small Scale Hard Rock

Hard rock mining is a definite skill that needs to be learned - especially where blasting is going to be required. There are special skills and methods that are used for surface and underground mining. I would urge you if you are thinking of conducting a small-scale hard rock mining operation and do not possess all the necessary skills yourself, that it would be an excellent idea to hire consultants or employees - they can help you with the knowledge and skills you may need to succeed.

Part III. Buying and Operating Used Mining Equipment

After environmental permits, one of the most critical tasks for the small commercial miner is acquiring the right equipment. You can't normally go down to the local prospecting shop and simply pick up a good trommel based placer plant off the shelf. Brand new equipment of this type does exist but it isn't cheap and there are often some seemingly good deals around on used equipment. However, you need to remember that mining is tough on equipment. Rocks and gravel are abrasive and many owners do to not give their equipment the full care and maintenance it needs. Used equipment is often stored outside were rain and weather can further damage it. Equipment that breaks down all the time has a hidden cost because of the profits which are lost while the equipment is being repaired.

Big mining companies, like big construction firms, usually buy new equipment - and there is a reason for it. When you consider the cost of the down time and costs of repairing used equipment, plus the tax benefits of the new, for most big concerns, the economics strongly favor buying new equipment. If you can afford new equipment, you should give it serious consideration. However, for many smaller concerns, the capital to buy all new equipment often just isn't available. For the individual who does not have the same tax situation and access to funding, quality used equipment, when it can be obtained, may be the better buy, but you need to be careful about what you purchase. Can the small miner really get what he needs at a considerable savings by seeking used equipment? Perhaps, but sometimes you are only buying someone else's headaches - equipment problems that may sink your project before it evens gets off the ground. You need to know what you are doing, because buying used mining equipment can be a rough sea to sail. However, if the right materials can be found, a considerable amount of money can be saved. I want to consider some basic rules for buying used mining equipment - ammunition to keep yourself from being taken or making a purchase that you will later regret.

A used Load-Haul-Dump unit like this may be a great bargain for underground mining, but will it hold up to daily tough use in a commercial mine? A close examination is necessary for all heavy equipment purchases before putting down your money.

First Rule: Always Inspect it in Person Before You Buy. If you are buying mining equipment sight unseen, you will never know what shape it's in until it arrives at your site. Condition descriptions from the seller can be misleading. Some of the most amazing twists and misunderstandings in the English language come in the communications between buyer and seller. The buyer may expect that "a little rust" means a few spots of rust here and there. The seller may mean that less than 20 % of all the iron in the piece has been converted to rust, so it's only "a little rust" because more than 80% of the original iron is still present as iron (even though the rust that is present may structurally weaken the equipment). I won't even try to describe the many ways the term "like new" can be interpreted (or misinterpreted). As in used cars, "like new" can mean very different things to the seller than it does to the buyer. Always closely inspect your purchase in person, communicate precisely and don't take things for granted. Take the time to travel to the person's house or business so you can see firsthand what you are buying. Then you can see for yourself what "like new" or "in good condition" really means to that seller. It is almost always well worth the time invested.

Second Rule: Beauty is Only Skin Deep, but major flaws run all the way through! Always inspect the wear points closely - they tell the real story about how much use and abuse that equipment has seen. These include foot petals, gear teeth, bucket edges, etc. Years ago I got a "behind the scenes" tour of the operations of a major heavy equipment auctioneer. The kind of heavy equipment he was selling, like bulldozers, front-end loaders and rock trucks were just the thing many mining or construction companies would be interested in. While they did repair some obvious and easily fixable mechanical problems, the major effort on each piece went into steam cleaning and painting. I have to admit it was amazing how much grease removal and a couple gallons of bright yellow paint improved the appearance of each item! The knowledgeable used equipment buyer however, is not taken in by pretty appearances. Simply giving well-worn equipment a fresh coat of paint does not actually make it "like new" again. As I noted above, wear points tell the true story.

Rule Three: If You Don't See it Run Correctly, Then You MUST Assume it Won't. Internal combustion engines are the source of power to operate most types of mining equipment and the engine is often the most complex and costly part to repair or replace. The condition of the engine is often the most critical item in getting good performance out of used mining equipment. If the seller tells you all about how he just repaired it and its operating great, but he has no fuel so he just can't start it now, don't offer to buy until he gets himself a few gallons of fuel and starts it up. Even better is to see it actually run and process some material.

Commercial operations offer a chance for larger production, like this 60 ounce gold dore button.

Rule Four: Fully Consider What the Word "Complete" Really Means. Look carefully over every piece and consider the condition of all the included equipment. You will need it all when you want to make it run. Make a mental note of all the items that are missing or in such bad shape that they will have to be replaced. Another thing a used equipment buyer needs to watch out for is when things have been specially "customized". You may not be able to get oddball or specially made, hand fabricated parts. Do you have the ability to fabricate your own? Always make sure you get all the various smaller pieces and accessories for your equipment before you drive off. Sometimes with a big load it's easy to forget stuff even if it's unintentional.

Rule Five: You Have to Count the Cost of Maintenance. Most used mining equipment will need some maintenance work or replacement parts to make it right. You have to count that cost in figuring the economics of your used equipment purchase. You just have to do the math here and figure what's worthwhile to you. If the used purchase price plus cost of repairs are going to be at least $20,000 and a brand new piece of the same equipment costs $22,500, you are clearly better off with the new equipment. If the used equipment does not make good economic sense, you have to walk away. You also have to consider the cost of the down time when buying used equipment. Down time is real, it's costly and it's a major disappointment to every miner. Know the general cost of replacement parts before you buy. Know how much that new engine will cost. Know in advance how much those new conveyor belts or new dozer track tread will run. That way you can really count the true and whole cost of your used equipment when buying. Once again, never buy something that is more than you can afford to fix all the potential defects and make it right. **A Special note about ebay:** Ebay has become the world's permanent garage sale. Everything from used sluice boxes to bulldozers can be found there from time to time. The quality of mining equipment for sale on ebay runs the full gamut from excellent to junk/scrap. Ebay is a great place to look, but it's very easy to overbid.

Rule Six: Let the Buyer Beware! – Don't Let Yourself Get Ripped Off.

One of the biggest keys to getting a good deal on used mining equipment is to know well the equipment that you are buying! Experienced dozer operators can see the typical wear points - even through a fresh coat of paint. If you don't know all about the equipment you are buying, then find someone who is knowledgeable and willing to check out the equipment with you. All I can say is really pay attention to what you are buying. Never let yourself be pressured into buying someone else's garbage by a fast talking, high-pressure seller. Count all the costs of putting the equipment into full working order and walk away if the seller won't take a fair price.

Summary of Small Commercial Operations:

Commercial operations are not for the faint of heart or those who are not in a position to take big financial risks. It's possible to make big money with a well-planned small commercial mining operation, but it's just as easy to lose your shirt. If you need help or advice – go out and get it. There are consultants and other folks out there with the experience and knowledge you might need – don't be like the guy who drives around in circles because he is too proud to ask for directions. None of us knows everything and the right information can make all the difference in the world.

When going commercial you have to remember that it's not for fun anymore – you will be making your prospecting interest into a job. If things start to go south and you start losing money don't make the mistake of throwing good money after bad - control the situation and shut down if you have to, so that you can minimize your losses.

Always remember the old proverb that - "He who fails to plan, plans to fail". So plan well and count all the costs – you need to have a large cushion of extra financing for safety and profitability. Things don't always go as expected in most mining operations; in fact they normally go worse. However, if you do it correctly, commercial mining operations done on a small-scale can indeed be highly profitable.

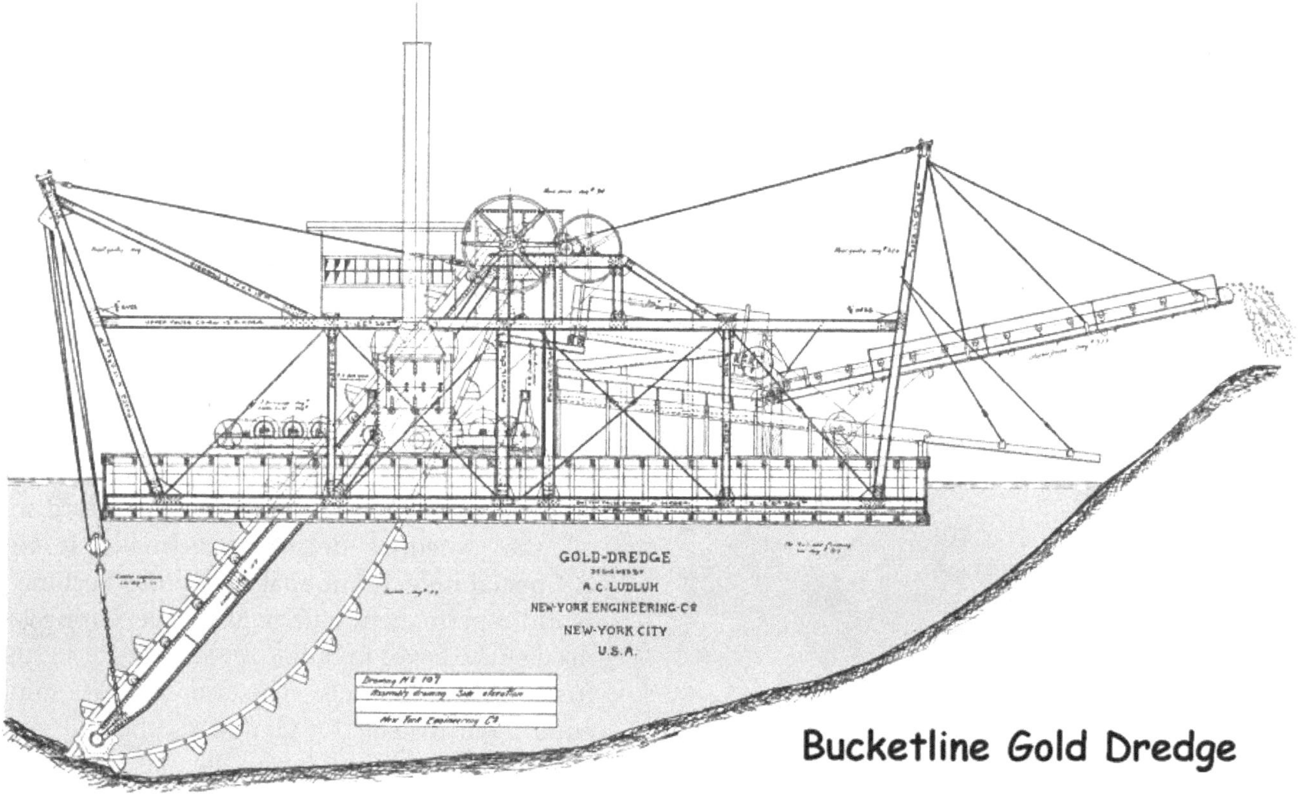

Bucketline Gold Dredge

Build Your Own Homemade Mining Equipment

I am an individual prospector and a dyed in the wool do-it yourself enthusiast as well. I have a degree in Mining Engineering from the Mackey School of mines, and for years I wrote a monthly series of articles for a Mining Magazine covering various aspects of prospecting for gold, so I have a lot of experience with prospecting equipment. (my YouTube channel, Chris Ralph Professional Prospector is also a great source of videos on how to build your own prospecting equipment). Over the years, I have built quite a number of items for myself, including my own dredge. A lot of the equipment on the market will seem a bit pricy to the new prospector, and probably pretty much the same to old prospectors as well. "Can I build my own equipment and save money?" - is a commonly asked question. In this chapter, I will give you my honest thoughts on that topic.

As I've mentioned previously, there are some types of equipment one can build at home. Simple sluice boxes come to my mind, but I have seen a variety of equipment, including well-built high bankers, dry washers and even dredges. I have put together design and planning information for a number of homemade mining equipment projects that you might attempt. Some guys will want to build larger equipment and others smaller, some motorized and others manual - there are just a huge number of possible combinations. I did not want to try to include detailed plans for 1,000 possible versions of stuff, but there is enough info here for you see what is involved in building your own mining equipment. This discussion is enough to get you started on the construction of your project.

A simple sluice box will allow you to process a lot more gravel and it is something that even a beginner could try to build. Prospectors have been building their own sluices for centuries. A sluice box can be built of wood, but things like dredges and dry washers are a bit more complex. If you are going to attempt something more advanced, you ought to have good welding and metal fabrication skills, access to the necessary metal working equipment, and a very good understanding of how the device works. This is because I have also seen homemade equipment doomed to fail because the maker either cut corners or did not fully understand the nature of how the device really operates. Examine closely the commercial units that are for sale -- this can tell you lots about how you ought to construct any homemade device. How are your fabrication skills? Many guys who are trying to save money don't have the right tools and so their homemade equipment doesn't hold up in the field. It is NOT true that you can build anything in the world using nothing but duct tape and bailing wire (although some guys seem to think that). You need the right tools to do the job the way it is supposed to be done. The truth is that with care, the proper tools and the skills to use them, you can make some worthwhile mining equipment for yourself. Take a look at the following projects and see if you think these are within your abilities to construct. I buy my project materials from the junk yard, Home Depot, the local prospecting shop or wherever I can purchase the items that I need at a good price. If you are interested in this, there is much more information that you will want to view on my website, including fabrication and tools information, and as I create additional, new "do it yourself" projects, with photos and information, my website is where they will be posted. I also have some comments about buying the equipment, tools and materials you need to complete these projects there. Each of the four projects which I have listed below has a lot of useful information, suggestions and description of what you need to do, so make a selection and take a look......

Project One:
Build A Homemade Gold Sluice Box

Project Two:
A Homemade Suction Dredge

Project Three:
Build Your Own Homemade Dry Washer

Project Four:
A Homemade High Banker Sluice

Project One:
The Homemade Gold Sluice Box

A sluice box lined with riffles is one of the oldest forms of gravity separation devices still being used today. They are simple and have been employed all across the world for thousands of years. A sluice is really nothing more than an artificial channel lined with devices to catch gold through which water flows moving the lighter materials such as clay, sands and gravels out of the sluice. The heavier materials remain behind, trapped by the riffles. For many years, most sluice boxes were homemade affairs designed and built by the gold prospector himself. To this day, in the gold bearing regions of third world countries, prospectors design and build sluice boxes out the most unusual items - sometimes whatever materials are available locally. You don't really need any special sluice box plans - the exact size is really not all that critical. They come in all sizes and a can range from small, portable aluminum models used for prospecting all the way up to large sluice boxes hundreds of feet long, which are used at fixed installations in commercial operations.

Making your own gold sluice is actually a very good beginning project for new prospectors in my opinion. Just take a close look at the sluices being offered by the manufacturers, and that will show you how to build your own sluice box. It won't be difficult to get some ideas to make your own plans. Sluice boxes can be made out of wood, aluminum, plastic or steel. Injection molded plastic is not really an option easily available to the do-it-yourself prospector, and steel has a tendency to rust, so wood and aluminum are the preferred options.

In developing plans for a homemade sluice box, the more time you spend thinking about your design, the better. You don't want to have to buy any parts you don't need, but on the other hand your sluice box needs to work and catch the gold efficiently. Quality materials and a good basic understanding of how a sluice box traps gold are important to your design. I think using miners moss underneath your riffles is a real important

A homemade aluminum metal and steel riffle sluice box made by the author – this is a big one, it's 20 inches wide.

item for capturing that fine gold. That is why miners moss is used in the sluices of nearly all commercial suction gold dredges. Having a liner underneath the riffles is an important aid in catching small gold dust, and is very worthwhile. I went with miner's moss under all the riffles in my sluice, and I strongly recommend it for you.

The typical wooden homemade sluice is made of boards and varies in width from 8 to 18 inches, usually with a depth of 6 inches to a foot. A typical length would be in the three to 6 foot range. Make sure the wood planks fit together well and caulk them so gold and other material can't escape. Riffles can be made from half inch square dowel nailed about every 6 inches down the length of the sluice. The section without riffles in the top of the box about a foot long is often left for the spot where material shoveled in. This type of sluice box does catch gold, and is easy to build, but is hard to clean out at the end of the day. In addition, the gravel will beat up the wooden riffles over time. It is also possible to create steel riffles that fit inside a wooden sluice, and in that case you should also use miners moss or some similar material to line the bottom of the sluice underneath the metal riffles.

Homemade sluices can also be made from lightweight aluminum. Wooden sluices tend to become waterlogged and increase greatly in weight after they have been in the water for a time. It's amazing how heavy they can become! Sometimes a good paint job can help slow down the waterlog effect on the wood. This gives aluminum quite an advantage and it is certainly preferred in the construction of the homemade sluice. The trough portion of the sluice, whether it is made of aluminum or wood, is usually roughly the same size.

This shot shows the low, flat riffles that only stick up just a bit for a typical hand fed sluice. Generally large, tall riffles are just not needed.

For those interested in making their own home made, hand fed sluice box from aluminum with steel riffles as a do it yourself type of project, I can say if you have any metal fabrication skills, you will find this an easy project. A little welding, a little metal folding and the project is done. If you purchase fairly thin aluminum sheet it will be possible to bend it yourself into the trough shape as a single piece (just don't go too thin, because it has to be sturdy). The prospector's portable gold sluice box has an advantage that it can be quickly taken up and moved to a new location, as the prospector searches for a spot with good gold. The grade slope when the sluice is set up usually ranges from about one to 2 inches per foot.

The riffles are really the most important part because they are the part of the sluice box that catches the gold. Riffles are often assembled from small slats of steel set at about 45° angle held in place by a vertical rail on either side. They are best assembled by being welded together. Take a look at the pictures in this chapter and you can see how I have welded my riffles together into a single easily removable piece. Keene and many other sluice box makers do the exact same thing. This makes the riffles easy to remove when it is time to clean the sluice out and recover the gold. Removable riffles, no matter what they are made of, are held down with bolts or wedges to keep them from moving around. Now my welds aren't pretty, but they do the job - they just need to be sturdy and hold the riffles in place. If you are going to do your own projects, I strongly recommend that you consider buying an inexpensive 110V welder. I recommend them so much that I have done up a whole web page on it. If you are seriously considering building your own sluice box, check out the DIY and fabrication part of my web page. If you would like to view some more information on how to operate a sluice box, be sure to check out the chapter of this book on using a sluice box to find gold.

A photo of the author's partly homemade dredge.

Project Two:
A Homemade Suction Dredge

Suction dredges are very handy and useful pieces of prospecting equipment, with the potential to produce some really nice quantities of gold. No question that many gold prospectors would love to own a dredge, but the prices for new ones are really high. The truth is however, that in the earliest days of dredging, there were no manufacturers, and all small suction dredges were handcrafted units made in someone's garage. There's no question that a good dredge can be made at home, and I know you can save some significant money doing it, because I've done it myself.

Building a dredge is a big project with a lot of plans and design decisions to be made. Take your time and think about what you really want to build. Think about what materials you have on hand or what you could easily acquire, then build a list of what you need to construct your dredge. Unfortunately, a simple set of dredge plans that would work for all sizes and styles of suction dredges is just impossible, so I haven't tried to prepare anything like that. However, you can do up a specific design for your own project. If you really sit down and think about things, and use measurements taken from the commercial dredge makers you can design your own set of plans based on tried and true designs.

Of course you will be building on the cheap to save money, but you don't want to shortchange yourself too much. You don't want your dredge to be rickety, fall apart, or to fail to function. Dredges get knocked around and they need to be functional, durable and sturdy. The time you spend sorting through design concepts, deciding what you will build and how you will build it will be well spent. Think about what you want and what you need then weigh those together with what you can afford. Do up some drawings and lists. Perhaps the best thing I can suggest is that you study the designs of the well-known dredge makers like Keene and Pro-line. These manufacturers have done quite a bit of research to produce good quality products, and they have tested different options and have developed efficient pieces of equipment that do the job well. Check out their web sites as most have good photos of their dredges and the individual components that make up these dredges - you can get a lot of information from their web sites. If your local prospecting shop has a dredge set up, take a close look and even measurements or photos if you can. Another great possibility is to join a prospecting club whose members actively dredge, and then go out to the claims and check out the members while they are dredging. Take some pictures of the dredges while they're in operation. The club members may even let you have a few minutes behind the nozzle so that you can get a feel for the whole experience. The more general knowledge you have about dredges before you begin your design, the better your construction plans will be.

When you are sitting down to plan the designs of your own home made gold dredge, probably the first question you will want to ask yourself is "What size of dredge do I want to build?" The bigger your dredge is, the more material you'll be able to process, but the more it will cost and the more it will weigh. So weight and cost are the trade-offs for the volume you can process, and practically speaking, the depth of bedrock you will be able to reach. The bigger the dredge, the more material it can handle per hour, but they go up fast in weight, and carrying them into your location can be a big job. You might want to look at the dredge section in this book, especially the part that considers the various sizes of small suction dredges. It is a general rule of dredges that it is not practical to dredge more feet of overburden greater than the inside diameter of your hose in inches. So for a 2 inch dredge you don't want to go through overburden deeper than about 2 feet, for a 4 inch dredge you don't want to go through more than about 4 feet of overburden, and the 6 inch dredge would be limited to around 6 feet of overburden, etc. Of course it is possible to dredge a 10 foot deep hole with a 2 inch dredge, but it will take forever and be highly inefficient. This little rule of thumb does not count the depth of water, only the depth of overburden to bedrock. Another piece of design information to consider in your planning is that a

Old style dredge with crashbox style sluice and truck tube flotation out on the water. This style flotation is easy to build for yourself.

four or five inch dredge is about as big as one guy can handle efficiently by himself in my opinion – others may disagree. If you're really big, tough and young you may be able to handle a 6-inch by yourself, but it will tax you. In my opinion, a 4-inch is probably the best size for a single guy operation, considering a trade off of weight and cost, and the abilities of the individual to conduct an operation on his own.

Will you need compressed air for diving? If so, you will need an air compressor, an air receiving tank and related diving equipment. Air is really handy to have, but it's optional, and you can always add air later if you plan well and allow space for the compressor on your dredge. At a minimum, your homemade dredge will need a motor and a pump.

A simple homemade dredge frame built from scrap aluminum purchased at a local scrap yard - viewed from underneath as it is mounted on my Garage ceiling.

Do you have the proper tools? Fabricating your own suction dredge takes tools – as I mentioned, this is not a little job. Things like a hack saw, pop rivet gun, drills, wrenches and the like are the minimum. Having the ability to do some arc welding, or to bend and shear cut metal are also handy abilities to possess. My little 110V arc welder, while underpowered, has helped me complete some serious fabrication projects. Some of these welders are cheap and if you don't already own a welder, this project might be just the excuse you needed to buy one! I so strongly recommend that you consider buying one of these welders that I have done up a whole set of web pages on welders and other project fabrication tools at my website. It's in the section with the rest of the DIY project information. It's also important to consider if you actually have the experience to build such a project. The more tools and fabricating skills you possess, the more likely it is that your home made dredge project will be a success.

Count the cost and figure what is realistically possible: It's not free! You can save a lot of money building your own home made dredge, but it still won't be cheap. I strongly recommend that before you actually begin to buy materials that you sit down with your design plan and figure those costs. Be sure you have enough money to complete your project. I decided to buy some parts from Keene, because I didn't think I could fabricate an item as good as what they had to offer. On the other hand, I bought some used dredge components as well. All in all, my 4 inch dredge cost less than 1/3 of what a new one would cost. Here are some of my comments on deciding what parts of a dredge project you might construct for yourself and what parts you might be better off buying:

Near Impossible to Make (But You Might Buy Them Used):

Engine:
Hey, I've seen some really nice homemade dredges, but no one forges the steel to cast the block and make their own engine. You may be able to purchase a used engine and save a few bucks, but no one makes their own engine from scratch. Be sure any engine you buy has the proper bolt holes and threaded drive shaft set up for attaching

the water pump - they come in various bolt patterns and drive shaft sizes, so be sure you are getting the right one (although a pedestal mounted pump could be used with any bolt pattern). Select a motor size to match the pump you will use. The manufacturers have certain motor sizes they use with certain pumps on dredges of a certain hose size. Plan to use the same sizes they use. Your engine is the source of all power on your dredge, and dredging is tough on these little engines, so you might want to seriously consider getting a brand new one.

Water Pump: Again, I've seen some really nice homemade suction dredges, but no one casts and machines the aluminum to make their own water pump. The centrifugal pumps used by all the dredge makers are well designed for this purpose. A good efficient water pump is important. I recommend you purchase one of the pumps designed by Keene. I got my pump used at my local prospecting shop. Don't use a trash pump and think you can pump the sand through the trash pump, saving the need for a jet. This is simply a very bad design idea, and the heavy load of sand and gravel going through the trash pump will destroy it in a short time. They are also quite heavy and a larger motor is required than would be needed if a centrifugal pump and jet were used.

Air Compressor: I own a compressed air pump, but I have not gotten around to mounting it on my current dredge. I have dredged for years with nothing more than a snorkel, but it does limit the areas that you can work. Long arms are a real asset. Be sure if you do get one that you design your system so that it has room to fit in on the mount with the motor. Don't use a pump or hose that is not intended for producing breathable air - compressors for tools and paint systems add oil that you don't want to be breathing....

Dredge Suction Hose: There are alternatives for hoses, but you need to be careful. Weak, thin walled materials will wear out very quickly and you will not save money with them. Transparency in a dredge hose is important so you can see where plug ups occur and take care of them. The type of material used by the various dredge manufacturers is very suitable and durable and will get the job done.

These are Do-able With Some Effort:

Sluice Box: A sluice box is probably the biggest and most important thing that you can make for yourself on a dredge. It's a key and critical item - hey, if your sluice box doesn't capture the gold, then your whole dredge project is a lot of wasted time. I've actually done up a whole separate portion of this section about designing and building the sluice box for your home made dredge. Its project 2, part B. Those pages will show you just what I did build my sluice box.

Header / Crash Box / Flare: This is one where you have to decide what you want to do. You can go with a crash box style head and is relatively easy to make. A crash box is just a box that sits at the head of the sluice, directing the flow of water and gravel into the sluice. The other option is to go with a flair header which is much more difficult to make, but gets a lot better gold recovery. I chose to go to the expense and buy a plastic molded flare header from Keene for my dredge. It's well designed and works fine. I have also seen guys build flares out of orange plastic traffic cones. I recommend using a damper (a sheet of heavy rubber) to help control the water and slurry flow out of the box. Dampers are good no matter which design you choose to use.

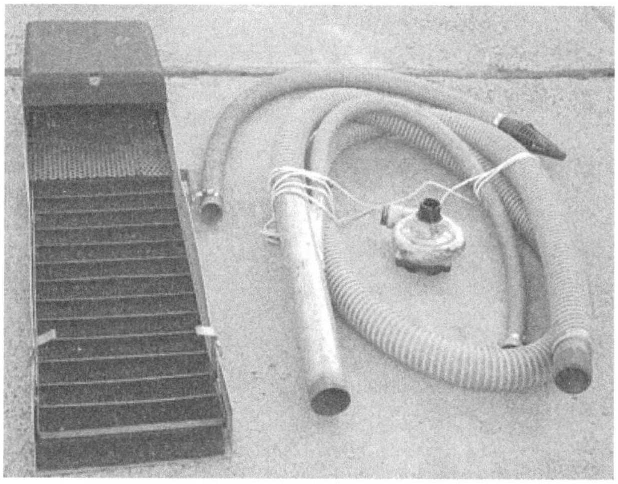

Various important parts and pieces of a dredge can sometimes be picked up separately at a good price. Many prospectors accumulate extra parts for various reasons. Ask around at your local prospecting shop or club meeting for who has extra parts that they might want to sell.

Jet:
This is a little difficult but not too bad if you have the right welding skills. First you have to decide if you want a suction nozzle or power jet type. The suction nozzle puts the high-pressure water in right at the nozzle. A power jet puts the high-pressure water in just before the sluice box. A suction nozzle is best in very shallow water; a power jet produces better overall suction. Most dredges are power jets, but some smaller units, 3 inches and smaller are suction nozzles. I recommend the power jet. Take a close look at the jets made by the manufacturers like Keene.

Nozzle:
Here's an easy project: really all you need is a piece of pipe about 10 inches long and of the right diameter to fit in your dredge hose. Simply take one end of the pipe and saw cut the pipe about every quarter inch. Take a hammer and gently tap the pieces down to form a cone on the tip. Weld a bead around the edge of your new nozzle and you're done. You can even weld on a handle to the nozzle if you wish (I personally don't like them, but some guys do).

Frame Structure and Flotation Systems:
This is another part of this project where you can really save some money if you want to. There are a number of different flotation types which you might select to fabricate for your dredge project. In the early days of portable suction dredges, most of them were floated on large truck inner tube tires. These large inner tubes are both light weight and inexpensive. This is the option I chose for building my home made dredge, but these large truck inner tubes are becoming more and more difficult to purchase. Few of the trucks still on the road use them and they are difficult to purchase even at tire centers specializing in servicing large diesel trucks.

Nowadays, most commercial suction dredges are built on plastic pontoons. They are more durable but they are also a lot heavier. There are a number of options in this direction as well. I have seen homemade dredges built on a variety of plastic pontoons. One interesting option is five to 10 gallon sized plastic shipping containers or barrels. Sometimes these can be purchased quite cheaply, but they are much heavier than truck tubes. The support frame you build has to be designed to go with these float units (which ever ones you choose). Another option is to purchase the standard pontoons from the manufacturer, and in general these are quite expensive, but they are sturdy and do a good job. The other part of the flotation system is the frame or mount for the sluice box and motor, allowing them to sit in a stable manner on top of your flotation units. I made my frame out of scrap aluminum in various sizes which I purchased at the local junkyard for only a few bucks. I bolted and welded the pieces together to form a rigid frame. It was inexpensive and not at all difficult to construct. Because it was flat, my frame sat easily upon my truck tubes. I use standard bungee cords to hold the tubes securely to the frame.

Foot Valve: The foot valve is an important but often overlooked item. You can put one of these together from a piece of pipe and some heavy screening with just a bit of bailing wire and some tack welds. You would also need to build in some sort of flap valve to allow priming of the pump with water (check out the commercial units). Sucking pure sand can ruin your pump in a fairly short time, so don't try to run your dredge without one - all the manufacturers supply them with the new dredges. They also keep out any twigs and leaves from getting sucked into the pump as well.

The author's homemade 4 inch dredge while in operation in California's mother lode country.

Sources For Parts and Pieces:
I have found that the junk yard / recycling yard can provide some important pieces that you may use at low prices. I suggest that once you have

good plans for the dredge you want to build, take your purchase list and go look through the local scrap yards - you can get stuff there a whole lot cheaper than you would at someplace like Home Depot. You may even find a suitable used engine there (remember you need an engine with a bolt pattern specially set up for pumps).

I suggest that once you have assembled all the pieces you need, and built your dredge, you need to test it. I suggest that you test it with a couple dozen pieces of small lead shot. Flatten them, and paint them red or some other bright color. Then suck up some gravel from the nearest gold bearing stream and put the shot in with the gravel you are processing. Be sure to take in a good bit of gravel both before and after you've sucked up the shot. When you clean up the sluice, count how many of the shot you have recovered and compare that to the number you started with. You should not lose more than one or two at the most. If you lose more than three or four you need to adjust your dredge or make some changes to improve it so you can be confident you are not loosing gold.

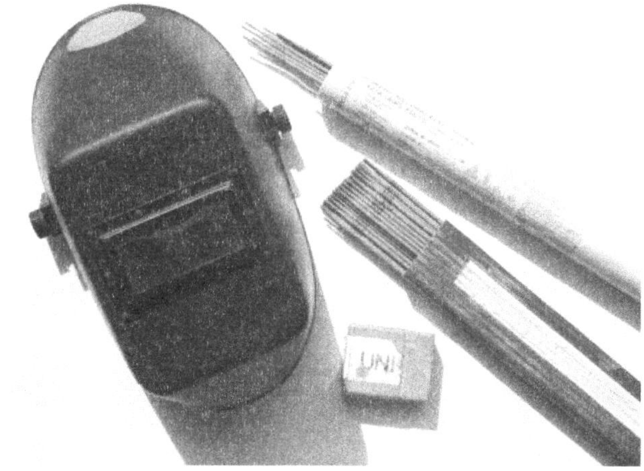

The proper fabrication equipment, such as the welding rod and helmet shown here can go a long way when attempting a mining equipment fabrication project.

A big part of designing and building your own reliable and functional dredge is understanding exactly how a dredge is supposed to work. If you would like to look at some more info that I have done up on operating a gold dredge, including a diagrams showing how a dredge works, take a look at the suction dredging chapter of this book.

You will also need a wet suit of some sort, although the actual type depends on where you will be dredging for gold. Even in California in midsummer the water gets awfully cold after an hour or so if you do not have some sort of protection. In Alaska, much more serious protection such as a well-designed dry suit is needed so that you can stay in the water while operating your dredge.

I hope you have found this information on planning and designing your own home made dredge useful. It is not an easy project, but it can be done. I wish you the best of luck with your own homemade gold dredge project; take care and remember the fabricator's old motto to "measure twice and cut only once"!

Here is the completed homemade dredge, with a new homemade sluice box.

Project Two, Part B: A Homemade 3 Stage Under Current Dredge Sluice Box

I have a mostly homemade 4 inch dredge which I have used for many years at various locations in the California mother lode gold country. A few years ago I changed out my old dredge pump for one with higher capacity - the plan was to get greater suction. It worked, but I noticed a significant drop in my fine gold recovery. I know that the old crash box style dredge sluices are not very good at recovering fine gold, but when you up the water speed too much, well, it really makes things even worse. I knew I had a problem, but while dredging in 2002, noticed a small nugget had worked its way down my sluice to the 6th riffle - it was on its way out of the box. I later weighed it and it came in at 8 grains, so I figured if gold that size could work its way out, my fines recovery was really bad and I needed to do something. I just could not put it off any longer, and I began to consider my options. Mostly because of the cost, I decided to make my own home made sluice box. I did not have any design plans or specific directions, but I am a mining engineer by background, and I own or have access to the necessary metal fabrication equipment, so the project did not seem beyond my reach. Besides, the 49ers designed and made their own sluice boxes, so why couldn't I make my own? I also figured I'd save some money in doing it.

Nearly all dredge sluices have Hungarian type riffles, and I wanted my homemade sluice to have them as well. The concept of the Hungarian riffle (also called the Lazy L) is that the water and gravel flow over the riffle and a little eddy sets up behind the riffle. Lighter materials continue down the box and out, but the heavier material like gold gets caught in the eddy behind the riffle, settles out and stays in the box. Basically these riffles are just a short bit of steel bent over at the top. They are nothing too complex to build. In planning the design for my own box, I took a look at sluice boxes of various designs at the local prospector's shop here in Reno. I also spent some time at a GPAA gold show here in Reno looking at Keene's display of all their equipment, with a special eye on how their sluices were designed. I decided that a flare and box design with under current riffles for catching fine gold would be the best one for me. This is basically what Keene is using on all their new dredges (except some of the smallest ones). I first thought about making my own flair, but decided to purchase a Keene flare. They are made out of plastic, which is a big part of why I chose to buy one rather than fabricate my own. I suppose you could make a homemade one out of fiberglass, but that would be heavy and a difficult project all in itself. So in purchasing the flair for my project, that meant I would be going with a standard 16 inch sluice width, which I knew would work fine with the rest of my equipment (which was designed for a 16 inch wide box). My old box was 53 inches long, so the new sluice had to be able to work with my existing flotation frame. Because of this, the new box is 54 inches long, 6 inches longer than Keene's 3 stage (for a 4 inch dredge). I spent a couple lunch hours going through local scrap yards, to see what I could get for building the box. I came up with a design, based partly on what I knew about sluice box operations and partly on what I could get in the scrap yard (or already had on hand in my garage from other projects). Scrap yards are a great way to get materials for metal fabrication projects very cheaply. One of the items I lucked out on was a box made of 3/16th inch diameter punch hole steel. It had been used for ventilation on some sort of electronics equipment. For me, it would be perfect for the undercurrent screen - just the right size. I got together all my metal from the salvage yard, and I just paid a couple bucks for all the steel, and only about $10 for the aluminum.

A work in progress: These are the parts and pieces of my homemade sluice box, prior to assembly and welding. Almost everything is laid together and put in its relative place. The metal slats on the lower right are the riffles. At this stage, you could really see my sluice project coming together.

In developing plans for a homemade sluice box, the more time you spend thinking about your design, the better. You don't want to have to buy parts you don't need, but on the other hand your sluice box needs to catch the gold efficiently. A good plan and a good understanding of how a sluice box traps gold are important to your design. I think miners moss, which is used in nearly all commercial gold dredges, it a real important item for capturing that fine gold. I went with miners moss under all the riffles in my sluice. In a three stage sluice, most of the gold gets caught in the upper part. As the gravel passes down the box, it goes over a screen, which in my sluice has 3/16 inch holes. The fine gravel and smallest gold goes down through the screen into the lower or underneath section of the sluice. There, the slower water allows the fine gold to settle out. Material that is not small enough to go in the 3/16 screen goes over an additional section of riffles before it leaves the box. This is a standard type of undercurrent sluice design.

A friend of mine has sheet metal shears and a brake. I cut the sheet steel to size on the shear and I bent the riffles on the brake. They are just like the standard Keene riffles, a modified Hungarian type. To make them I'd grab about a quarter inch of the metal and bend the rest up at a 45 degree angle. This is all there is to the basic Hungarian or "lazy L" riffle. If you have access to a sheet metal bending brake the riffles are nothing to make. The riffles on the first (front) set are 3/4 inch high, the riffles on the upper deck of the under current are 1 and 1/4 inches, and the lower under current riffles are 3/8 ths of an inch high. I used standard mild 1/16 th inch steel for all the riffles. The sluice box itself is just a big sheet of aluminum bent into a trough, and was no big deal. It only took about 2 hours to do all the shear and brake work (and it only took that long because I was being careful and taking my time).

A project like this really does require tools like the brake and shear, and some sort of welding equipment. My sluice would have been nearly impossible without them. After cutting and bending everything, the next step was to weld up all the steel into the designed configuration. There is a lot of welding to assemble the riffles. Nothing too difficult, just a lot of spot welds. I am sure Keene and the other manufacturers have pre set jigs for making up riffle sets, but I had to make do and set things up for my self so it took me a bit longer. This was by far the most time consuming part of the project. My welds aren't pretty, but they do the job - they just need to be sturdy and hold the riffles in place.

The upper riffle section of the sluice box, all completed and assembled with the Hungarian style riffles and small expanded metal section in the front.

Flat steel strips are welded on the ends of the riffle to hold them in the proper position. The wider part of the riffle is turned over and set at a 45 degree angle to the bottom of the sluice. In that position, the narrow bent part is parallel to the bottom of the sluice box. The picture with this chapter shows at set of riffles, all welded up and painted, ready to put in the sluice box. For people with the right equipment, and some metal fabrication skills, this is really not a difficult project, but could be tough for a beginner. While lots of guys would like to try it, as I have noted, you need the right tools and skill set. If you don't have the experience fabricating metal items, you might want to seek out a friend or relative to help or advise you. I probably had 10 hours of actual metal fabrication work, but a lot more time in design. There is no question that I invested the majority of my time in planning and designing my sluice project. In order to use my existing flotation frame, I needed to modify the frame slightly to lift the engine up about 2 inches in order to accommodate the new flare that now passes underneath it. My flotation is still based on rubber truck tire inner tubes. These systems work fine, and I have been using them for years. The frame itself was mostly flat, but it was not too difficult to cut and bridge across the motor mount supports. Keene used to make systems with truck tire flotation long ago, but all their sluice designs were crash boxes back then.

The first time I started up my new dredge, I noticed how much of a difference the new box made. No more frothing white water flying down my sluice - just water running down nice and orderly. The first clean up confirmed what I had hoped for. I was getting much better fine gold recovery. The sluice was doing exactly what it was supposed to do. I never had any problems with the under current section loading up and becoming plugged. A few folks have had problems with properly adjusting the undercurrent intake on Keene's 3 stage sluice box so that the material flows on through without plugging.

The authors homemade dredge sluice box after some of the construction has been completed. The project is starting to look like a sluice box.

When I cleaned the under current (lower level) riffles, I noted that only a little gold showed up there - however I think I was loosing nearly 100% of this fraction before. There was loads of black sand, and only a little fine gold. Keene advertises that 90% of the visible gold stays in the first set of riffles, and my box performed in a similar way. A slight amount also showed up behind the riffles of the upper deck of the undercurrent sluice, so I put in miners moss there. The final clean up showed a very significant increase in the total amount of fine gold over what I would have previously expected. In my opinion, the first most important thing to improve your gold recovery is to get rid of the old crash box. The flare slows down the flow and spreads the material out, greatly increasing your gold recovery. Previously, my sluice box was full of froth and whitewater, and fast water is another gold recovery problem. It's a wonder I got any fine gold at all with my previous set up! The box held together fine, one spot weld broke in the field, but it did not affect the sluice's operation. A quick spot weld when I got home and it was all fixed before I put it away for the year. The revised and elevated motor mount worked just great.

For those interested in making their own home made sluice box in a do it yourself type of project, I can say if you have any metal fabrication skills, you will find this an easy project. A little welding, a little metal folding and before you know it, the project is done. If you are just making a sluice to shovel gravel into, there is no need to have an underflow or two level box. In that case, one level is fine. Making your own hand fed gold sluice is a good beginning project for new prospectors in my opinion. On the other hand, if you are making your sluice box for a dredge, then I do recommend the two level or undercurrent type box. It is what Keene and some other dredge makers are using on their new dredges. Just consider this chapter, take a close look at the sluices being offered by the manufacturers, and you can make your own plans. I do not have any written plans that I offer, but I think you can make a plan for yourself easy enough. Just go out and take a good look at a friend's sluice box, or go down to the local prospecting shop and take a close look. You can see how they are made and I think you will find that if you can do a little welding and metal folding, that you will have no problem making a box for yourself. I do suggest you use miners moss under your riffles; this stuff makes a big difference in catching fine gold. Here is a summary of the construction costs for my homemade sluice box:

Scrap Aluminum:	$ 10
Scrap Iron:	$ 5
Paint (rattle cans)	$ 4
Miner's Moss	$ 56
Welding Rod (small)	$ 8

TOTAL	$ 83.

I think a new dredge sluice box in this size is well over $500. It's funny that the biggest expense was the miner's moss. I really do believe that this stuff makes a big difference in fine gold recovery.

Photo of an old time homemade desert placer dry washer. In the old days, all dry washers were made by the miners who used them. This is a wooden, bellows driven, portable machine.

Project Three:
Build A Homemade Dry Washer

In a number of places, placer deposits exist in desert and other regions with little or no water. The most practical way for the individual to work these types of deposits is with a dry washer. Have you wondered if using a dry washer would help you find more gold? Here is how you can plan and build your own homemade dry washer......

A dry washer separates gold from sand by pulsations of air coming up through a porous cloth medium. It is the gold prospecting equipment item of choice when working in desert areas. The vibrations and flow of air replace the function of water in the normal sluice, allowing the gold to settle downward. Screened gravel passes down an inclined riffle box with cross riffles. The bottom of the box consists of a thin, light weave canvas or some other fabric through which moving air passes upward.

The riffle section of a dry washer. Note that unlike a wet sluice, the gold is caught ahead of the riffles, as opposed to downstream. The large flat area is basically a landing pan where the material drops into the sluice area.

Many folks believe that a dry washer is a piece of gold prospecting equipment that you just have to have if you're working in the desert, but the price of a new one can be expensive. At least for the bellows type dry washer, they are not that hard to make. It really has only been in the last 30 years that you could buy pre-made dry washers quite easily. For a hundred years before that generally most dry washers were homemade affairs designed and built by the gold prospector himself. If you'd like to be a part of that great old tradition, read on.

First, the bulk of what homemade dry washers are made of is wood, so standard home wood working tools are sufficient for all but the riffle tray. I prepared a generalized plan for an average puffer type homemade dry washer. The image is a couple pages further on. You can adopt and use these plans, or use them as a starting point and design your own dry washer. To build one, you need to think about how you will build your particular unit, and what design ideas you want to use on your own homemade dry washer.

The first design concept you need to think about is the air source. Do you want to build a bellows puffer, or a powered blower type of unit. Both work well, but the puffer type is plainly easier to build. Because of this, I recommend a puffer type dry washer if this is the first one you are building. Remember that dry washers are such that you could always modify your unit in the future to switch from a puffer type to a powered blower type if that's what you eventually decide you really want to do.

The directional flap on the inside (upper portion) of the bellows. It allows air in, but not out.

If you decide to go with a power blower type of design you will need an air blower unit and some source of power for it. I have seen units like this run off electric leaf blower motors such as one might buy at a hardware store. It would be very possible to run something like that off a little gas powered electric generator. Gas powered leaf blowers work fine as well. No matter what your design, you will need the following construction materials: wood, metal, cloth, screws and screen. I suggest 3/8 inch weave for the material screen - half inch would probably be OK, but I suggest 3/8 inch as best. Too large a size and those larger rocks will help push gold out of your sluice, and if too small you may lose a nice nugget if you dig one up.

I mentioned that you will also need standard wood working tools. See my web page for more thoughts about obtaining the tools you will need. One of my best suggestions is that if you want to start fabricating mining equipment (and keep the stuff you have in good repair) consider purchasing a small welder. This type of unit would come in handy if you were fabricating the riffle tray. I know that mine has come in very handy for me a number of times both in making equipment and around the house.

The dry washer consists basically of two boxes, and upper box and a lower one. The upper box is really just a screened hopper you can feed the gravel into. The material too large for the screen falls off the screen as "oversize". The exact dimensions of the upper box are not that critical, but you don't want it to be too big or too small. The lower box holds the bellows or blower as well as the riffle tray set up.

The air hole pictured from the outside (below) the bellows. It allows air in, but not out.

You shovel onto the screen until the upper box is full, then with a puffer type, you step to an upwind location and pull the cord to make the bellows puff. The material then works its way down through the hopper, and across the sluice riffles. When the hopper is empty, you repeat the cycle and start shoveling again until it's full.

Make sure the screen you purchase for the upper box is durable and that on the hopper it is well supported from underneath. If you toss too many big rocks onto a poorly supported screen it will simply bend and break. That is something that can be fixed, but when you are out in the field, it can be very inconvenient. A couple quarter inch diameter steel rods can be very helpful support for the screen.

The gold catching riffles in the lower box of a dry washer are different than those on a water sluice. The design on a water sluice box is that the gold is retained downstream of the riffle. On a dry washer, the gold is designed to be held on the upstream side of the riffle. The riffle tray is one of the things that really need to be put together by welding. I suppose small bolts would be a possibility, but welding would be much superior. Welding up the removable tray does not take that much work. On the underside of the riffles is a light weave muslin cloth that allows the air to pass upward through it. The riffles need to be a movable tray, so that periodically (usually every hour or two of run time) the riffles and the concentrates can be removed and put in a bucket for later processing. In addition, with the puffer type, dust gets sucked into this lower air box and it needs to be cleaned out from time to time. The sluice box must sit high enough to allow for tailings to drop off the back end.

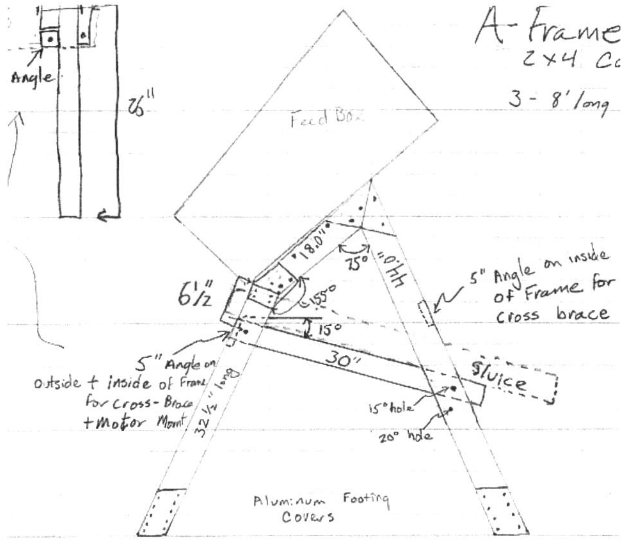

A hand sketch by the author made for the construction of a dry washer. Sketches like these are useful for anyone who wishes to build this type of project.

On the underside of the lower box is where the air comes in. If you use a blower, this is where the duct enters. If you are making a puffer / bellows type dry washer, this is where the bellows will go. Good heavy canvas makes a good bellows. Material that has been rubberized or made waterproof is best. You also need a flapper so that air can go in and go out only through the riffles. The one way flow flapper is usually just a piece of 1/8 inch thick rubber covering a hole about 4 inches in diameter. The bellows are hinged on one side and a pull cord is attached to the end opposite the hinge. Pulling on the cord puffs the air through the riffle cloth. See the photos for further detail. It's helpful if the support legs for the dry washer are adjustable in some way, as the ground surface in the gold fields is often very uneven. Understanding the basics of how dry washers work is important if you are building one.

Take a close look at the commercially available units to see how they are built. If you would like to look at some additional info on dry washers and the basics of how they operate, take a look at the chapter in this book on dry washers and the general occurrence of dry placers.

Project Four:
Homemade High Banker Sluice

The highbanker has become a popular piece of prospecting equipment. It is really just a sluice box set up with a pump that enables it to run at a location away from the water's edge. A close relation of the dredge and river sluice, the highbanker has the advantage of being able to work bench gravels left high and dry by river erosion. The disadvantage of a highbanker is that it needs almost as much water as a sluice, and needs a good sized motor and pump, to move the water supply up from the river source to the highbanker unit. The higher and farther away your gravels are, the larger the motor and pump you may need. The main types of deposits for which a highbanker is a good choice are gravels away from the water, such as bench gravels and residual placer gravels.

There are two main elements of a high banker: they are the feed box and the sluice box. Here is a discussion of the function of these two parts, to better help you design you own home made highbanker sluice: The sluice box part of a highbanker is really is not that different from a regular sluice. The riffles may be a bit taller, but the riffles and carpet or miners moss underneath the riffles is virtually the same. Gold gets caught in the riffles and moss in the same way. Besides the pump system, the only other important difference between a regular sluice and a sluice box for a highbanker is that the highbanker sluice has adjustable legs for support. In fact, many hand fed sluice boxes can easily be adapted with a pump and legs for use as a highbanker. For more information about building a river flow sluice box, see project number one on designing and building your own hand fed sluice box.

The feed box is basically a box with a grizzly screen and water sprays. Water is normally sprayed from plastic pipe with a series of holes drilled into it. A simple water valve ahead of the spray pipes controls the amount of flow into the pipes. A slat type screen allows water and the smaller material to fall through and go into the sluice box. The slats should be set on an angle so that much of the material will slide off the slats. The material too large to fall between the slats is washed with the spray and then rejected.

The feed box of a high banker has a set of slats to screen off the coarser rocks and pebbles.

The operation of a highbanker consists basically of shoveling material into the hopper, where it is washed by the water sprays. The water, sand and smaller materials drop down into the sluice box area. Enough water must be pumped into the system to run the sluice, although the sluice box section of a high banker is usually set at an angle much steeper than a hand fed sluice in a river is set.

It is very possible to convert an existing sluice box to a highbanker. If you have low, flat type riffles as are found in a hand sluice, you may need another set of riffles that are a little taller.

The modern high banker power sluice in action. Water flows down the sluice box section.

To make the conversion, the sluice will need to be modified to have legs and a feed box mounted on top of it. You'll need a baffle piece to close off the front end of the sluice, so material does not go flying out the front of the box.

Building the sluice portion is the most important part of the project. Here is where you will need your metal fabrication skills, but you will find the sluice box portion of the highbanker is really just like building a hand fed river sluice box that I described under project one. Your highbanker will need the same standard riffles, with miners moss as a lining underneath them. One difference is that you will also need adjustable legs for the high banker sluice to sit on. There are a variety of adjustable leg designs, but the legs need to be stable and strong. Because of the weight of the weight of the feed box on top, the legs and support structure need be able to handle it. Take a look at what the commercial manufacturers build. Because of the weight of wet wood, I strongly advise using sheet aluminum for your high banker. The sluice will also need to have some prevision for mounting the feed box on top of it. If you purchase fairly thin aluminum sheet it will be possible to bend it yourself into the trough shape as a single piece (just don't go too thin). For more related information, see project number one on building a hand fed sluice box.

This different, tall design for a high banker runs essentially the same as other designs.

Building the feed box portion of the highbanker is not too difficult either. It's simply a box you can shovel material into that has a set of steel rods set about 1 inch apart to allow fines to pass through, while the coarse rocks will stay on top or slide off into the reject pile. An opening cut in the box allows the oversized rocks to just slide out. Some designs have this rock screen on top of the hopper. The steel rods or slats that screen the material have to be strong and durable so they can stand up to having material shoveled onto them (so don't buy little thin ones).

To run a highbanker, you need a system to pump water to it. The choice of the right pumps and hoses makes a difference. A 5 or 6 HP engine with a pump similar to the Keene 180 centrifugal pump should do. It's important to note that all pumps deliver less water as the highbanker location is moved uphill. Pumping uphill increases the resistance on the pump, and as you move uphill, eventually there is a point where the pump will not provide sufficient water to run the highbanker. I have a web page on my site about pumps, motors, hoses and other component equipment, but take a look at the pump / motor combinations which are used by the commercial makers - they will do for most situations.

The grade slope on the high bank sluice box is set up a little steeper than a normal hand fed river sluice box, and usually ranges from about three to four inches per foot. It's good to be able to adjust the drop so you need a range of possible settings on the adjustable support legs. Be sure to make the riffles easy to remove so that it will be no problem when it is time to clean the sluice out and recover the gold. The riffle set needs to be firmly held down with bolts or wedges to keep them from moving around. I recommend you build them from steel. If you build your riffles with steel, your welds don't have to be pretty; they just need to do the job. They just need to be sturdy and hold the riffles in place. As with all of these projects, the more tools and fabricating skills you possess, the more likely it is that your home made dredge project will be a success.

Above: Brand new high banker units ready to purchase for running your placer gravels.

Final Clean-up: Capturing the Gold Out of Black Sands

Every placer miner has to deal with it in one form or another and some locations have it far worse than others, but nearly every placer has at least some of it: black sand. It collects in our concentrates and gets in the way of recovering our fine gold. Some prospectors worry about getting every speck of gold and save it in buckets until they have tons - others take all the gold they can conveniently remove and then toss out the rest. So what is this mysterious stuff and what should we do about it? Let's take a look at black sands and the best ways to extract all the gold that is reasonable to get, and our options for recovering that gold.

The sand, rocks and gravels in the placer deposits we work have a density of about 2.5 to 3. Gold on the other hand, has a density of 19.3 when pure. That difference is what makes all the gravity methods of capturing gold work - whether it's a simple pan, or a sluice box, dredge, dry washer, table or jig, that big difference in density is what makes working with placers pretty easy. However, these gravity-based methods are not just specific to gold - they capture all the heavy materials in the gravel and concentrate them, including black sands. The frustrating thing about black sand is that while it is relatively easy to separate the normal sand and gravel from the gold, it much more difficult to separate the small sized gold from the black sand and other concentrates. There are many methods to accomplish this job - some a lot more efficient than others.

Black sands are heavy, which is why they collect in your pan, dry washer or sluice. They have density of about 5, meaning that those black particles are about 5 times heavier than water. Perhaps the first question about black sands is along the line of "what is this stuff?" Most of black sand is made up of two crystalline oxides of iron: magnetite and hematite. Hematite has a density of 5.3, and magnetite a density of 5.2. It's not surprising that two iron minerals dominate the concentrates, as iron is the fourth most common element in the earth's crust. Chemically, magnetite has a formula Fe3O4. Its crystalline structure places it in the spinel family of minerals. It melts at 1538 C and has a density of 5.2. Its structure makes it highly magnetic. Hematite is not as chemically resistant as Magnetite, and its chemical formula is Fe2O3. The little-known secret about hematite is that it is actually very slightly magnetic. Nothing like magnetite is, and most magnets will not lift it, but it can be picked up with certain extremely strong magnets and we will talk more about that important fact shortly.

Black sand, like the concentrates shown in these pans, is a challenge to many placer operators.

In addition to magnetite and hematite, black sands may also contain small amounts of a lot of different heavy minerals depending on the rock types in which you are working. Heavy concentrates may or may not contain at least some of the following: chromite (an ore of chromium), Diamonds, platinum, gold bearing amalgam (lost by early day miners), lead (from fishing weights and bullets), floured mercury, cinnabar (a natural mercury ore), sheelite (tungsten ore), zircon (a rare gem with a density of 4.7), garnets, rutile, ilmenite, pyrite and barite. The proportions that you may find of these minerals or man-made products depend on the specific location where you are prospecting. Some of these things, like diamonds, platinum and amalgam, are valuable in their own right. Mercury and amalgam may contaminate the gold you recover in your sluice box.

The amount of black sand and other heavy minerals, which you find associated with your placer gold, varies a lot - there is not a simple direct correlation. There are places that are loaded with black sand but have no gold. Conversely, there are also places that have good quantities of gold but very little black sand. It seems like at least a little black sand is found practically everywhere, so perhaps the next question is "where is all this black sand coming from?" As I noted, iron is the fourth most common element in the earth's crust, and some iron-bearing minerals are found in most rock types. They are especially common in metamorphic and igneous rocks. Since gold deposits are often associated with igneous and metamorphic rocks, it's easy to see why we nearly always find at least some black sands in our placers. So now that we know a bit about this stuff and where it comes from, what are we supposed to do to get the gold out of it?

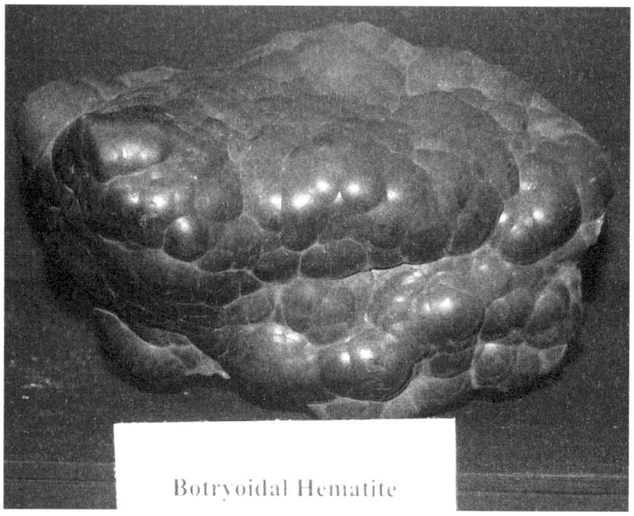

Tiny particles of Hematite, an oxide of iron, is one of the major components of black sand

Many new prospectors, believing that it somehow contains large quantities of "hidden" precious metals, find black sand mysterious and incomprehensible. Often this leads to the promotion of expensive and highly questionable treatments to "unlock" the "invisible" values of black sand - including "special" assays, strange leaching techniques, and "magic" fluxes. The truth is that most of the values in black sands are tiny free gold particles, which can be hard to recover by standard gravity recovery methods. Any values truly locked into the black sand, surrounded by minerals such as magnetite so that it cannot be recovered by normal gravity separation systems, are typically very small. The amount of any truly "locked in" gold can be scientifically determined by very meticulously recovering all the free gold possible from a sample of black sand, making sure to carefully examine it under 10x magnification to see that all the free gold has been removed, and then take that clean black sand and run a standard fire assay on it. Get an assay test before you invest money in any expensive treatments, because in nearly all cases, this assay will show it is not economically worthwhile to pursue any further treatment of the black sands, especially for the small operator. The answer as to how to best deal with black sand depends on how much you are generating. If you are an individual prospector who dredges or dry washes during weekends or while on vacation, you might generate 20 to 200 pounds of the stuff each year. If you just snipe with a pan or use a hand operated sluice box, the amounts will be even less. On the other hand, if you are a large commercial operator, running a dozer or front end loader to feed your sluice, you may produce tons of black sand and the use of more sophisticated equipment such as shaker tables will be justified. Let's take a look at both possibilities:

Processing Small Amounts of Black Sand: This is the category into which most prospectors and weekend or seasonal miners fall - including me. I used to spend a week or two per year dredging, dry washing and sluicing. I don't dredge since it was outlawed in California, but the considering the combination of all of my prospecting, I generate something around 20 to 40 pounds per year of black sands that need to be processed. Here are the techniques I use to go from concentrates fresh out of the sluice box, all the way to clean gold. At the end of each dredging day, I clean up the front section of my 3 stage sluice box. About every other day I clean out rear top section, and about once per week I clean out the undercurrent section of the sluice box. Many new miners, worrying about loss of gold, clean up too often and this takes away from the time they could be dredging and finding more gold. Sluice boxes on newer dredges are well designed, and if your sluice is set up so that the gravel is not accumulating so much that it buries the rifles nor running so fast that it sweeps the riffles clean, then once a day clean up is all you should need on most commercial dredges. When operating properly, your sluice will have about one inch of carpet showing between the gravel of the middle riffle of the sluice and the one below it.

After dredging all day, I empty off my sluice into a large plastic wash tub. I then screen those concentrates using a quarter inch screen to get rid of the small rocks. Any gold that fails to go through a quarter inch screen is a nice nugget. After a quick sort to find any coarse nuggets, the reject rocks and pebbles are discarded. The next step is to pan through the remaining concentrates. The point of this step is to remove the light sands which are in those concentrates. Once I have panned through the fines so that the lighter sand is removed, and I take out any pickers, flakes and small nuggets large enough to be worth removing by hand. After all these larger pieces have been removed, all that remains is a small amount of black sand, which contains all the fine sized gold and other heavy mineral concentrates. I then put these concentrates into a container - lately I am using the new plastic coffee cans. Don't use metal cans to store wet black sand as the cans rust. This will cause the black sand to harden up into a cemented, solid mass. I save and accumulate this fine black sand and gold concentrate all season, waiting to process it all at once at the end of the year.

This is a blue bowl concentrator, an inexpensive and simple to operate device for separating fine gold from black sands. It runs on the flow of water from my garden hose. This one belongs to me and I use it to process the black sands I recover from dredging and sluicing.

Spiral Pans that spin and collect the gold toward the center are also popular. They work just fine, but are a little more expensive.

Dry washing concentrates are processed in a similar manner, depending on where I am camping. In locations where it is easier to haul all the concentrates around than it is to haul all the water needed to process them, I sometimes simply take all my dry wash concentrates home and process them in a tub at the house. The clean black sand and fine gold is added to the black sand fines obtained by dredging, sniping, or sluicing for final processing at the end of the prospecting season.

At the end of the prospecting year (which for me is late October or early November), I have accumulated something between 20 and 40 pounds of black sand concentrate. In my final clean up, I use a blue bowl hooked up to my garden hose. Careful screening is the key to success with the blue bowl. First I screen the concentrates with a 30 mesh screen. I pan down the coarse matter that is larger than 30 mesh, until there is only a few cups of this material. The minus 30 mesh material is then processed in the blue bowl. Processing this material in the blue bowl is not fast, but it does work. You can relax while you are doing it as not much attention is required once it is set up. I will add about a cup of black sand, then run the bowl for a few minutes, shut down the water flow and add another cup or so, etc. until I have gone through the entire amount of -30 mesh black sand. I do not take the material down to pure gold in the blue bowl, but it does remove the majority of the black sand. At this point, I am normally left with a cup or so of coarse material I've panned down, and about 2 cups of the minus 30 mesh fine gold and black sand material after it was processed through the blue bowl.

The next step requires a very strong magnet - you will need one of the new rare earth super magnets, and a good sized one with about 50 pounds or more of pull. Now your normal horseshoe magnet made of iron alloys simply will not do for this. The magnets sold by the most of the major prospecting shops to separate out black sands will not do. Those old style magnets will pick up magnetite, but not hematite. The new powerful rare earth super magnets are necessary because they will pick up the hematite, which is only very slightly magnetic. The old magnets will pick up around half of the black sand, but the super magnets will typically remove more than 95% of it - and that makes a big difference! These magnets are so strong that you need to be careful. You can get your fingers pinched and smacked as these super powerful magnets attract steel objects from surprising distances and with great speed. Put the fine sized concentrates in a plastic pan with about an inch or two of water - and spread them out into a thin, even layer. Carefully pass the magnet over the black sand from about a half inch above the black sand concentrates. On this first pass, do not touch or stir the concentrates with the magnet. The trick is to make the magnetite and hematite jump up through the water to the magnet leaving the gold behind. Remove the black sand that sticks to the magnet and repeat. Eventually, when nearly all the magnetic minerals are removed, you can stir the magnet through the remaining sand in order to pick up as much of the material as possible. This is the part of the process where you need to examine the material on the magnet very closely as a few particles of gold may be attached to magnetite and stick to the magnet. Repeat the process until no more magnetic minerals can be removed. You can examine the material you removed in the last part of the process with a magnifying glass to see how well you did. I normally see a few tiny specs of gold may only be less than a dime's worth, so there is no great need to worry too much about them. This process is done separately for both the 30+ mesh size and the -30 mesh size. The processing of the coarse material with the magnet typically goes much faster. The next step is to dry both fractions separately.

After removal of all magnetic minerals, only a few items normally remain. Gold, a few bits of lead, a garnet crystal or two, etc., may remain in the 30+ mesh size. This fraction is easily sorted with your fingers when dry on a white sheet of paper. The trash is discarded and the gold, plus any other valuables like platinum pieces are saved. For more information on platinum placers, see the chapter on platinum in the appendix. The smaller fraction may have some non-magnetic black sand (commonly Chromite in many parts of California) and some grey-white heavy sand of a very fine size (commonly zircon). When dry these can also be separated on a white sheet of paper - I normally use two sheets and with a little careful blowing to separate the gold from these smaller final concentrates. With care and a little effort, using this technique you can produce a clean gold product ready for sale or shipment to a refiner. By the way, that 20 to 40 pounds of black sand typically yields at least 2 or 3 pennyweight of fine gold, an amount well worth my time and effort at today's current high prices. There are of course, a number of ways to go about doing this final cleanup. Some folks use mercury to get the small gold. I have done this in the past, but with the advent of the super magnet and the method

A large commercial operation like this bucket line dredge will generate very large and substantial amounts of black sand.

outlined above, in my opinion, mercury is simply no longer necessary for the individual prospector. I like to stay away from the use of mercury, because the regulatory agencies are so crazy about it. All in all, it is faster not to use it, and the few specks I loose with the magnet are hardly worth my time. There are spiral panning wheels, micro sluices, tables and other devices for final clean up, they can do the job and all have their proponents, but I use a blue bowl, mostly because it is cheap and it works. Sometimes (actually fairly often) my gold needs to be treated to remove mercury stains that either happen in the river or during recovery and processing. I prefer to use nitric acid for this, but some guys also remove mercury by heating. Mercury vapors are toxic, so if you do this with heat, be sure you do it outside and be careful. Also be sure all the lead has been removed before heating, or you will have to use nitric acid to separate the melted lead from the gold.

Handling Commercial Amounts of Black Sand: Even small commercial operations may generate more than 50 pounds of black sand per day, and because of this, large scale processing equipment beyond the scope of the individual prospector is normally justified. These operators commonly make use of shaking tables, large spiral wheels, mercury amalgamation and other commercial equipment to handle these larger volumes of black sand concentrates. It is always important for the commercial operator to keep close tabs on his recovery equipment and do regular testing to make sure it is working properly. If this is not done, the operator may well be losing significant fine gold in the black sand. During the early 1980's I had the privilege to inspect and evaluate the operations of St. Joe Minerals' bucket line dredge operation at the Yuba Goldfields near Yuba City, California. At the time, they delivered their entire finished black sand product to the property owner after they had extracted all the gold that they believed was present in the material. It turned out that they were not keeping good controls on their processing. The property owner simply had one of his employees re-run the black sand over a finishing style shaker table, and he recovered additional gold, in the amount

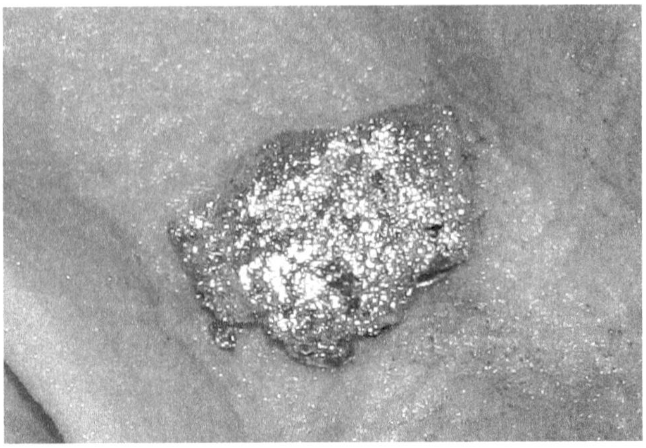

Small chunks of amalgam like this can still hold significant gold. Most prospectors, however, have moved away from using mercury.

of about 3 ounces per day. So don't forget that when you are dealing with commercial qualities of black sand it pays to stay on top of things and be very careful.

Large operators who generate many tons of black sand each year may wish to consider additional steps beyond tables or spiral wheels such as reprocessing after crushing the black sands or leach extraction by cyanide. However, these options should only be considered after assaying the finished tailing material to see if cost and effort of any additional processing is justified. With the continued regulatory concern over mercury, it has lost much of its traditional place as a method of recovering fine gold, but it is still used, especially in less economically developed countries. For commercial operators in the US, mercury can be a significant environmental liability, as even comparatively small spills may be quite expensive to clean up.

Amalgamation has been used for thousands of years to recover small sized or fine gold. Historically, it was used in many operations to increase gold recovery. In spite of this I recommend against using it in almost all applications. Especially for the small prospector, there are many better options available for separating your fine gold from black sand. Many prospecting shops no longer even sell mercury. In spite of its potential danger, I am presenting some information about using mercury to recover fine gold because of its historical significance and general interest to the individual prospector and miner. If you decide to use mercury with your prospecting operations, it is mandatory that you take care to learn all the environmental regulations, and necessary safety precautions. Follow them carefully in every step of your work, because mercury can be dangerous if not handled properly. Basically, amalgamation is the practice of bringing free gold particles into contact with mercury. When clean gold comes into contact with mercury, the two substances mix to form a compound called amalgam - an amalgam is simply an alloy of gold and mercury. The gold literally is dissolved into the mercury. This allows the collection of very small sized particles of gold. At the end of the operation the mercury and gold amalgam are collected after which the two are separated, and the mercury recycled. Now larger gold and nuggets will not be completely converted (except over a very long time) and only a thin coating of amalgam forms. However, large nuggets should not be captured with mercury; it should only be used in recovering the fine sized gold. Personally, I strongly urge individual prospectors to use techniques other than mercury to recover the fine gold. I just find the alternatives safer and easier. Environmental regulators go nuts if they find the stuff in a public place. Mercury is still used by some small prospectors, and if you decide that you want to use mercury for yourself, I recommend further research on the topic of amalgamation. Be sure to research mercury use in chemistry books, safely hazard references, environmental regulations and mining reference texts. You must take all the necessary precautions. The use of mercury involves some inherent risk to your health if not handled properly. In the final step, the mercury is distilled in a retort in order to recover the gold and mercury. The fumes are toxic, and mercury chemicals can be absorbed through the skin. Once mercury is absorbed into the body it is retained for a long period of time and is not excreted out. It attacks the central nervous system. I used to use mercury to recover my fine gold, but the advent of the rare earth super magnet has convinced me it's unnecessary to use mercury on my scale of prospecting efforts to separate my beautiful placer gold from those nasty black sands. For more information on this topic, see my website page on using mercury.

Selling Your Gold To Refiners, Specimen Collectors and Tourists:

Everyone knows that gold is a valuable commodity. Even small amounts can bring some good money. But that is just the question: Just how do you get the best money when it comes time to sell your gold? That is what we are going to discuss here. Almost half the battle in getting the best price for gold is not being in a rush when it comes time to sell. A lot of guys save up their gold, but then a crisis hits and they are forced to sell at whatever price they can get.

Beautiful sprays of natural gold like this specimen sell for far more than the value of the metallic gold they contain.

Unfortunately, the emergency "I have to sell right now" price is not normally as good as the price you will get if you can be choosy and wait till the right buyer comes along. So remember that one of the most important things about getting the best price for your gold is good planning. Make sure you have enough money in the bank to cover all those little emergencies and unexpected things

Nuggets like this crystallized specimen piece from Rye Patch, Nevada, command a high price.

that come along. If you don't have a safe reserve of money in the bank, you ought to sell some gold until you reach that point.

Many miners strongly prefer to keep their gold as a stash of money. However, as I've noted, unlike a bank, you can't just go out and make an instant ATM withdrawal as you can from your account at the local Savings and Loan. In addition to being a stash of money, the gold you keep becomes a trophy of the hunt, filled with great memories of good or difficult times and the efforts involved in finding and collecting the gold. I know that for me, the gold I save is filled with great memories of all my experiences prospecting out in the field - the desert, the pines, the mountains. For me, I never want to sell it. It would have to be a very dire financial emergency for me to decide that I need to go out and sell my gold - especially the nicer pieces.

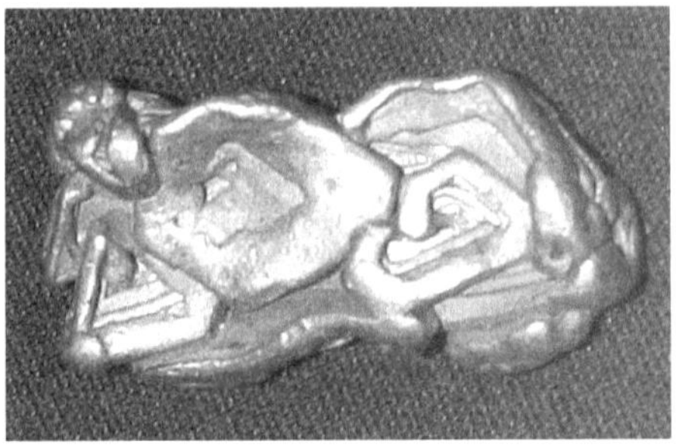

Here is another fine crystalline gold piece that would command a high specimen price.

There are a lot of different forms of natural gold and of course some of them will bring you better prices than others. Beautiful crystalline and leaf gold specimens as well as large nuggets will bring premium prices, so it's important to understand and know what kind of gold you have. Let's identify a few of the different classes of gold you might find, with the purposes of selling in mind:

1. Small flakes and dust.

2. Small nuggets at least two grains in size.

3. Medium sized nuggets, one pennyweight to 1 ounce.

4. Large nuggets greater than an ounce in size.

5. Good-sized gold Quartz specimens.

6. Crystalline gold specimens

Each of these categories has its own special market that it is best suited for. A real big part of getting the best price for your gold is knowing what markets are available for what you have and what appropriate prices are. Part of the problem for me in writing this chapter is that I know the price of gold will change, and so if I recommend a certain price for your gold, and the price changes, it will make what I recommend simply wrong. So for purposes of reference in this chapter, the price of gold as I am writing is about $850 per ounce. You can re-figure things based on the difference between that and the price of gold when you are reading this chapter. There is a significant collector market for nice gold nuggets. In order to be selling your gold to collectors, you need to have something special and collectible, not just gold dust and tiny flakes. This usually means at least medium sized nuggets, or some special crystalline gold specimens. Tourists and some casual buyers may be interested in smaller nuggets, but we'll consider that possibility a little further on in this chapter. You may even try mineral specimen shops - they sometimes may make you an offer, especially if your gold is crystalline or has other desirable properties for mineral collectors. Finding and connecting with collectors has always been the problem with getting a premium price when selling gold or other mineral specimens. The Internet of course, is the great equalizer and there are several prospectors who are regularly marketing their gold through Internet websites where they can get full retail prices. Creating and maintaining a website is a whole additional proposition, but if it is your cup of tea, then it is a

Small nuggets like these find a ready market for a variety of purposes, mostly related to making jewelry that features natural gold nuggets.

great opportunity to reach and market your gold to people who are interested in buying your gold at a reasonable price. Nuggets of a pennyweight and more up to about half an ounce, should command a price at least 35% greater than the gold weight. Nuggets of a half-ounce or better should command a 50 to 80% premium. Nuggets greater than an ounce command a larger premium, but is variable depending on how spectacular the piece is, but the premium is often in the range of 100% (double the gold price) and up. These prices are figured for average pieces which are not crystalline nor reflect any special shapes or other notable characteristics. Also remember that for American buyers, American gold pieces are rare and therefore command a premium price.

Jewelers will use smaller to medium sized nuggets in making natural nugget jewelry. In some cases they will also use gold quartz material that has been cut up and polished into gemstones. In order to market to jewelers, you will need these types of materials. Not all will be very interested in what you have, and may only make a low grade offer to buy your gold cheaply. Try to look for jewelers who offer or even specialize in natural nugget jewelry - these are the ones who are most likely to know the true value of your gold and offer you a fair price if they are looking for that type of gold. Some of the most valuable gold around is found in crystalline gold specimens. Some have little trigon shaped crystals, but leaf and wire forms are also well known. These pieces can easily sell for five times the value of the gold contained in the nugget. Be sure that if you have nice crystalline gold that you do not let such beautiful items go cheaply out of your hands. They are rare finds in most gold districts, and therefore should command a premium price for their sale. Of course the value of a specimen is always in the eye of the beholder, and it may take some time to market these special high dollar pieces, but it is worth it. For more information on crystalline gold, see the minerals chapter in this book.

There are also gold containing specimen nuggets and attractive quartz specimens. Quartz pieces containing a high percentage of gold are sometimes sold simply as nuggets. Quartz pieces containing only a very small percentage of gold are perhaps best crushed and the gold extracted. The experienced prospector learns the best marketing potential of his finds. I have prepared a separate webpage on my internet site that goes over the procedures for handling specimens as well as crushing them.

Gold Bearing quartz can be made into beautiful jewelry. This attractive jewelry pendant is set in gold and surrounded by emeralds. This gold quartz was recovered from a mine near Grass Valley, California.

Quartz with a medium amount of gold can be cut up to produce slabs to be made into gemstones for use in gold-quartz jewelry. The goal should be to produce material that has roughly 20 to 40 percent of the surface covered with gold. This of course requires a thin kerf diamond bladed saw set up for cutting hard materials like quartz. Such equipment is commonly available at many rock hound shops, as they can also be used to cut gemstones as well.

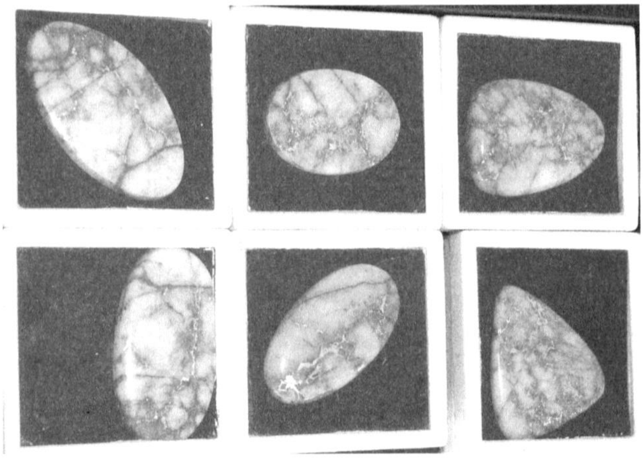

Gold quartz can be made into some striking gemstones. These gems were very popular in the late 1800s, and are still desirable today.

Finally, quartz with significant gold content that is crystalline or leaf shape within the quartz matrix can sometimes be etched or treated to produce beautiful gold quartz specimens. There is an entire art to treating gold quartz with hydrofluoric acid to produce these spectacular pieces. Hydrofluoric acid is a dangerous material that is capable of dissolving glass and quartz. It will also destroy your bones and the tissues of your body. Even the vapors are hazardous and serious precautions are necessary to use it safely. Pieces of gold quartz are usually sawn with diamond blades of the type mentioned above to produce a piece of the proper size and shape that the artist wishes to develop into an attractive gold quartz specimen. Hydrofluoric acid treatment together with chipping and further sawing are used until the piece presents the proper level of beauty. In many ways this art is a trial and error process, and not every piece works out perfectly as planned. For those who would like to delve more deeply into this subject, I have to state that it is beyond the scope of this book and further research would be required.

Getting premium prices for gold dust and small flakes takes a little more work. In historic gold mining areas where there is significant tourist traffic, you can usually do well selling small nuggets and flakes to tourists. Your placer gold is weighed into small lots and put into vials filled with water. With this type of setup, you should charge at least double the spot price of your placer. The difficulty here is in finding a spot where you can easily sell to tourists. If you have to sit all day and wait for people to walk by, you're not really getting much back for the time you're investing by sitting there and selling.

Another possibility for marketing fine gold at a premium price is to sell "unprocessed" gravel on ebay or other locations, and allow people to pan out their own real gold. Some operators like Felix Paydirt of Alaska have created an entire business out of this concept. Felix Paydirt actually advertises in magazines, and sells this type of product through the mail. Some folks just have no access to real gold bearing gravel and want to see what it is like. To make up this kind of material for yourself, you would select a spot where you know there is some gold present in the gravel. Dig up some of this material and screen it to remove the larger rocks and pebbles - anything larger than a bean. Then weigh out a fixed certain amount of the gravel, say about 1 pound. Add an amount of gold flakes and dust, perhaps four dollars worth if you are going to be selling the finished bag for $10. Some operators produce a “large size” lot of gold bearing material with 15 to 100 pounds of gravel product and more. The bag of gold bearing gravel plus the small amount of fine gold and flakes is sold as being unprocessed gravel, guaranteed to contain gold. In this way, because you used gravel that you knew was gold bearing there probably will be a bit more gold than you put in and exactly how much will be determined by the buyer who purchases your bag of gold bearing gravel. This type of product is also very popular on eBay. Additionally, this is another possibility where a website might be helpful in reaching the market that you are trying to sell to.

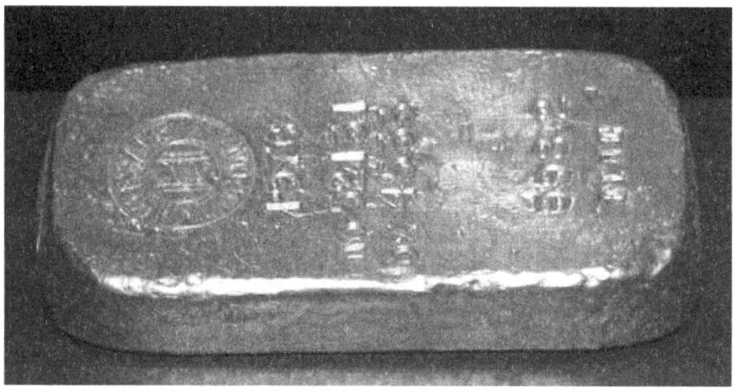

Refiners take raw gold products and purify them, turning them into bars or other forms of bullion and them selling that material to those who fabricate gold products like jewelry.

Using An Industrial Precious Metals Refiner: Last, but certainly not least, is the possibility of sending your gold off to a refiner. A refiner will not pay you anything extra for beautiful specimens or nuggets; they will simply pay you for the inherent scrap value of your gold metal. As a result the best things to send them are fine flakes and dust as well as scrap or amalgam sponge. Also, because of fixed fees and minimum charges, it can be expensive to send small amounts of gold to a refiner. Most refiners will not accept amalgam that has not had the mercury removed. The refiner is in business to make money for his company, he is not in business to help you. Most refiners are honest enough and will provide you the opportunity for your material to be tested by umpire assay if there is any dispute. Most miners who have experienced frustration in their dealings with refiners are disappointed by the results because they did not have a good handle on the purity of the gold they were selling. While many placer miners will claim that their nuggets are 90 or 95% pure, that is normally a highly unusual case. Overall impurities of 75% to 85% are far more common. If your gold has any significant quartz or iron oxides within it, the percentages may be even lower. It is not uncommon that small miners feel cheated or at least heavily confused by the results of the refiners assay. There are some good things about using a refiner but there are also some serious disadvantages as well. With the refiner you know that you will get paid on a fixed schedule. However, once you ship the gold, your material is basically out of your hands. You must trust that the refiner will deal with you honestly and process your material, giving you your full due. When you are dealing with a refiner, be sure to learn all the rules and requirements for shipping and policies that the refiner enforces. Package and ship your materials according to their policies. It is worthwhile to check out the refiners website or ask around to people you know about their experiences with that refiner. When your material begins the refining process, it will first be logged in to their records and then melted down into a single bar or button. An assay will be taken of that material and you will be notified of the weight of your material and the result of the refiners assay. With most refiners you have a certain period of time in which to respond and/or possibly dispute the result of their assay. If you dispute it, a second assay will be taken and run by a different laboratory and the results of the two checked against each other. Refiners have different rules for how any discrepancies between the two assays are dealt with. Once you agree with their weight and assay results, a refiner will pay you based on his schedules. Fees are charged for the assay and refining processes and these are deducted from any payment you are to receive. Fast payment for your gold usually also incurs additional fees, so if you can wait a little bit and be patient, you may make a little more money on your gold. While you are accumulating your gold, it is important to carefully and safely store the gold you find until you sell it. It is not safe to have large amounts of gold just sitting around your house in a box or a jar - it's just too valuable. Many prospectors and miners have had their nuggets stolen because they failed to keep their gold in a

Never send large nuggets to a refiner unless you are in dire need of quick cash – you will be paid only for the gold value, and nuggets are worth more than spot values.

safe location. Gold is valuable, and it attracts thieves. If you have more than an ounce or two of gold around the house, you really should purchase a safety deposit box in the vault of your local bank and make arrangements to keep your gold in the safety of their secure vault.

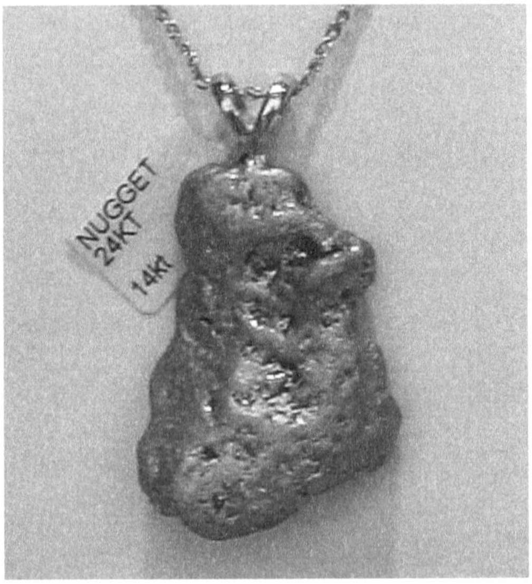

A loop bail is easily soldered to the top of a nugget, and presto, it's a valuable pendant! Though marked 24K it is more likely about 21K.

Making Up Your Own Nugget Jewelry:
Another popular way to get the greatest dollars out of your nuggets is to turn the larger ones into jewelry. Well-designed natural gold jewelry can be extremely beautiful and it is popular with many folks, both men and women. Tie tacks, earrings and pendants are simple to do. Rings and some other jewelry can be a bit more complex. With the pendants, all that's necessary is to carefully solder on a jump ring or other type of jewelry bail onto the nugget. These sorts of mountings are available from a number of commercial jewelers, as is the low temperature gold solder needed to attach them. Tie-tacks and earrings are created by attaching the nuggets to pre-made pads on tie tack or earring settings. Rings paved with small nuggets are a bit more difficult, but still are a common project.

Although it takes some practice and skill to learn how to do this, soldering is really not that difficult, you just need to learn the techniques of using the torch. The solders melt at a temperature much less than the nuggets. Classes in this type of jewelry work are offered through the art departments of many colleges, including community colleges that are easy to get into. Torch supplies, fluxes and cleaners are available from many of the same jewelry supply firms that sell the settings. Some prospectors end up really getting into this aspect of the craft.

Melting Down Your Gold:
Many prospectors acquire fine sized placer gold in one form or another. Fines from panning, sluicing or dredging, fine gold from amalgamation, electro-winning or other chemical process, or even scrap gold collected from metal detecting in parks are all categories which prospectors may wish to melt down to pour into a solid bar or button form for storage or sale to a refiner. Melting down gold is not particularly difficult, but there are a few things that it is helpful to be aware of concerning this procedure, so that you do not get into problems.

Many natural forms of gold contain impurities, so a flux is required to collect them and remove them from the gold. The more impurities which are present, the more the amount of flux that is required. Some small amount of flux should be used even when the material appears clean. Fluxes also help to hold together fine gold particles and keep the torch from blowing them away. Never indiscriminately bring a torch down onto fine powdered gold in a crucible as you can easily blow hundreds of dollars' worth of gold right out of the crucible. Wind speeds from a torch in a crucible can easily blow tiny gold around, so be sure to add a protective layer of pure flux, such as borax, on top of the fine gold flux mix. Then bring the torch down slowly, allowing the torch to slowly and gently melt the top flux layer so it will spread out and protect the gold bearing layer below from blowing out of the crucible. In addition, heating the crucible too fast may also damage it by cracking, so take it easy and don't go too fast in your heating. Pure gold melts at 1060 C or 1940 F, and this is a good bright red heat. More or less of other metals alloyed with your gold such as silver or copper may raise or lower the melting point a bit. 14 K Gold, varies because there are many 14 K gold alloys, but a typical melting temperature for 14 K is about 1650 F. A fairly powerful torch, with an oxy-acetylene or MAP gas

flame will be required. Melting can also be done in a furnace, so long as the requisite hot temperatures can be reached. A propane torch like those sold at hardware stores will not do the job. Be sure to do you work in a well-ventilated area. Some gold is contaminated by mercury, mostly by man, but some is natural. You don't want to breathe those mercury vapors.
There are many formulas for flux mixtures, but for simple melting of clean gold fines, a mixture of borax and sodium carbonate works fine. In some cases, a small amount of very fine silica sand can also be added, but that is usually unnecessary. I would add about two big pinches per ounce with clean jewelry scrap, more for dirty scrap. I'd use 4 pinches with each ounce of clean placer scrap, more with dirty placer materials. Any cover layer of flux you might need would be in addition to that. Also, don't add any lead compounds as would be added for assay fluxes. This is not an assay, so lead is not needed.

Molten gold can be poured into various molds.

The crucibles used for melting your gold must stand up to the fluxes being used, and crucibles made of Graphite carbon are the most common type used. Borax fluxes can dissolve and eat through the walls of many ceramic type crucibles and you don't want to spring a hole in the side of your crucible when your gold is molten and red hot. That situation could be both very costly and highly dangerous! Graphite crucibles of the right size are available at many rock shops which have supplies for jewelry casting. Folks who melt metals to cast gold and silver jewelry use the same types of crucibles and fluxes as you will want to use. There are many possibilities for shapes and molds to pour your gold into. The simplest are open molds such as brick or bar type molds. Some folks use the depression of a fire assay cupel to create small buttons. The cupels will absorb any excess flux which was used, and are just the right size for pouring buttons.

Lost Wax Casting:
It is even possible to use the lost wax casting method to pour your melted gold into various shapes, or to make jewelry. Things like little miners, eagles, gold picks made of real gold are all possibilities - and of course the more standard jewelry products like rings are possible as well. The art of lost wax casting is a whole field of study in and of itself, and has been used for many centuries. A wax model of the item to be cast is made and then a sprue (one or more) is attached and the whole thing set in a special type of plaster of Paris. The wax model is set such that the sprue comes up to the surface. When the plaster and wax set up is heated, the wax melts and vaporizes, leaving a void the shape of the model. Gold or silver is poured into the sprue hole such that it flows down and fills the void in the plaster, making a gold or silver casting of the original wax model. Even today, this is still the technique by which most gold jewelry is made.

High purity gold, such as placer fines should not be cast or used in jewelry without the addition of copper and silver to increase hardness and durability. In addition, it is necessary with most raw gold mixes to add extra flux during melting to clean, capture and remove impurities. Any gold alloys which are used in jewelry and other lost wax projects must have good flow properties so that they can be properly cast. I have done a little bit of this type of casting myself, but as I noted, it's a whole art form unto itself, and requires a whole set of tools and equipment, knowledge, etc. Some photos of the process and more detailed information are available on my website if you would like to investigate and learn more about jewelry making and casting, but there are also many good books written on this very interesting topic.

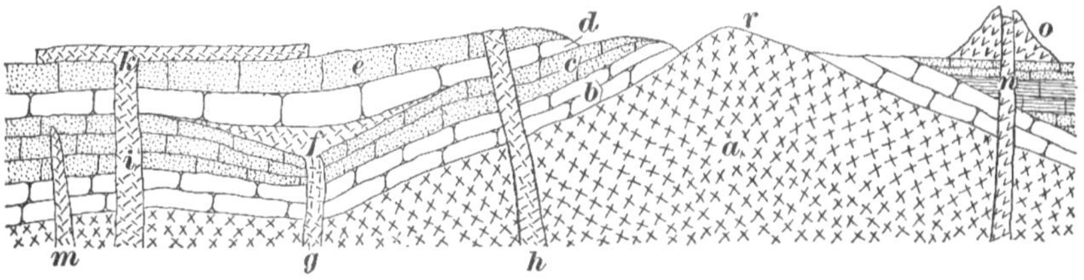

PART III -BASIC GEOLOGY FOR PROSPECTORS

Understanding Minerals:

Many individual prospectors would like to know a little more about the geology of gold deposits. There is no doubt that in the final analysis, no matter how well you operate your dredge, metal detector or other prospecting equipment, unless you can find the places where gold is located, your equipment cannot help you recover it. Prospectors know they need that background knowledge, but it can be almost impossible to find the right kind of information. The "how to" prospect books commonly offer a few words on geology, but it is very difficult to find good information that goes beyond the most basic beginner level, yet is oriented to the individual prospector. This portion of the book will give you what you need to know, without too much unnecessary extra material.

I don't want to be too technical, but I want to thoroughly cover the basics of geology with an eye toward the individual prospector that is principally interested in precious metals like gold and silver. We will touch on other metals and gemstones to a lesser extent. We will first cover the topic of valuable minerals and then take a look at the geologic formation of common rock types. After that, we will go into a little about mountain building processes and the plate tectonics forces that cause so many of them. We will then take what we know about these topics and take a look at how hard rock gold deposits form, and finally how placer deposits form from these hard rock deposits.

Basic geology texts in use at local colleges spend little time on the subject of economic geology - deposits of valuable minerals like gold. My son's college introduction to geology class text had all of one little paragraph to cover all types of placer deposits - gold, gemstones, everything. There are also some advanced textbooks on gold deposits, but these are really directed at the professional geologist, reaching far over the head of most individual prospectors.

Misunderstandings of how and where gold deposits form can lead to wasted time prospecting, and a view that gold could be anywhere (and who knows where that might be). The often-repeated saw that "gold is where you find it" does not really mean that the location of gold is some great incomprehensible mystery that can never be fathomed. Yes, sometimes you have to really think about why gold is found in a certain location, but a better understanding of geology and mineral occurrences can be of a significant benefit to the prospector. You don't need to have a PhD to be successful, but it does take some learning and a significant effort. I honestly think that this is one of the important reasons why lots of folks get interested in prospecting, but then get discouraged.

The characteristic texture with concentric circles and bright green color identify this mineral as malachite, a secondary copper carbonate mineral that is commonly associated with gold veins and other precious metal deposits at some locations.

So what should the prospector know about minerals? There are thousands of identified and named minerals, and it's certainly not necessary to know a large number of these, but it would be worthwhile for the individual prospector to know and recognize a few dozen of the most common rock forming minerals and a few other types which are commonly associated with gold or other valuable deposits.

Basically, a mineral is a stable combination of elements (sometimes just a single element) that is found naturally in the earth. When studying minerals, geologists define them scientifically as materials which possess all of the following special characteristics. So minerals are: 1) Naturally occurring; 2) Inorganic (not the product of a living organism); 3) a solid material with a definite chemical composition (though some chemical variability does occur) and 4) Have an orderly internal crystal structure (although there are some exceptions, including things like opal).

POCKET LENS.

A magnification lens is a handy tool for identifying minerals in the field – one every prospector should own.

Minerals are formed when the elements (chemicals) that make them up come together in the earth's crust to form a stable substance as the result of some geologic process. Minerals tend to be the most stable forms of the various possible different combinations of these chemicals at the temperatures, pressures and other conditions present at the time of formation. There are a huge variety of geologic processes which form minerals, from the cooling of molten masses to circulating solutions of water that may cool and form crystals. When the geologic process forming the minerals is slow, it sometimes allows the growth of large crystals, but when cooling or the change in the geologic conditions is fast, crystals are usually small. Some minerals show a change in crystallization, so one can recognize what temperatures they formed at. Silicon dioxide crystallizes as quartz, but at higher temperatures, it crystallizes in different forms - same chemical makeup - but a different crystal form. The two high temperature forms are called cristobalite and tridymite. When a geologist finds cristobalite, he knows the range of temperatures the rock must have formed at because below a certain temperature or above it, cristobalite won't form (it will either be quartz if cooler or tridymite if hotter).

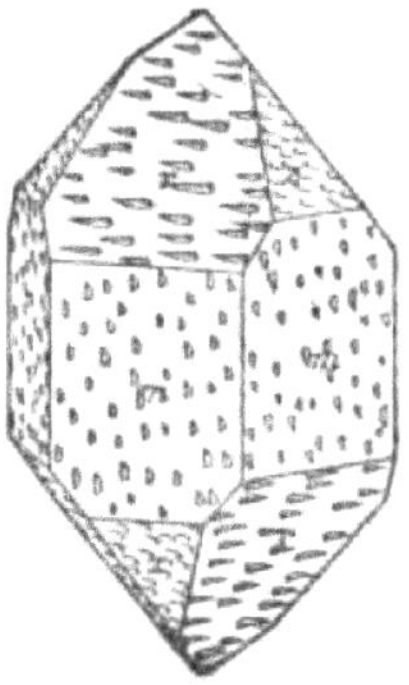

The common six sided quartz crystal is perhaps the most common mineral crystal the prospector will come across. Most grow from a base of vein quartz or other rock and are not terminated at both ends as this example is.

Because minerals are naturally occurring, the elements which are the most common in the earth's crust tend to make up the most common rock forming minerals. Since the two most common elements in the crust are silicon and oxygen, it is no surprise that the mineral world is dominated by silicates, minerals that contain silicon and oxygen. Iron and aluminum are the next most common, and these are commonly found in many rock forming minerals as well. Water also appears in the makeup of certain minerals. The relative abundance of other elements in various minerals depends in part on their percentage of the earth's crust. The most common basic rock forming minerals are quartz, feldspars, micas, clays, amphiboles, pyroxenes and calcite (all are silicates except calcite). These seven mineral groups are made of common elements and form the bulk of most of the rocks found in the earth's crust. These very common minerals are among those the prospector should learn to recognize. For elements that occur in only tiny quantities in the earth's crust, minerals containing them only occur where some special geologic process has produced an unusual concentration of these elements. These concentrations of the less common metals such as gold, silver, copper or platinum are the kinds of deposits that we seek as prospectors.

The prospector wants to focus his energies on finding valuable minerals that can be sold or processed for a profit. Valuable minerals also include gemstones such as diamonds, emeralds, tourmaline and others, as well as metal bearing minerals that may contain gold, silver, copper or other metals. Many of the most sought after metals occur as sulfides, or more rarely as tellurides. Sulfides often have a metallic type of appearance and are present in many valuable ores. Most gold - quartz veins have at least some sulfides like pyrite (an iron sulfide) present. Other sulfides like acanthite (silver), galena (lead), chalcopyrite (copper), arsenopyrite (arsenic), sphalerite (zinc), cinnabar (mercury) and stibnite (antimony) can also form depending on the chemistry of the rocks present and the chemistry of the mineral solutions. Sulfides are an important group of minerals to the independent prospector, and one group he would want to be familiar with. So what is the difference between a mineral and an ore? Well, ores are mineral matter that can be mined and processed at a profit, producing some valuable commodity. If the valuable product cannot be extracted at a profit, the stuff may be mineralized rock, but its not ore. Valuable minerals can come in many forms including a wide variety of metal containing minerals and even gemstones as well.

Minerals can be broken down into a number of different classes for study, but one of the most common is primary vs. secondary origin. This breakdown is most commonly used when speaking about ore minerals. Primary minerals are the minerals formed directly by geologic processes in the earth's crust. This is often some sort of circulating heated water that will form a primary deposit as the waters cool or the pressure containing these fluids is reduced and the

This ore specimen from the rich silver mines of Virginia City, Nevada contains many complex dark colored silver bearing sulfide minerals in a gangue of quartz and adularia feldspar.

minerals in the solutions drop out to form various minerals. Examples of primary minerals include metallic gold, silver, or electrum (a natural combination of metallic gold and silver), various sulfides and even tellurides. Silver most often occurs as silver sulfides, or as a minor element within other sulfides, such as galena or sphalerite. Gold most often occurs as a native metal, though it can be found combined in the form of tellurides, or as tiny bits of metal within pyrites or other sulfides.

Secondary minerals are formed by the weathering action of air and water on primary minerals. Minerals of secondary origin are an important class of ore as the weathering can concentrate the metals or other valuable constituents of the primary ores producing ores of much higher grade than the original deposit. Weathering partly (or completely) dissolves certain elements and can re-deposit them as new minerals. This process can form rich concentrations near the water table, and below oxidized and leached surface deposits. An example of this geologic process is the iron oxide deposits called gossans, but a gossan is not necessary for secondary minerals to form. For base metals, the secondary minerals are different than the primary minerals. The secondary minerals are other types of things like carbonates, oxides or sulfates. Primary lead often forms as galena, but secondary lead minerals are things like cerrusite (lead carbonate) or angelsite (lead sulfate). Primary copper forms as a variety of sulfides including chalcopyrite, but secondary copper minerals are things like chrysocolla (a silicate) or azurite and malachite (both carbonates). The most common secondary zinc mineral is smithsonite. With Silver, one might see chlorides or bromides as secondary minerals.

Gold is difficult to dissolve, but where the conditions are right, secondary metallic gold, usually fine grained, is the most common form. Secondary gold enrichment requires the presence of manganese, sulfides, chlorides and a semi-arid environment. The sulfides oxidize in the presence of air and water to form sulfuric acid. Air also oxidizes manganese minerals to form manganese dioxide. Bringing sulfuric acid and dissolved chlorides in to contact with the manganese dioxide will release small amounts of chlorine and free chlorine will dissolve gold. The environment that a secondary deposit forms in needs to be semi arid, so the water can carry the dissolved elements down and redeposit them - if there is too much water, the elements will be washed away before secondary minerals can deposit (they usually deposit through concentration by drying). Because chlorine is not stable and coarse gold does not dissolve well, secondary gold enrichment is not nearly as efficient as it is with base metal mineralization. This is actually lucky for us, because if gold was easily dissolved and chlorine easily formed, there would be no placers for us to detect - the gold would dissolve away before it had a chance to roll down the hill into streams and form the placers we work!

Gangue minerals are the minerals which are not valuable that are found associated with the ore in both primary and secondary deposits. After processing, these are usually sent to a tailings pile. Quartz is the most common gangue mineral found in gold bearing veins, but a large variety of others including pyrite, calcite, barite and adularia feldspar can commonly be found in gold bearing veins as well. Gangue minerals can often tell us much about the conditions under which a vein or other type of deposit was formed. Gold ores that are considered as "free milling" or said to contain "free gold" are those in which the gold is in a form that it can be easily separated from the gangue minerals, usually meaning that it is uncombined and coarse enough to be captured by gravity based extraction methods. In some ores, the gold is tightly bound up inside the sulfides and not easily separated from them. These ores are called refractory or rebellious ores.

Quartz (white) and pyrite (striated metallic colored cubes) are two of the most common gangue minerals found in gold bearing veins. This fine mineral specimen came from a famous California mother lode gold mine.

Accurately recognizing minerals often depends on a number of tests. There are a number of mineral identification charts which, once the basic characterization tests have been performed, one can use to look up and identify the mineral in question. These tests are necessary to evaluate the important characteristics of the mineral, and include:

Color - While color is the most obvious feature of a mineral, it can be misleading as many minerals can be both transparent or opaque and occur in a wide range of different colors. The perceived colors of a mineral depend on the transmission, extinction and reflection of light striking it. When light passes through a mineral, it is transparent. When a mineral absorbs nearly all the light reaching it, the mineral appears black; on the other hand, if it reflects all of the different colors of white light, the mineral will appear white. Again, if it reflects only the red waves of ordinary light and

extinguishes all the others, it will appear red; this goes for the other colors as well.
Some crystals possess an unusual property of displaying different colors when viewed in different directions by transmitted light, and this property is called pleochroism. It is easy to view when present, and is commonly used to identify some types of minerals. Tourmaline is a well known mineral which possesses this property.

Pointed six sided crystals with characteristic growth striations along the hexagonal faces are indicative of quartz. This transparent purple quartz material is best known as amethyst.

Streak color tests - Many minerals are dark or intensely colored, and it may be hard to tell a very dark red material from a very dark green one as these minerals may both appear black to the eye. Rubbing a piece of the mineral on a white piece of unglazed ceramic like the back of a ceramic tile produces a streak of powdered mineral making the color analysis easier.
Luster – The type of exterior luster or shine a mineral shows may be used to identify it. Some well known types of luster are:
Metallic - The ordinary luster of metals. Gold, iron pyrites, and many sulfides have a metallic luster;
Vitreous - The luster of glass. Quartz, feldspar and many other minerals are examples of a vitreous, luster;
Resinous - The luster of resin. Sphalerite and Sulfur are examples;
Pearly - The luster of a pearl. Talc, brucite, and some mica minerals like sericite have a pearly luster.
Silky - The luster of silk. This luster is peculiar to minerals having a fibrous structure. The fibrous form of gypsum, satin-spar, and some asbestos minerals are good examples of a silky luster.
Adamantine - The luster of a diamond. This luster is due to high refractive optical properties and is shown in only a few other minerals like cerrusite.
Hardness tests, Hardness is one of the most important tests in identifying a mineral – perhaps the first which should be made after observing the color and other optical properties. These tests are often administered by scratching with a set of known minerals to determine where on the Moh's scale an unknown mineral lies. If one mineral scratches a second one, and the second cannot scratch the first, the first mineral is harder than the second one. If both are capable of scratching each other, they are considered of essentially the same hardness. For more detail, see the chart below.

MOHS SCALE OF HARDNESS:
A scale ranging from 1 to 10;
1. Talc
2. Gypsum
The typical fingernail is between 2 and 3.
3. Calcite
The copper cent has a hardness between 3 and 4.
4. Fluorite
An iron nail has a hardness between 4 and 5.
5. Apatite
Most types of glass have a hardness between 5 and 6.
6. Orthoclase feldspar (a potassium bearing feldspar)
7. Quartz
8. Topaz
9. Corundum (includes ruby and sapphire)
10. Diamond

Cleavage - many minerals have an internal weakness which exists along certain fixed planes caused by the internal bonding patterns of the atoms in the mineral, and these planes of weakness are a property of that mineral's internal crystal structure. Cleavage is often a great assistance in determining the nature of minerals which closely resemble one another in their crystalline form, and in other external characteristics. The number of cleavage planes, their angles in relation to one another and how easily they form are all used in mineral identification. While one can often see evidence of cleavage in a mineral surface, in testing to open up a cleavage on a specimen, it is often necessary to use a sharp knife or razor, and a strong blow. Minerals may cleave in one, two, three, or more directions, but normally cleavage in one direction is generally obtained with greater ease than the others. When such a break in the mineral is obtained easily and cleanly, it is known as a perfect cleavage. Minerals such as halite (rock-salt), galena, calcite, fluorite, gypsum, topaz, mica, hornblende, and feldspar all afford excellent examples of cleavage.

This obsidian specimen has broken showing a perfect chonchoidal fracture pattern – the same pattern shown in broken glass.

Fracture – minerals have a tendency to break a certain way. The character of a mineral's fracture, as displayed on the broken or chipped surfaces of minerals can be used to identify it. It is important that the viewer distinguish between the smooth, flat surfaces, resulting from the cleavage of a mineral, and the irregular surfaces characterizing true fracture surface. Fracture surfaces may be conchoidal, like glass; even where the fracture surface is flat or nearly flat; uneven when the fracture surface is rough by reason of minute elevations and depressions; and irregular, forming a hackley type of breakage.

External Crystal forms – Based on their internal atomic structures, minerals also form an external crystal shape. Crystals are mineral bodies bounded by regular surfaces, usually flat, arranged upon a definite plan which is an expression of the internal arrangement of the mineral's molecules. The crystal faces represent growth planes on the crystal. Crystallography is the scientific study of those crystals. The crystal shape can also be used to identify a mineral. As an example, quartz has a very distinctive crystal formation, which makes identification of quartz crystals easy. Many other minerals have similar characteristic crystal forms and these are often used in mineral identification.

The crystals on this ore specimen are brassy colored, but their triangular crystal faces tell us they are certainly not pyrite, but tetrahedrite, a copper and antimony sulfide that often contains considerable silver.

These known crystal shapes are classified into specific crystal systems. The recognized crystal systems of minerals are as follows:

Cubic - Featuring three equal axes at right angles. This form includes the cube, octahedron and dodecahedron as well.

Tetragonal - Featuring two equal lateral axes, one vertical axis, at right angles.

Hexagonal - Featuring three equal lateral axes, making angles of 120 with each other, a vertical axis at right angles to the plane containing the lateral axes.

Orthorhombic - Featuring three unequal axes, all at right angles.

Monoclinic - Featuring three unequal axes, one vertical, one at right angles to the vertical axis, the third making an oblique angle with the plane containing the other two.

Triclinic - Featuring three unequal axes, none at right angles.

Each of these crystal systems are then divided into a number of subgroups.

Twinned Crystals: At times, the growth of crystals can be partly altered, producing growth in new directions off the original crystal growth face. These are known as twin crystals, and in them one or two axes of one individual are parallel to the corresponding axes in the other individual. In twin crystals one part is in reverse position to the other part, and the twin may be viewed as being produced by the revolution of one half through an angle of 180 degrees.

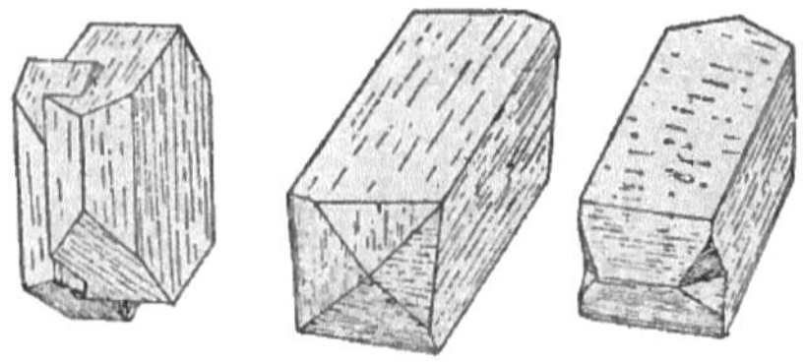

Carlsbad Twin. Baveno Twin. Manebach Twin.

Several examples of twinned crystals of the potassium bearing feldspar mineral orthoclase.

Specific Gravity **-** The specific gravity of a mineral is the ratio of the weight of the body to that of an equal volume of water. The weight of water varies a little with the temperature, but since its variation is, very small, it is neglected in determinations of the specific gravity of minerals. Since the specific gravity of water is considered to be 1, that of silver 10.5, the meaning being that silver is 10 and a half times heavier than an equal bulk of water.

The specimen to be tested to determine its specific gravity should be free from foreign matter and from decay, unless it is the specific gravity of such a specimen which is actually wanted. It should also be free from cavities, or these all need to be filled with water during the measurement under water. The cardinal principle employed in most specific gravity determinations
is that the loss in weight of a body immersed in water is the weight of a volume of water equal to that of the body.

When taking specific gravity measurements with a balance, a device to hold the mineral needs to be constructed (usually out of thin wire or thread). The weight of this holding device is subtracted out of the total, usually as a tare weight. The mineral to be weighed should be suspended from the device, and the mineral weighed both dry and under water.

After carefully weighing the suspended mineral in the air, a tumbler of water should be placed so that the mineral is well immersed beneath the surface of the water, taking care that no bubbles of air are adhering to, or imprisoned by, the mineral. The bubbles of air which attach themselves to the rough surfaces of minerals when suspended in water cause them to displace more water than they would otherwise; consequently, the specific gravity is apt to be understated. To avoid this, the bubbles should be shaken off or otherwise removed. If the weight of the mineral in air is then divided by the difference of its weight in air and water, the result will give its specific gravity.

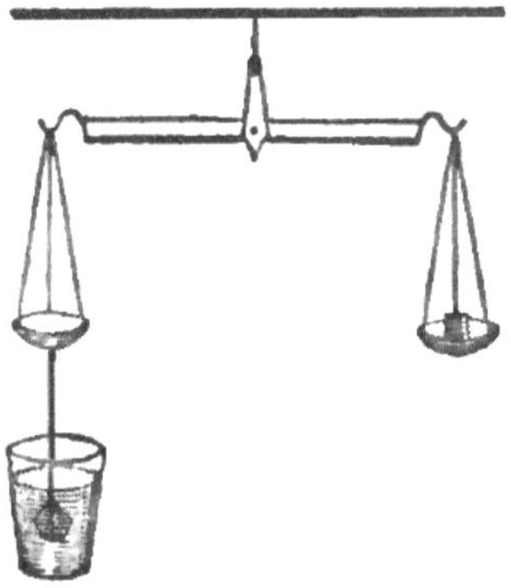

Many scales can be adapted to measure the specific gravity. The specimen must be weighed dry and then again while suspended in water.

Blowpipe treatment – Some minerals give off emissions which color a flame or melt when heated, and although uncommon, high heat is a method sometimes used in the field to identify minerals. The blowpipe flame is a simple test for the presence of certain elements, which are then used in determining the nature of the mineral.

X-Ray diffraction pattern – This is not something that can be done by the amateur, but scientists can use this technique to identify very small amounts of a mineral in a powder form. This technique is commonly used to confirm identification of newly discovered minerals, as each mineral has it own unique x-ray diffraction pattern signature – it's like the fingerprint of minerals.

Chemical Classification of Minerals:

One of the most common ways to classify minerals for study is by their chemical makeup. Most of these groups share a variety of characteristics because of their common chemistry. The most commonly recognized categories of minerals with similar chemistry are as follows:

1. **Native elements:** These are minerals consisting of elements uncombined with other substances. These include minerals like gold, native sulfur as well as diamond and graphite (both are carbon).
2. **Sulfides:** These are minerals that combine one or more metals and sulfur. They are fairly common and include a number of important ore minerals such as pyrite, galena and sphalerite.
3. **Sulfates:** These are minerals that combine a metal and the sulfate ion. This class includes common minerals such as gypsum and barite.
4. **Oxides:** These are minerals that form as a combination of a metal and oxygen. This class includes minerals such as hematite and magnetite - which are the iron oxides commonly found in black sands. This also includes corundum which is the mineral that makes up both ruby and sapphire.
5. **Carbonates:** These are minerals that combine a metal and the carbonate ion. This class includes the common rock forming minerals of calcite and dolomite which form limestone, marble and other related rocks like dolostones.
6. **Silicates:** These are minerals that include both silicon and oxygen - other than the various forms of silica, they also include one or more other metals as well. This class of minerals includes the most common rock forming minerals. Examples include quartz, various feldspars, micas, pyroxenes, amphiboles, the tourmaline group and a great many others. Silicate mineral structures include a wide variety of crystalline arrangements including single silicate tetrahedra, single chains, double chains, sheets and three dimensional frameworks.

These are the most common types, but there are actually many other less common mineral classes as well. These less common chemical classes include borates, urananates, hydroxides, Halides, tellurides and phosphates as well as others with only a handful of minerals in each category. These mineral groups are rarer, but also include some types which are economically very valuable. Telluride minerals can be extremely valuable because they can contain large amounts of gold and silver.

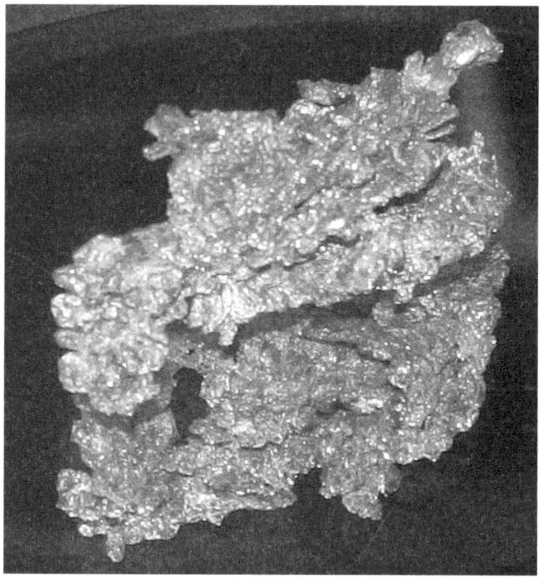

Elemental gold is found in many forms, some crystalline as shown here, and some smooth and water worn. Because of its many unusual properties, gold is not hard to identify.

It is beyond the scope of this book to provide complete listing of minerals or a full table for identifying minerals based on the types of tests noted above, but there are a number books available which can be purchased from the usual outlets which do have listings and tables of may rare unusual or valuable minerals. One can determine the hardness, luster, crystal shape and other features and use those facts to identify the mineral specimen. The best of these guides have excellent mineral photos and can be very handy in the field. Some basic information along these lines is also available over the internet - I have a set of mineral pages on my website with color photos and the basic identification information for many common ore minerals.

Once you have collected all the necessary information on hardness, color, cleavage, crystal form, etc. one can simply look up those characteristics in a mineral handbook or guide and find out what the mineral is. It is strongly recommended that the prospector own one or more of these handbooks. As you get more experienced with seeing a number of different minerals, you will begin to recognize the common ones fairly quickly. That's something that just takes some time to learn. With practice the prospector will learn to recognize quartz, feldspar and the other common rock forming minerals, as well as common ore minerals both in their secondary forms (like green copper mineral malachite), and in their primary sulfide forms (like the brassy chalcopyrite or silver-gray galena). Although we cannot take the time to look closely at a large number of different types of minerals, we will stop and take a serious look at a few very important minerals. Because of their importance to prospectors, first we will take a look at gold, then the sulfides and tellurides.

Gold as a Mineral:

Pure gold is a bright yellow metal, unaffected by the atmosphere or by the impurities it might normally contain. Because of its beauty and other properties, gold has been used as money and a store of wealth since pre-historic times. It is still an important and compact form of wealth, being especially prized in politically unstable parts of the world. In many cultures, it is normal to keep a significant portion of one's wealth as gold. In those parts of the world where banks and other financial institutions cannot be reliably trusted, many individuals will store their wealth in the form of gold. Even in the most advanced countries, when politicians use inflation to reduce the value of money for their own purposes, gold still retains its value. The color of pure gold is attractive, but may be varied considerably by alloying it with other metals. A mixture with copper produces an alloy with a reddish tinge, while a certain percentage of silver, notably 15 per cent, yields a distinctly green color. Finely-divided gold is purple by reflected and green by transmitted light. The metal is exceptionally malleable, and can be drawn into very thin wires or beaten into exceedingly thin sheets. Its density when perfectly pure is 19.3, but this density is, of course, never attained in nature.

Natural gold nuggets always exhibit a lower density because they contain certain amounts of silver, copper, iron and sometimes other metals as impurities. The melting point of pure gold is approximately 1,063 C. Gold is very resistant to any corrosion or chemical attack. Ingots lost at sea and sitting for centuries on the ocean floor are recovered, but show no visible sign of corrosion. Neither sulfur nor selenium are known to combine directly with gold; and few common acids have any effect on it - including sulfuric and acetic (vinegar). It is dissolved by a few unusual chemical mixtures, the best known of which is aqua regia, a combination of nitric and hydrochloric acids. It is also dissolved by free halogens (fluorine, chlorine, bromine and iodine) in moist environments. Tellurium vapor attacks gold, yielding a crystalline telluride. Gold is also attacked by perchloric acid.

One of the questions commonly asked by new prospectors is how to recognize gold. I covered this under the beginner's questions section, but we will go ahead and take another look here.

Occasionally, gold forms nice cubic crystals as displayed in the beautiful nugget pictured above. This is definitely one nugget that you don't want to hit with a hammer to see if it really is gold! Nice crystal gold specimens like this fine piece have a premium value approaching twice the spot value and should be protected.

Fortunately, gold is usually easy to identify. Few minerals have as intense a metallic luster, or form in the same types of shapes. It is heavy and soft, and deforms like a soft metal when struck or cut. Two minerals commonly confused by beginners as gold include pyrite and mica. Mica breaks into sparkling little sheets when rubbed between the fingers and just does not have a true metallic luster. Mica is also fairly light - it does not have the density of gold, and it will not be found in the heavy concentrates when panning or dredging. Pyrite, when fresh and unoxidized, does have a nice metallic gold color and though not as heavy as gold, is heavier than most other rock forming minerals. Because of this, it will be found in the heavy concentrates. It also commonly occurs with gold in quartz veins. That's why it gets the well-deserved name of fools gold. It often forms cubic to angular shaped crystals, which are usually easily recognizable. Luckily, pyrite also rapidly oxidizes in the air and the surface changes to a rust brown mineral, usually limonite, and in this form it is not often mistaken for gold. If it is hit by a hammer, pyrite shatters because it is brittle, but gold will flatten out in a manner similar to lead. This is a quick and easy test, but should not be preformed on things that could be nice gold specimens, as it destroys their collector value (which can be significant). I always tell folks that once you've seen enough real gold, it's hard to mistake mica or pyrite for the real thing. My website has a good number of clear photos showing natural gold in a variety of forms if you'd like to see some.

This photo shows a fine chevron crystallized gold nugget from Rye Patch, Nevada.

The beauty of natural gold is widely appreciated by both prospectors and mineral collectors, but the pinnacle of beautiful natural gold specimens are the spectacular crystalline forms. As a result, these specimens of golden scales, plates, ribbons and crystals are normally saved and treasured. Unfortunately, the natural growth conditions that form gold deposits only rarely create these very attractive specimens. At most mines, native gold is more normally found in the form of irregular masses and blobs in quartz or sulfide veins, or as impregnations in the country rock adjacent to mineralized fissures. In fact, in most modern mines, the gold is in the form of very tiny particles, too small to easily see with the unaided eye. In placer gravels, the wear of the erosion process means gold normally is pounded into the outwardly rounded forms of scales, grains, slugs, and nuggets. While large alluvial nuggets are beautiful and valuable in their own right, the crystalline forms are still the ones most highly valued. In fact, the value of crystalline gold in an attractive matrix specimen can exceed the intrinsic metal value of the gold by a factor of 10 to 100 fold in some cases.

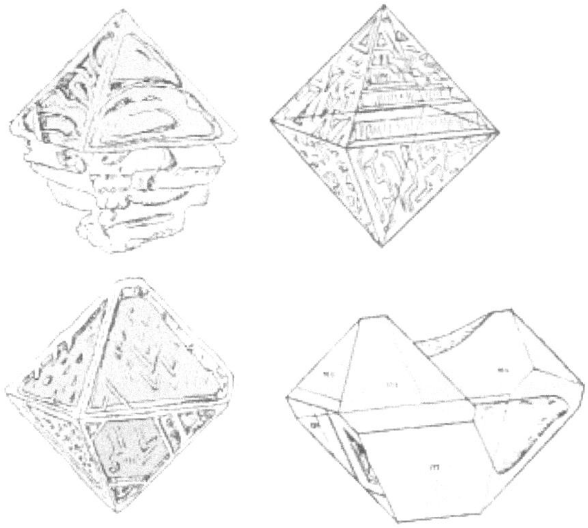

Here are some drawings of gold in various forms of etched, twinned and deformed crystals.

Although gold only rarely shows its attractive crystalline forms, where the geologic conditions are favorable, as in cavities or other locations where growth and expansion is not hindered by a lack of space, gold obeys the natural laws of its crystal growth and crystallizes in various isometric (cubic) forms. This partly explains the rarity of fine crystalline gold specimens. This is because open voids are uncommon in and of themselves. Normally, the walls of the cavity limit the lines of crystal growth, or growth is cut off by the concurrent growth of another mineral, such as calcite or quartz. Often any voids that do exist are later filled with quartz or other materials, including clays. In addition, large single crystal growth requires stable conditions that support slow growth - at least growth slow enough to lead to the formation of a few large crystals rather than millions of tiny ones. Faster growth will support the development microscopic pieces or at least mossy and dendritic forms. Gold crystallizes in the isometric system, and usually forms crude octahedrons, but specimens showing dodecahedrons, cubes and trapezohedrons are also found. Crystals are rarely perfect, and are normally irregular, sometimes exhibiting unusual wiry, reticulated or dendritic shapes. Many forms also show some sort of distortion of the crystal, and some show extreme distortion. Crystal twinning is common in gold.

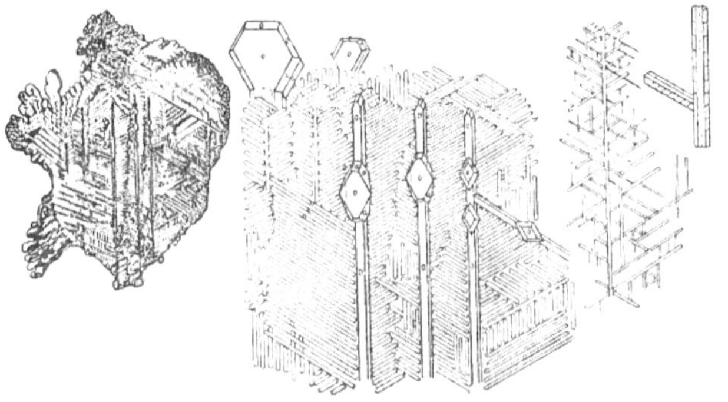

This drawing shows the herring bone pattern that forms chevron nuggets, as is seen in Nevada.

Between the combined effects of all the different crystal forms, twinning and a range of possible crystal distortions, gold can be found in a very large variety of crystalline forms, all of which have their own attractive appearance. The number of potential combinations is so large that all the possible formations are hard to fathom. Beautiful wire forms are found in a number of locations, with one of the most famous locations being Farncomb Hill in the Breckenridge District of Colorado (beautiful leaf and other forms are found there as well). Many of the dendritic forms are also beautiful, and one of the most beautiful dendritic forms of gold are the herring bone style dendrites, with some of the finest examples of this being the "chevron" gold of northern Nevada. Many of the nuggets from the classic locations in Pershing and Humboldt counties show a grooved V or chevron shape that is characteristic of the crystalline gold from this area. Not every nugget from the "Rye Patch" region shows this formation, but they are fairly common. California produces some specimens of ribbons and sheet gold, some with attractive crystal patterns on their surface, and these are also highly sought after.

Crystalline gold specimens have been recovered from both hard rock and alluvial deposits. Gold crystals from placer deposits may exhibit good crystal form so long as they have not traveled too far in their alluvial journey. After an extended period of travel in a stream, in time the pieces become rounded. Some hard rock specimens must be treated with hydrofluoric acid (a dangerous operation) to reveal the gold contained within them. Hydrofluoric acid will not react with gold, but does dissolve quartz. The most common single crystal form for gold is the octahedron. Many of the most perfect specimens of this type have been derived from the gravels of Victoria, Australia – however some fine examples have come from other locations including Romania in Europe. Large crystals from this location, especially the octahedrons, often possess deeply recessed faces and prominent growth on the crystal edges, indicating a comparatively high local concentration of gold in the surrounding solution. These recessed faces indicate the absence, at least in the immediate vicinity of the crystal, of solutions that have only just passed the point of saturation. The normal equilibrium process of deposition in these solutions of crystal growth form flat surfaces on the octahedral faces. The concentration of dissolved gold sufficient to create growth of crystals does not necessarily exist uniformly throughout the whole bulk of solution flowing through a vein-fissure during its formation. Geologists have found that the chemical makeup of the solutions that form gold deposits vary quite a bit over time during the deposition process.

At times, the amount of gold contained in the geothermal waters during vein formation must be relatively high, because this situation causes rapid deposition and results in the formation of small sized gold. The gold of quartz veins is often so fine and widely dispersed that it is not visible to the eye on normal examination. As I noted, many modern mines process gold too small for the eye to see. The Carlin type mines of Nevada are well known for their rich deposits of microscopic gold. Portions of some Carlin type deposits are formed by boiling of hydrothermal solutions and rapid deposition of the gold these solutions contain. This results in extremely small gold particles. When tiny particles of gold are dispersed uniformly through a portion of quartz vein, the vein material acquires a greenish hue, a feature that is most distinctly observable in specimens from the richest free-gold zones of certain epithermal deposits.

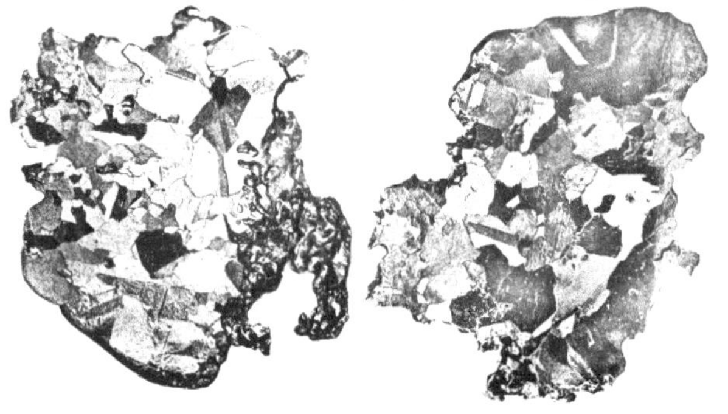

This image shows the interlocking crystals found on the interior of virtually all gold nuggets.

While alluvial nuggets with their rounded, smooth surfaces may seem to have a non-crystalline, blob like texture, the inside of these pieces reflect a very different situation. Investigation of the internal structure of nuggets has shown that polished and etched sections of rounded and water worn pieces of alluvial gold almost invariably show that they possess an interlocking internal crystalline structure.

Simple gold crystals that demonstrate the ideal octahedral model shape are rare. The crystal forms of gold, and particularly specimens from Romania, Australia, the Ural Mountains, as well as those from Western North America, have been closely studied by a number of scientists. In gold crystals, a considerable distortion from the ideal forms of isometric symmetry is the norm, as is shown in the drawings and photos of this book. Abnormal forms are explained by repeated twinning parallel to the octahedron and by elongation or flattening along a di-trigonal axis of symmetry (a cube diagonal). Twinning combined with elongation may produce what appear to be rhombohedral crystals. On the other hand, flattening along an axis normal to the octahedral face may produce forms that simulate very closely monoclinic plates. Other faces beside those of the cube may also be striated. Moss-gold, wire-gold, and dendritic forms indicate growth and crystallization along the di-trigonal axes of symmetry, accompanied by abnormal elongation along those axes. In some cases, as in certain unique specimens of wire-gold from the Santa Isabel mines of Colombia, gold wires are formed by elongation accompanied by twinning. Specimens from this mine were, curiously enough, terminated by well-defined crystals of chalcopyrite, crystals of which were attached like buds to the ends of the wire-like threads.

This is a fine museum quality crystal gold specimen composed of a series of octahedral crystals joined together.

Gold occasionally shows crystalline skeletal growth as parallel octahedra united on the octahedral faces. Sometimes it occurs as successive cubes formed along an axis of ditrigonal symmetry. This last form, with its bundled wires in a twisted corkscrew shape is common enough for skeletal growths of native silver, being best known in specimens from Kongsberg, Norway. Unfortunately this beautiful form has been noted in only a few gold specimens. These are commonly termed "Rams Horn" forms by collectors. One of the few localities that have produced this type of specimen is the Ground Hog Mine at Gilman, Colorado.

All natural gold contains some impurities, and the amounts vary quite a bit. Other metals typically alloyed with gold in various natural deposits include silver, copper, mercury, iron and palladium. More rarely traces of bismuth, lead, tin and zinc can also be found naturally alloyed with gold. By far the metal most commonly alloyed with gold is metallic silver. The percentage of silver mixed in with the gold appears to have very little influence on the form of crystallization, as might be guessed from the fact that native silver has the same isometric crystal form as gold. Nevertheless there are some grounds to think that the purer gold is, the greater its tendency to adopt the simpler forms (octahedra and cubes) when crystallizing rather than the more complex forms like trapezohedra, hexoctahedra, etc. These forms are found most commonly in electrum and lower purity gold. This interpretation is complicated by the fact that crystals of the purer gold are, as a rule, derived from deeper and hotter mesothermal deposits, while those of electrum (a natural

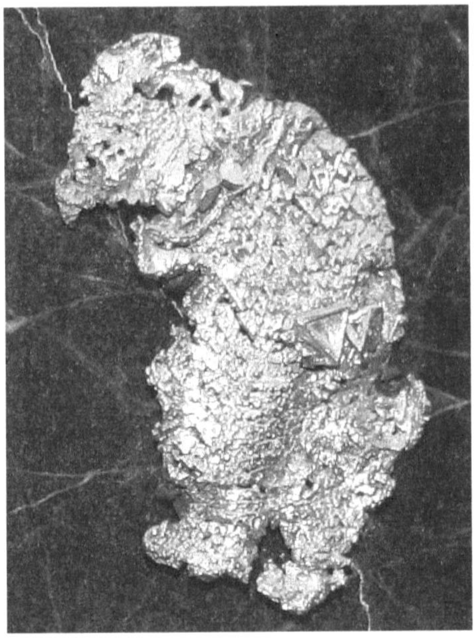

This crystalline gold nugget shows the shape of a bear – and was originally found in California.

gold-silver alloy) are derived mostly from epithermal fissures that formed in cooler, near surface deposits. Any affect of the silver on the crystal form is hard to separate from the changes caused by the physical conditions of formation, and not from inherent crystal forming properties of gold and gold-silver alloys. The purity of placer gold varies quite a bit, but even the purest native gold contains some silver. California placer gold typically contains about 10 to 15 percent silver, while Nevada placer contains that much and sometimes more. The purest natural gold yet recorded is that from the Great Boulder mine,

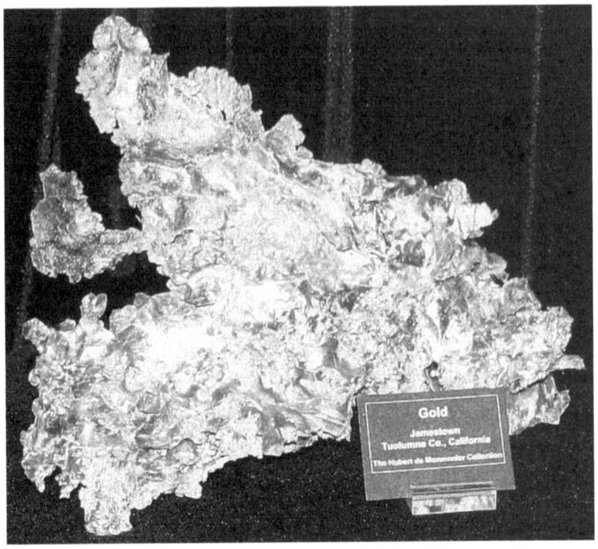

Another form of gold: Ribbon gold. This piece is from Jamestown in the California Mother lode.

Kalgoorlie, Western Australia. It was apparently formed from the decomposition of auriferous tellurides and was .999 fine. A small quantity of gold from the Pike's Peak mine, at Cripple Creek, Colorado, also showed an assay fineness of .999, and was probably formed the same way. A very large portion of the outcrop gold of the famous Mount Morgan mine in Queensland, was .997 fine. Mount Morgan has furnished probably the greatest bulk of naturally fine gold, as the two first-mentioned results were obtained from assays of small picked samples, rather than from the mass of the gold recovered on a commercial scale in the mines mentioned above. Alluvial gold directly derived from the degradation of auriferous veins is typically higher in quality than the gold of the parent vein, since meteoric waters act on all sides of the liberated gold particles to remove much of the more soluble silver and copper. It is commonly noted that alluvial gold-dust from the lower reaches of any given river is much higher in fineness than the coarser grains and nuggets of gold found in the upper reaches of the same stream. This increase in purity comes from the long exposure to leaching as it travels downstream. In the case of the finer sized dust, it has a grater surface area for the action of water and air to remove the silver and copper. Testing has been done that demonstrates that the surfaces of larger alluvial nuggets contain less silver and are of higher purity than the gold in the interior of these rounded alluvial nuggets. Again, this occurs because of leaching away of the surface silver in wet alluvial environments.

There are several natural alloys of gold, and some have been given names. Let's take a look at these natural alloys, and how they figure into the world of nuggets and gold crystal specimens. Perhaps the best known of these alloys is electrum, a natural alloy of silver and gold. Electrum was known to and named by the Romans. Pliny, the ancient historian, defined electrum as a natural alloy containing at least one-fifth of silver, and it was probably named on account of its light yellow color. The term is still used by modern geologists and other authors to describe natural gold-silver alloys rich in silver. Electrum is found most commonly in the Tertiary andesitic epithermal goldfields of North, South, and Central America, New Zealand, and Hungary. Its color varies with the percentage of silver present, from yellowish white to a pale yellow. The greater the percentage of silver that is present, the paler the color will be. Electrum's specific gravity normally ranges between 12.5 and 15.5. While sometimes the proportions of gold and silver present represent specific molecular ratios, it is most probable that in the majority of occurrences, electrum is composed of a solid solution alloy mixture of gold and silver. The occurrence of electrum or low-grade gold in the veins of the younger volcanic deposits alone may be regarded as evidence of a near surface origin for nearly all electrum.

Mercury is also an important alloy agent. Miners have long used mercury to amalgamate gold in the process of its recovery, leaving behind mercury in many gold districts that gets into rivers and commonly contaminates the gold that is present. In placer mining areas that have long been known and worked, the great majority of the amalgam now found is due to past mining efforts, resulting from the union of gold with quicksilver lost by the old miners. While man caused gold amalgam may be fairly common, it also occurs naturally in small amounts. It contains a variable proportion of mercury, and like the man made material, is of indefinite composition. Amalgam with 57 per cent mercury is reported from the alluvial placers of Colombia. Natural amalgam from Mariposa, California, yielded, on analysis, 61 per cent mercury, and had a specific gravity of 15.5. Native amalgam is found in the gravels of the Pek River, in eastern Serbia, where it is termed "zivak," and in Victoria, Australia, in the quartz of the German Reef at Tarrangower.

Pyrite: The most common sulfide mineral.

Sulfide and Telluride Minerals

Having considered gold in detail, let's take a look at the sulfides (and the tellurides and selenides that go with them). Besides metallic gold, they are the primary ore minerals of most precious metal mineral deposits. Gold, silver, copper, nickel and so many others - valuable metals are very commonly found in or closely associated with sulfides. We've talked a little about primary and secondary minerals, so we can think about sulfides as the original primary ore minerals that formed deep down within the earth. The early day miners never liked to see loads of sulfides in their deposits, because the oxidized surface ores, with their secondary minerals were a whole lot easier to work. Nature had already done some of the processing and extraction for them. Yet it happened a lot more often than they would have liked: the easily worked rich bonanza ores they discovered on the surface gave way as they mined downward, turning into a vein full of sulfides, base metals and other nasty, rebellious and refractory ores. In many locations where they had initially been recovering 90% of the precious metal assay values processing oxidized surface ore, with the rebellious sulfide ores they were recovering less than 50%. The remainder of the precious metals in their ore simply flowed uselessly out from the mill and into the tailings pond. Many rich and profitable mines suddenly became money losers when the miners got down to those unoxidized, primary ores. That's why the miners called them rebellious.

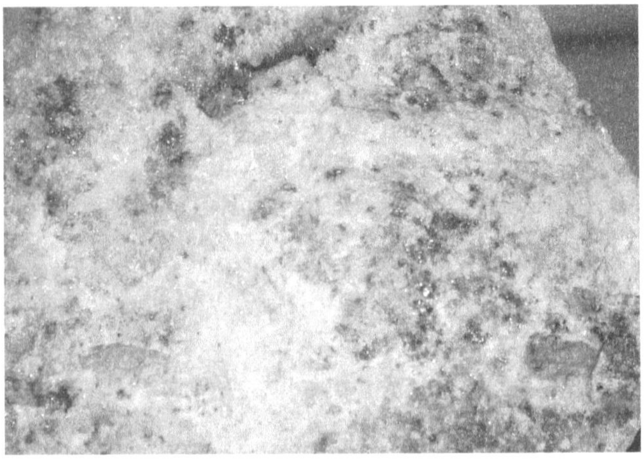

The complex silver sulfide minerals sprinkled through this Nevada ore specimen makes it harder to process for precious metals.

A great example is found in the silver deposits of Austin, Nevada. Here, the first miners found rich oxidized silver chloride ores on the surface, some of them running as high as 1000 ounces per ton. That surface ore presented no real problems in milling, so the miners installed the Washoe pan process already in use at Virginia City and started to pull the silver out of their ores. Unfortunately, at only about 75 feet down, the miners struck the water table and the ores below that were unoxidized and full of silver bearing sulfides like enargite, polybasite, tennantite and other rich but rebellious minerals that refused to surrender their values in the mills of the day. New and expensive roasting processes had to be installed, and as a result, the mills charged $50 and up to mill the ore and only paid 80% of the assay values back to the miners. By contrast, charges for processing free milling ore in Virginia City at the time was only around $20. Sulfide ore from Austin that didn't contain at least 70 ounces per ton of silver had to go out on the dumps, because it was unprofitable to ship them to the mill! This sad story was repeated all over the Western US and actually all across the world. Over the years, miners worked to develop new ways to handle these difficult ores, and they made progress as time went on, but even cyanide doesn't normally do as well in processing sulfide ores without some special additional efforts. Sure, today we've got some new technology, and there is no doubt we're a lot better off, but sulfide ores still require a lot of extra treatment and handling to extract their values - which adds a significant cost. Let's take a look at some of those devilish minerals that make us work extra hard to extract the values they hold.

Sulfur, selenium, and tellurium are all members of the same chemical family, and as a result they form similar mineral compounds when they react with many valuable metals. This is why I am addressing the three of them together. So what are these nasty problem causing minerals? Well, sulfides, selenides and tellurides are simply minerals which contain sulfur, selenium, and tellurium combined with some metal. Usually this combined metal is the valuable one we are interested in, but sometimes other valuable metals can be mixed in as trace elements. This is the case for silver, copper and many other metals, they actually form chemical combinations with the sulfur to form silver sulfides, etc. Gold however does not combine with sulfur (as it's so chemically inactive), but tiny pieces of metallic gold are sometimes surrounded by sulfide minerals like pyrite. As a result, the gold is hidden inside the sulfides so that it is difficult to extract and capture. So when gold occurs disbursed through a crystal of pyrite, while the pyrite itself is not particularly valuable, if significant amounts of gold are contained within it, the gold particles inside are valuable.

Sulfides

Sulfides are the most common member of this mineral grouping. Pyrite, galena, sphalerite, and chalcopyrite are all common sulfide minerals found in precious metal veins and other hard rock deposits. Both pyrite and galena are formed in the shape of cubes. When metals combine with sulfur, the resulting mineral is called a sulfide. Most have a reflective metallic sheen, like lead sulfide, also known as galena. Some however commonly take on a more earthy appearance. Acanthite, which is silver sulfide, commonly appears as a sooty looking earthy black material. There are also sulfur minerals which contain more than one metal such as parargyrite (ruby silver), bournonite (peacock copper ore) and tetrahedrite. Most sulfides contain an excess of sulfur. As an example, the chemical formula for galena is PbS. That means there is one atom of sulfur to each atom of lead. However in reality, galena crystals usually contain considerably more sulfur than a 1 to 1 ratio with the lead would call for. The same is true for pyrite.

These gray colored cubic crystals are galena.

Sulfides are generally fairly dense and heavy, so they can be recovered by simple gravity methods. Vibrating tables are sometimes used for this purpose and the sulfides will usually show up in the table middlings. However, in most commercial milling operations, sulfides are recovered by a method called flotation. This technique employs the fact that sulfides are not easily wetted by water. As a result, in a froth of tiny bubbles, they will tend to stick to the air bubble. The froth of tiny bubbles with their concentrate of sulfides can then be separated off and treated. Different minerals have a preference for different types of conditions in the flotation cell, so by planning the chemicals used to make the froth, the ph of the water and other factors, it is possible to separate differing sulfide minerals out of ores which contain more than one type of sulfide. This is extremely important in base metal ores which contain multiple valuable metals. Just why are sulfides such a big problem? The first problem is that they contain locked in values.

That was the biggest reason sulfides presented so many problems to the old time miners. Gold and silver contained inside some sulfides amalgamates poorly, and in those early days, amalgamation was the chief recovery method for hard rock ores. The old timers developed the pan processes where they ground up the ore and amalgam mixture for hours at a time (the Washoe Process), and for some ores that did the job to oxidize and open up the sulfides. Sometimes an extended grinding period did not help very much, and some sort of roasting was required prior to amalgamation for it to be successful.

The second problem with sulfides relates to the difficulties that these minerals present in the cyanide leaching processes. The first difficulty with cyaniding sulfides is that in a wet environment with oxygen, sulfides tend to oxidize to form sulfuric acid. Acids will destroy the cyanide and create poisonous gases. It is extremely important when treating high sulfide ores with cyanide to keep a sufficient protective base in the solution. Many sulfide minerals also contain things like copper or antimony that can react with the cyanide, preventing the cyanide from leaching out the gold and silver.

Sulfides are relatively unstable in a number of ways – they are easily broken down by oxygen. In most gold or silver bearing veins, the sulfides in the surface outcrops are already broken down by oxygen in the air. However, once you get down close to the water table, sulfide minerals start to appear (as happened at Austin).

The old timers took advantage of the instability of sulfides by simply heating them to burn off the sulfur. Almost all sulfides will oxidize quickly when they are heated to a high temperature. The problems with sulfides and the fact that most of the problems can be removed by roasting, led to the development of a variety of ore roasting systems that were installed in the mills of that day. Heating ores to elevated temperatures (roasting) causes the oxygen in the air to convert the sulfur to sulfur dioxide, which is a gas and is emitted to the air. Unfortunately, it is also a pretty nasty air pollutant. This is why there are very few smelters still operating and doing this type of process in the US. However, this "burning" or roasting of the sulfide ores does work to convert them into a more free milling state, allowing the values to be extracted. Exact temperatures used in roasting vary, and excessive temperatures cause problems. One of the unusual things about sulfides is that when heated to high temperatures they will melt. The product of melted sulfides is called a matte, and it is a highly corrosive liquid.

This silvery colored sulfide is soft and slippery – it is molybdenite, and is commonly used as a lubricant in grease.

Over the years, a number of small scale miners have sought to treat small lots of ore by burning off the sulfur as sulfur dioxide. Unfortunately, sulfur dioxide is rather pungent and offensive stuff. It's a serious air pollutant and can be toxic if you inhale a lung full. If you start choking and coughing, that's a real good sign you're getting too much. As if that's not enough, any arsenic in the ore will be vaporized and emitted as poisonous arsenic trioxide - and those are some very toxic air pollution fumes. This is a concern because arsenic is actually very common in many gold ores. To roast sulfide ores, it takes a good red heat and a good amount of air, but not too hot as you don't want to melt them. Even if you get everything right and you're successful, the neighbors may well call the police on you! That's why I can't recommend "at home" roasting methods to small miners.

On the other hand, the way the Spanish did it - the old "patio" process - is still a possibility for small miners. Actually more recently the big mining companies have come around to adopting a modernized use of this ancient practice, because it's less polluting. The method is to grind the ore up finely, put it in some sort of tub and keep it moist. Stir and turn it over from time to time to allow the air access to the material. Natural bacteria found in sulfide ore will do the rest. Add a tiny bit of lawn fertilizer to the water you mix in to help them along. Don't submerge the ore, just keep it wetted down. In this kind of environment, the natural bacteria will break down the sulfides and convert them, using the oxygen in the air, to sulfates. It's slow and may take months to complete. There will also be a little bit of stink, but nothing like roasting. Even in spite of this, your oxidized sulfide solutions will be strongly acidic and may be full of toxic heavy metals like lead, antimony, cadmium and copper. They may well legally constitute a hazardous waste, so you can't just toss them in the trash or dump them in your yard. As a result, any treatment of sulfide ores needs to be done with an eye toward compliance and any applicable environmental regulations. If you want to try something like this, be sure to do some through research beforehand.

In the modern equivalent of the patio technique, large heap leach operators have cultivated strains of these natural bacteria that like to eat sulfides. Their sulfide ore heaps are first leached with the bacteria "soup" and small amounts of diluted fertilizer solutions are added to help the bacteria grow. It takes some months, but the sulfides are broken down and when the heap is later leached with cyanide it produces much more gold because of the work of the bacteria.

The newest technology to provide answers to dealing with sulfides is the autoclave. This is a high tech and more environmentally friendly way to oxidize sulfides so that their metals can be recovered. It is most commonly used with precious metals extraction. The autoclave is a

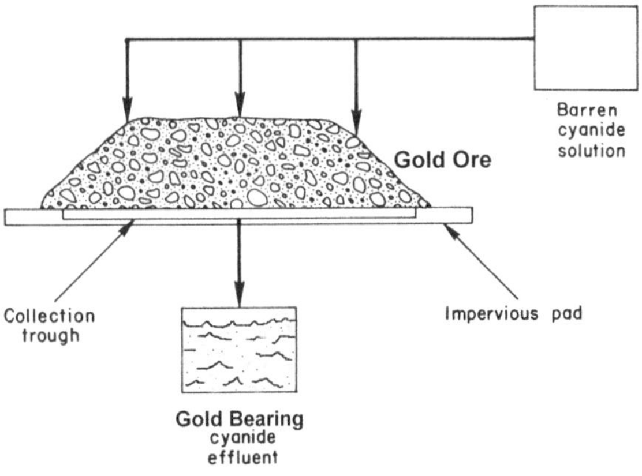

Working low grade ores by heap leaching is a standard method of recovering gold from ores.

large sealable vessel into which gold and silver ores are placed for treatment. Autoclaves provide a heated, wet, high-pressure environment which allows the oxygen present to quickly break down the sulfide minerals. Often strongly oxidizing chemicals like hydrogen peroxide are added to help supply oxygen to the mix.

Because many sulfide ores are rich in base metals, and because recent price increases have made base metals valuable once again, miners are asking about how to convert these rebellious sulfide minerals into the valuable base metals that they contain. The processes used vary depending on the particular metal to be produced, but the first step in most of these techniques involves getting rid of the sulfur in some sort of oxidizing roast process. Smelting is the next step. For lead and zinc, the smelting process involves using coke carbon to reduce the ores to metal. Zinc is also produced by an electrolytic process which involves dissolving the roasted ores in acid. With copper, oxygen from the air is capable of stripping out the sulfur out of a molten copper sulfide matte. The raw, unrefined metals produced by the smelters is then further refined and cast into ingots for sale.

Selenides

Many selenide minerals have often gone unnoticed in years past, as in many cases selenium is present as traces in normal sulfide minerals. Only in recent times have mineralogists routinely recognized the difference between the standard silver sulfide acanthite (argentite) and the lesser-known silver selenide, naumannite, which to the eye appears very similar. Both often take on a sooty black appearance in veins. In recent years, naumanite has been found to be an important mineral in low sulfidation epithermal bonanza grade ores in Nevada. The rich Nevada ore deposits at the Midas mine at Gold Circle, the bonanza deposits of the Sleeper mine in Humboldt County and the deep epithermal veins at the Holister mine all contain naumannite as an important ore mineral. Many silver bearing deposits in Nevada previously noted to contain sooty black deposits of silver sulfide, may actually be rich in silver selenides, especially naumannite, because of that mineral's similar fine grained black appearance. As far as environmental considerations, it is important to consider the impacts of any treatment you are doing. While sulfur and its associated chemicals are toxic enough, selenium is even worse. You must understand the risks is any chemical treatment you do.

The mossy calaverite on this rich Cripple Creek, Colorado ore specimen contains plenty of gold. Old time miners sometimes missed telluride ore.

Tellurides

Tellurides are a series of minerals containing the element tellurium. While it is sometimes found in nature as the native element, tellurium is one of the very few elements that chemically combines with gold to form natural stable minerals. While tellurides are not common on a worldwide basis, locally rich telluride deposits have formed some very famous deposits. Gold bearing tellurides were important in some of the richest ores at Goldfield, Nevada - high grade ore that contained upwards of 10,000 ounces of gold per ton. If you have tellurides in your ore, it's important that you be aware of that fact, because it affects how you

This crystal specimen shows calaverite up close.

will process the ore. Tellurides are leachable by cyanide but can take an extended processing period, as the telluride minerals take time to dissolve completely in the leaching solutions. They have a metallic appearance similar to sulfides. There are several telluride minerals that are of interest to prospectors and miners - here are a few of them:

Calaverite: AuTe2: Calaverite is the name of a tellurium bearing mineral that occurs in certain low temperature hydrothermal veins, but is also sometimes found in higher temperature veins. It was first discovered in Calaveras County, California, but is also important at Cripple Creek and Telluride Colorado, as well as Kalgoorlie, Australia. It is brittle and has a brassy gold to silver white metallic sheen. Its appearance sometimes confounded old time prospectors who confused it with pyrite or arsenopyrite. No miner wants to toss out super rich telluride ore thinking that it is just pyrite or some other base metal sulfide - but that has happened in a number of instances. It commonly occurs in the presence of sulfide minerals. Unlike pyrite, calaverite contains very significant amounts of gold - about 40% gold by weight. With a 40 percent gold content by weight, you can see it wouldn't take much calaverite to make for some very rich ore.

Sylvanite: A rich telluride of gold and silver. After calaverite, it is probably the next most important telluride mineral. It can assay as high as 30% gold. It is found with other telluride minerals in some precious metal ore deposits.

Petzite: A gold and silver telluride that is mostly silver. The Silver content is usually about three times that of the gold.

Nagyagite: This is a complex mineral containing gold, tellurium, sulfur, lead, antimony and bismuth that occurs in epithermal veins.

Hessite: This mineral is a silver telluride.

Coloradoite: A rare telluride of mercury, but it commonly occurs with deposits of precious metals.

Altaite: This is a rare telluride of lead, but can contain a fraction of precious metals.

Above: An early day hydraulic mining scene in the California mother lode.

Early Day North Carolina Underground Gold Miners

Recognizing and Understanding Common Rock Formations

Now that we have taken a look at individual minerals, we will turn our attention to rocks. Rocks are simply an aggregate formed from the accumulation of minerals stuck together into a single mass. Minerals form under various conditions, and when the minerals come together in a solid material, they form rocks. Some rocks are formed principally of one type of mineral like quartzite, which is a rock formed mostly of compressed quartz sand. Other rocks have many different minerals, and a good example of that is granite which may contain small amounts of a dozen or more minerals. The scientific study of rocks is known as petrology.

There are three basic categories of rocks, which are Igneous, Sedimentary, and Metamorphic. Simply put, these classifications are based on the origins of the various rock types. Igneous rock types come from deep within the earth. They have been melted during at least one point in their existence. Sedimentary rock types form when weathering by wind or water remove minerals from existing rock masses and deposit them elsewhere. Metamorphic Rock types form when significant heat or pressure changes the nature of existing rocks. Usually these changes take place far below the earth's surface. As with minerals, there are many named types of rock, and it's not necessary that the individual prospector know all of them. However, it's worthwhile to know the basics and be familiar with the most common types and especially those that are commonly associated with gold and other mineral deposits.

There is actually a cycle of rock renewal where igneous or other rocks weather and accumulate to form sedimentary rocks and eventually are buried themselves. These sedimentary rocks are in time changed by heat and pressure to metamorphic rocks and if even more heat and pressure is applied, the rocks become semi-molten and are renewed as igneous rock again. We will take a look at each of these three main types, starting with igneous rocks.

IGNEOUS ROCKS

Igneous rocks are important to the prospector as many types of gold and silver deposits are created with igneous rocks as the source of heat that drives the formation of the deposit. The word igneous means fire-formed rocks - it is used for rocks that were molten at one time. The rocks that form this group are the ones that crystallize from that molten material. Below the Earth's surface the molten material is known as magma, but once it erupts onto the Earth's surface through a volcano or crack it is called lava. Lava, of course, cools quickly because it is cold on the earth's surface. The cooling rates determine the texture of the igneous rock. Quick cooling means the rock will be fine grained like most lava, and the minerals formed will be tiny, even microscopic. Slow cooling will result in the growth of coarse grains easily visible to the eye like those in granitic rocks. Igneous rocks erupted on the surface are called igneous extrusive rocks and those that do not come to the surface until after they cool (so they are allowed to cool slowly) are igneous intrusive rocks. Igneous rocks are categorized by the chemical makeup of the original molten magma "soup" that makes up the mix out of which the rocks form.

This unusual specimen shows a lighter salt and pepper granite with a darker diabase inclusion.

The first category is granitic or felsic rocks. These rock types include: rhyolite, the extrusive (lava) version and granite, the intrusive (slow cooled) version. They are dominated by silicon and aluminum (Si, Al). They are usually light in color, are characteristic of locations on the continental crust, and form a stiff (thick) lava or magma. The magma of Mt. St. Helens is of this type, and because of their stiff, thick nature, volcanoes with this type of lava are prone to explode, just as Mt. St. Helens did. Krakatoa, a south Pacific volcano that exploded in 1883, also produces lava of this type. When rhyolite cools very quickly, if it contains little dissolved gas, it forms obsidian or volcanic glass. If the magma contains large amounts of gas, the gasses form a froth, resulting in a rock called pumice. Granite consists of interlocking crystals which have crystallized slowly from a molten mass. In examining a piece of granite, the most conspicuous crystals which meet the eye are usually those of feldspar. They are commonly pink, white, or yellow, and break along smooth cleavage planes which reflect the light like tiny panes of glass. Mica may be recognized by its glittering plates, which split into thin elastic scales. A third mineral, often translucent and waxy looking, but harder than steel, breaks along irregular surfaces like broken glass, is quartz. Granite usually only shows a small sprinkling of darker colored minerals.

This is rhyolite, a fine grained, light colored volcanic rock that is chemically similar to granite.

The second type of igneous matter is the andesitic rock. The andesitic rock types are intermediate in composition between felsic and mafic types, still containing significant silicon and aluminum, but also much more iron and magnesium than the felsic rocks. As a result, they are also intermediate in color between felsic and mafic rocks with lots of gray tones ranging to sometimes even purple-brown colors. They are also characteristic of continental crust materials. Rocks from this category include: andesite, the extrusive (lava) version, and diorite, the intrusive (slow cooled and coarse grained) version. Diorite, which generally has no visible quartz, is composed mostly of hornblende and feldspar, with the feldspar being of a different variety than the feldspars found in granite and syenite.

This granite hillside is part of Rich Hill in Arizona. The horizontal and vertical jointing in the rock results in naturally rounded boulders. This is common in many igneous intrusive rocks and is called sphereoidal weathering.

The third type is mafic rocks. These rocks contain abundant ferromagnesian minerals (magnesium and iron silicates). They are usually dark in color (typically dark gray to black). They are characteristic of volcanoes, like the lava of the Hawaiian volcanoes, or the Earth's oceanic crust. This material forms a runny (low viscosity) lava that flows smoothly. The rock types that form this category include: basalt, the extrusive (lava) version, and gabbro, the intrusive (slow cooled) version. gabbro is a dark, heavy, coarsely crystalline aggregate of feldspar and augite (a dark hornblende mineral). It often contains significant magnetite (the magnetic black oxide of iron) and olivine (a greenish iron-magnesium silicate). A rock cooled at moderate speeds, and intermediate between basalt and gabbro is diabase.

Although all igneous extrusive (lava type) rocks can display this characteristic, most basalt contains some "air holes" which geologists call vesicles. These are formed when steam and other gasses from deep in the earth are trapped within the lava, and as it cools and hardens, the trapped gasses form bubbles, which remain as vesicles when the rock solidifies. Under certain geologic conditions, larger crystals can grow within a finer grained volcanic rock matrix. The larger crystals are called phenocrysts, and the rock itself is said to have a "porphyritic texture" sometimes the rock is just called a porphyry. Under rare circumstances, these larger crystals can form valuable gemstones. These include such gems as the sunstones near Plush, Oregon and the sapphires of Montana and certain places in Asia.

The heat required for formation of gold and silver deposits is not extremely high, but some heat is necessary to create the natural convection currents which make the mineralized waters flow. In many situations, the heat of igneous rocks, either intrusive or extrusive, supplies the heat needed for the formation of precious metal deposits, and it is very common across the western US and many other places that gold and silver deposits are located near to or actually within igneous rocks.

Abundance of the Elements in the Earth's Crust

Name of Element	Symbol	Lithosphere Per Cent of Whole
Oxygen	O	47.33
Silicon	Si	27.74
Aluminum	Al	7.85
Iron	Fe	4.50
Calcium	Ca	3.47
Magnesium	Mg	2.24
Sodium	Na	2.46
Potassium	K	2.46
Hydrogen	H	.22
Titanium	Ti	.46
Carbon	C	.19
Chlorine	Cl	.06
Bromine	Br	—
Phosphorus	P	.12
Sulphur	S	.12
Barium	Ba	.08
Manganese	Mn	.08
Strontium	Sr	.02
Nitrogen	N	—
Fluorine	Fl	.10
All other Elements including Gold, Silver, Platinum, Arsenic, Copper, Lead, Mercury, Nickel, Tin, Zinc, Radium, etc.	—	.50
Total		100.00

METAMORPHIC ROCKS

Metamorphic rocks are another important class of rocks. These are materials that form as one type of rock and through exposure to heat and/or pressure the original rocks are changed (metamorphic means "changed form"). One of the best-known examples of this is the formation of marble - when limestone is subjected to pressure and perhaps a little heat, the limestone is re-crystallized into marble. The forces of heat and pressure can cause changes in both the texture and mineralogy of the original rock. Rocks that become deeply buried after years of having other materials deposited on top of them may be subject to a considerable force of pressure and heat. The heat comes from the fact that once the near surface conditions are left behind, the heat of the earth increases as one moves deeper and deeper. The effect of the compaction and heat can cause re-crystallization of the minerals present, and change the texture of the rocks. This heat and pressure may also cause a change in the types of minerals present; for example, garnet is a common result of metamorphism, as various elements present in the rock may combine to form garnet, since it is more stable at higher

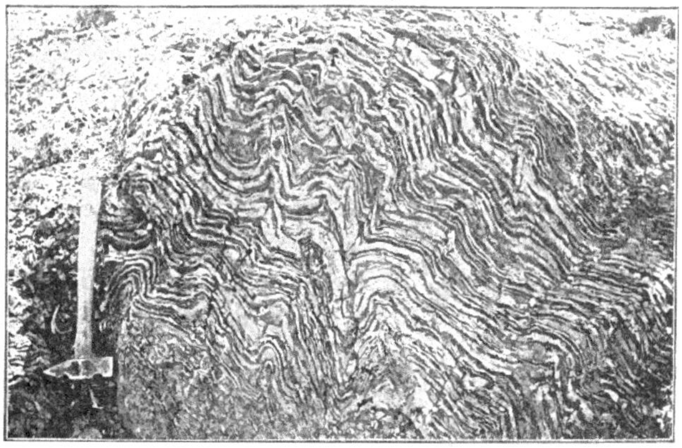

These wavy dark and light lines mark the rock known as gneiss – it is strongly metamorphosed.

temperatures and pressures. Depending on the heat and pressure a rock is exposed to on burial, rocks like limestone can be changed into marble and sandstone can be changed into quartzite or shale into slate.

As I noted above, pressure forces increase with depth due to the weight of the overlying rocks. A cubic foot of granite weighs 167.9 pounds. Burial to a depth of only 1000 feet would result in pressures approaching 1,200 pounds per square inch. Many metamorphic rocks have been buried at depths far greater. Heat also increases with depth. Heat may come from large bodies of molten rock rising under a wide geographic area. Intrusions of hot magma can literally bake rocks as it intrudes them. Lava flows can also bake rocks on the ground surface. Lava or magma in contact with other rock can also cause changes in minerals. Even if no igneous rocks are present, the heat of the earth's core becomes significant as burial becomes deeper. An area-wide exposure of rocks to increased pressure and temperature with depth causes regional metamorphism. This regional metamorphism occurs at burial depths of 5 - 40 km. Tectonic pressures associated with convergent plate boundaries and continental collision can also cause Regional Metamorphism. Directional pressure along fault zones causes an effect called Dynamic Metamorphism, resulting in the crushing and ductile flow of rock. Foliation is a broad term referring to the alignment of sheet-like minerals, which may leave a rock susceptible to breakage along certain planes. Types of foliated rocks include: schist and slate both of which are derived from the action of metamorphic forces on shale. They are typical of regional metamorphic forces. In these schistose (or foliated) rocks, the planes of division are determined by the parallel arrangement of crystalline grains of some cleavable mineral, which is very commonly mica, but can also be other types such as hornblende, talc, or chlorite. This structure is characteristic of many of the metamorphic rocks. In some cases, the parting plane is the result of the original stratified arrangement of the material in a sedimentary rock, but in other cases such a parallel arrangement appears to be due to pressure or shearing, causing a rearrangement of the materials of the rock. A schistose structure may, accordingly, be developed in rocks of igneous origin, resulting in greenstones and greenschist.

In regional metamorphic rocks, the temperature and pressure the rock was subjected to can be estimated using metamorphic index minerals. This occurs because heat and temperature causes certain minerals to become more stable and others to become less stable. Minerals then convert over

Mica Schist like the rock above, easily splits into flat shapes – it is commonly associated with gold.

to the more stable mineral forms. Lower-level metamorphic forces (low temperatures and pressures) - say about 200 degrees C, will be forming rocks like slate and hyalite. The typical minerals present would include chlorite, muscovite and biotite. As the temperature increases, intermediate grades of metamorphic forces would produce rocks like schist, and include typical minerals like garnet and staurolite. High levels of metamorphic forces, including temperatures approaching 800 degrees C, result in rocks that have been on the verge of melting. Typical rocks would include gneiss and migmatite, and minerals like sillimanite. When exposed to even higher levels of heat, complete melting of the material occurs and it becomes full-fledged igneous rock magma.

It is worthwhile for the prospector to be able to recognize rocks of slate and schist rocks as there are a number of different types of gold deposits commonly associated with these types of rocks. Schist and slate both represent the finer types of foliation in metamorphic rocks, with thin, parallel, crystalline parting planes which fit together almost like the pages of a book. In schist, these are often slightly crumpled in form. These folia can be distinguished from the laminae of sedimentary rocks by their lenticular form and lack of continuity, and especially by the fact that they consist of platy, crystalline grains, and not of particles rounded by wear.

Mica schist is the most common of all schists, and in fact, it is probably the most common of all metamorphic rocks. It is composed of mica and quartz in alternating wavy folia. The waves are small but visible. All gradations between this rock and phyllite can be found. Schist rocks are normally created through the metamorphism of slates and shales. It can be distinguished from slates and shales by its visible mineral grains and wavy foliation, and although both of these are small, neither of them are normally found in slates and shales. Mica schist is widespread in the goldfields of Arizona and California. In fact, in a study of seventeen of the worlds most important placer regions, 17 of 17 were associated with some form of foliated rocks like schist, slate or phyllite. Not every placer is associated with these rocks, but many are. Other types of schist include talc schist, consisting of quartz and talc; hornblende schist, a rock which sometimes results from the metamorphic alteration of basic igneous rocks; and quartz schist, which is produced from quartzite by the development of fine folia of mica along planes of shear.

Schist, marble and quartzite are all examples of metamorphic rocks that were formed out of sedimentary rocks. However, metamorphic rocks can also be formed by the effects of heat and pressure on igneous rocks. One of the most common of these is greenstone. Greenstone is formed from rocks like basalt, diabase or andesite. These rocks owe their color to the presence of green colored minerals like chlorite, actinolite, or epidote (or a combination of them). Now not every green colored rock is a "greenstone". I have seen green colored andesites, quartzites and other rocks which are green colored, but are not greenstones. The true greenstone rocks are also sometimes called metavolcanic rocks. They are also closely related to greenschists, which are foliated (easily parted) rocks that form from the metamorphosis of basic rocks like basalt, diabase or andesite by the effects of heat and pressure.

Greenstones and greenschist are of great interest to prospectors, because they are associated with gold in many locations around the world, from California to Canada, and from Alaska to Australia. Gold deposits, especially those of the mesothermal type, are associated with greenstones. This is because under heat and pressure, these rocks release water, and that water circulates in the formation of gold deposits. In many districts, gold veins form at the contacts of greenstones with other rocks, especially metasedimentary rocks like schist, slate or quartzite. I will discuss this process in more detail in the chapter on hard rock gold deposit formation.

There is also a type of metamorphic action that is formed when the high heat associated with igneous rocks comes into contact with other types of rock. This is called contact metamorphism. Certain types of tungsten deposits and even some gold deposits occur where rocks are changed at the contact of hot igneous rocks with other materials. During the contact metamorphic process, new elements can be introduced into the rock. Segregation of certain elements can also occur under the exposure to heat and pressure. Contact metamorphic deposits often contain high concentrations of garnet and epidote.

Sometimes heated geothermal waters will rise right to the surface as in this natural vent at Steamboat Hot Springs near Reno, Nevada. It is thought the Comstock Lode once looked like this.

Another specialized category of "changed" rock that the prospector will want to be on the lookout for is those rocks that have been changed by hydrothermal alteration. Hydrothermal alteration can affect rocks of all categories, igneous, metamorphic and sedimentary, and at times the changes are so strong that it is difficult to tell what the original unaltered rock was. Geothermal fluids cause alteration of rocks as passing hot water fluids flow through them changing their composition by adding or removing various chemical components. Temperatures of these fluids can range from weakly elevated to several hundred degrees. The chemical composition of these fluids is extremely variable, so that there are many different results from these types of changes. Several sources of these hydrothermal fluids often combine into one flow affecting the chemical make of the fluids and the resulting rock alteration. Water can come from the magmatic rocks themselves, because many magmas do contain significant water that is released during the final stages of cooling. As noted under the discussion of greenstones, certain metamorphic rock changes can release water as the minerals change and re-crystallize. The other main source is near surface groundwater or "meteoric" water.

Porous and permeable rocks allow the waters to circulate more readily, which may allow large scale mineralization and alteration as is found in the Carlin type deposits of Nevada. Some types of rocks are extremely impermeable and may create a dam or ponding effect behind the impermeable barrier. Fluid movement can also be facilitated by the presence of lots of thin layers along which the water can flow. Hydrothermal fluids also commonly circulate along fractures and faults. This is why altered rocks are frequently associated with mineral bearing veins that form where these fluids flow through larger, open space fractures and faults. Zones of altered rock may form an envelope around vein structures and can be an excellent guide to deposits that may be found in the same area.

Hydrothermal alteration may be found as bleached rocks where the weathering of sulfides that have been introduced into a rock forms light colored clays. It can take the form of silicification where large amounts of silica are introduced into the rock, sometimes completely changing it into a silica rock, often called jasperiod, because of its similarity to jasper. In addition, depending on the chemicals in the hydrothermal fluids, a variety of minerals such as sericite (a mica), adularia (a feldspar) or alunite (a sulfate) may be introduced into the rock. Many of these new minerals will cause a color change in the rock and can be recognized by the individual prospector.

The heat of the metamorphic environments is important in the formation of many types of metallic ore deposits. These are the kinds of conditions that commonly lead to the formation of mineral deposits, including gold. This is why most gold and other precious metals deposits are associated with metamorphic or igneous rocks. Deeper-seated forms of ore deposits such as mesothermal gold deposits are often associated with metamorphic rocks as mesothermal types of conditions are present deep in the earth where metamorphic rocks are formed.

The Grand Canyon of the Colorado River cuts down through many thousands of feet of layered sedimentary rocks, including limestone, sandstone and shale.

SEDIMENTARY ROCKS

The third major class of rocks is those of sedimentary origin. Sediment means an accumulation of loose particulate material such as clay, sand, gravel, etc. This group includes all rocks made of sediments, their materials having settled either in water upon the bottoms of rivers, lakes, or seas, or on dry land, as in the case of deposits made by the wind and by glaciers. Sedimentary rocks are divided into the fragmental rocks--which are made of various types of rock fragments, either coarse or fine--and the different types of rocks which consist of chemical precipitates.

The sedimentary rocks are sometimes divided according to their composition into the following classes: 1.) The arenaceous, or rocks made principally of quartz, including beds of loose sand and gravel, sandstone, quartzite, and conglomerate (a rock made of cemented rounded gravel or pebbles); 2.) The calcareous, or lime rocks, including limestone and a soft white rock formed of calcareous powder known as chalk; 3.) The argillaceous, or clay rocks, including muds, clays, and shales. These three classes grade by mixture into one another. Thus there are limy and clayey sandstones, sandy and clayey limestones, and sandy and limy shales.

This sandstone shows layering in two directions. One was laid down, then tilted, then the second layer that is still nearly horizontal was laid on top.

Most sedimentary rocks tend to have a layered type of structure, which is caused by the environments of their formation, because sediments are often put down in periodic flooding or drainage events. Loose sediment becomes a solid sedimentary rock through a process called lithification, which involves compaction, cementation and recrystallization (of carbonate sediments).

Terrigenous sedimentary rocks are those that are derived from the weathering of pre-existing rocks, which have been transported to a location such as a basin, where they are deposited. They have a clastic (broken or fragmental) texture consisting of: 1) clasts (larger chunks of rock, such as gravel); 2) matrix (mud or other fine-grained sediment surrounding the clasts) and 3) cement (which is the binder holding it all together). Cements that bind terrigenous sedimentary rocks include minerals such as calcite, iron oxide or silica. These rocks include conglomerate or breccia, sandstone, siltstone and shale. There are also sedimentary rocks formed from chemical precipitation in some kind of depositional basin (such as oceans or lakes) from chemical components of rocks that become dissolved in water. These chemicals may be removed from water and made into rocks by chemical processes, evaporation, or with the assistance of biological processes (such as shell growth). In some cases, it is difficult to sort the two out (in carbonates or some siliceous rocks, for example), so they are put together as the chemical/biochemical sedimentary group. These rocks include the evaporates (like salt deposits), the carbonates (limestones and dolostone), and the siliceous rocks. Other examples of these rocks include gypsum and travertine.

Another type of sedimentary rock is the carbonate group. These sedimentary rocks are formed through both chemical and biochemical processes. They include the limestones (many types) and dolostones. Two minerals are dominant in carbonate rocks: calcite (calcium carbonate) and dolomite (calcium and magnesium carbonate). Still another type of sedimentary rock is those which are derived from organic origins - these are the coal group. This group consists of rocks which form by the build up of organic matter (mainly plant fragments). Because of this, they lack minerals (which must be inorganic, by definition). Different types of coal form depending on the depth and pressure the material has been subjected to. In order of increasing depth of burial (temperature and pressure): 1) peat (porous mass of brownish plant fragments resembling peat moss); 2) lignite (more black than peat, but still very crumbly); 3) bituminous coal (dull to shiny and black; sooty; layers may be visible, usually fairly solid); 4) anthracite coal and jet (extremely shiny and black, may have a slight golden shine; not sooty). Technically anthracite is a metamorphic rock due to high temperatures and pressures to which it has been subjected.

One final specialized form of sedimentary rock is the ironstone. At a certain time in the distant past, increased oxygen in the atmosphere led to a massive precipitation of iron out of lakes and oceans. Formations of this type are found worldwide associated with rocks of that age and have been a considerable source of iron ore. Ironstone is also used as a more general term for zones, bands and segregations of iron oxide minerals that are sometimes found associated with schist rocks and gossan deposits. These latter types are commonly associated with gold deposits, both in the US and Australia. In general, there are comparatively few gold deposits associated with sedimentary rocks as they are not often associated with the heat needed to create ore deposits. However there are some notable exceptions. The huge disseminated Carlin type deposits in Nevada occur mostly in sedimentary rocks. There are also some ancient placer deposits that have become incorporated into sedimentary rocks - their erosion contributes to the gold in modern rivers and streams. The Tertiary River deposits of California would fall into this latter category.

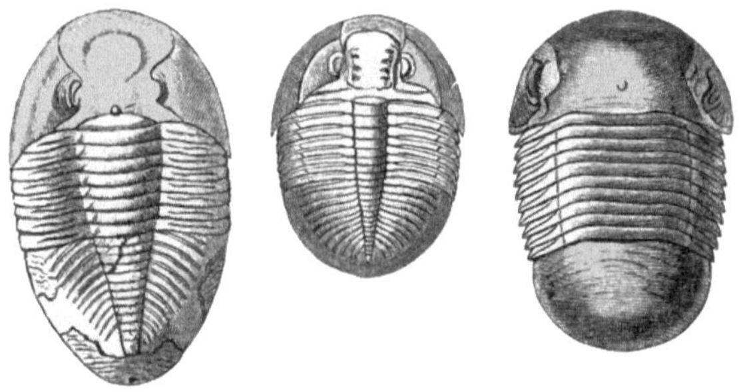

Trilobites are a popular fossil that come in many types but the animal disappeared long ago.

One of the things many prospectors would like is to have a better ability to identify and recognize different types of rock in the field. While a geologist might saw the rock into thin sections and look at it under a microscope, the individual needs to be able to recognize things right in the field. It takes a little training and practice, but this is a worthwhile skill, and probably the best way to do this is to get a good geology handbook. These types of books are available at larger bookstores with good coverage of natural history topics. A good field handbook will have clear pictures and descriptions that will assist you in identification. Rocks are identified by their mineral makeup and texture (size) of the minerals. In order to successfully identify rocks, one needs some feel for the minerals that make them up. A good geology field handbook will help you understand the observable characteristics of a rock and then assist you along in the process of identifying it.

Finally, I'd like to cover a little bit on the geologic column. The geologic column is basically a system that gives names to certain periods of time in the geologic past. For the most part, the various times are differentiated from each other by the types of fossils that are found in the sedimentary rocks of that time period. This is because each individual body of sedimentary rock, each formation, constitutes a record of the time when it was laid down. The structure and the character of the sediments of each formation tell whether the area was land or sea at the time when they were spread; and if on land, whether the land was part of a river plain, or a lake bed, or was covered with wind-blown sands, or by the deposits of a glacier. If the sediments are marine, we can determine also whether they were laid in shoal water near the shore or in deeper water out at sea. The stratum of each formation essentially writes a separate local chapter about the geological history of the earth at that point.

From the popular movie Jurassic Park, most folks could guess that the Jurassic period includes a part of the times of the dinosaurs. The dinosaur extinction at the end of the cretaceous period is marked by various dating methods as having occurred about 66 million years ago. Quaternary is the name for the current geologic time period. Various other names are given to other time periods, and longer periods are broken down into various sub-periods. It is probably not all that important for the individual prospector to memorize all the names for these geologic periods or how many millions of years ago they occurred, but it is worthwhile to understand their nature and the basic concepts behind them. This includes the concept that younger rocks generally lie on top of older rocks, which is important.

While it's not terribly important to know, just for reference the periods of geologic time are as follows: Precambrian time is divided into three eras, Paleozoic time into six eras, Mesozoic time into three, and the Cenozoic into two.

The eras of Precambrian time are divided into:
1. Hadean – 4600 to 3800 MYA.
2. Archaen – 3800 to 2500 MYA.
3. Proterozoic – 2500 to 544 MYA.

The eras of Paleozoic time are the following:
1. Cambrian – 544 to 505 MYA. In this era, the animals were exclusively marine invertebrates
2. Ordovician – 505 to 438 MYA, In this era appeared a few Fishes, Insects, and terrestrial plants. The Cambrian and Lower Silurian eras may be called the Age of Invertebrates.
3. Silurian – 438 to 408 MYA. In this era, fishes, insects, and land plants became more abundant.
4. Devonian – 408 to 360 MYA. In this era, fishes showed a further increase in diversity of type. Amphibians seem to have made their first appearance. The Upper Silurian and Devonian eras are known as the Age of Fishes.
5. Carboniferous - 360 to 286 MYA. Amphibians became abundant, and a few reptiles appeared in the closing period of the era. It is divided into an older portion (the **Mississippian**) and younger (the **Pennsylvanian**).
6. Permian – 286 to 248 MYA. The Carboniferous and Permian eras together are called the Age of the Amphibians.

The eras of Mesozoic time are the following:
1. Triassic – 248 to 205 MYA. In this era, reptiles first became abundant, and the earliest mammals made their appearance.
2. Jurassic – 205 to 140 MYA. In this era, reptiles became still more abundant, and presented a greater diversity of type. Birds made their first appearance. Gymnosperms were the dominant type of vegetation.
3. Cretaceous - 140 to 66 MYA. Modern angiosperm plants and fishes of the modern type (Teleosts) become abundant.
The eras of Cenozoic time are the following:
1. Tertiary- 66 to 1.6 MYA. The Tertiary era is known as the Age of Mammals.
2. Quaternary – 1.6 MYA to the present. Man himself, and other existing species of Vertebrates, make their appearance. The era may be called the Age of Man.

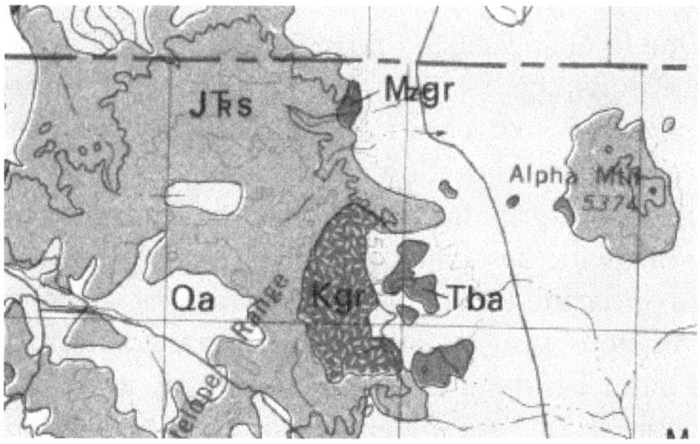

This section of a geologic map near Alpha Mountain in Northern Nevada shows Jurassic/Triassic slate/shale (Jtrs); Cretaceous granite (Kgr); Tertiary basalt (Tba) and Quaternary (recent) alluvium gravels (Qa).

It's also helpful in reading geologic maps to be able to see the comparative ages of rocks where one type overlies another. If the mineralization in a certain area is found in rocks of a certain age, the rocks much younger than that are usually not favorable for deposits as they were not around when the period of mineralization occurred. In that example, a prospector would do best to focus his attention on outcrops of the rocks that are from the favorable time period. Additional information on the geologic column can also be found in any good geology handbook, which is something every prospector ought to own. Probably the most important thing the independent prospector needs to know about different rock types is that the rocks that are most typically associated with gold deposits are those of the igneous or metamorphic types. This is because of the association of heat with mineral deposits. Taking the time to learn to recognize these important rock types in the field is worthwhile. Next in this series, we will take a look at how the geologic processes related to igneous rocks, volcanoes and mountain building lead to the formation of mineral deposits.

Plate Tectonics, Volcanism and Mountain Building for the Prospector

The formation of gold, silver and many other types of deposits depend on a natural segregation of the elements from the earth, so that gold or silver (and other minerals) become enriched enough to form a valuable ore. The heat and pressure that is usually necessary to cause these natural concentration processes, is often associated with mountain building and/or volcanic activity. Because of this, it's useful for the individual prospector to know a little something about plate tectonics, and how plate tectonics relates to volcanic activity, faulting and mountain building.

Plate tectonics is a topic on which many books have been written, and the average prospector may have only a passing interest in it, but a basic understanding of the concept can help understand the large regional geologic forces which lead to the formation of gold deposits. The basic theory of plate tectonics is that the surface of the earth consists of a series of comparatively solid plates, which "float" so to speak, on the semi-liquid mantle below. This semi-liquid layer circulates due to convection and other forces. Single "hot spots" of upwelling magma create isolated volcanoes like those responsible for forming the Hawaiian Islands. In the middle parts of the Atlantic and Pacific oceans, long and consistent lines of upwelling material from within the core creates rift zones that push up basaltic rock, which cools and spreads out the two oceans from their centers. The spreading forces in the Atlantic and Pacific are so strong that it causes the plates on the surface to push up against each other, and some plates actually get pushed underneath others. The continents are made of plates of lighter, less dense rock that "float" over the denser semi liquid mantle which is made of basaltic rock. Both the collision and sliding under processes (called subduction) cause mountain building and volcanic activity. The Himalayan mountains - the highest on earth - have formed as the result of the collision of the Indian sub-continent plate as it presses into the southern part of the Asian plate, because neither plate has slid underneath the other, that collision has created the uplift of a huge mountain range. On our part of the planet, the North American plate has been colliding with or sliding over plates on the Pacific Ocean side of the continent for many millions of years. The long collision of the Juan Del Fuca and Pacific plates against the North American Continent has led to a long and repeated series of mountain building episodes (orogenies) of the Cordillera. It has also added a considerable amount of material to the western edge of the continent, as deep-water sediments were thrust up or accreted onto the western shores. This is how rocks that formed deep in the ocean are now lifted thousands of feet above sea level. These older deep-water sedimentary rocks cover much of Nevada and large parts of California. Regional low angle thrust faults across Nevada are the scars of the powerful forces of these colliding plates. The

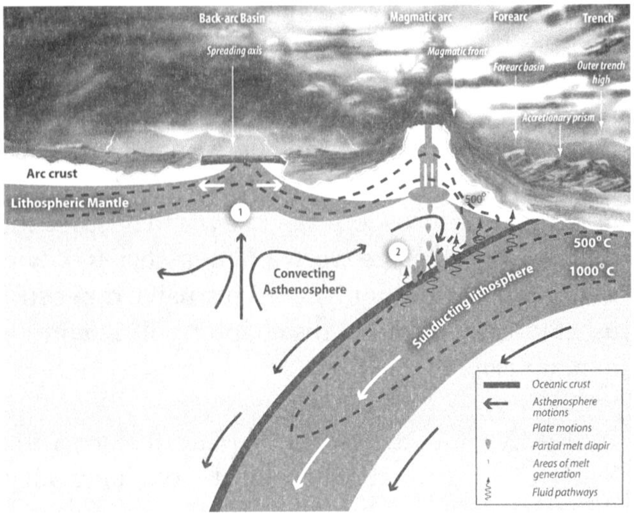

Subduction zones along the earth's crustal plates cause volcanic action and form igneous intrusive rocks, commonly resulting in gold deposits.

buckling and folding force of the two plates against one another created the costal ranges in California, and the resulting volcanism and igneous activity has created both the still active Cascade Range volcanic system, as well as the older Sierra Nevada Range. To this day, it causes the continued activity on the San Andreas – Queen Anne fault system, one of the planet's largest faults. Other shallow to deep water ocean formed sedimentary rocks that are common in nearby western states, including the Colorado plateau area of Arizona, Utah and Colorado were added to the western part of the continent by the same process.

Where crustal subduction slides one plate down below another, the resulting melting of the subducted rocks allows for segregation and the lighter, less dense (higher silica) portions tend to move upward, toward the surface. This means that the parts of the rock higher in silica tend to rise back toward the surface. This is the process that is thought to have led to the formation of the Sierra Nevada batholith – a large block of granitic type rocks that form the backbone of the Sierra Nevada Range. This mass of molten material brought heat from the deeper parts of the mantle, miles below the surface upward closer to the surface. The hot material becomes the heat source and driver for the formation of many mineral deposits. In the case of the Sierra Nevada range, the heat of the upward moving igneous rock became the source that created the wide spread gold deposits of the mother lode.

So all this may be well and good, but what interest is it to the individual prospector? Many experienced prospectors know that precious metals deposits are very often associated with metamorphic and/or igneous intrusive rocks such as granite or diorite. This chapter will attempt to explain *why*.

Indirectly, it is the forces of plate tectonics that lead to the formation of most gold and silver deposits. The mountain building processes resulting from plate collisions brings the heat of the mantle closer to the surface, creating the forces that form metamorphic rocks, and lead to the fissures and faults that are commonly the conduits for ore bearing fluids.

Volcanism:

Volcanoes are formed when hot molten material from deep within the earth is able to rise to the surface and erupt out. If the magma moves too slowly to the surface, it will solidify and become igneous intrusive rock (like granite or gabbro) rather than an extrusive rock (like basalt, which is a lava-type rock). It is no accident that the "ring of fire," a ring of active and ancient volcanoes, is located around the rim of the pacific plate encircling where other plates come in contact with it. The edges of these plates are breaks in the crust along which deep magma can move quickly to the surface. There are a number of types of volcanoes; the following is a brief listing of the best-known types:

Shield volcanoes are the largest of all volcanoes on Earth. The Hawaiian volcanoes are the most famous examples of this type. These volcanoes are mostly made up of basalt, a type of lava that is very fluid and flows easily when erupted. For this reason these volcanoes are not steep sided. They are only explosive if water somehow gets into the vent, otherwise they are characterized by lava fountaining that forms cinder cones and spatter cones at the vent. This explains why Kilauea volcano on the Big Island of Hawaii is safe enough for everyone to hike in and see. It's safe because the basalt flows easily and smoothly - it just oozes out like red-hot toothpaste.

The author on the slope of Pu`u `O`o volcano on the big island of Hawaii. Molten lava is flowing down the hill a few hundred yards behind me.

Strato volcanoes are the type that makes up the largest percentage (about 60%) of the Earth's volcanoes. They are characterized by eruptions of cooler and more viscous types of lava. The usual lavas that erupt from strato volcanoes are andesite, dacite, and occasionally rhyolite. These more viscous lavas allow gas pressures to build up to high levels (they are effective "plugs" to the magma plumbing); therefore these volcanoes commonly suffer explosive eruptions. Mt. St. Helens is of this type. Mt St. Helens is very dangerous when she is about to erupt because her rhyolitic lavas often don't flow well, so it's not uncommon that they just plug up the volcano until the pressure builds up to the point where the volcano explodes - killing everything for miles around.

Large rhyolite caldera complexes are the most explosive of Earth's volcanoes. These are volcanoes that often don't even look like volcanoes. They are usually so explosive when they erupt that they end up collapsing in on themselves rather than building any tall structure. The collapsed depressions are called calderas, and they indicate that the magma chambers associated with these eruptions are huge. One of the rocks formed by this kind of explosion, called a rhyolitic ash flow tuff, covers huge areas of Nevada - there were a lot these types of volcanic explosions in the distant past in Nevada. These caldera volcanoes are important to prospectors because they are commonly associated with the formation of mineral deposits along what are known as ring fractures.

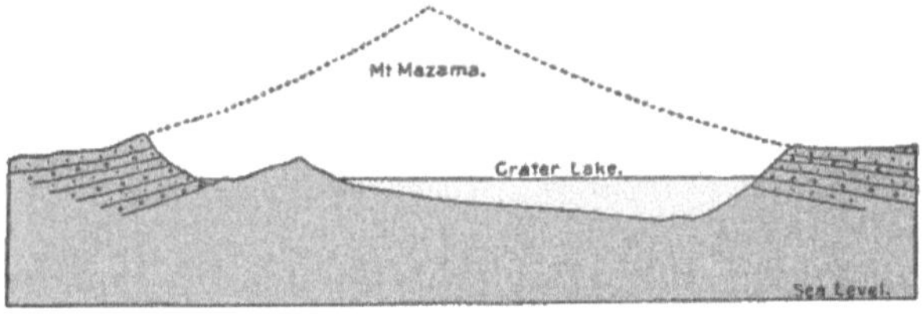

Crater Lake in Oregon is the remains of a Strato volcano that exploded long ago similar to the way Mt St. Helens did in May of 1980. The outline shows the part blasted away and the lake.

Lesser-known volcano types include: **Monogenetic fields,** which are a collection of sometimes hundreds to thousands of separate small vents and flows spreading all its separate eruptions over a large area;
Flood basalt provinces, these large areas are covered by thousands of square kilometers of thick basalt lava flows - sometimes more than 150 feet thick, and extending for hundreds of miles;
Mid-ocean ridge volcanism occurs at plate margins where oceanic plates are created. This is where the basalt intrudes to the surface and pushes or spreads the plates apart.

Mountain Building:

Mountain belts are typically formed by plate tectonic activity, specifically plate collisions and igneous activity related to plate subduction. An orogeny is the name geologists give to a time or episode of mountain building activity. The formation of the Cordillera (the western mountains, including the Rocky Mountains, Cascades, Sierra Nevada, and Coast Ranges) occurred over a number of episodes of geologic time, and each of these orogeny periods has their own name. The process that uplifts and produces mountain belts is called orogenesis. The same forces that uplift mountains create related events such as rock folding; thrust faulting (which occurs where horizontal forces thrust one block of rock over the top of another); metamorphism and igneous activity.

There are a number of types of mountains, including: volcanic mountains built up from a series of eruptions; folded and thrust-faulted mountains built up from the buckling of significant horizontal forces; and fault-block mountains where one block is uplifted relative to another.

Because rock is a fairly brittle material (it is strong under compression, but very weak under tension) mountain building and igneous activity (volcanoes or emplacement of intrusive rocks like granite) often results in considerable faulting, shearing and breaking of rock. The fissures created by the broken rock can then become conduits where geothermal fluids can flow and eventually form mineral deposits. The igneous activity, uplift and broken rock all occurring in the same close location is the cause as to why so many mineral deposits are associated with mountain ranges.

A break or fracture plane where geologic forces cause the rock on one side to move relative to the rock on the other side is known as a fault. Faults can be anything from very tiny - only a few feet long and less than an inch of movement - all the way to huge regional faults like the San Andreas

Fault system. They can be sharp and narrow or form wide zones of broken and crushed rock. Faults can bring two different types of rock together, or take a portion of a productive vein and move it to another location. Many productive veins have been lost in the past due to movement caused by faulting which occurred after the vein was formed. Movement after the vein is formed is called post mineral faulting, and movement which occurs before the flow of mineralized, geothermal waters is pre-mineral faulting. Pre-mineral faults often become the conduits for mineralizing geothermal fluids that form veins and other deposits. Faults are geologic formations of interest to the prospector. They are associated with contact zones, dikes, sills, and batholithic intrusions, as well as silicification and mineralization of rock from fluids moving along them. Larger scale faults are the ones best suited for moving lots of heated water and forming significant gold deposits.

Formation of the Sierra Nevada Goldfields:
As an example, let's take a closer look at the geologic relationship of the Sierra Nevada range, the Sierra Nevada granitic Batholith, Tertiary volcanic activity in the Sierra Nevada and the famous mother lode gold deposits.

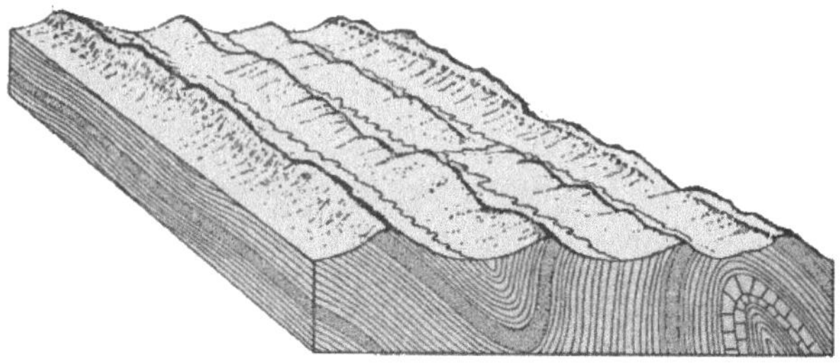

The Sierra mountain range block before uplift, consisting of mostly folded, gold bearing metamorphic rocks, with streams draining mostly in a North – South direction.

The formation of the Sierra Nevada range began in the Jurassic period, approximately 150 million years ago. At that time, the spreading tectonic forces of the pacific ridge pushed an off shore island arc much like present day Japan into the West coast of North America. This collision pushed up a set of mountains, a geologic event called the Nevadan orogeny. These hills were the first formations of the Sierra Nevada range. The pressure from this event produced metamorphic rock from the miles-deep sediments that were lifted out of the ocean. Most of the land that forms California and Nevada was added onto the North American plate during those periods of collision. At roughly the same time, a subduction zone started to form at the edge of the continental plate. As a result, the eastern edge of the Pacific Ocean plate started to dive beneath the North American plate. The lighter fractions of the magma from the melting oceanic plate separated and rose to create bodies of igneous rock deep below the surface. These plutons were formed at various times, from 115 million to 87 million years ago, cooling slowly at a depth of about 6 miles below the surface. For the most part, they have a granitic composition. Heat from these deep bodies circulated mineralized fluids up through deep seated cracks and faults in the metamorphic rocks that were formed when the rocks of the ocean floor were shoved onto the North American plate. These circulating fluids formed the deep mesothermal gold ores of the mother lode country and the western Sierra. Continuing geologic action uplifted this overlying metamorphic rock along with the rest of the Sierra Nevada. Over time, large portions of the overlying rock were removed from the area by erosion, exposing the gold bearing veins. By 65 million years ago, this early Sierra Nevada was worn down to a range of rolling low mountains, a few thousand feet high. The wide and flat rivers that drained these hills were rich with gold placers formed from the erosion of gold bearing veins in the mineralized metamorphic rocks.

Continued subduction of the Pacific (Farallon) plate underneath the North American plate led to an extended period of volcanic activity that began in the late Cretaceous and early Tertiary period. A series of large volcanoes started erupting explosively and spewing forth volcanic ash. Andesitic lavas erupted from fissures and dikes in the Sierra Nevada. This volcanism buried the gold bearing rivers in ash and mudflows. Flows of andesite lava on top of the mud flows further capped the river channels with a weather resistant top. The most southern of these volcanoes still visible are the Sutter Buttes, near Marysville, which has long been dormant. The next in this series of volcanoes is Mount Lassen which is still active - the chain extends toward the north into what is now called the Cascade Range. In time, the subduction ended and the pacific plate began to slide northwest against the North American plate along what has become known as the San Andreas fault zone. The once huge volcanoes south of Mt Lassen on the western side of the Sierras that were associated with plate subduction stopped erupting, became dormant, and were worn flat.

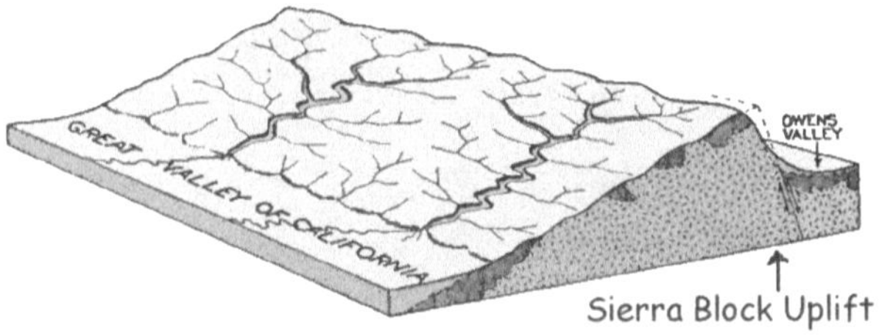

The uplift of the Sierra mountain range block, tilted up mostly on the east side, caused drainage that goes mostly in an East- West direction.

About 25 million years ago, the Sierra Nevada started to rise and tilt toward the west. Rivers eroded away much of the volcanic ash capping the area and cut deep canyons on both sides of the range. New and often rich placers formed along these modern streams, taking gold from both the eroded veins as well as reconcentrating the gold from the older Tertiary channels where the modern streams cut through them. This is the geologic environment which resulted in the legendary mother lode placers discovered by the 49ers.

Nevada's Epithermal Gold & Silver Deposits

Since it is partly related, let's also look at how tectonic activity also led to the formation of Nevada's epithermal gold-silver deposits.

Most of Nevada's important mineral deposits are Tertiary in age and are related to Tertiary volcanics. About 45 million years ago, new volcanic activity began to sweep across Nevada, moving out of the Northeast in a southwest direction. New volcanoes began to form and erupt, blanketing Northern Nevada with what was mostly ash flow tuffs. After reaching central Nevada about 33 million years ago, the spread of volcanic activity began to slow, finally dying out around 19 million years ago. Much of this volcanic activity was expressed through the explosive eruption of large rhyolitic caldera complexes. Calderas are structural depressions that form from the catastrophic explosive eruption of magma causing collapse of the chamber roof. These explosions normally result in considerable quantities of rhyolitic ash flow tuffs - rocks which are widespread across Nevada. This process usually results in extensive fracturing of the area rocks - often having a ring-like structure around the edge of the caldera depression. They are also the sites of long lived near-surface heat sources that may drive hydrothermal and hot springs systems to mineralize local rocks through the caldera's system of fracture faults. Many of Nevada's epithermal deposits are thought to be associated with caldera volcano systems. They formed at depths much shallower and temperatures much lower than the Mother Lode veins of California.

As the volcanism slowed, the magmas moved through the highly fractured Walker Lane area resulting in pervasive alteration of the area rocks, the formation of hot spring systems and widespread shallow epithermal mineralization. In fact, the majority of Nevada's most important bonanza type epithermal deposits are located within the Walker Lane province including those at Tonopah, Goldfield, Aurora and Virginia City. There are also dozens of smaller districts and isolated mines. As in California, this volcanic activity was also related to the subduction of the Pacific (Farallon) plate beneath the North American plate, where melting of the plate material led to the formation of lighter magmas, which rose and were extruded on the surface or erupted as ash flow tuffs. The activity slowed and eventually ended as new plate movement began to be taken up by sliding along the San Andreas Fault rather than continued plate subduction. After the long period of subduction, the tectonic forces in California transitioned to a sliding type regime, but the Great Basin began to experience a period of extension, which many geologists believe was caused by the force of upwelling magma from below. Geologists call this a rifting process, bowing up the plate as the upward flowing magma from the mantle underneath pushed upward. This uplifted the whole region, with the highest point in the middle. As the upwelling force of the magma raised up the state, a force stretching apart the region was created, and large V-shaped blocks of rock within the region slid downward forming the now familiar basin and range province. Nevada still has a gentle bow shape - as the central part of the state has the highest elevations. This period of extension across Nevada which created the basin and range topography, also created deep faults that tapped down to molten rock below. This led to a time of revived volcanic activity, called the bimodal volcanic period, because of the fact that the lavas that erupted consist of two modes - a high silica rhyolite and a basalt. There are some important metallic deposits in Nevada that are associated with this later period of volcanic activity, including precious metal deposits at Seven Troughs and Sleeper in northern Nevada and the Bullfrog district in the south.

Summary:

So, in the end, the topic of plate tectonics and the related mountain building, faulting and volcanic forces it creates are very important to the prospector. It's the pushing force of the spreading center that causes the tectonic subduction. These forces are important to goldfields all over the world - from Australia to Alaska to the Ural Mountains in Russia, and even the Appalachian goldfields of the eastern US. I merely chose California because it is a decent example as it's so much more recent, and it's easy to see the causes and resulting effects as many are still in operation. The mid-Pacific rift is spreading and pushing against the North American plate - it runs N-S and is still currently spreading.

Material has been subducting under the North American plate for a long time. Faults form generally parallel to the direction of the source of pushing from the rift. In California, most veins run N - S, actually a bit NW-SE of true N-S. This is parallel to the mid-Pacific rift fault which also runs N-S, and perpendicular to the force the rift exerts. In Australia, the "Continental Island" has been pushed around all over the place since Precambrian times. The subduction related gold deposits of Victoria and Western Australia have the same geology but bear no relation to the current, modern spreading zones, as they are related to forces of collision that are much older. In spite of that age difference, they have much in common geologically. The force back then might have been in a different direction - perhaps E-W, or some other different direction, but rotation of the Australian continent over millions of years has changed this and now has the current spreading oriented in a N-S direction. Anyway, veins or reefs form in faults, and deep faults form parallel to the tectonic spreading line, and roughly perpendicular to the force generated by the spreading zone.

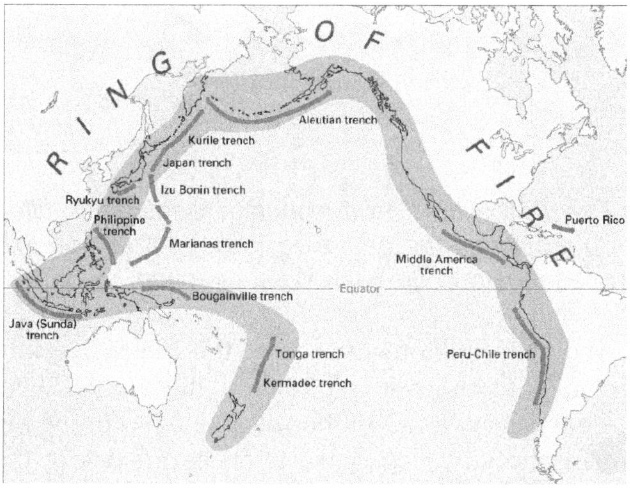

The Pacific Ring of Fire: a zone of volcanic and tectonic activity surrounding the Pacific Ocean. (Diagram Courtesy USGS)

In the Ural Mountains of central Russia, there are many rich and famous gold fields that are also related to tectonic collisions. During the Carboniferous period, the Siberian and European plates collided. Neither plate subducted under the other, and as a result, the great Ural mountain range was lifted up at the zone of collision. The two plates have been stuck to one another ever since. The mountain building and deep faulting underneath created mesothermal goldfields of a type that is geologically similar to those found in Australia and California. Erosion of this old mountain range has created vast placer fields that still produce gold. The same deep faulting also brought up rocks which created the platinum placers for which the Ural Mountains are famous.

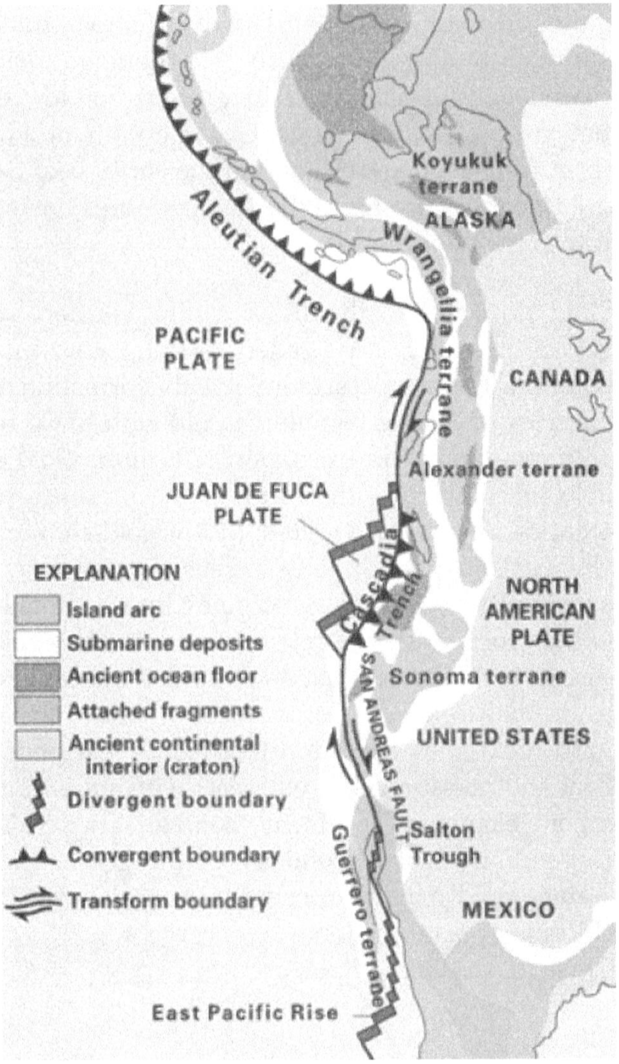

This closer view shows the "Ring of Fire" faulting and rift zones along the US Pacific coast. This activity formed the goldfields of California, Oregon and Nevada. (Diagram Courtesy USGS)

Even the southern Appalachian gold belts of the southeast US are related to mountain building activity caused very long ago by the subduction of a part of the eastern Atlantic plate and a later collision of North America with the continent of Africa which occurred between 500 and 270 million years ago.

The subduction process has a tendency to form off shore island arcs, similar to what Japan looks like today, and similar structures have been added to the pacific coast of the north American plate over the years. When the basalt of these island chains and the floor of the spreading ocean are pushed onto the continent, they create a situation where near surface lavas become deeply buried and subject to regional metamorphic temperatures and pressures. The rock formed by heat and pressure on basalt and similar materials is a metavolcanic rock, and the most common example is called greenstone. Greenstones are commonly associated with gold deposits in California, Australia and elsewhere. In some places in Australia, these metamorphosed volcanics (greenstones) contribute much gold to the vein formation process. However, they also contribute water to the system. Most geologists think the water is actually taken out of the minerals in the rocks. Many minerals include what is known as water of crystallization - water incorporated right into the very crystal lattice of the minerals. Under high pressure and some heat, the water comes out of the crystals and is available to circulate, leaching gold and minerals out of one location and re-depositing them elsewhere. The water needs an opening to flow, hence flow along faults creates the veins (or reefs) we see in the field.

Some other unusual types of rock which are associated with these deep subduction zone systems are the ultra-basic intrusive rocks, like dunite and peridotite. These rocks are often emplaced along deep fault systems which allow bodies of iron rich, silica poor rock to rise up close to the surface from their original position deep within the earth's crust. Another intrusive rock associated with these deep faults is diabase. The dunite and peridotite are easily altered, and form a rock known as serpentine. In addition to being associated with some unusual rocks, these deep fault zones are also closely associated with gold mineralization. These ultra basic intrusive rocks also sometimes contain metallic platinum, and give rise to platinum placers. See the platinum chapter in the appendix for more details on the formation of platinum placers.

Geology of Hard Rock Gold Deposits

In this chapter, we are going to take a look at the formation of hard rock gold deposits, and the geologic processes that lead to those formations. This chapter is probably the most technical one in the book, but I think it's important that the experienced prospector understand how primary gold deposits are formed. It may not really be for beginners, so feel free to skim through this material, but as you gain more experience in the world of prospecting, you ought to come back and read it through carefully if it seems a little bit unclear on your first reading through it. I tried to use as little technical language as I can get away with while still being clear about the subject at hand. Don't let a little geology lingo intimidate you - being a good prospector is in a large way about the knowledge you posses and a higher level of knowledge will help you find more gold in the long run.

Geologic processes within the earth can either concentrate or disperse gold and other elements from adjacent rocks. It is the results of the concentration processes that the prospector is most interested in, because those are the ones that create valuable mineral deposits. The process of mountain building, volcanism and intrusions of molten rock continue to work at various places across the planet. Many of these geologic forces are the same ones that in one way or another lead to the formation of gold deposits. Waters heated by magma and other rocks deep in the earth circulate and continue to create gold deposits below the surface.

Since most primary hard rock gold deposits form from circulating waters within buried rock deep down in the earth, openings such as fault zones are needed to allow the fluids to flow and circulate. The fluids can circulate through almost any kind of rock as long as there are openings or the rock is porous enough. Sometimes the water circulates along the contact between different rocks, or along folds. The temperatures are actually not as hot as many would think. Most gold is deposited from solutions within the range of 200F to 500F, although a few are higher. These are temperatures similar to those we use in cooking food. Occasionally temperatures are higher or lower, but most fall in this range. Gold deposits do not form from molten gold within the earth, because gold deposits are formed at temperatures far too low to melt gold.

Gold does not easily dissolve into solution, so special processes are needed to bring gold into solution. These processes involve the formation of complex molecules combining gold with sulfur or chlorine - and on rare occasion, tellurium. Gold is able to combine with sulfur in the presence of chloride ions to form gold-sulfur or gold-chloride chemicals that can dissolve in heated water. I have specifically decided to avoid going into the details of the geochemistry of gold solutions in this whole discussion, because I just don't think it's necessary to do that to learn what we need to know about hard rock gold. The important fact is that under heat and pressure, gold will react with sulfur and other elements to form soluble chemicals. Depending on the chemistry of the geothermal waters, gold may form chemical complexes with chloride or bisulfide ions.

Geothermal heat from deep within the earth dissolves gold and creates valuable deposits.

The gold bearing chemical complexes that are formed in hydrothermal systems are not particularly stable, and under the right conditions they break down and release metallic gold. Where the rare element tellurium is present, the situation is a bit different, and because gold - tellurium chemicals are more stable, they result in the formation of gold telluride minerals. As I said, these gold-sulfur chemical complexes are not all that stable, so that when the chemistry changes or the waters cool and the pressure drops, the chemicals decompose, releasing the gold to form free gold deposits. These same solutions also dissolve quartz when it is present in the rocks, and because quartz is so common, it is also common in the solutions that form veins and other types of deposits. Other minerals which are not valuable that form along with the gold in a deposit are known as gangue minerals.

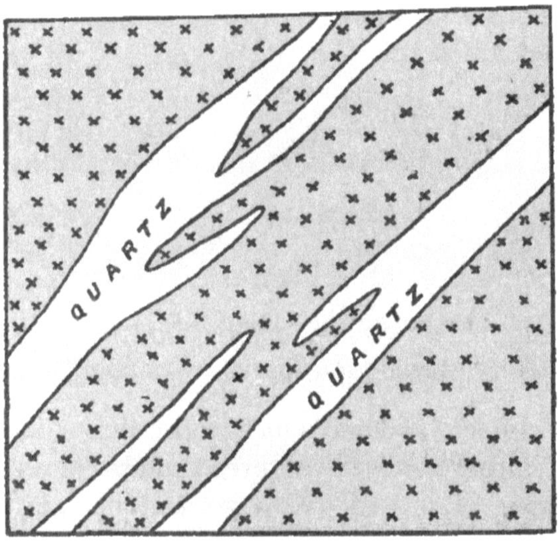

Quartz veins are not always one solid strip – they often branch, pinch and swell.

Its hard to imagine things like gold or quartz being dissolved by water solutions, but if the water is hot enough, the pressure high enough, and the chemistry is right (with acids, and other elements like sulfur and chloride ions present in the mix), then gold, quartz and other things you don't expect to see dissolving will go into solution. The solutions usually move by natural convection (the fact that hot things tend to rise), and natural flow from high pressure (deep) to low pressure (surface) areas. As they rise, the waters start to cool as they move farther away from the heat source in the ground and closer to the surface. As the heated waters rise, they cool, and things like quartz and sulfides become less soluble, so they come out of solution often in the form of veins and other deposits.

Its important to note that although quartz is often present with gold, is not required to dissolve the gold – it's the sulfur that's required. Sulfur is very common in geothermal waters (as in hydrogen sulfide - the odor of rotten eggs). Most natural hot springs have a very strong sulfur odor, and most gold - quartz veins have at least some sulfides like pyrite present. Sulfides are a chemical combination of sulfur with a metal of some type. Pyrite, the most common one, is a combination of iron and sulfur. A few other sulfides which are common in gold bearing veins include: acanthite (silver), chalcopyrite (copper and iron), galena (lead), arsenopyrite (arsenic and iron), sphalerite (zinc), and cinnabar (mercury). These and other sulfides can form along with the gold depending on the chemistry of the rocks and the solutions present. This also explains why at times gold forms with little or no quartz present, but only very rarely with no sulfur or iron present.

Gold is chemically one of the least active elements and only rarely combines with others. Gold occurs in only a few forms in the typical deposit. The three most common forms are free gold, gold encased in sulfides, and gold tellurides. Free gold is intimately associated with quartz, calcite, alunite, and other nonmetallic gangue minerals. This free gold is easily separated from the gangue minerals for recovery. Gold encased in sulfides is still in a metallic form, but becomes much more difficult to recover. The particles of gold inside sulfides like pyrite, galena or chalcopyrite are tiny and often completely surrounded by the sulfide mineral. This means that the sulfides must be somehow broken down before the gold can be recovered from inside them. The old timers called this type of ore "refractory" or "rebellious" because of the difficulty in processing it. A third type of gold occurrence is the gold-telluride minerals. There are a number of minerals in which the elements gold and tellurium combine. These minerals can usually be processed by cyanide, but it takes a little longer than normal to process them and it's important that the mine operator working ores that contain these minerals be aware of their presence in the ores. Tellurides are not that common, but they do occur in a number of locations across the globe. Gold ores can have a combination of these types of occurrences, and gold can be present in all three modes in some ores. It is quite common that a lot of quartz veins contain a certain percentage of free gold with the rest of the gold being encased within the pyrite and other sulfides present in the vein.

The Depth of Formation:

The depth of formation of a gold deposit has a great effect on its nature and characteristics. While quartz is fairly common at all depths, other minerals characterize the depth the formation. As an example, one would expect to find cinnabar and stibnite at or very near the surface. A little further down perhaps a thousand feet up to about one mile, one might expect to find things like opal, chalcedony, quartz, calcite, sericite, chlorite, and adularia as gangue minerals. In addition, barite and fluorite are also sometimes present. The common ore minerals would include such sulfides as chalcopyrite, galena, sphalerite, and arsenopyrite. In this zone of formation, most gold bearing deposits contain a very significant percentage of silver. These are termed epithermal deposits. The depths of the ore shoots in formations such as these can be significant but are often notably shallower than those deposits formed at deeper depths.

At depths approaching one to 2 miles the chemistry will change again, the predominating gangue minerals are quartz, calcite, dolomite, and ankerite, more rarely siderite; fluorite and barite are also occasionally of importance. At this deeper depth of formation most of the gold deposits have comparatively little silver. These deposits are considered in the mesothermal class. It is also important that in these deeper levels of formation the ore bodies typically extend over a wide vertical range. Miners sometimes work these veins to depths many thousands of feet deep. Higher temperature systems form at even deeper depths, and here a different suite of high temperature minerals are found. The high temperature gangue minerals that might be found in some gold bearing veins include things like pyroxenes and amphiboles, the various garnets, apatite, pyrrhotite and schorl tourmaline. The gold vein systems of Appalachia in the southeastern US fall into this category, and it is estimated that most were 3 or 4 miles below the surface at the time of deposition. Geologists have sophisticated ways of estimating the depth of formation for various vein materials, but a quick scan of the gangue minerals that are present in a vein will usually give a very good rough estimate of the depth that it was formed at.

A large open fissure left by fault movement in Arizona. Deep underground, veins could form.

The heated waters which form these deposits come from a mixture of three main sources with a number of related subcategories. The sources of water for any particular type of deposit may be complex and in some places are still hotly debated by geologists. Some of the water is magmatic (meaning it came from water contained within the hot volcanic or granitic rocks inside the earth). Igneous intrusive rocks, which may release their contained water (called "juvenile" water) during the final stages of cooling. In regions of metamorphic rocks, a potential source of the fluids is dehydration reactions which take place during the metamorphic event. With increasing temperatures and pressures of metamorphic change, early, low temperature, hydrous minerals may reform and recrystallize into new, higher temperature, anhydrous minerals. The excess water is released and will circulate through the surrounding rocks where it may scavenge and transport metals to sites where they can be precipitated as ore minerals. Near surface groundwater is another source of water (called "meteoric" water). This type of water comes from rain that has percolated downward from the surface.

In general, the higher the sulfide and base metal content in a deposit, the higher the percentage of magmatic type waters that were present in the mix during its formation. Many mesothermal (deep) types of gold deposits are thought to be formed from water coming out of water bearing minerals that are under heat and pressure. That is why these types of deposits are nearly always associated with metamorphic rocks and typically contain only a small percentage of sulfides. Evidence from some ore deposits suggests meteoric waters may mix with juvenile or metamorphic waters during late stages of mineralization.

Underground in a stope exposing the North Star vein at Grass Valley, California. This vein has a shallow dip, but was mined to great depths.

Many different kinds of rock naturally contain tiny traces of gold - usually between 5 and about 100 parts per billion. The circulating hot waters can leach these traces of gold out of the rock, then re-deposit it in other locations, concentrating those trace levels of desirable metals to the point where the deposits become valuable ore. Gold can deposit in a hydrothermal system at various locations, including points where cooling occurs, where chemical changes in the wall rock occur (like changing from one type of rock to another); where there are pressure changes occurring (like boiling), and where there are sharp changes in the vein (like bends, folds or where smaller veins split off from the main vein). There are other situations or changes as well, but the ones mentioned above are the most common.

Because what we are discussing is a general description of how gold deposits are formed, it could be applied to a number of different types of deposits. It is also somewhat of a simplification of the complexity of the chemistry and nature of gold geology, but it will provide us what we need to know as prospectors. The exact nature of the chemistry, water sources, porosity of the rock, temperature, and water flow rates determine what type of deposit is produced. These factors vary from gold district to district. Variations and changes in these important parameters can yield a variety of different types of primary gold deposits.

Gold is often found associated with quartz, partly because silicon and oxygen, the two elements that make up quartz, are the two most common elements in the earth's crust. Although we don't think of quartz as being a water-soluble material like sugar or salt, at high temperatures and pressures, quartz is slightly soluble, and like the traces of gold, the heated geothermal waters can pick up the quartz and re-deposit it elsewhere. Usually when gold is found with little or no quartz, there are signs that sulfides were originally present such as limonite and other iron oxides. Even though its common to find gold and quartz together, sulfides or chlorides are the chemicals actually needed to dissolve and then deposit gold - the quartz really isn't needed to carry or deposit gold, the quartz just comes along for the ride.

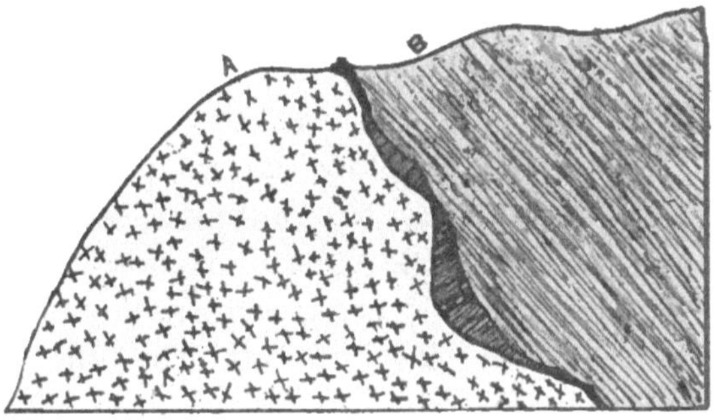

The contact of different rocks, as shown here, are a good place for gold veins and deposits to form.

Gold and quartz are found together in an infinite continuum of gold -quartz mixtures from pieces that are nearly pure quartz with just traces of gold all the way to pieces that are all gold metal. Generally, pieces with less than 50% gold by weight are called specimens rather than nuggets, but there is no hard and fast rule. Much rests in the eyes of the beholder, especially when selling nice natural gold pieces. Gold deposits can be found in many types of rocks, and while metamorphic formations are common hosts for gold veins, they can occur in igneous rocks like granites or rhyolites, and even sedimentary rocks like limestone. That's why it's important to know what the favorable geology is for the district or region you are exploring.

Hydrothermal Brine Fluids:

Hydrothermal fluids can cause chemical alteration and change in adjoining wall rocks by the contact of these fluids with the rocks. They are often acidic, and the changes in the rock composition occur when the fluids add, remove or redistribute elements within the rock. Sometimes the gold mineralization actually extends into the wall rock for a significant distance, and sampling must be done to determine the width of the profitable ore. The water temperatures can range from weakly elevated to maximum of about 800 hundred degrees Fahrenheit. The chemicals within the fluids are extremely variable. The fluids may contain various types of dissolved gases like carbon dioxide, salts (briny fluids), water, and mineral acids. Sodium chloride (common table salt) in amounts up to a few percent is a common constituent of these fluids. Mineral deposits are usually built slowly with large amounts of water being involved depositing only a little bit of quartz and gold at a time. A recent scientific study estimated it would take about 8.5 trillion gallons of water to deposit a single ton of high grade gold ore. In order to allow large quantities of hydrothermal fluids in plutonic settings to circulate, a large-scale conduit is needed. Often these convective paths occur in the nature of faults of other porous rock formations. The flow of the fluids is similar to the water circulation within a pot of boiling water. The hottest water rises fastest directly above the heat source, the water cools, and the flow direction changes to horizontal, and finally downward along the sides of the pot. In a similar manner, hydrothermal fluids circulate upward and outward from an igneous intrusion at depth. Porous and permeable host rocks (those containing lots of interconnected pore spaces) allow the fluids to flow freely, for example, in coarse-grained sandstone or an acid leached limestone layer. Some types of rocks, like shale or slate, are extremely impermeable. A layer of shale or clay can cause damming or ponding of the hydrothermal fluids, which can lead to a concentration of mineralization behind the impermeable barrier. Therefore, contacts of impermeable layers on top of favorable porous layers below are favorable sites for concentrations of mineralization. Fluid migration can be also facilitated by the presence of lots of thin layers, as long as some are permeable.

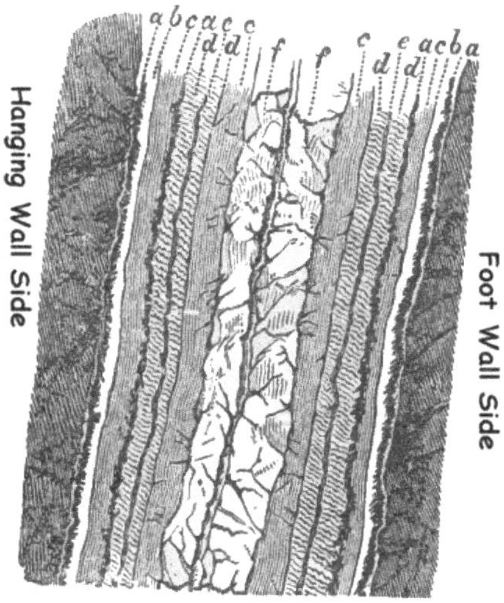

Ribbon type layering is formed by changes in the fluids over time. When the chemistry of mineral fluids change, different materials are deposited.

Normally, during the long geologic life in which a gold deposit forms, the mineralization will vary over time. This means that for a time the fluids will be rich in one set of minerals, and then change and deposit another set of minerals. As an example, the first fluids into a fault zone might be rich in silica leaving behind an initial deposit of quartz along the walls of the vein. The fluids might then change to be rich in sulfides, leaving behind a zone of pyrite mixed with galena. Perhaps then a pulse rich in gold would be produced, and gold would be deposited in the vein. Sometimes there are multiple pulses of quartz and other minerals. Where a regular series of pulses comes one after another, this produces what is called ribbon veins. This type of situation is very common in many types of veins. It is also quite common that gold is among the last things to be deposited. This is not always the case, but it is very common. Hydrothermal fluids also circulate along sharp folds, fracture zones and faults. Practically any rocks that have a well-developed fracture system which creates openings may serve as a host rock. At times, the intrusion of igneous intrusive rocks like granite will fold and fracture the adjoining rock, allowing circulation of mineralized geothermal fluids. This is one reason why contacts of those igneous intrusive rocks with brittle rocks like shale or schist are favorable sites for mineral deposits, even hundreds of feet away from the actual contact, because the folding and fracturing can occur throughout a wide zone along the intrusive contact. Zones formed in this way continue to yield good-sized gold nuggets to metal detector operators who understand the geology of these types of occurrences and know how to search for them.

Strong folding of rocks can create faults and other openings that lead to the formation of gold veins.

As I've noted, most common conduits for these hot water solutions are natural fault and fracture zones, because they are excellent places for fluids to circulate and precipitate mineralization. This is why most veins are shaped like fault zones, which are long and narrow sheets. It is also why these deposits are sometimes called fissure veins - they form in fissures. Faults are not perfectly flat plains and they usually have some sort of wavy character to them. Movement along faults tends to open up voids which allow places for the fluids to flow. Faults can move up and down or side to side or a combination of these, but no matter the direction of movement along the fault, openings where water can flow are created. Veins form where the faults allow hydrothermal fluids to flow through larger, open space fractures and precipitate mineralization along the walls of the fracture, eventually filling it completely. Where the walls of the fault are made of impermeable rock, the vein will have a sharp contact with the wall, but were the wall is of porous rock, the vein may grade into the wall rock and some of the wall rock may even constitute profitable ore. This is the process that forms nearly all gold-quartz veins.

Veins also commonly form at the meeting of two different rock types, also for the reason that on the contact where the two rock types meet is a zone where water circulates better than through the solid rock itself. Veins that form along intrusive dikes are great examples of this. Folding can also give solutions a place to flow, because strong folding of rock also involves movement, fracture zones can develop at the top bend point of a strong fold. Sometimes a zone of broken or fractured rock develops with no strong single fault shear, and in that case a series of small parallel veins may develop. These are known as stockwork zones, and it is not unusual that coarse gold would form in these situations. Sometimes strong single vein systems will fray out into a stockwork system at the ends of the vein, or where it crosses into a different rock type. Under some conditions, pipe or plug shaped deposits may form where geologic conditions favor tube shaped openings. At times the circulating mineral waters can go though wide zones where a whole area of rock is porous, and if there is enough solution moving and depositing gold, you might get a big disseminated deposit, like those mined by open pit methods here in Nevada.

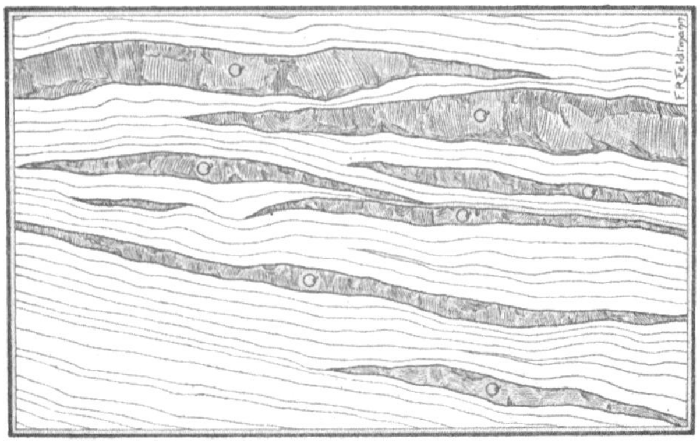

Stockwork vein systems are a made up of a series of narrow veins all in a zone.

While many types of gold deposits occur in veins, what actually constitutes a gold bearing vein is highly variable. Veins can be found in all sizes and lengths with a wide variety of appearances. They can be a tiny fraction of an inch to more than 50 feet wide. They can penetrate down into the ground for a few feet or for thousands of feet. They can extend along a length of a few feet or be up to thousands of feet long. At times they may outcrop 10 or more feet above the ground or be completely hidden under the soil. Veins can be nearly 100% quartz or contain no quartz at all. Sometimes a vein will contain large amounts of broken country rock cemented together by vein matter. Other times, a vein may contain none of the country rock that surrounds it. Sometimes a vein will be no more than just a zone of fractured rock impregnated with silica. Veins can have sharp walls totally different from the vein, or they can grade slowly, transitioning bit by bit from what is clearly vein to what is clearly wall rock, with no sharp change in between. They can be snow white or all shades of gray, red, yellow, brown and black (plus some other colors as well). So with all these possible variations, how does one distinguish what constitutes a vein?

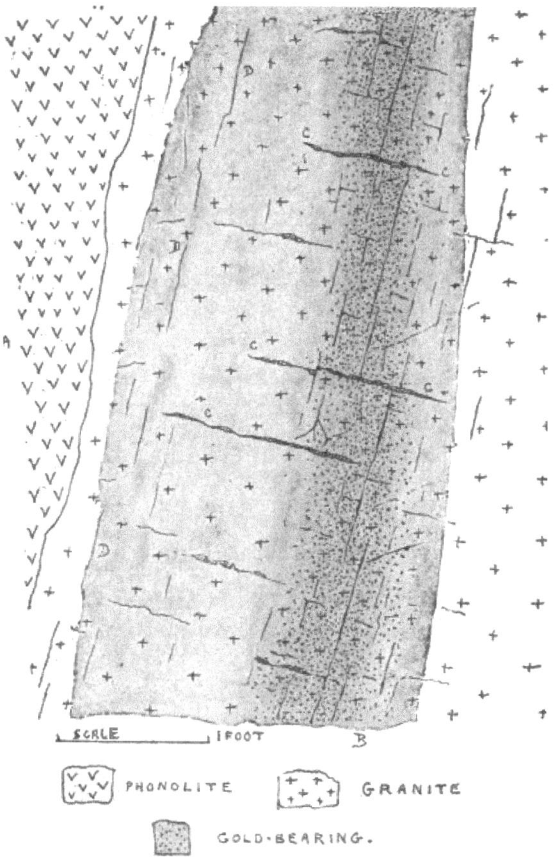

Rich mineralized fracture zone along the contact of phonolite (a volcanic rock) and granite, Cripple Creek District in Colorado (not a true vein).

I have to admit that under certain conditions, it is not always easy to distinguish a vein from the surrounding rock. While veins made of pure quartz that crop well above the ground can be as plain to see as the nose on your face, subtle zones of silicification or other altered rock can be hard to spot. There were certain situations where unrecognized veins fooled miners for a long time. One of the richest gold districts in the United States, the Cripple Creek District of Colorado, fooled miners for the first couple years because they didn't recognize the vein zones and other deposits of the area. The first prospector to announce he had struck rich gold deposits there was considered crazy because other people had examined the same area and saw nothing worthwhile. I have included drawings of a number of different kinds of veins in this chapter to help you as a prospector keep in mind all the different possibilities. Perhaps the best way to know what to look for is to know what occurs in your particular area. Each gold district has its own characteristic types of mineralization, so by doing research and knowing what occurs in your area, you can know what to look for and be prepared.

The most common geochemistry required to dissolve gold involves sulfur, which is why sulfides like pyrite, galena and arsenopyrite are so commonly associated with gold veins. In most gold bearing veins, the gold does not normally form into well-developed crystals. Unlike quartz which often forms good crystals, gold requires a very specific set of circumstances to grow into good crystal shapes. Few gold deposits produce good crystal specimens, as most gold forms naturally as irregular masses and lumps. Many of those irregular masses have a form suggesting that they might have been melted at one time, but that is virtually never the case. However, where the right conditions are present, gold can form in a variety of crystal shapes, including dendrites, leaf, deformed octahedrons and cubes. The placer field at Rye Patch (Majuba) in Nevada is one of the places where the conditions were right for the growth of crystallized gold. This location has gained a reputation for some nice-looking specimens - the term "chevron" is locally used to describe the typical shape. Not all Rye Patch gold has the chevron shape, but the chevron shaped nuggets are fairly common there. The chevrons I think are a combination of the dendrite and the octahedron forms. Some unusual set of chemistry, temperature and pressure conditions present at Rye Patch led the gold to grow into these attractive V shaped crystals.

It is very common that gold is emplaced during the later stages of the mineralization, which is why rich concentrations often form along either the middle or outside walls of the vein - as these are often the last locations to fill up. Faulting may develop breccia and gouge, which is often favorable for replacement style mineralization (this is where the mineralizing solutions literally replace the country rock by dissolving it). The forms of mineralization and alteration associated with faults are highly variable, and may include massive to fine-grained, stockworks of veinlets, and occasionally vuggy crystalline textures in some breccias.

Gold Pockets and Ore Shoots

No large vein deposit is uniformly rich; in fact most are highly variable and fluctuate greatly in value from point to point along their length. Parts may be rich, but in most deposits, the bulk of the vein material is fairly low in grade. In certain districts, the ore richest in gold occurs in only pockets or "bonanzas" as the miner describes them. In the pocket oriented mining belts, miners drive workings along the direction of the main vein hoping to encounter a "bonanza" but never know when or where the next one will be found. Whereas the normal concentration of gold within a vein might be less than a quarter ounce per ton, a pocket would be represented by ore containing several thousand ounces per ton. It often occurs that drifting along the vein must be carried for long distances in veins that do not pay for the extraction of the metal before such a pocket is encountered. This is why working pocket mines underground normally does not pay. Testing the vein exposures and surface outcrops with a metal detector might be a much more profitable proposal. There are many types of locations within a deposit that can show the significantly increased values of pockets and bonanzas. These include the intersection of veins, places where the vein bends or pinches down, and also points where chemical changes in the wall rock occur, such as the crossing of zones with high iron concentrations. Bonanza style enrichment can occur where the deposit encounters favorable beds in sedimentary rocks that cause chemistry changes, or near pre-mineral faults. Miners use their knowledge of the area geology to assist them in finding these rich ore concentrations, because these places are where the ore that is economic to mine occurs.

As I mentioned earlier in this chapter, the gold forms complex bisulfides to get into solution. Often a reaction with iron minerals is what causes

Drawing of a bonanza gold pocket from the Rathgeb mine in Calaveras County California. The D-E Crossing is a rock zone rich in iron.

the gold to come out of solution. The quartz and iron are carried in water solution as well as most of the other sulfides in the ores you see. High iron rocks or zones in the country rock can create rich pockets of bonanza ore. These high iron zones were known as "iron crossings" by the old time California miners and indicator beds in Australia. There are mines with occurrences like this in Nova Scotia as well. All of these locations were the sites of rich pockets of free gold, some of them containing many thousands of ounces of gold. These smaller ore bodies are known as bunches, pockets, or kidneys; however, in gold-quartz veins these small spots may be exceedingly rich. Rich pocket mines like the 16 to 1 mine in Allegheny have produced millions of dollars from very small areas within the vein. This mine has been producing pockets like this from the 1800s right up to the present time. Another cause of gold dropping out of solution is vein intersections. Often where small side or feeder veins cross with a major vein, significant chemical changes can occur where the two fluids mix, resulting in the formation of rich pockets of free gold.

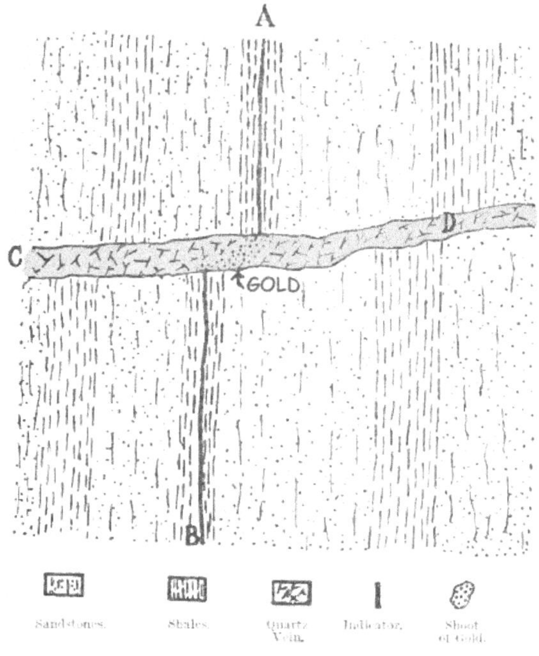

Rich Pockets of gold, like the previous California example, formed where a quartz vein crosses a special indicator bed, Victoria State, Australia. Like the California example shown previously, the special crossing bed is very rich in iron minerals.

Even in the vein systems which do not have rich pockets of coarse gold, there are nearly always zones and areas where materials of higher grade have been deposited. The zones are known as ore shoots or ore bodies. The length along the vein of these ore shoots varies quite a bit, from less than 100 feet to more than 1000 feet long in some large, low grade vein systems. It is a general rule, with only a limited number of exceptions, that ore shoots of extremely high-grade tend to be small and discontinuous. Conversely, ores of moderate grades tend to be more continuous and extend over longer lengths.

On the other hand, when the grade of the ore within the productive shoot is low to moderate, then it is typical that the ore shoots will be large and extend over significant areas. In underground mines, it is these larger vein sections with a moderate grade of ore that are at the most profitable to dig and process over the longer term, and these constitute the types of ore bodies which large mining companies extract. They are simply more reliable and easier to follow, therefore more optimal for a mining operation to work on. In the end, no matter how large they are, ore shoots do not extend on down the vein forever, and they all come to an end eventually. One of the reasons that miners stop working on an ore body is simply that they lose them. Post mineral faults can cut off and move a vein a considerable distance. A miner then may have difficulty in relocating the vein if he does not know which direction the fault has moved, and the direction and distance it has moved. Many good mines have closed down because faults have cut off the vein and the continuation of the ore was lost.

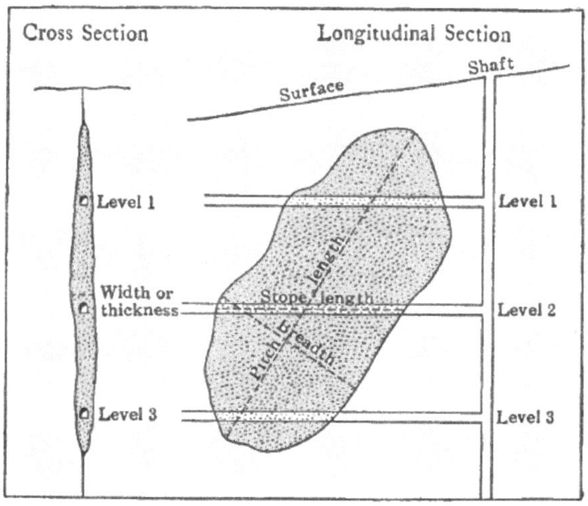

Ore shoots within a vein have a definite shape and size. As shown here, they often pitch downward at an angle. This is the portion of the vein which miners extract in a working called a stope. In a mine working a vein, the stope is tall and long, but only as wide as the vein itself.

When miners encounter a section of vein with ore that is profitable to dig out and process, they will extract it in a working that is known as a stope. The rest of the vein is left behind because the grade of the rock is too low to process at a profit. Ore stopes will often be only as wide as the width of the vein, but may extend hundreds of feet both along the length of the vein, and vertically up and down it. Sometimes these higher-grade zones will run vertically up and down the vein, however more commonly they will pitch downward to one side or another related to the direction of the flow of the original mineralized waters. I've included a drawing (above) that will illustrate this and show how an ore body pitches at an angle.

A stope is the part of a mine where miners extract the ore, leaving a hole where the vein was.

Depending on the width of the vein that the miners are taking out, they will normally leave behind pillars to support the walls as well as wooden stulls or other shoring to support the remaining wall on either side of the vein. The narrower the vein, the more simple these support systems are. Drilling and blasting are nearly always required, and miners will work to design their systems such that gravity can be used to move the ore to a central collection point where it can be hauled to the surface. Close sampling of the vein is often done, as miners do not want to extract any areas where the ore does not meet their minimum standards to be profitable. The miner certainly does not want to send rock to the mill which will cost more to process than the metal values contained within it. Where small areas of low grade material occur within a larger body of higher grade material, these zones of low-grade are not mined and left as pillars to support the walls on either side. Only when he really is forced to do so will a miner leave high grade material behind as a support pillar. The walls on either side are known as the hanging wall (above) and the footwall (below). For more information on this terminology, see the definition section of the appendix.

In many vein systems, it is very common that there may actually be one or more parallel veins, separated by a few feet to perhaps a couple hundred feet of barren rock. The veins will tend to run parallel to each other. This is why many old time mining operations would crosscut from a known vein out into the barren rock searching for parallel vein structures. Modern miners do something very similar to this using diamond core drills to look out into the barren rock for any parallel veins.

Working in a stope was a dangerous assignment, as it was often easy to fall a long way. Wooden stulls supported the walls left behind, and the old time miners worked only by the faint illumination of candlelight.

Because of the chemistry of gold deposits, there are several elements that dissolve and precipitate in a manner similar to gold and can commonly be found associated with it. These include mercury, arsenic and antimony as well as some others, depending on the chemistry of the particular gold deposit. Some of these elements are a bit more soluble than gold and can disperse to a wider zone of mineralization. In this situation, these other metals essentially form a geochemical halo around the deposit. Because of this, geologists look for these elements in their exploration for gold as they may indicate the presence of gold ores nearby. In Nevada, metals like mercury, arsenic and antimony form halos around large low grade disseminated type gold deposits. Analysis of these rocks by atomic absorption (usually just referred to as AA) testing methods has replaced the 500-year-old standard fire assay in some types of exploration applications. Many of the assay shops here in Reno do as many AA type assays as they do fire assays. The standard method for AA analysis of a sample is to treat it with an acidic leach to take into solution a variety of metals. A small amount of the liquid is injected into a clean flame. The AA detects specific wavelengths of light given off by the metals when heated in the flame. The AA unit is then able to quickly give metal contents for a variety of elements typically including gold and silver but also heavy metals like mercury, arsenic and antimony and base metals like lead, zinc and copper. All these results are taken from one sample and the costs are only just slightly higher than a standard fire assay for gold and silver only. Geologists sometimes explore for large deposits by sampling stream sediments and then use an AA type of assay where they are looking not just for gold, but a full range of associated elements. This is a typical method that has been used here in the US to prospect for large low-grade hard rock deposits (like the Carlin type deposits of Nevada).

Alteration of the surrounding rock on the surface is also common near mineral deposits, as the geothermal waters can either leach or add new elements to the surrounding rock, often changing the existing minerals in the rock. Geologists call these zones alteration halos. Understanding how rock alteration works can lead to some good ideas for exploring around old mining districts for new veins. Often, being able to recognize the alteration of rocks and the sometimes subtle changes created by the hydrothermal waters can show which areas on the periphery of known districts will be favorable for exploration and prospecting. In certain locations, rich coarse gold may lie below or otherwise near on the outskirts of low-grade deposits. Across a major fault from the low grade (but huge) gold mines at Carlin, Nevada are placers and small veins with coarse gold. The gold in the open pits is microscopic but there is a huge amount of it. The placer deposits and veins with coarse gold have produced much smaller amounts by comparison, but still significant. In other locations, different metals may deposit above or below one another vertically - this knowledge has also become an exploration tool.

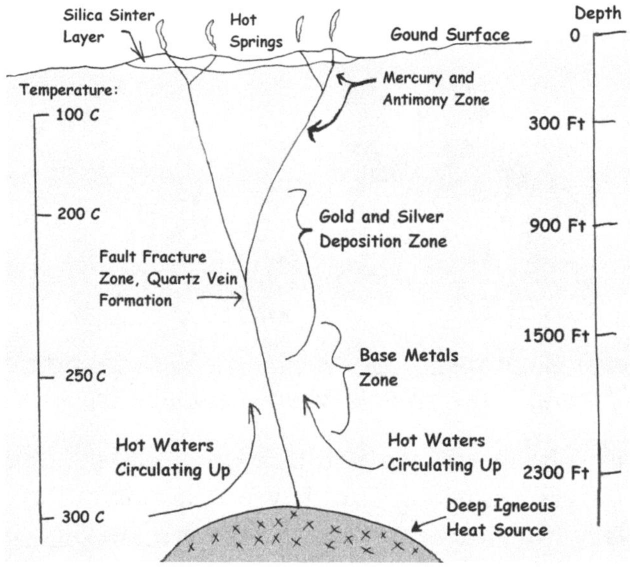

This vertical geologic section shows how different types of epithermal deposits are related based on the depth and temperature of formation. Epithermal gold and silver ores form at moderate depths and temperatures.

In recent decades, geologists have found that in many epithermal systems, mercury and sometimes antimony will be deposited at or near the surface. These are places where a hot springs system may form as the heated water is released up on the surface of the earth. From 1000 to 2000 feet down below the surface would be the general location of the zone which is optimal for the deposition of rich gold and silver ores. Further down below the gold silver zone is the location that would be expected to be rich in base metals such as copper or lead, and sometimes including silver as well. This knowledge has been used for finding new ore deposits in Nevada and elsewhere. An excellent example of this type of occurrence was found at the National District in northern Nevada, where the higher elevation deposits are principally mercury and antimony, mid-depth are rich gold deposits and at the lower depths are lead and silver veins. These are helpful ideas to have in mind when looking for gold around the boundaries of old productive districts.

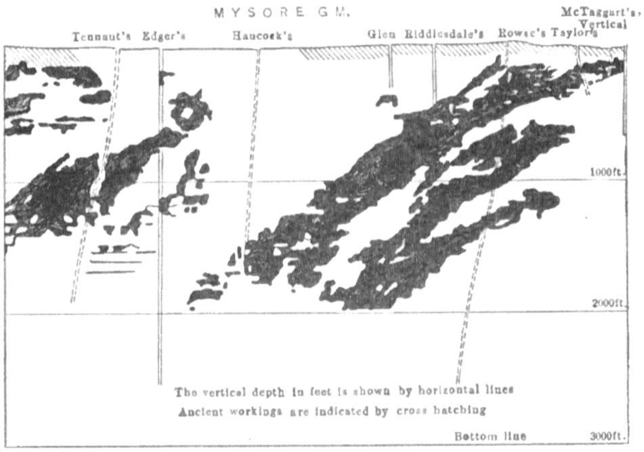

Map of the ore bodies in the Kolar Gold District of India, showing the slanted pitch of the ore bodies.

Secondary Gold Enrichment:

Certain types of metals, most notably silver and copper, form important enrichment zones near the surface of their deposits. This happens because the copper and silver bearing sulfides dissolve in sulfuric acid. Small amounts of sulfuric acid are formed when the sulfur that is found in sulfur bearing minerals like sulfides is exposed to rain and air. Especially in arid regions, the copper and silver are dissolved and carried downward into the soil in and around the vein. Secondary minerals like cerargyrite (silver chloride or horn silver) and malachite (copper carbonate) are formed. This process can form some very rich deposits. Gold however is not soluble in sulfuric acid and is not so easily dissolved. Gold can be dissolved in the presence of sulfuric acid, if a chloride and manganese dioxide are also present. However, even under favorable conditions, gold is generally dissolved less readily than copper or silver and precipitated much more readily than either of them. As a consequence, ores enriched in gold are likely to be found much nearer to the surface, and the secondary gold formed in this process will be very small in size.

The most common primary manganese mineral which is found in gold bearing veins is rhodochrosite or calcite which contains a certain percentage of rhodochrosite. Rhodochrosite is a mineral commonly associated with epithermal gold and silver veins. Because of the requirement that significant manganese be present for gold enrichment to occur, we would expect that enrichment in the gold values would occur almost without exception, in veins which contain a significant amount of manganese minerals. Geologic investigations have confirmed this. Additionally, where gold bearing veins are rich in manganese, we almost never have placer deposits, because the gold will be dissolved in the presence of manganese. The flipside of this is that both gold placers and outcrops rich in coarse gold are generally found in connection with non-manganiferous deposits.

Because of this, essentially all coarse placer gold nuggets originate from the erosion of primary hard rock gold deposits. Some placers are re-concentrated from older placers, but at least some time in the past, these older placers came from primary deposits. A small amount of coarse gold has probably resulted from some very unusual secondary enrichment processes, but the majority of the coarse gold which has been found originally grew in primary deposits from circulating hot water solutions under pressure. In the past, many geologists have considered that large nuggets might grow through cold, secondary processes that occur very near the earth's surface. Recent scientific testing has shown that this is not the case, and that coarse sized gold is formed almost exclusively through hot circulating waters deep within the earth. A more detailed discussion of placers and how they relate to secondary enrichment processes, including the recent scientific research I mentioned, can be found in the placer geology chapter of this book.

The Tonopah gold and silver district is a fine example of desert secondary gold. The Tonopah district in Nevada did have secondary gold enrichment, as the near surface ores were much richer in both gold and silver than the ores deeper down. The ores from this area do contain significant manganese and sulfides as well. The mining operations which worked downward for years showed that the near surface ores were almost universally higher grade in both gold and silver. Silver was present in the near surface, oxidized ores as chlorides, and the gold as free gold. So given the geology and the grades encountered, there is little doubt that there was some significant secondary gold enrichment at Tonopah. On the other hand, the gold in both the primary (deep) and secondary (near surface) gold from Tonopah was very fine grained and no placers were formed at Tonopah, even though something like 2 million ounces of gold was mined there.

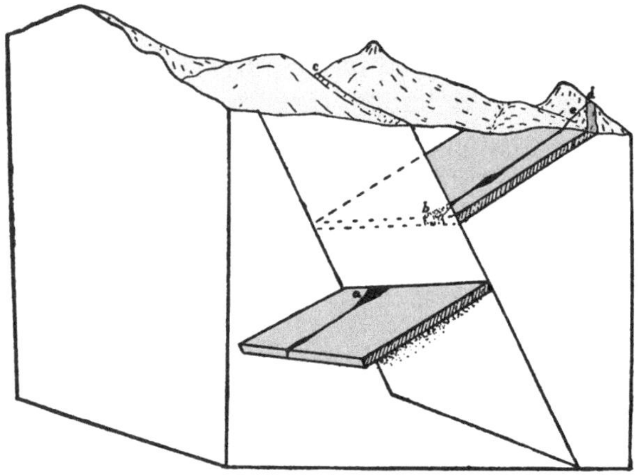

This diagram shows how miners can loose a vein due to offset faulting movement. Unless they knew which direction the fault moved, they would not know where to look for it.

While the gold formed in secondary types of deposits is normally very small in size, the gold which is formed in the primary hard rock and vein deposits can be virtually any size from microscopic up to several hundred pounds. Many prospectors are interested in lode deposits chiefly because they are the source of placer gold, so let's take some time to look at the geologic processes that lead to the formation of larger sized gold in hard rock deposits that eventually turns into placer upon erosion.

The Formation of Coarse Gold

The prospector using a metal detector is primarily interested in coarser gold as larger nuggets are easier to detect, and of course, more valuable! Growth of larger gold requires some special circumstances, and it turns out that coarse gold most commonly forms in small veins (for the sake of argument, let's say for our discussion that coarse gold is stuff a Pennyweight or larger). Big veins which have a width and length large enough to be mined commercially underground tend to produce mostly small gold - wires, small pieces, etc - even though they may be fairly rich. Of course, there are a number of exceptions to this rule, but generally speaking, it is true. So why, in general, does big gold tend to form in small veins, while mostly only small gold forms in big veins?

Growth of large, coarse minerals from water solutions requires stable conditions where large crystals or nuggets can continue to grow over an extended period of time. It takes some time to grow big gold, so conditions must be fairly stable to allow this to occur. This is pretty much the same for quartz, tourmaline, mica or gold – and most other minerals that commonly grow from water solutions. Rapidly changing conditions, including rapid drops in pressure or temperature generally lead to the formation of fine-grained crystals - the minerals are forced to come out of solution so fast that large crystals cannot form.

Often, the source of the heat is some sort of cooling magma (lava, still molten or solidified but still hot). This kind of heat source throws off so much heat energy that the rock may well be warm to within 100 feet of the surface, even if the source is over 2000 feet down. The temperature does not drop off suddenly once the solution gets a short distance from the source. The drop in temperature as you move away from the heat source is gradual, and usually fairly linear, at least to within a few hundred feet of the surface. So the temperature of the rock (and the water solutions as well) becomes directly related to how far they are from the heat source. The rate of cooling then becomes a simple function of how fast the fluids move upward.

With that in mind, the small feeder veins, and splits off the main vein are sort of the quiet backwaters for the solutions forming veins. Smaller openings just don't allow as much flow volume, and the slower flowing solution can take its time to cool and release its mineral load. The small flows of the small veins move upward slowly and have a chance to take their time to cool. This creates fairly stable conditions in the growth zone where decreases in temperature and pressure are such that minerals, including gold, can grow to a good size. These smaller veins are still spotty in their values, with some parts rich in coarse gold, and other parts low in grade and nearly barren. Usually these veins are far too small for any commercial underground mining operations, with widths of up to a few inches and strike lengths of 5 to perhaps 50 feet. Some of these small veins were formed at cooler, epithermal temperatures, and some at warmer mesothermal temperature ranges, but those that contain coarse gold all had the stable conditions that allowed for coarse growth. These types of small veins are also sometimes called feeder or stringer veins. The exact temperature of formation may be important to geologists, but is probably not of that much interest to the individual prospector.

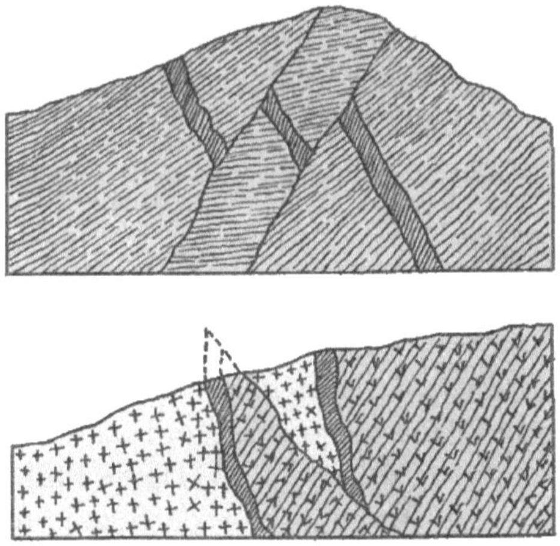

Examples of how faults can move a vein around. The lower shows one vein made to look like two.

Wide fault zones and other large openings in the rock generally lead to more solution flow, and greater quantities of quartz, sulfides and gold. This means bigger veins have more commercial potential, but also it means more rapidly changing conditions, which means rapid deposition. This generally leads to growth of fine-grained minerals. In larger openings, the larger flows move quickly upward and the solutions become over-saturated, so the minerals in solution must drop out quickly, forming fine-grained minerals of small size. In fact, it is common that, where there are large fault zones, which conduct the fluids quickly upward, the waters are still fairly hot when they actually reach the surface. This often results in the formation of a hot springs. Steamboat Hot Springs, just south of Reno, is a classic example of this.

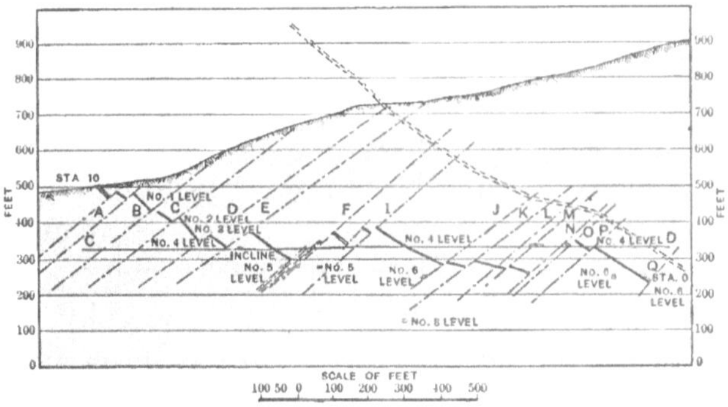

Actual example of a badly faulted vein broken up into many segments – Berlin, Central Nevada.

Most productive hard rock veins are not uniformly rich, but have rich sections, which are called ore shoots as I mentioned earlier in this chapter. These are usually oriented down the dip of the vein, and represent places where rapid changes in pressure or temperature caused the metal bearing solutions to drop much of their metals content. For the average underground mining operation, the fact that the mineralization is fine grained is no problem, for them the advantage of the big veins is in greater quantities of gold deposited in them. Even on the occasions when coarse gold is formed in a larger vein system, it most commonly occurs where the vein pinches down, bends or splits out into smaller veins. Coarse gold in large veins also can occur where the vein undergoes a change in the wall rock, and the change in the chemistry of the enclosing rock can cause coarse gold in what is otherwise an unexpected place for it to form. The tendency for coarse gold to form in smaller veins is not a hard and fast rule and there are some large veins, which have produced very coarse gold (like the 16 to 1 mine in the Allegheny District of California). However, the formation of big gold in small veins is by far the most common scenario for the growth of coarse gold. So it is true in general - just allowing for a few exceptions here and there.

Let's take a look at some field examples of gold deposits, with respect to coarse and fine gold. In some districts, small vein openings are all that occurs. The top of Rich Hill in Arizona is an intensely jointed granite. Hot solutions circulating through the joint cracks in the granite were all of the "small vein" type. There were no large faults that allowed larger volumes of fluid to circulate. Erosion over the years concentrated the coarse gold, which grew in these small veins, which in turn produced an extremely rich placer deposit. In the same way, the placers at Gila City (Dome) in Yuma County Arizona produced some considerable gold, though no rich veins were found at that location. The Arizona Bureau of Mines came to the conclusion that "many pockety or small low grade veins supplied the gold". At Gila City, the accumulation of gold from those many small sources resulted in a rich placer. Still another example would be the placers at Rye Patch (Majuba) in Nevada. The gold is derived from a series of small quartz veins, which occur in shale in a 2 mile wide belt on a pediment east of the range front along the Majuba Mountains. Many of Nevada's mountain ranges have large fault zones, which parallel the mountain ranges at the edge of the range, where the valley meets the range. The displacement on that fault allowed igneous rocks to come near the surface, creating the heat source which formed the veins. This range front fault trends north south and little if any gold is found west of the fault zone in the mountains themselves. The quartz veins in the mineralized zone are not large and have not been mined as veins underground, as they are just too small. Most of them have a strike length of perhaps 15 to 30 feet. In some districts, fine gold is all there is - and as a result, there are basically no placers formed. This is the main reason why some rich hard rock gold districts don't have much in the way of associated placers. The most productive gold vein district in Nevada (excluding the Carlin type deposits) was Goldfield. This district was discovered about 1903, and produced nearly five million ounces of gold, much of it from extremely high-grade ores. The smallest veins were several feet wide, but the largest were more than 30 feet wide and stretched to thousands of feet in length. In spite of the high-grade ores found there, no significant placers were associated with the Goldfield deposits and all the gold in the veins was very fine grained. The hot fluids circulated through pre-existing faults, which provided large openings for the waters to circulate quickly through. The Jarbidge District in Elko County, Nevada is another example from Nevada of a district with good gold veins, but only fine gold and little or no placer. So when planning a trip to detect around old hard rock mines, be sure to find out, not just if the gold found was "free gold", but just how coarse the gold from that district was - it makes a big difference!

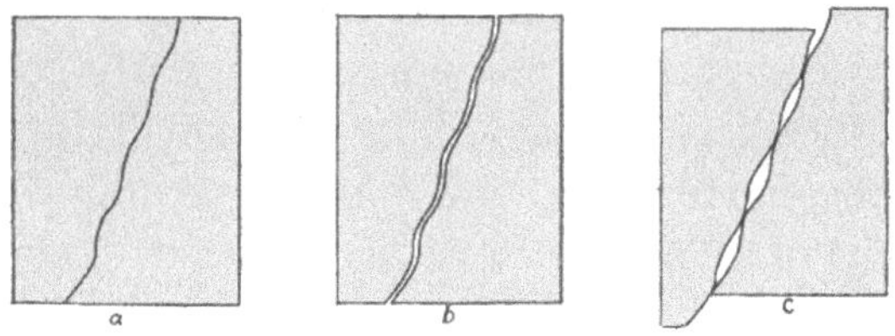

Examples showing how movement along a fault opens up void space in which mineralized waters can flow and deposit gold.

In some districts both large and small veins occur. At Randsburg in southern California, rich veins were mined underground yielding over one million ounces of gold. These veins are located in and around the town of Randsburg, though a few of the vein deposits trended off to the south of town. Mineralization in and around two of the largest vein systems (the Yellow Aster and Baltic) have been mined as an open pit deposit in recent years, yielding over a million ounces of gold. The placers, which yield gold coarser than what was found in the veins, are for the most part located on the pediment north and south of town - away from the productive larger veins. The Kofa placers of Yuma County; contain both larger veins which were mined underground and smaller veins which produced coarse gold. USGS Bulletin 620 says the following about the coarse gold found in these placers: "It has evidently been derived from the disintegration of auriferous veins in the metamorphic rocks as it is much coarser than the gold contained in the North Star and King of Arizona veins". The North Star and King of Arizona veins were two larger veins, which were mined underground. The Vulture placers of Maricopa County, Arizona also fall into this category. The Arizona Bureau of mines states in Bulletin 168, "The origin of the placer gold, in Red Top basin at least, appears to have been the small quartz veins in that vicinity. The gold of these veins, like that of the adjacent placers, appears to be coarser than that of the Vulture vein". The Vulture vein was a large and productive deposit, but the gold from the productive vein was small.

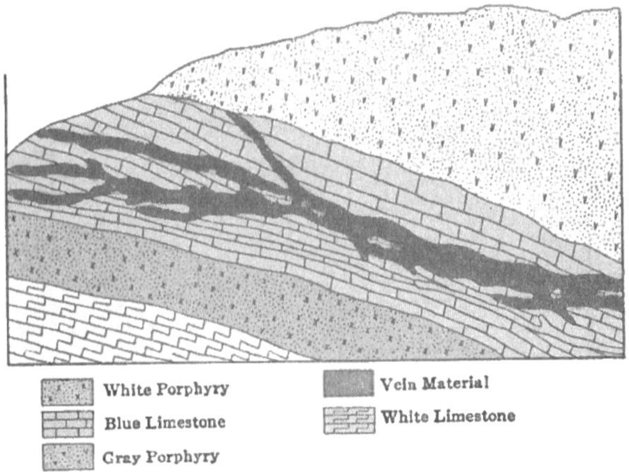

Replacement deposits tend to be irregular – they often replace limestone with sulfides, but they can be very rich in gold and silver as with this example from the Leadville district of Colorado.

Placers formed by small, pockety epithermal deposits form an interesting class of deposits for the prospector armed with a modern metal detector. As a group, they are not well prospected for except in a few places like Rye Patch in Pershing County, Nevada. Most of these are residual placers (depending on where one draws the line between residual and eluvial placers). Much of the Nevada placer tends to be high in silver. Electrum is a natural high silver content gold alloy. Some of the placer gold found by prospectors in Nevada is actually electrum. Generally gold deposits formed under lower temperature epithermal conditions are higher in silver, and the gold derived from them also higher in silver. This is why, generally speaking, Nevada placer nuggets are higher in silver than nuggets from the mother lode deposits of California (which were formed under the hotter mesothermal conditions). While it is true that epithermal deposits form some good residual placers, this is also true for other types of geothermal sources as well. Even in the mother lode region of California, there are many small residual placer deposits. This is in spite of the fact that these are considered mesothermal type deposits. There were many places in the mother lode area where old time "pocket hunters" worked for decades after the initial big "rush" of prospectors. These folks had no metal detectors and used only a pan and shovel to trace the gold back to the residual placer source pocket. The old timers had a tough time locating them because of the spotty nature of these deposits and the limited effectiveness of their tools. We still have the same kinds of problems as with the old timers, because these residual placers are spotty and sometimes they can be hard to find, even for the modern nugget-shooter armed with a good detector (but that is the reason they are still there).

Basics of Hard Rock Gold Prospecting:

Serious prospecting for hard rock gold is something that not many of the folks reading this book will ever really get involved with. In spite of that, I'm going to take some time and present the basics of hard rock prospecting in two phases. This part here at the end of the section on hard rock deposits will cover the basics, with a more detailed look at hard rock prospecting to be presented in a later chapter of this book. There is actually good reason to do hard rock prospecting, as there truly are very good opportunities in this area for the individual who's looking to find some gold or make some money. Really only a very few of the most serious individual prospectors ever get into this part of the prospecting game. The main opportunities are in the finding of rich gold pockets and in the locating of larger sized deposits that could be claimed and then leased to professional mining companies. The downside is that hard rock prospecting is far more technical than placer prospecting. To be successful you will need to know the basics of gold geology and understand the favorable indicators and details of how gold occurs in the district or region where you want to prospect. Many gold deposits do not consist of the classic "quartz vein" - they have a different appearance and you'll need to know what to look for in your region. For the hard rock prospector, the time spent doing research is every bit as important as the time spent in the field.

Outcrops of quartz veins are the first thing prospectors look for in a new area, but the old time prospectors were looking for the same thing and tramped all over the hills searching for them. It's pretty darn hard for the modern prospector to find something in the way of an outcropping quartz vein that the old timers have not already found and sampled. Because of this, the modern prospector has to look for things that are harder to find. This includes deposits that don't outcrop or would have otherwise been missed by the old timers. The truth is that many gold veins outcrop only very poorly or not at all. Veins which do not outcrop well are of course much more difficult to find. There are several reasons why a vein may not outcrop clearly. Part of it has to do with the amount of overgrowth in the area and how quickly soil builds up from the plant growth. Another part has to do with how easily the rock around the vein weathers. A deep layer of soil can easily cover over a quartz vein and hide it. In spite of this type of obstacle, a metal detector can be used to see rich chunks of gold bearing quartz through the soil. The classic methods of loaming or tracing gold deposits by their dispersion fan are also important. The details of tracing gold dispersion fans (also known as loaming) will be covered in a later chapter.

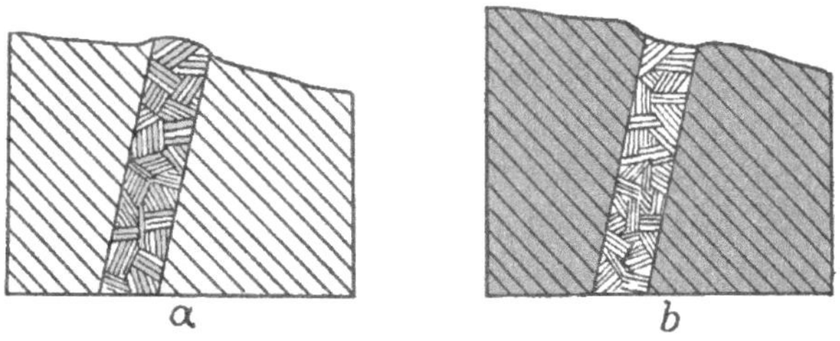

A - *vein more resistant than the country rock = good outcrop.* **B** *–vein less resistant than country rock = no outcrop, vein likely hidden under soils.*

Another possibility occurs when the veins break up into a series of small parallel veinlets, which together are called a stockwork deposit. Sometimes these small veinlets outcrop very poorly. They may leave some evidence as traces of float quartz strewn downhill from the deposit, but are not usually weather resistant enough to crop out above the level of the soil on the ground. Keep this kind of thing in mind when prospecting old hard rock districts. Quartz is very resistant to natural chemical weathering and concentrations of quartz float scattered around an area are strong indication that one or more veins may be present up hill from that location even if none are outcropping. Perhaps the first step for those who want to do hard rock prospecting and have located a spot showing good looking quartz or other silicified rock, is to sample it. The tried-and-true old-time way of doing this is to start with taking samples of average looking material and then crushing it up and panning it by hand.

This 5 foot wide quartz vein is buried under about 6 feet of soil – it is exposed only in a road cut.

Gold panning is an excellent method of concentrating a sample down to its heaviest components. If you see a decent amount of free gold in your sample then you're on the right track. Even if you don't see a lot of free gold but there are significant sulfides, you may wish to proceed to the next step, which is having material assayed by a reliable operator. There are a number of reliable assay companies, and having a standard fire assay done isn't cheap, so you'll want to be sure of the potential of your material before you send it in. At this time, a typical fire assay runs roughly $30. Sometimes people suggest they have a cheaper method. My thoughts about trying to cut corners by getting them done on the cheap with some sort of fast and dirty process, is that if you're going to invest significant time and money or ask someone else to invest significant money based on the results of those assays, then they need to be done accurately and done right. Once you have found an area of valuable float, then you'll need to trace it up hill to its highest point. There you might be lucky and find a few spots of outcrop, but in many cases the vein will be completely buried. The way to locate a buried vein is by trenching. The old timers cut trenches in the ground by hand, but we have more efficient tools like backhoes and bulldozers. Trenching should be done in a direction that is perpendicular to the likely direction of the vein. Obviously, if you go to use a backhoe, bulldozer or other heavy equipment on government land, you will need to obtain all the necessary government permits beforehand.

This is a small quartz vein outcropping in the desert gold country of Nevada; this type of outcrop in a remote area would be a worthwhile spot to check with a metal detector.

Valuable precious metal deposits can also form as the hot waters dissolve or remove the existing rock minerals and replace them with sulfides and other ore minerals. These types of ores are called replacement deposits. Beds of host rock such as limestone that can be dissolved by the acidic hydrothermal fluids are favorable sites for replacement deposits. Many replacement deposits are marked on the surface by exposures and outcrops of leached iron oxides and boxwork shaped quartz called gossans.

Gossans are derived from the surface oxidation of bodies of rock which are very high in sulfides, often mostly pyrite. As a result, they typically show on the surface as concentrations of yellow to orange to red iron oxides. Significant quartz outcrops may or may not be present along with the iron oxide-stained soils.

Gossans are interesting. I have found some as small as maybe 10 feet in diameter. In places these gossans can yield gold coarse enough to find with a metal detector, but in other places gossans are chiefly base metal deposits (NE Nevada is like this - many of the gossans there are primarily base metal sources with only traces of gold). You just have to know something about the geology of the district you are hunting to know the likelihood of gold being present in significant amounts.

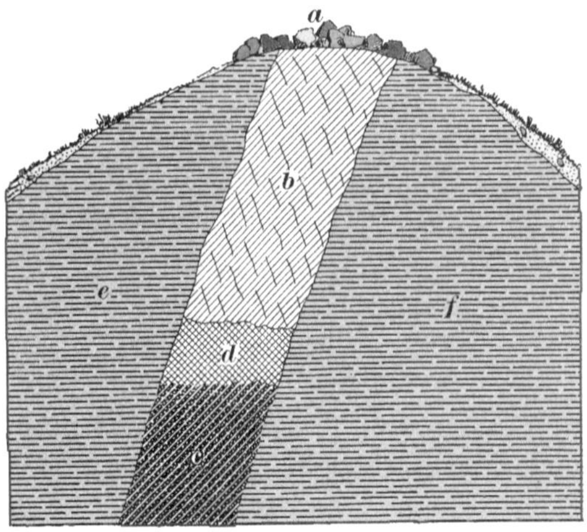

***Gossan Diagram**: A: outcrop material – leached low grade; B: Leached gossan zone of low grade; D: enriched high grade zone; C: Unaltered primary zone; e and f: country wall rock.*

Veins Verses Disseminated Deposits

As I mentioned before, not all gold deposits consist of a simple and classic gold-quartz vein. Some gold deposits have a little bit of mineralization spread out over a large area, instead of concentrated down into a single vein or set of veins. These spread out deposits can contain huge amounts of gold. Deposits containing more than one million ounces of gold are not unheard of. There are several varieties, but in general they are known as disseminated gold deposits, because their values are disseminated through large areas of rock. These deposits are often formed in large fracture zones where the fluids that form the gold mineralization can spread out to a larger area. These fracture zones also allow a large amount of fluid to flow. Another situation that leads to large disseminated deposits is the case where the geothermal fluids encounter favorable strata. This may include things like limestone beds where the fluids, because they are acidic in nature, can literally dissolve away the limestone in the bed. Layers of very porous rock can sometimes have the same effect. In these cases silica, gold and other related minerals replace the limestone or other openings within the favorable bedded rock strata. Large disseminated gold deposits are very specialized topic that will be covered in more detail in a later chapter.

Nevada is famous for its large disseminated deposits of microscopic gold. The Carlin trend by itself has produced over 60 million ounces of gold and continues to produce millions of ounces each year. Many of these types of deposits are associated with significant fault zones and favorable beds within sedimentary rocks. The tiny sizes of the gold particles are caused by rapid deposition, often within zones where sudden changes of pressure and boiling of the hydrothermal fluids occurred. Sometimes these deposits are marked by bodies of jasper-like silicified rock known as jasperioid.

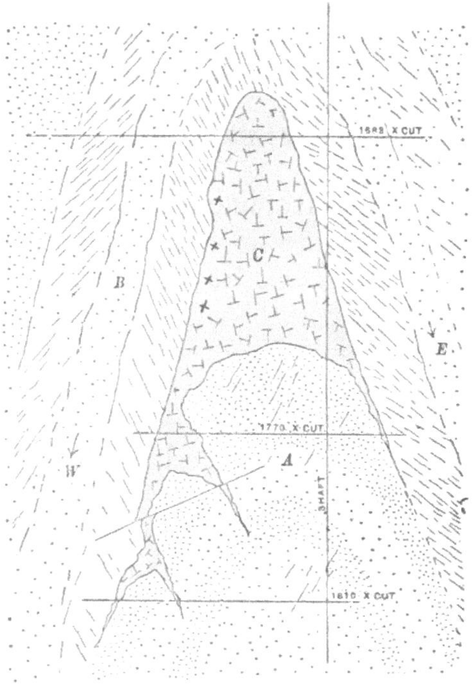

Saddle vein structure, as found in the mines of Bendigo, Australia (Victoria State). This shows a vein filling a fold in older sedimentary rocks.

As another example of gold deposit formation, let's take a look at some of the veins of the Australian goldfields (what we call veins in the US are known as reefs in that county). In the Australian state of Victoria, the veins occur principally in Upper Cambrian and Lower Silurian metamorphic rocks; while in New South Wales they traverse the beds of the Upper Silurian and Devonian systems. During the Devonian period in Australia, or at its close, great upheavals took place, and in these areas granite in many places was brought to the surface and the enclosing rocks were fractured and faulted along a number of lines, the direction of which depended upon the lines of upheaval of these granites. The fissures that were formed in this process conducted geothermal fluids and became mineralized. Eventually these deposits were uplifted a distance of about 3000 feet until they reached the surface, and from the time which erosion began, the formation of alluvial deposits commenced.

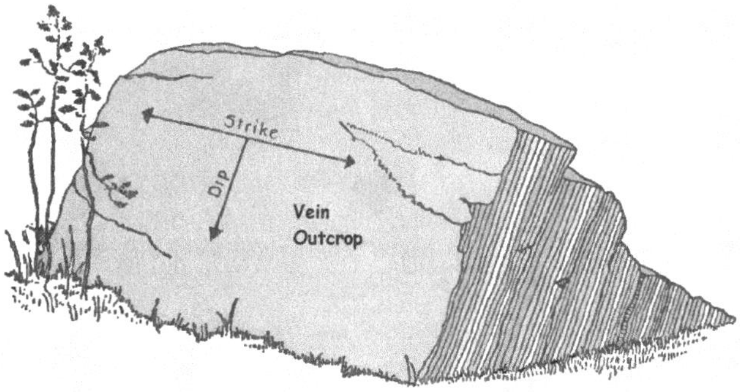

Example vein outcrop showing the vein strike direction and dip angle down into the earth.

At the same time, there are many veins in that country which are of much later origin. The Silurian rocks, for instance, are traversed by dikes of diorite, some of which are on such a massive scale as almost to merit the term of bosses, and these dikes have been found to be very closely associated with the occurrence of gold. Within the folded Ordovician rocks, the quartz veins are abundant and generally follow the strike of the strata, being massed along certain productive "reef lines." Frequently they are conformable between shale and sandstone, but some of them cut across their strike.

The saddle reefs of this area constitute an interesting class of deposits, in which masses of quartz fill cavities produced at the tops of sharply folded anticlines. The richest parts of these unusual deposits are said to be at the caps of these reefs where rich geothermal solutions may have been ponded by the impermeable rocks above; the two side branches (which are called the eastern and western legs respectively) are normally relatively poor. Although sometimes one of these legs will pay to work for some short distance downward, while the other one is barren.

Mineral deposits are often classified by geologists using the temperatures they were formed at. This is reasonable because deposits formed at the same temperatures often share a closely related chemistry, associated elements and similar minerals. In general, epithermal types of deposits are formed nearer the surface and tend not to persist to great depth, though there are some exceptions to this rule. Mesothermal deposits are formed deeper in the ground and at higher temperatures, and tend to be more persistent, both at depth and on strike at the surface. However, as with the epithermal deposits, there are exceptions to the rule, and some small, pockety, mesothermal deposits do exist, and in fact certain parts of the mother lode in California were very famous for small but rich pockets of lode gold.

Mesothermal quartz vein, this photo was shot underground at the Bullion River mine, French Creek in Shasta County, California.

Geologists also classify gold deposits using descriptive "models". These are detailed descriptions of known types of deposits, including the geology, mode of occurrence, and associated minerals. Geologists use these models to help them look for other occurrences where similar geology may give rise to similar deposits. There are literally hundreds of these models for various kinds of deposits, though some are rare and found in only a few unusual situations.

For prospectors who are seriously interested in learning more about prospecting for hard rock deposits, it would be worthwhile to learn about the different geologic deposit models (they will be discussed in the next chapter). It can be hard to put all this geologic information into practice in the field, but it's worth the time to learn and try it. It will help you do better research and find more gold in the long run. Good research is really very important in the hard rock prospecting game. Because of this, I will be covering the basics of some of the more important geologic models in the next chapter of this book.

The old timers got the bulk of the easy stuff, but there is still plenty of good gold still out there today and it's worth looking for, especially larger deposits of gold too small to be easily seen without magnification. Considerably more information on hard rock prospecting, including a discussion of leasing large hard rock deposits to mining companies who will pay you for them, can be found in this book in the chapter called "Prospecting for Large Commercial Scale Gold Deposits".

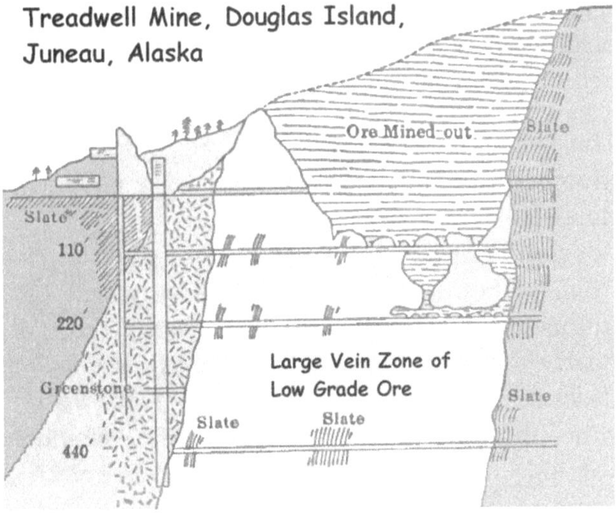

A century ago, the Treadwell mine at Juneau, in Alaska produced many millions of dollars of worth gold from a very large low-grade body of ore that was located between the contacts of greenstone and slate country rocks. Rock contacts are frequently fault zones and can host valuable gold ores.

Understanding Geologic Deposit Models

Geologists have long noted the similarities of certain types of mineral deposits all across the globe. Similar minerals, similar country rocks, and similar grades of mineralization may occur in several different parts of the world. Silver deposits in Nevada have important geologic similarities with silver deposits found in Mexico and South America. Gold deposits of Ballarat and other districts in Australia's Victoria state have much in common geologically with the mother lode gold deposits of California. Understanding how these deposits are similar and their essential characteristics can help a prospector or exploration geologist in seeking new deposits and evaluating the well known ones. Geologists have developed extensive descriptions of certain classes of mineral deposits that show these similar relationships. The descriptions of mineral deposits belonging to the same category are known as geologic deposit "models".

The U.S. Geological Survey has long been involved in assessing and classifying mineral deposits and resources, including the development of these mineral deposit descriptive models. Some of the earliest modern work done in this line was produced by Waldemar Lindgren, nearly a century ago. A "mineral occurrence" as described in these models is a natural concentration of minerals that is considered valuable and is normally evaluated in terms of some economic commodity, such as copper, silver or gold. As an independent prospector, being able to recognize and evaluate a few of these types of deposit models may be helpful and lead to more and better discoveries (especially for those types which are found in your area).

The attributes or properties of a mineral occurrence consist of all the geology, mineralogy, origins and other significant features that are exhibited by a specific kind of mineral deposit. When applied to develop a model, the characteristics of a deposit are sorted and classified into a specific group that represents similar deposits. The "model" then becomes a set of unifying concepts that explains or describes the essential geologic phenomenon that lead to the creation and nature of a mineral deposit. The purpose of these models is to communicate information that helps mankind find and evaluate mineral deposits. They are extensively used by modern geologists for planning exploration programs, the interpretation of sampling results and resource assessment. Of course not every deposit fits perfectly into these models, and every deposit has its own special and unique characteristics, but it is surprising how well these models are able to describe certain types of deposits.

The descriptive models also include the geological environment, describing the types of environments in which the deposits are found, such as favorable host rocks, as well as the source rocks believed to be responsible for hydrothermal fluids which may have led to the formation of these deposits. Terms like the "Tectonic Setting" in a model description would cover the major regional features that could lead to the formation of such deposits.

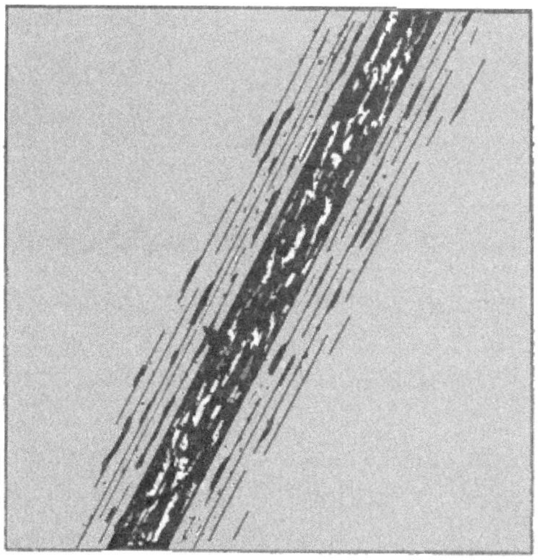

Sometimes vein faults develop parallel breakages, which also get filled with mineral fluids. This creates what is called a sheeted vein.

The "deposit description," lists the identifying characteristics of the deposits themselves, particularly emphasizing aspects by which the deposits might be recognized through their mineralogical, geochemical and geophysical anomalies. The mineral deposit models developed by the USGS are based on the study of over 3,900 well-explored, well-characterized deposits and include estimated pre-mining tonnages and grades for the deposits. Over 60 different deposit models are recognized, and descriptions for all of them can be found via the internet. One of the other interesting features of these models is that when the grades and tonnages of the various deposits in a particular class are plotted, they often adhere to a well-developed curve for that particular type of deposit. In general, the higher the grade the smaller the overall deposit. Deposits which are suspected to be small or having a very low grade are seldom sampled well enough to be characterized in terms of grade and tonnage, thus they are not normally included in the grade - tonnage portion of the models. Graphs of this grade-tonnage distribution have been prepared for many of the models by the USGS.

As geologists continue to study and scientifically explore mineral deposits, these models continue to be refined, and further detail is added to their descriptions. The amount of exploration and knowledge gained each year varies considerably from one deposit type to another. While a great majority of the valuable mineral deposits of the world fit well into one category or another, some famous deposits simply don't fit well into any known model and may be essentially unique. Examples of this include the Coeur d'Alene Ag-Pb-Zn veins, the massive Zn-Mn-Fe oxide/silicate bodies at Franklin and Sterling Hill in New Jersey, the Cu-Zn-Pb-Ge ores of Kipushi and the Tsumeb pipe copper ores. These deposits remain geologic enigmas despite, especially in the instance of the first two, extensive study and research spanning many years.

The well understood models have become refined enough that geologists can use them to estimate the potential for deposits that are as yet undiscovered, perhaps lying under a shallow cover of recent alluvium gravels, or otherwise hidden. The presence of existing deposits already discovered in a certain region confirms that specific ore-forming processes have occurred within those geologic environments. That region is then believed to be favorable for the occurrence of certain deposit types, and suggests potential for the presence of other deposits of that same type which are formed by similar or related processes.

So having considered the basic concepts of mineral deposit models and how they are used, let's take a look at some examples of well known deposit types and see what we can learn from these models. Let's first look at the model for "Comstock Epithermal Veins" named for the famous Comstock Lode in Nevada.

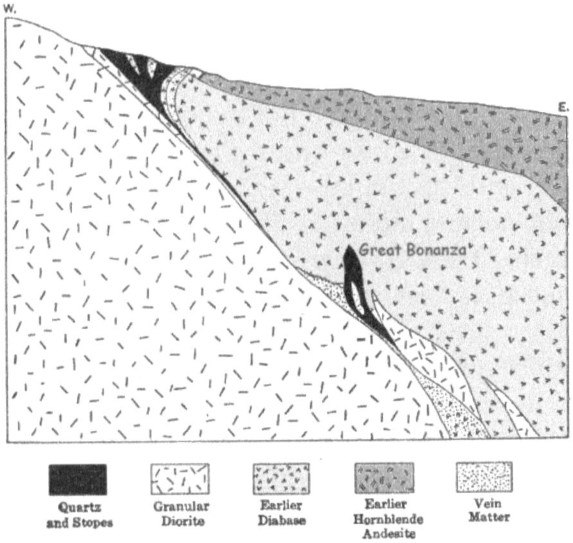

Geologic section through the Comstock Lode. The lower black area was one of the richest silver deposits in the US. It was known as the "Great Bonanza" and produced several million ounces of gold, and nearly 100 million ounces of silver.

The typical Comstock type epithermal vein contains gold, electrum, silver bearing sulfides, and acanthite (silver sulfide) as its most valuable minerals. Base metal sulfides such as galena, sphalerite, chalcopyrite and arsenopyrite are moderate to sparse, and the valuable minerals in the vein often constitute no more than few percent of the total vein matter. This type of deposit generally has overall low base-metal grades. Silver is characteristically much more abundant than gold by weight. The veins are vuggy and consist mostly of quartz and adularia (vugs are open holes often lined with tiny quartz crystals and adularia is a type of feldspar). Typical gangue minerals are quartz with pyrite and adularia, but calcite, sericite and chlorite are also commonly present. Other well known gangue minerals such as barite, fluorite, rhodochrosite, kaolinite, and montmorillonite are moderate to sparsely present. The normal host rocks for these deposits are light to medium colored volcanic rocks such as andesite, dacite, quartz latite, rhyodacite and rhyolite. Because these deposits are associated with volcanic action, many are associated with countries which share boundaries on the Pacific "ring of fire". The country rocks in the area of the deposit are often heavily bleached and altered by the fluids which formed the deposit. The altered rocks typically exhibit iron oxides such as limonite, jarosite and goethite; acidic alteration products like alunite; and the feldspars in these rocks are argillized with the formation of clays like kaolinite. The formation of the veins is related to local volcanic action, and the veins are formed in large and continuous fault systems, including major normal faults, fractures related to doming, and ring fracture zones related to calderas. Some of these deposits are believed to be related to hot springs systems. As a result, they often exhibit a distinct variation in mineralization going downward in depth. Nearer to the surface they are sometimes much richer in minerals containing arsenic, antimony and mercury. Other types of deposits which may be associated with these epithermal systems include placer gold (derived from vein erosion) and alunite-gold deposits similar to those found at Goldfield, Nevada.

Mineralized fractures rich in gold, but without any vein matter; Cripple Creek District, Colorado.

Well known examples of deposits in this category include, of course the Comstock Lode at Virginia City, Nevada, but also the deposits at Tonopah and Aurora in Nevada, Oatman in Arizona and Calico in California. The majority of the US deposits are hosted in Tertiary volcanic rocks. A number of the important silver deposits of Mexico also fall into this type, including Dolores, El Rincon, and Guanajuato. The great silver mining districts of Peru in South America are also of this type, and they have also been found in Japan as well.

If a prospector wanted to search for Comstock type epithermal vein systems, he would use a geologic map to identify regions of favorable volcanic rock types and fault or fracture zones within them, then use satellite photography to identify areas of bleached rocks within those favorable areas.

Another famous deposit type well worth considering is the low sulfide gold-quartz vein, also known as Mesothermal quartz veins, or mother lode veins. This is a common model and a significant number of gold veins throughout the world can be classified under this type. The most famous example of this type of deposit in the US is the mother lode vein system of California. In these deposits, gold is found in massive and persistent quartz veins located mainly in regionally metamorphosed rocks.

Favorable rock types for this kind of deposit include regionally metamorphosed volcanic rocks like greenstone (metavolcanic) belts; oceanic metasedimentary rocks like slate and schist; as well as graywacke, chert, shale, and quartzite. Dunnite, gabbro and serpentine and other basic to ultrabasic intrusive rocks may also be present. These veins are found mainly along shear zones (faults) in metavolcanic and volcaniclastic rocks of the greenschist facies as a result of regional metamorphism and are often related to deep-seated igneous plutons.

The veins are normally persistent for long distances along deep, regional, high-angle faults, or joint sets within granitic rocks. The best deposits overall are located in areas with greenstone. High-grade ore shoots can be found locally at metamorphic rock - serpentine contacts, which are often located along fault zones.

Disseminated ore bodies are sometimes found where these veins cut granitic rocks.

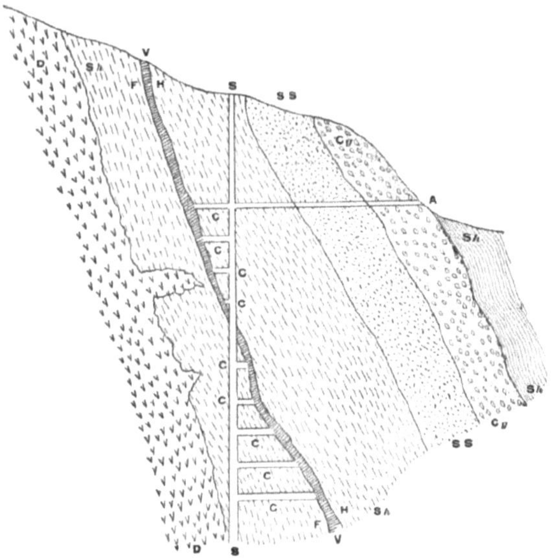

Here is the typical method of working a vein by a shaft downward with many levels for access.

In addition to forming along large faults, saddle reefs form in zones of heavily folded rocks. The veins commonly exhibit a ribbon quartz structure, with few if any voids, as any open-spaces are commonly destroyed during vein deformation which occurs after deposition.

These deposits are associated with continental margins, island arcs and accreted margins. The veins are generally emplaced after the regional metamorphic forces have changed the local rocks. They locally cut older granitic rocks. The tectonic setting for these deposits is one produced by regional compression, and these forces develop the faults and joint systems on which the veins are emplaced. This compressional tectonic setting is normally the result of colliding crustal plates with the subduction of one of them.

The veins consist principally of quartz with native gold, together with minor amounts of sulfides which may include pyrite, galena, sphalerite, chalcopyrite, arsenopyrite and pyrhotite. Additional minerals, some of which may be important locally, include tellurides of gold and silver, scheelite, native bismuth, tetrahedrite, stibnite, molybdenite and fluorite. The productive ore-grade quartz may be colored grayish or bluish in many instances because of the presence of dispersed fine-grained sulfide minerals. In addition to quartz, common gangue minerals in these veins include calcite, dolomites, and siderite.

The quartz in these veins commonly contains fluid inclusions (bubbles) rich in carbon dioxide, and where the veins occur in rocks rich in Ca and Mg, such as basalt, greywacke or serpentine, they are enveloped by carbonate alteration zones. Alteration around the veins commonly shows carbonate alteration with siderite and possibly with ankerite and albite. In many vein systems the carbonate alteration shows a halo extending outward into the wall rock. Green chrome bearing mica, together with dolomite and talc can be found in vein systems bordering ultramafic rocks. Sericite and disseminated arsenopyrite and rutile are found around veins in granitic rocks.

Abundant quartz float is commonly found around the veins on the ground and in the soil. Free gold particles can often be recovered from soil samples merely by panning. For geochemical testing, it has been determined that the best pathfinder element is arsenic, but silver, lead, zinc and copper have also been used for this purpose. Other deposits commonly found associated with this deposit type includes placer gold-PGE, kuroko massive sulfide deposits, and Homestake type gold deposits.

Well known examples of deposits in this category include, of course, the Mother Lode gold belt in California, including all the important gold districts of the Sierra Nevada, including Alleghany, Grass Valley and Sierra City. The important districts of the Klamath-Trinity River region such as those at French Gulch also belong to this class. Many of the important gold deposits of British Columbia in Canada, including the Bralorne-Pioneer area also belong in this category. In Australia, the famous deposits of Victoria's golden triangle including Ballarat, Bendigo and Sailor's gully also are of this type. Most of the important hard rock gold deposits of New Zealand are of this type as well.
Not all of the models are made for hard rock deposits – there is a model for gold and platinum group placers as well. Not all of these contain both gold and platinum, some contain only one or the other, but there are some with both and they were

all combined into one model because of their related nature. As described in the placer model, these deposits consist of elemental gold and/or platinum-group alloys in grains and nuggets in gravels, including sand, silt, and clay, and their consolidated equivalents, as found in alluvial, beach, eolian, and (rarely) glacial deposits. Placer deposits form in high-energy alluvial environments where gradients flatten and river velocities lessen, as at the inside of river bends, below rapids and falls, beneath boulders, and similar locations. Gold placers commonly derive from various gold vein-type deposits as well as porphyry copper, copper skarn, and polymetallic replacement deposits. Famous examples of gold placers are found in many locations, including the Sierra Nevada range in California and the golden triangle of Victoria, Australia.

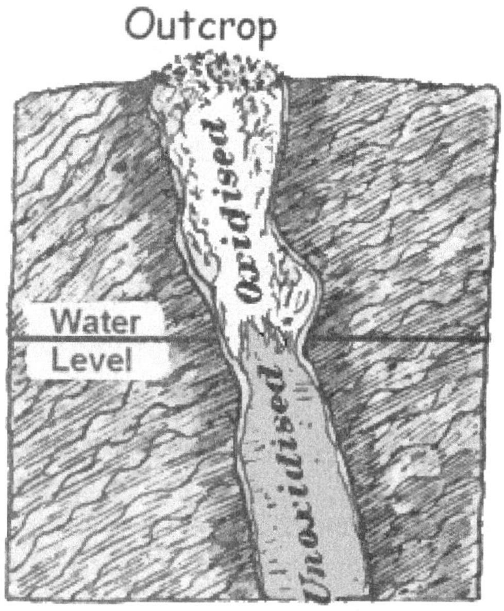

Polymetallic veins are often subject to secondary surface enrichment in the oxidized zone, with the unoxidized ores being of much lower grade. The oxidized level usually extends down to the water table line, below the water table is unoxidized.

A model similar in some ways to the mesothermal gold quartz deposit is the polymetallic vein. They are found in proximity to granitic intrusive rocks. Like mesothermal deposits, the veins are largely quartz and/or calcite, and form in fault and fracture zones. They may be valuable for their gold, silver or base metal contents. Generally, they have values much higher in both silver and base metals than the standard mesothermal gold-quartz vein.

Nevada is famous for its sediment hosted gold (also known as the Carlin type deposits). While a significant number of these have been discovered in the state, only a handful have been found outside it. Sediment-hosted gold deposits characteristically contain microscopic gold grains localized along thin fractures and disseminated throughout sedimentary host rocks. The ore typically contains pyrite, stibnite, realgar, and orpiment but only rarely contains gold large enough to be visible to the naked eye. Hydrothermal alteration associated with the gold includes replacement by jasperoid, argillization, leaching of carbonates, and locally, introduction of carbonaceous material. In Nevada, many of these deposits are located on or near to the Roberts Mountains thrust fault.

Beyond the few models we have taken a close look at here, there are dozens more that may be of interest to the individual prospector. Some of the other models that may be of interest to the prospector include: Alaskan type platinum group elements; diamond pipes (kimberlites); porphyry copper-gold; epithermal quartz-alunite gold; hot spring gold-silver; and Creede type epithermal. There are actually many others that also may be of interest to local prospectors in certain areas. Considerable information on descriptive models of minerals deposits is available over the internet, and USGS Bulletin number 1693 by Dennis Cox and Donald A. Singer is the place to start. This document is available for free download from the government over the internet. Much of the information presented in this chapter was derived from this publication. While this report was originally published in 1986, and there have been some considerable updates and refinements of the models since then, USGS Bulletin 1693 still makes a great place to start your deposit model research. USGS Bulletin 2004, Developments in Deposit Modeling (Bliss, 1992) contains some updated information on the topic and is also worth researching.

Geology of Placer Gold Deposits

Placer deposits are the focus of many small-scale prospectors, as they can be worked on a small scale with comparatively inexpensive equipment. In this chapter, we will take a look at the geology and formation of placer deposits, and how a better knowledge of placer geology can improve the individual prospector's take. Placer deposits are natural accumulations of valuable minerals concentrated in loose materials such as stream sediments or in beach materials. In brief, the origin and formation of placers is all about the forces of weathering and erosion. The process of erosion is the break down and removal of the rocks and minerals that form the earth's crust, also commonly known as weathering.

The weathering process is divided into two basic types - chemical and mechanical weathering. Chemical weathering occurs when rocks and their constituent minerals are decomposed by the dissolving forces of surface waters that are generally acidic in nature. Soluble minerals, such as calcite, are particularly affected, but many other minerals, including silicates (such as feldspars) and sulfide minerals can also be dramatically affected. Common chemical weathering reactions include dissolution, oxidation, and hydration. These separate the minerals themselves into their individual chemical constituents. Some elements are leached away and removed by the waters. In contrast, mechanical weathering involves physical forces such as freezing which disintegrate rocks into constituent minerals, but do not dissociate the minerals into their elemental components. Mechanical erosion is extremely important to placer formation, because not only does it loosen and wear away material, but it can also sort, trap or concentrate minerals as it transports them to a new location.

The fact that placer gold is directly derived by mechanical processes of erosion from vein deposits or other hard rock occurrences is absolutely certain, and examples of convincing character are present all over the world. This does not imply that every primary deposit can always be worked at a profit, but they may still form placers even if not suitable as hard rock deposits.

The gold in the outcrops of quartz veins or other gold deposits is freed from the parent body by erosion and passes either at once into a stream-bed or gets there after a slow journey down the hill-slopes. The progress of the gold towards the valley bottom is dependent on the supply of running water and on the "creep" of the hillside gravels moving downward. Where gold lodes are exposed, it is reasonable to expect placers, and where placers have been found the search for the source of their gold suggests interesting possibilities for prospecting.

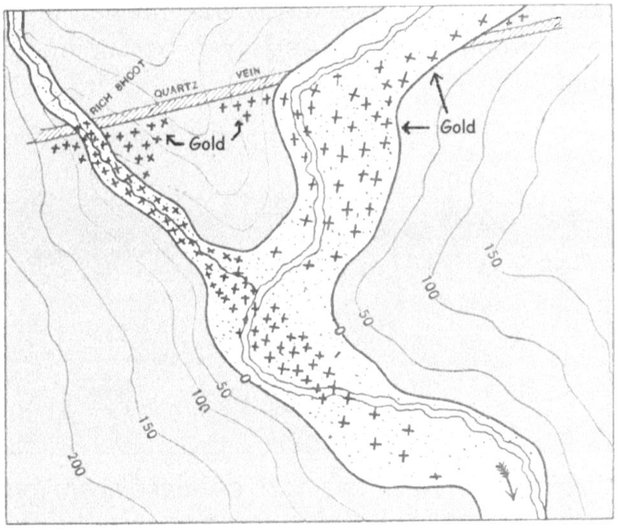

This illustration shows how a stream crossing a gold bearing quartz vein picks up gold. Often many small veins contribute gold into a placer.

Gold is extremely resistant to chemical attack and natural weathering processes; its density causes it to resist most movement by mechanical forces of erosion. As a result, gold tends to be left behind as large quantities of rock are removed through the process of erosion. Any gold that has weathered out of its original rock matrix, be it a quartz vein or another source is called placer gold. Once it is freed from the vein source, any accumulation of that gold is called a placer deposit. Other minerals such as platinum or certain gemstones are also heavy and resistant to weathering and can be similarly concentrated in placer deposits. There are several different kinds of placers. They are classified based on how far the gold traveled, its origin, etc. The four most common types of placer deposits are: 1) **Residual** - where the original vein has weathered, but the placer gold remains more or less "in place" and still within a few feet of the original source; 2) **Eluvial** - where the gold has traveled a short distance down from the source, but has not made it into streams and other drainages - these are often called hillside placers, and can be difficult to distinguish from residual placers; 3) **Alluvial** - Where the gold has made it into area streams and rivers. These placers are sorted by running water and usually the gold lies mostly on or near the bedrock (these are by far the most productive); 4) **Beach** placers occur where small gold particles make it all the way down river to the ocean or a large lake. Wave action on the beach can concentrate the heavier fraction of the sand, producing black sand layers containing fine gold.

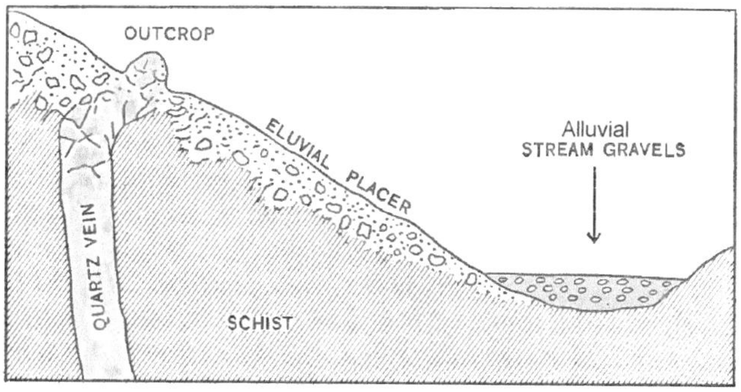

This diagram illustrates how residual placers form at the outcrop, eluvial placers are on the slope and Alluvial placers are in the stream bed.

It is important to remember that even in rich placers the gold is much more concentrated in certain locations than others. The natural concentration of gold in certain portions of rivers, streams or beaches is called a paystreak. These concentrations are extremely important to the individual prospector. Field recognition of good placer areas and the points at which paystreaks are formed isn't always easy. The experienced prospector is able to "read" the river with its bends, blockages and drops and predict where these paystreak concentrations will occur. A general understanding of present day (or past) hydrologic environments is necessary in order to predict where concentrations of placer minerals (paystreaks) have been deposited. Reading a river in this way is mostly about seeing where the water would be flowing when the river is running at flood stage and raging through its canyon bed. These are the times when the gold is moving. Other types of placers, like beach or residual deposits, have their own techniques that a prospector uses to "read" them. This skill can take a lot of experience because what is a good sign in one area is not necessarily good or even present in another area. To read a river for paystreaks, one must envision what the river looks like and what it is doing along the bottom when it is at its highest flood stage. Desert drainages are even more extreme in this regard as they are affected by "flash" flooding where huge amounts of water flow for short periods down a normally dry ravine. The prospector needs to see how obstructions, bends and changes in gradient come into play and affect gold deposition. Excessive turbulence and fast water at flood stage may prevent paystreak formation at sites that would otherwise seem favorable. It is not always a simple task, but with time and experience, the individual prospector will learn to see and read the locations where placer deposits form. Stream borne placers are frequently found at some distance from lode mines, and while there are many potential causes, one must remember that in the larger watersheds exceptional floods that occur only once in a thousand years can move great quantities of material long distances downstream. When conditions are right for gold concentration, including the right stream gradient and/or other conditions that will permit heavy minerals to settle, placer deposits will form. Nicely riffled gravel, often called a shingling of the bars, is conducive to placer formation. Coarser sized rocks and gravel are normally associated with deposits of coarser gold.

Glaciers often act in just the opposite way that a stream concentrates gold, as glaciers destroy accumulations of placer gold because they randomly mix up the material previously concentrated by stream action, so unless the primary vein deposits giving rise to the glacier gravels were unusually rich, these gravels are normally of little value. Under certain circumstances waters flowing off glaciers or streams, or drainages cutting through older glacial material can re-concentrate the gold material that has been randomly mixed up by glaciers to form good placers.

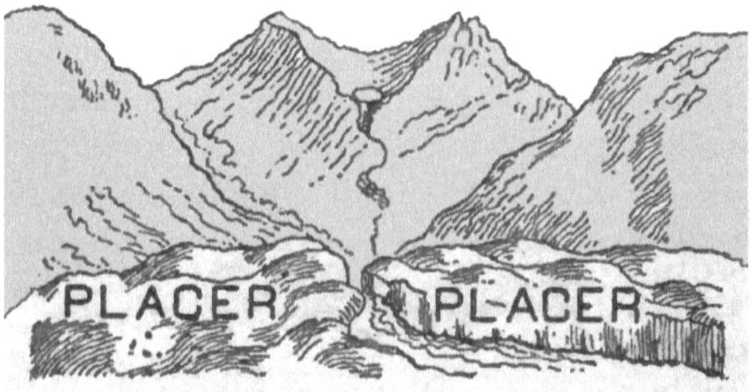

While some gold will lodge high up in the steep part of a drainage, most gold will deposit lower down, where the gradient is more moderate.

Placer geology focuses on the various forces that concentrate the placer gold into economic deposits. A number of factors influence the ability of a mineral to become concentrated in a placer, including factors such as the settling rate, both of the desirable materials as well as the gangue that is hopefully carried away. The settling rate is a function of the grain size, grain shape, specific gravity, surface roughness and electrostatic charge. Larger grains, because they are heavier, settle faster than small ones. Some particle shapes are more likely to catch currents and be whisked away, while others are able to "hunker down" and resist flows to stay in place.

Associated Heavy Minerals

Placer deposits may contain a number of heavy minerals, which settle out with the gold in the sluice boxes. The minerals and metals occurring in gravel deposits consist of those found naturally in the rock traversed by the streams forming the placers, but there are certain minerals almost always present. These include pyrite, magnetite, hematite, ilmenite, (these last 3 make up most black sand), garnet and zircon (heavy white sand). Other rare concentrate minerals can include monazite (heavy yellow sand), cassiterite, platinum, osmiridium, rutile, corrundum, epidote, topaz, barite and even diamonds. Pyrite or marcasite may actually form in the gravels from flowing waters. Other occasional associates of gold, probably derived from its primary deposits, are silver as nuggets (Alaska and Arizona), native bismuth (Queensland and Alaska), native amalgam, palladium-gold, native copper, and cinnabar (see the appendix at the end of the book for more information on platinum group metals).

While the quartz pebbles so abundantly found in many gold-bearing gravels do not ordinarily contain visible gold, there are many instances of such occurrences.

Placer gold occurs in many shapes and sizes, all the way from tiny specks to chunks weighing a couple hundred pounds. Larger pieces (generally bigger than 10 mesh) are called nuggets, and smaller, flat pieces are called flakes. "Colors" are the tiny particles (generally < 0.001 oz), which are often found by panning or sluicing. Placer gold is never fully pure, but instead is a mixture of gold and other native metals (usually silver, copper, mercury, iron or bismuth). Though sometimes as much as 98% pure, most placer runs from 75% to 90% gold. The purity of placer gold is normally referred to as the "fineness", which is essentially the weight percent of gold stated in parts gold per 1000. As gold particles travel further downstream, the metal impurities are leached out, typically causing an increase in fineness as the gold moves downstream.

This very large heart-shaped gold nugget weighs in at 33.85 ounces and was found at the Ganes Creek mine located near McGrath, in Alaska.

Recovering large, coarse gold nuggets is a very exciting experience, and the quest for large gold is often the "holy grail" of the independent prospector, no matter what the tools or equipment that are being used. Hefty, good-sized gold nuggets are the stuff dreams are made of and looking too closely at them can really give a person a bad case of gold fever! Deciding the minimum size for coarse gold is a relative thing, but for our purposes here, we will consider nuggets of a pennyweight or larger as constituting "coarse" gold. Coarse-grained gold and other heavy minerals are usually found associated with coarse sized sediments, such as pebbles, cobbles, boulders and coarse sands. The coarser sediments, due to size and weight, drop out of suspension in the same types of hydrologic environments that deposit the smaller heavy minerals like nuggets. This is why it is not unusual to see a large accumulation of boulders at a bar of gravel on the inside bend of a river. Particle size deposition changes as a flooding period ebbs and flow speeds decrease. Lighter materials such as sand or silt may be deposited as beach sands on top of coarser deposits with valuable heavy minerals underneath.

Sharp edged, angular gold nuggets like this have not travelled very far from their original source.

Normally, coarse sized gold has seldom traveled very far. This statement, however, is relative only and implies that coarse gold is generally nearer its source than fine. When the alluvial gold in a district is all coarse, it does not always follow that all the gold that came from the source veins was equally coarse, but only that the forces of concentration have separated the fine sized gold and washed it away. Some prospectors think that gold breaks down from larger to smaller in a manner similar to hard minerals like quartz. However gold tends to smash down into flakes without breaking because gold is not brittle and does not easily break down into smaller pieces. Gold is a very malleable material, and tends to get caught in the upper parts of a river system. Platinum is less malleable, but like gold is still very resistant to disintegration.

The texture of gold is also an indication of the distance downstream it has traveled. Rough, angular texture is generally considered to be an indication of close proximity to the source. Rolling in a river for long distances rounds gold, so angular pieces cannot have traveled far. However, gold can be formed in the rock in both coarse and smooth forms. This happens because the geothermal solutions in which gold pieces grow can both deposit gold as well as dissolve it. There are nuggets which, after forming are subject to partial etching and erosion by the geothermal fluids. These can lead to the formation of rounded nuggets and blob shaped particles of gold that, in spite of their appearance, have not traveled far from the source.

Types of Placers:

I've mentioned the main classes or types of placer deposits already, and I'd like to take some time and delve into each of them. We will take a real serious look at the details of each type, including tips, suggestions and thoughts about hunting them. Three geologic conditions are usually necessary for the formation of placers: (1) The occurrence of gold in primary rock deposits to which erosion has access; (2) the separation of gold from those primary rock deposits by weathering or abrasion; and (3) the transportation, concentration and deposition of the eroded gold bearing material. Placer deposits are categorized based on the various geological processes which form them, so let's get started with our detailed look at each of them as follows:

Eluvial or Residual Placers

These are placers which are at or very near to their original host rock, having moved little or no distance away from the vein or other source. They typically originate in regions where gold bearing rocks are subjected to deep weathering, during which the gold may undergo more or less concentration, and also migrate down slope to some extent. These residual placers are deposits in which have never been transported by flowing waters, so the concentration of gold takes place only because of rock weathers and the resulting soils are washed away. The gold grains are usually angular, as they have not been exposed to the wearing and smoothing action of streams. In the United States, this type is known in the southern Appalachians as "saprolite", and in California as "seam" diggings". They have also been found in a number of locations around the world, including Australia, Brazil and the Guianas in South America, Madagascar, some places in Europe, etc.

These gold placers are formed when the area rocks are subject to an extended period of deep weathering and secular decay at the surface. It has been thought by some that this kind of deep rock decay is peculiar to the tropics, but this is not correct. This same process has been active under more temperate conditions in the southern Appalachian States, in California, and even in Alaska. When the outcrops of gold-bearing veins are decomposed, a gradual concentration of the gold follows, either directly over the primary deposits or on the gentle slopes immediately below. When located on a hillside, the vein bends over and disintegration breaks up the rocks and the quartz, the latter as a rule breaking up much more slowly than the country rocks; the minerals in the rocks weather into limonite, kaolin, and soluble salts. These are slowly washed away, and the volume is greatly reduced, leaving areas paved with broken quartz and gold concentrations. Both quartz and gold are chemically resistant to weathering and remain behind. The country rock turns into a loose, deep red, iron rich soil often with significant amounts of quartz on the surface. The gold in these deposits consists of rough and irregular pieces and has a fineness essentially the same as the gold in the primary vein. Miners have given the term eluvial deposits to these hillside residual gold concentrations. They are ideal for the modern metal detector because of their near surface nuggets.

In the gold region of the southern Appalachian States, the decomposition of the country rock which generally is a schist, may reach a depth of 100 feet or more. The decomposed material of the gold-bearing veins slides downhill, mixing with the weathered rock. The old miners of the south termed this material "saprolite". This type of deposit has given rise to a peculiar system of mining by which the whole mass is washed by the hydraulic method and the more resistant quartz boulders are crushed in a stamp mill with a coarse mesh. This has been practiced in the state of Georgia at Dahlonega and is sometimes called the Dahlonega system. The Appalachian gold-bearing region is essentially one of mica-schists, but the detrital gold deposits are dependent not so much on the schists, as on a number of basic dikes which cut them. The gold is set free by the complete oxidation of gold-bearing pyrites. Similar deposits were worked in California, particularly in Eldorado County, and are here called "seam diggings" from the fact that the gold occurs disseminated in quartz seams traversing a certain belt of schists. These deposits sometimes create legal contests owing to the uncertainty whether they should be considered as placers or as mineral-bearing veins.

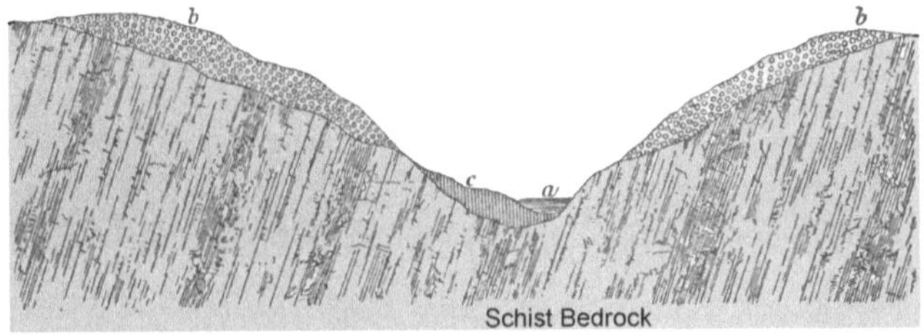

Eluvial hillside placers (shown as b) can be of a very large volume as compared to the river gravels (shown as c) and water (shown as a).

In parts of Brazil, the schists and gneisses are covered by gold-bearing soil accumulated in place. Another example is the "Tapanhoancanga" of the same country. This is a bed of residual or lateritic iron ore up to 10 feet thick covering the underlying hematite schist and containing gold throughout. The gold probably occurs in veinlets in the schists and the gold-bearing detrital material has been concentrated from a considerable thickness of schist weathering in place. Excellent examples of eluvial deposits are reported from the Guianas, though ordinary stream placers are the most common deposits in those countries. The lateritic deposits containing gold are more or less confined to the plateau and lesser hill- slopes occupied by diorites, diabases, amphibolites, epidiorites, and hornblende schists rocks which are regarded as the most frequent carriers of primary gold. Over a great part of this gold-bearing territory secular decay of crystalline rocks has resulted in a deep mantle of iron red, clayey laterite and in places the gold has been concentrated in this material below outcrops of gold-bearing veins. Many of the stream beds are also worked for placer gold, the detritus usually resting on the clayey surface of the compact laterite. It seems more probable that the gold contained in these deposits is derived from mineralized fracture or stringer zones in the local greenstones. It has been definitely proved in these districts that the local gneisses, granites, and porphyries are not the source of the metal.

Residual gold placers are important in Western Australia. This photo was taken over 100 years ago near the golden mile during one of the great rushes to the Australian outback.

In the nation of India, these deposits are of recent or sub-recent date, and rest on metamorphic and igneous rocks. Most of the detrital gold of India has been derived from quartz veins in the metamorphic rocks, and from chlorite and hornblende schists. In Madagascar, an iron red lateritic deposit is derived from rocks cut by auriferous quartz veins and yields gold in some quantity. Some large nuggets have from time to time been discovered there. These deposits are due to the breaking down and complete oxidation of masses of gold bearing pyrites, and are connected with a variety of rocks.

Dry or Eolian placers

In arid desert regions, where the rocks are disintegrated by long exposure on the surface, the lighter particles may be blown away while the heavier ones, including gold, remain behind. At most placer locations, the powerful eroding force of flowing water dominates the shaping of landcapes - canyons are cut and mountains are removed and relocated in the form of sands and gravels. However, in areas where little rain falls and terrains are flat or very gently sloped, a specialized and somewhat unusual type of placer is formed: the eolian or windblown placer. This is basically a subclass of residual placers where wind is a more important force to remove soils than water from rainfall. Prospectors that specialize in electronic detecting for nuggets are especially interested in this type, as it yields nuggets that are close to the surface of the ground, and therefore more easily found with a detector. In fact, eolian placers are generally superficial in nature and normally do not go more than a few feet into the ground.

Eolian placers are formed in comparatively flat terrains where the forces of water erosion do not have enough slope to remove the soils and other surface minerals - often these are desert environments where little water falls. These are also poor environments for plant growth, which makes things easier for the prospector, as the lack of plant life opens the surface to easier erosion by the wind. The rocks exposed on the surface become weathered as air and water convert feldspars and micas to clays over long periods of geologic time. Light sands, silts and clays are slowly removed by blowing winds. In time, materials resistant to chemical weathering such as quartz and gold become concentrated in a thin layer at the surface as the lighter materials and weathered soils are removed mostly by wind erosion. Smaller stones resistant to weathering may result in the formation of a "desert pavement" of stones covering over lighter clays and sands. In many eolian placer areas, a high percentage of the "pavement" material may be pieces of vein quartz.

There is also some removal of the soluble elements of certain minerals by leaching, and sometimes minor removal of minerals such as clays that are easily held in suspension by water solutions. Where there is no opportunity for liquids to drain away, water soluble materials may become concentrated as salt flats in the form of things like chloride salts. In the case of rocks which contain large amounts of calcium (like limestone) the soluble part may re-solidify just below the surface as cemented caliche gravel (referred to as calcrete in Australia). Gold in eolian deposits tends to accumulate in patches near its original source point and it doesn't move much, showing rough edges with little if any signs of wear. In fact, in these types of deposits, gold nuggets can sometimes be found right on the ground surface.

Residual placer deposits are normally formed at or very near the surface. Because of this, they make excellent targets for prospectors looking for gold using metal detectors.

This is why this type of deposit is of special interest to the electronic prospector. Though eolian placers can be formed by the re-concentration of gold within auriferous conglomerates or similar rocks, most commonly eolian placers are a form of residual placer where the gold has not traveled more than a few feet from its original source. Reading these types of placers are difficult, but the general areas where nuggets can be found are often marked by the presence of vein quartz on the surface. The gold does not necessarily come from the veins, but the mineralization that caused the quartz in that area is often the same geologic force that would place gold nearby as well.

Certain placers in Australia have been concentrated only through the agency of the wind, the lighter material having been removed. Placers in the Red Sea Hills of NE Sudan in Africa fall into this class as well. Deposits concentrated by wind (eolian) forces can, of course, be formed only in dry countries where long periods of surface decay has paved the way for the work of the dust storms; from the decomposed and crumbled outcrops of the lodes the winds blow away the lighter sand, leaving a mass of coarser and heavier detritus which contains the gold. Such wind-born placers have been noted near the outcroppings of the West Australian gold veins. Examples of this kind from the western states of America include some of the deposits in Arizona and northwestern Nevada, including the Rye Patch, Rabbit Hole and Sawtooth deposits. Perhaps the best method available to work these placers is by metal detector as they lie at or near the surface and commonly contain coarse gold.

Desert Pediment / Alluvial Fan Placers:

Placer gold and other heavy minerals can also accumulate in alluvial fan deposits, which are fan-shaped areas of unconsolidated, unsorted stream sediments that form where the streams come out of an uplifted area and meet a flat valley. The deposition of these materials is due to the sudden decrease in the stream gradient and consequent decrease in stream velocity and turbulence. Mixed gravels of various sorts and sizes are normally dumped at these locations as they become too flat to move larger rocks and of course, gold. Because larger gold takes more current to move it, smaller gold and quartz specimens are moved out farther on the pediment. Quartz and gold pieces are moved even farther because their overall combined density is less than that of larger, purer gold nuggets. The overall grade of these mixed placers is usually low because the gravels are not concentrated by normal placer processes that wash away lighter rock and gravel leaving the heavier gold behind.

Concentration of gold within desert alluvial fans and pediment areas is complex as these systems are subject to renewed uplift forces, erosion and reconcentration of gold previously deposited within the fan. In many ways, the slopes of these eroded pediments are very similar to desert dry residual placers and can become basically a form of that placer category. This happens by the same

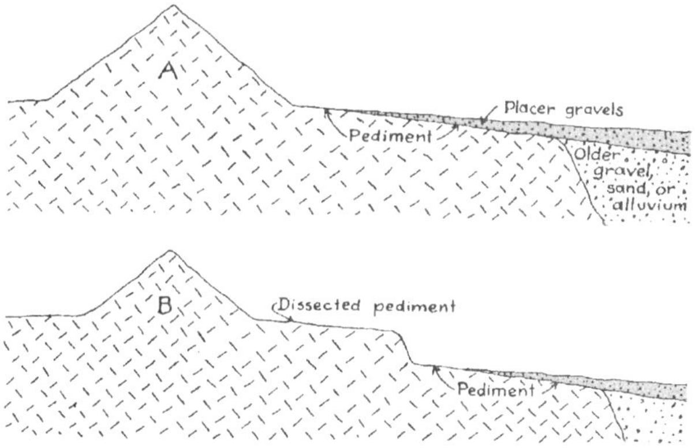

Pediments form important placers in many desert areas because the drainage slope decreases as the pediment is reached, so gold drops out.

forces in lower gradient areas, where lighter materials are washed away. This type of placer is important in certain desert areas of the US. In any areas where new drainage gullies cut through the older gravels, normal washing and concentration of placer gold will occur, because the alluvial environment will concentrate the gold. At times a shallow bedrock pediment may underlie the alluvial fan, and washed shallow pediments may concentrate gold in this shallow bedrock. Flats, drainages and slopes are all worth checking, and both dry washers and metal detectors can do well in these situations. All these processes take place in uplifted pediments, so you can see why I commented that these environments can be complex. Some geologists believe that the very rich placer gold deposits of the Witwatersrand District in South Africa are ancient fan deposits which have been buried and lithified into solid conglomerate stone.

When a stream is flowing, gold works its way downward, eventually becoming lodged in the bedrock cracks and crevices.

Stream Placers

Running water is by far the most important agency in the formation of gold placers, because flowing water is such a powerful agent in concentrating gold. As a result, this type of placer has produced by far the most gold, so I will spend the most time looking at these deposits. The mass of rock shedding off a mountain, even in a very rich area, is normally too low in grade to mine. However flowing streams are very good at pushing rock and gravel downstream, while retaining the gold, much like a sluice box. In fact, in this section on stream placers I am going to refer to that analogy many times as it is a good one to help people understand how a flowing stream concentrates the gold. So it is good to remember that stream placer deposits are the results of nature's vast sluicing process as she wears down the mountains. The formation of these stream deposits requires not only the gold bearing rock, but the power of rapid streams, flowing in a suitable channel. Under the force of flowing waters, rock is ground up into sand and gravel, then sifted by the power of the streams, with the lighter and coarser materials being carried away and the heavier particles lodging in the bed, just as they do in the riffles of a sluice box. In many mountain gravels, the sands of rivers coming from the mountains contain a certain amount of gold, but it is only under favorable conditions that the gold is sufficiently concentrated to render the deposit profitable to be worked by the labor of man.

Once gold moves down into flowing streams and rivers, it is subject to the strong sorting forces of water. Among the most favorable of these conditions is a comparatively narrow channel, with a hard and jagged bedrock and ridges or bends in its course, which, by causing a slowdown in the speed of the current, allow the heavier particles of gold to settle to the bottom and hold them there when once they have settled. Most of the concentration of heavy minerals can only occur when current velocity and sorting forces are at a maximum. In water, gold and gravel are subject to the forces of water buoyancy. This buoyancy force causes the difference in density between rock and gold to be far greater under water. In air, gold is 7.1 times denser than typical rock, but underwater, gold is 10.8 times denser than that same rock.

Crevices, ledges and other changes in bedrock, (Marked as A and B) that are commonly buried and hidden under the gravel, catch the gold and hold it much like the riffles in a sluice box.

Because of this, rock is moved away downstream much faster and more easily than gold, and the eroded gold particles tend to remain behind, getting concentrated in traps along bedrock and in zones of slowed water currents (velocity shadows).

So we know that during flooding, heavy minerals resist movement by the flows far more than the general river gravels. This causes the high-density materials like gold to move downward in the gravel bed until these minerals end up resting on the lowest streambed floor created by the scouring action of the flood. Streambed floors that are not the true bedrock are called false bedrock or flood layers. Periodically, larger scale floods scour the streambed completely down to the true bedrock, resulting in deposition and accumulation of the heavy minerals on bedrock. This is why the richest pay streaks of placer gold and other heavy minerals are usually found on or very near bedrock. Traps and crevices along the bedrock allow gold particles to drop out of the flow and "hide" from the forces of the high water currents, concentrating gold further in these spots. Any spot where the current lacks the force to move the heavy minerals out of the catch can be a trap. That's why it's so important to clean out those crevices completely.

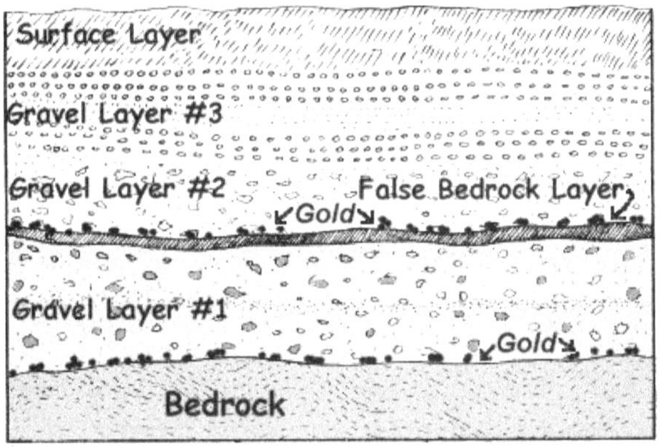

False bedrock or flood layers form when floods do not scour all the way down to bedrock. Paystreaks of gold can form on top of flood layers

Because a river concentrates gold only at certain locations, it makes a big difference exactly where you choose to dig for gold. This is because the placer gold deposits formed in rivers and streams are anything but uniform - with some parts rich in gold but most parts being very poor. Again, I am stressing to you, a river is very much like a sluice box where gold catches in certain parts but not in others – in fact, the old timers even called the bedrock outcrops of a stream by the name "riffles". Under the right conditions, a stream will allow the sand, gravel and rocks that enter it to pass right on through but it will retain heavy minerals like gold.

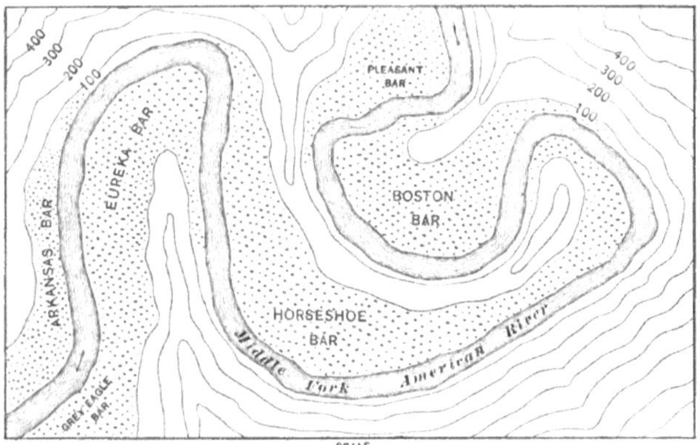

Sharp bends in a river create spots where gold is deposited. This is a set of sharp bends on the Middle Fork of the American River in California.

To be sure, a river or stream is not as efficient in concentrating gold as a well designed sluice, but it definitely does concentrate gold into rich deposits at certain points along its path. Just like in a regular sluice box, there are certain places in a stream that one would look for gold and others where one would not expect to find any concentration of the precious metal. Those rich, gold bearing sections of the river are called paystreaks or sometimes the "run of gold". The goal of the individual prospector is to find those rich sections and work them for their gold. Learn the skill of finding those rich spots and you will be well rewarded for your efforts. Because placer deposits are so spotty, you can spend about forever working the poor sections of a stream or river and find very little gold, while another guy working maybe 50 feet away can be producing ounce after ounce of nice nuggets. The critical question is: "How you find those rich sections where the gold is concentrated?" It isn't always easy, but the old timers call that skill learning to read a river - to see in your head how it flows when it's flooding, to look at all the boulders and outcrops and then be able to find the locations of it's paystreaks and concentration spots.

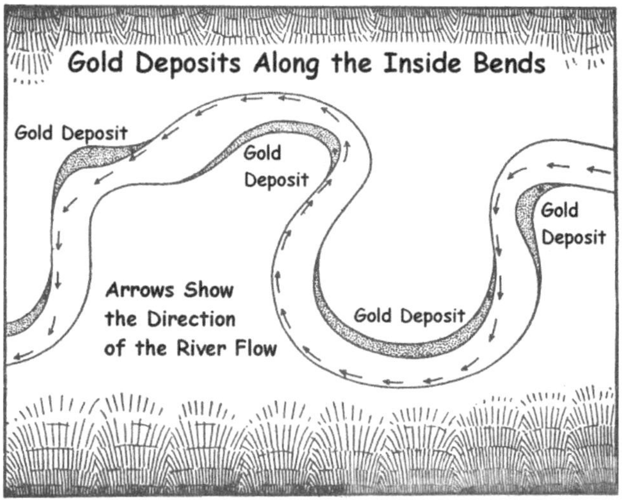

Gravel bars and paystreak gold concentrations form naturally on the inside bends of streams. They are an important target for the prospector.

The same resistance to movement by the force of the floodwaters also causes gold and the heaviest minerals to move in a line of decreased flow force (such as inside bends). The ideal environments for heavy minerals to drop out of suspension and form a paystreak are so called low pressure areas or "velocity shadows", where the current looses speed and force due to a stream obstruction or change in direction. It is hard to judge where these spots will form as the appearance of the river flow during flooding is much different than what is happening during low water flows. There is always a little bit of guesswork and estimation in learning to read a river, and it's a skill that takes time and practice to gain, but it can make all the difference in the world in your success of finding or not finding worthwhile quantities of gold. This is especially true for dredgers, but also true for snipers and other stream prospectors. Because reading a river is not a 100% perfect science, it is necessary to sample and test your conclusions as to where the gold will concentrate.

I've seen a number of different demonstrations showing how gold accumulates in a stream, but perhaps the most interesting was done by Dave McCracken on one of his videos. In this particular shot, he has a model of a river set up on an 8x4 sheet of plywood. The model is just a long narrow trough a couple inches wide, but along its length there are river bends, obstructions, places where the "river" widens and narrows, and other normal examples of the changes in a river. He set it up on a gentle incline with a constant flow of water coming in and then sprinkled in fine gold dust at the top. As the viewer watches, the gold lodges on the downstream sides of bends, at places where the river widens, etc - exactly where one would expect it to concentrate. Just as striking is the fact that for most of the locations along the length of the model river there was little or no gold. Real rivers are just like this - strong concentrations in a few places and very little in most others. As I noted before, I've seen several of these demonstrations, and it is always striking to see how gold will accumulate in some places and not at all in others along the model streambed. These demonstrations strikingly get the point across that it can make a huge difference exactly where you dig for your gold - you can be in a good paying deposit right on the paystreak or you can be slightly outside of it and getting little or nothing for your work. One of the other examples I have seen was just a simple trough with a rock in the middle that was about half the width of the trough in size. The rock did not sit perfectly flat on the trough bottom, just as most boulders would not match the bedrock surface perfectly. Gold was sprinkled in and some of it passed right on through and out the back end, but the portion of the gold that went toward the rock lodged underneath and behind it. To show another example, I have included with this chapter a drawing of one of these experimental water deposition demonstrations as done by the Canadian Geological Survey. This is a very complex model, using sized galena, garnet and quartz sands. The galena, marked in black on the model results shows where the gold would have been deposited. So how does one learn to read a river? Well, the first step is to learn a bit about placer geology. Gold is heavy and resists movement, so in most streams significant amounts of gold only get moved and deposited during times of extreme flooding and high water. Because of this, it is important to remember that when you are trying to read a river, you must see the river in your mind as it is during flood stage when the waters are high and raging. You have to imagine what the river would look like when it is many feet deeper and flowing much faster than you would see in a typical normal water flow.

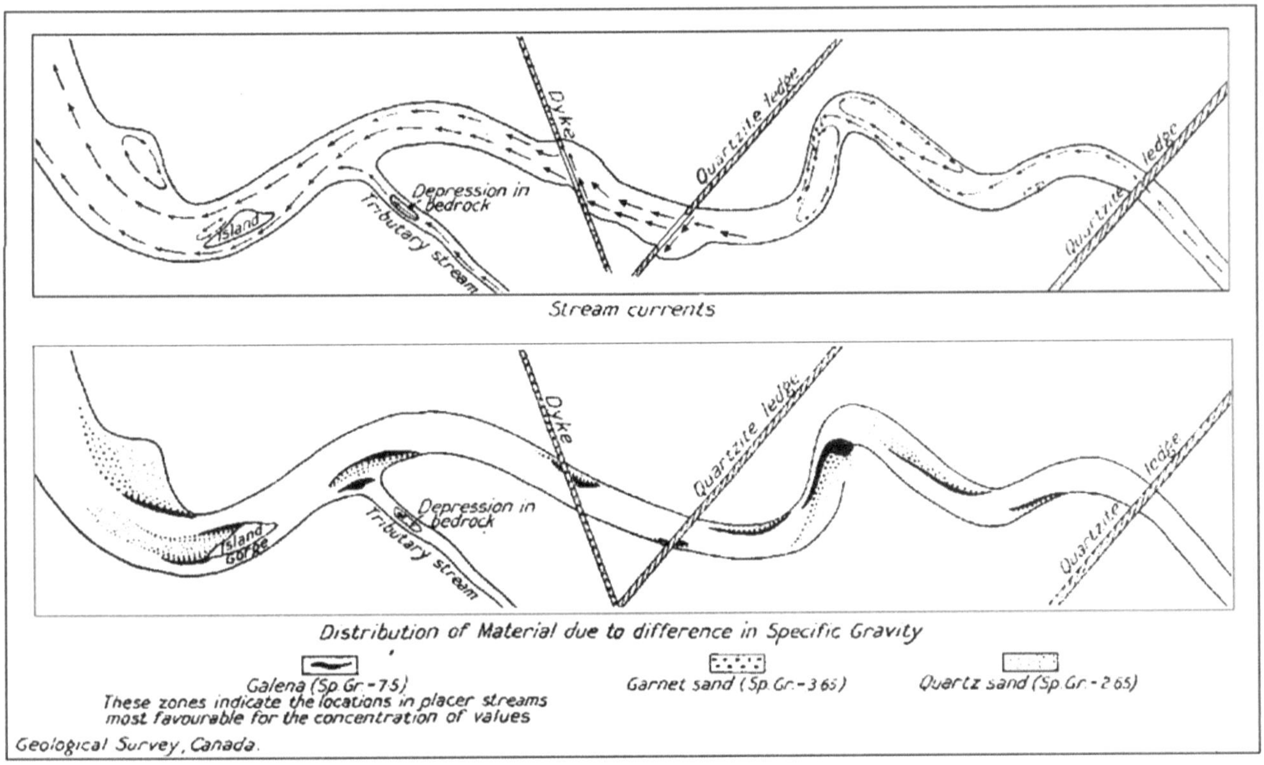

This diagram from the Canadian Geological Survey shows how gold concentrates along bends in a river. The deposits behind the island would be similar to those found behind a large boulder. In this test, they used various minerals instead of gold, but the results are the same. The galena (black) would correlate with coarse gold, the garnet with fine gold and the quartz sand with normal stream gravels and rocks. Note how the sharpness and length of the bend makes a big difference in the formation of the paystreak deposits.

Let's take a moment and imagine just what that looks like, as prospectors are normally out hunting for gold when the streams are low and the weather calm. At full flood stage, the river will be churning and boiling - the water will be flowing 5 to perhaps 15 feet higher than the low water level, and sometimes even higher. Instead of the soft sounds of the water as it normally trickles by in low water times, the river will be rumbling and making a noise like a freight train as boulders bounce off each other and roll along the bedrock. The speed of the river will be much greater as well, and the water would be flying by. It would be full of mud and clay as well. If you could somehow go below the water and see what was happening without being killed, you would see that the high-speed water has swept most of the sand and gravel out of the crevices in the bedrock. The heaviest items in them, however, such as gold or lead bullets would remain and these would fall down to the bottom of the crevice. Any gold that rolled over the crevice as it moves downstream would fall in and be caught. As the gravels churn and roll along, gold moves downward toward bedrock. Once it reaches bedrock, it has a tendency to stay down there, sliding along bedrock as it moves. Where bedrock crops upward like a riffle in the stream, the flows would be slowed and gold would drop out. Something similar happens where large boulders block the flow of the stream. Where the river bends or turns, the flow is fast on the outside of the turn, and slow on the inside. Where the water slows down enough on these bends, gold will drop out of the flow and be deposited with rocks in a gravel bar. Gold is very dense and has a strong resistance to movement. It takes far more energy to move a one pennyweight nugget than it takes to move a small one grain flake. In a similar manner, it takes far more energy to move a 1 ounce nugget than it takes to move a one pennyweight nugget. Gold will only move when the speed and force of the water are sufficient to push it along. As a result, it takes more speed and force to push a larger nugget than it does to push a smaller one. When the gold reaches a point where the speed and force of the

The dark water spot behind this large boulder shows how slow, quiet water forms behind an obstruction even in a raging fast current. This fact allows the gold to accumulate there.

water are not sufficient to push it, it will stop. That's why gold will tend to deposit where the speed of the water slows. This can happen in a number of situations, but as noted, the best known examples are outcrops or boulders, cracks and crevices of the bedrock, or bends in the stream.

Gold does tend to form a crude line of travel as it slides along, which exists between the fast water and the water too slow to move it. Because different sizes of gold take different amounts of force to more it, the "line of gold" has a width - the coarser sizes more to the outside part of the river where the water is faster and the finer on the inside where it is slower. This is a big part of

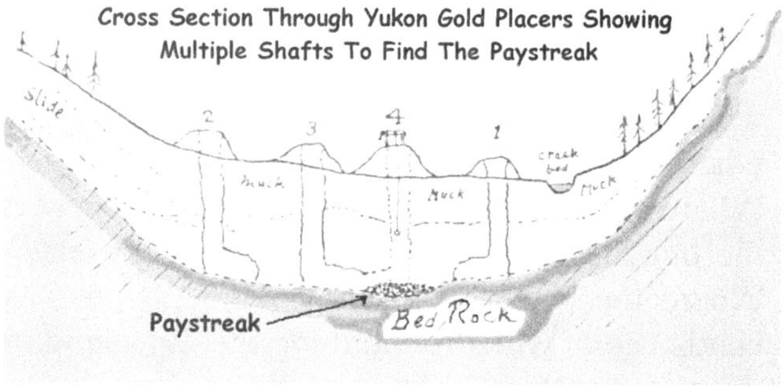

On the Klondike, miners had to dig test holes down to the bedrock to find the paystreak where the gold was concentrated.

why the outside part of a bend will tend to have more coarse gold than the inside. In long straight stretches with a constant grade, the "line of travel" will tend to disburse and open up more evenly across the whole river. Another result of the fact that gold resists movement is that the largest percentage of the coarsest gold tends to be closer to the source. The resistance of gold to movement also means that the gold will almost always be found on or near the bedrock and especially in its irregularities and crevices. Overlying gravels may contain some fine gold for a distance of a few inches to 15 or 20 feet above bedrock in thick deposits. Above the pay gravel there is usually a mass of boulders and gravel containing little or no gold. While the bulk of the gold usually rests on the bedrock or within a foot or two of it, there are some exceptions to this rule. In some gravels the coarser gold is occasionally scattered through the lower 4 to 20 feet on layers which are termed false bedrock, a clayey impermeable layer lying above the true bedrock. Because these false bedrock layers are often established during floods they are sometimes called flood layers.

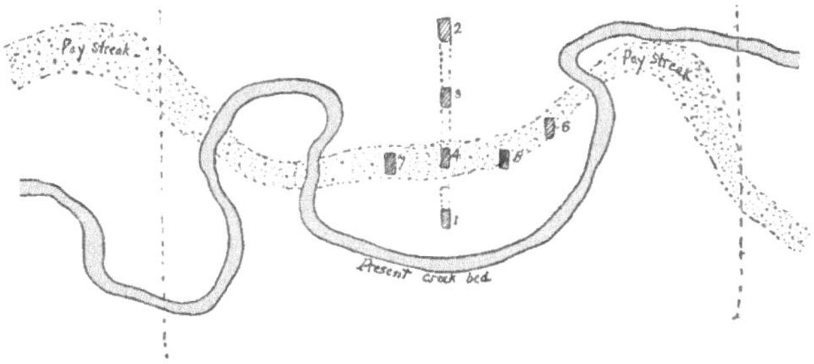

Sometimes the gold in a buried paystreak will be independent of the course of the modern stream. This example is from the Klondike gold placers.

When it does move, gold has a strong tendency to move downward through the gravel until it hits bedrock or some other surface that it cannot penetrate like a false bedrock made of clay or some other material. This movement does not happen after the gold has been emplaced, it happens only as the gravels are moving during a time of flooding. Gold works its way down into the bedrock in a most amazing manner. In hard rock, it will settle into the very smallest crevices, dropping down into spaces that are only just slightly larger than the nugget itself. Even hairline fractures can collect gold. The author once dug a quarter ounce nugget from a tiny hole in bedrock only slightly larger than the nugget itself. In soft rock, gold works downward to a depth of several inches up to 5 feet (depending on the depth of the cracks, holes and fissures in the bedrock), so that it is normally necessary to mine a certain amount of bedrock to get all the gold in larger scale placer mining operations. With only a few exceptions, the distribution of the gold through a gravel bed is far from regular. The vast majority of the gold is usually found on

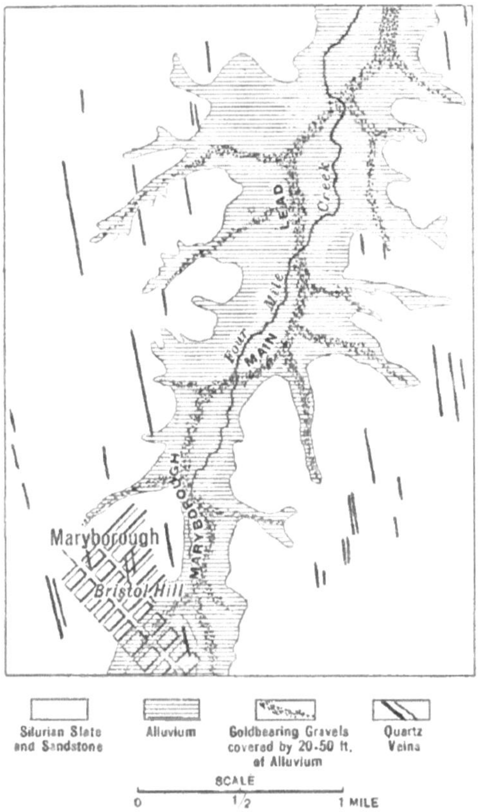

Paystreak lines from the ancient channels located just outside of Maryborough in Australia.

or near the bedrock, but even near bedrock the values are not evenly spread over the whole area of the stream bed. Normally the richer part of the paystreak follows a twisting and winding course, along the line of gold's travel. Its richness is distinctly affected by the character of the bedrock, sometimes splitting and re-forming, following first over to one side, then crossing diagonally to the other side. This richest run of gold is not necessarily found in the deepest depression or gutter of the stream - often the best concentrations of gold ride up on one side of the stream or the other. Even ancient placer gravels have strong, well defined paystreak lines, and sometimes these lines are clearly independent of the course of the present drainage systems. These include deposits like the bench gravels of the Klondike region, the deep leads in the Maryborough, area of Victoria in Australia and the Tertiary gravels of California. While this "run of gold" may follow a distinct and well-defined course on a gently sloping to almost level country rock, sometimes the run of gold may twist and follow a devious course. These pay streaks are formed during flood periods with well balanced conditions of erosion and deposition.

Just like the tilt of a sluice box, the grade of a stream greatly affects how gold will accumulate in it. From very steep to nearly flat, all ranges of grade can be found in watercourses which contain gold-bearing gravels. The water current velocity is faster in steeper areas, creating greater scouring forces. In small steep tributary creeks the drop may be many hundreds of feet per mile, but the placers in these places are usually very spotty. Conditions in these types of streams are high on erosion and poor for deposition. Often the gravel in the streams will only be a few inches deep on top of bedrock, or sometimes the bedrock will simply be swept clean and exposed. In and around powerful gold catches, or in short stretches where these steep creeks flatten out, some gold may accumulate, but in general they have only minimal amounts of gravel, and what gold accumulates in them is likely to be coarse nuggets. Because of this, the coarse gold with shallow overburden does in part balance out spotty nature of the gold in steep tributaries. However, the steep grades and fast flowing waters in steep tributary streams can cause unusual water flow situations which are much harder for the miner to read. When working these little streams, look for clear and powerful gold catches that would work even against very fast water flows. Deep, narrow crevices are the type of catch to look for. Faster currents also cause greater turbulence, which in turn increases the ability of the stream to carry heavy minerals in the suspended gravels. For this reason, where the stream gradient suddenly decreases is often a good area to look for paystreak placer deposits, as the gold will tend to drop out of the flow where the speed of the flow lessens significantly. Where the steep tributaries combine to form larger streams, their beds are less steep, but the bigger streams are still quick-flowing, and subject to floods with rapid rise and fall. Although bedrock can sometimes be seen along these larger streams, because of their flatter grade, a considerable quantity of gravel can also accumulated along their beds. It is in streams of this type that the most important alluvial placer deposits are formed, including rich gold-bearing gravel bars. The larger streams and rivers, which have grades from 50 to 150 feet per mile tend to produce much bigger and more consistent placer deposits (and ones which are much easier to read and evaluate). Many good-sized rivers have formed extremely rich gravel bars because their more moderate grade has allowed the gold and gravels an opportunity to accumulate.

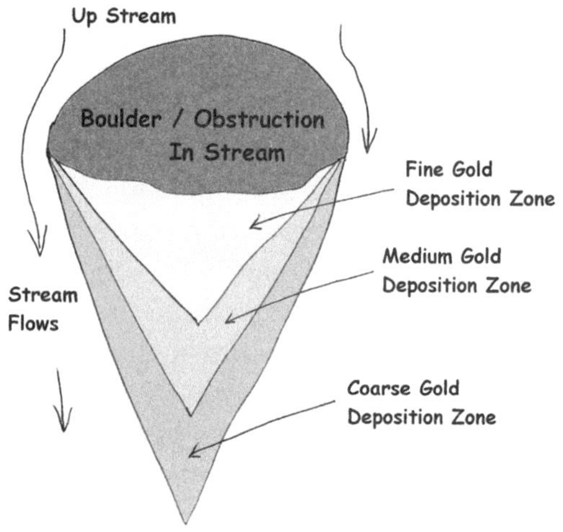

How paystreak gold deposits behind a boulder.

Here are some typical examples of stream gradients: In most California gold bearing rivers in both the Sierra and Klamath region are in the range of 50 to 150 feet of drop per mile. The Tertiary channels of the Sierra Nevada have much shallower grades, typically in the 30 to 50 feet per mile range. This reflects the influence of the Sierra uplift in increasing the grade since the Tertiary channels flowed. The old White channel, in the Klondike district, has a grade of about 30 feet per mile, but it's important to remember that the Klondike is an area of deeply buried bedrock where a large amount of material has been deposited on top of the gold. Many of the present Alaskan streams which have comparatively shallow gravels possess a grade of 100 to 150 feet per mile. In the principal Tertiary channels of Victoria, Australia, low gradients down to 20 feet per mile have prevailed. The best results of gold concentration are probably obtained in rivers of moderate grades, in the range of about 50 feet per mile, under nicely balanced conditions of erosion and deposition. This type of balance allows about as much gravel as is coming into the stream system to move on out of it, leaving only heavy materials like gold behind. That is the kind of concentration which is like a sluice box, and is exactly what the miner wants to see. Where the grade drops below about 20 feet per mile there normally is not enough force to move coarse gold along and only fine flaky gold can be moved down stream. When active deposition of deep gravels begins to take place in a river channel, the concentration of gold ceases.

This steep section of a gold bearing stream in California has captured very little gold. Both above and below, where the grade is more gentle, considerable gold was mined years ago.

The type of rock which makes up the bedrock can greatly affect how gold accumulates. Smooth, hard bedrock makes for a poor place to catch gold - the nuggets simply slide over it. Certain types tend to polish smooth with few breaks, cracks or joints. Granitic rocks, sandstones and some slates will yield this type of smooth and unproductive stream bed. Serpentine is yet another material which normally forms a smooth and unsatisfactory bedrock. Jagged and uneven bedrock is prone to forming gold catching pockets, crevices and cracks, creating a situation like the riffles in a sluice box. These types do a great job of catching gold. Schists and slates often make good bedrock, especially when their parting (foliation) strikes in a direction that is perpendicular to the stream channel and dips steeply. In some types of bedrock, the tendency to be polished smooth or form rough surfaces may be related to parting in the rock and the direction of the parting in comparison to the direction of the stream flow. This situation promotes the formation of narrow crevices. In the ideal situation, parting is perpendicular to the flow and the floor of a stream then forms an effective set of natural riffles, collecting gold as the gravels pass over. In the same way, bedrock that is partly decomposed with a portion converted to clay will sometimes yield good results as well. In some situations, compact clay is good bedrock (or false bedrock), also clay sandstones and clayey decomposed volcanic tuffs. The occurrence of this type of material was common in some of the California bucket line dredging grounds. In limestone, deep and irregular solution cavities may form crevices that contain the gold. This is why perhaps the very best bedrock for catching gold is limestone, although this is not a very common bedrock in the gold country of most regions, limestone sometimes erodes out very deep pockets that can descend to a depth of 50 feet or more. Of course crevices in bedrock don't have to be 50 feet deep to work well, although when gold gets down in them it is stuck there for a long, long time.

Bedrock potholes like this, where the gravels are packed in tightly can sometimes hold significant amounts of gold.

As I mentioned, coarse gold resists movement, and the shape of the gold itself affects that resistance to movement - for equal weight pieces, flat shaped gold tends to lie down flat and resist movement that way. Round pieces may have a tendency to roll and therefore move a little easier for their size. No matter what the shape however, gold tends to move to places where can get out of the strongest, fastest flow of water. In a sluice, that typically means in behind the riffles. In a river or stream, you will also be looking for those obstructions and other places where the flow slows down and gold can be caught. Gold's resistance to movement also causes gold to tend to be pushed into a common line of travel as it moves down a stream. It is along this line of gold's travel that pay streaks are most likely to be found. In long stretches of straight streambed, the gold will be scattered across the width and any line of travel will be weak and spotty.

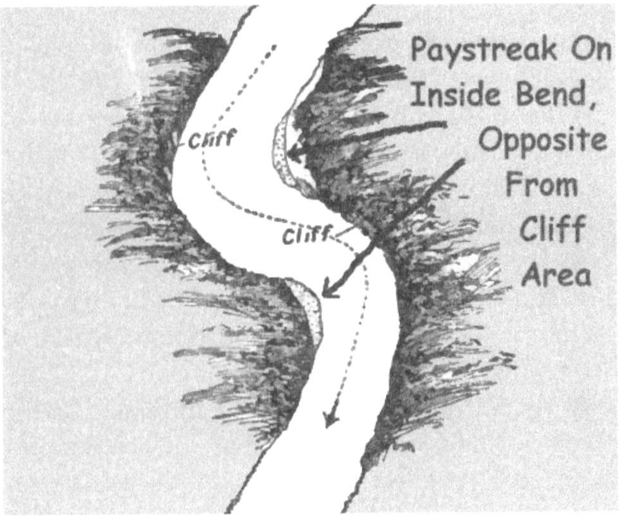

Paystreaks of concentrated gold deposits will accumulate on the inside bends, at the inside edge of gravel bars in larger streams.

Perhaps the best known of the places for paystreak formation are locations where the current makes a significant direction change. The fact that rivers never flow long distances along perfectly straight lines is chiefly due to the varying hardness of the rocks along their course. When a river is in flood stage, the water is deep, and the speed is sufficient to carry boulders, cobbles and gold along its course. When the rains cease or the snows have melted, the river levels begin to drop downward toward its normal rate of flow. The point at which the reduction of speed will be first noticeable will be the inside parts of the bends. As both gravel and gold are deposited by rivers at all points where the current is slowed by any means, it follows that, as the river subsides, gravel bars will be formed along the insides of these bends, and, if the

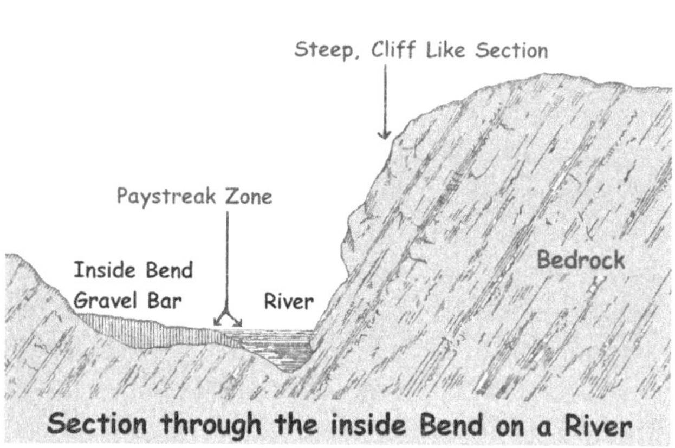

On a bend in the river, the outside part has strong erosion and often forms a cliff or steep wall. Gold bearing gravel will deposit on the inside, low erosion part of the river bend.

country rock is gold bearing, then gold will be deposited with the gravels. On the inside part of the bend, the water must slow down to make the direction change, and the gold, resisting the flow of the faster outside parts, will move toward the inside of the bend and the slower water. The heaviest particles drop out first, with the lightest last. Many large and rich paystreaks have formed at the bends of a gold bearing river, so they are worth looking for. This also means that in a paystreak, the coarsest gold accumulates toward the head and outside parts of the inside bend, while the smallest is found on the inside and toward the downstream end. The sharpness of the turn can greatly affect how gold accumulates and the concentration that is formed in the paystreak. Although there are many variables, sharper turns tend to produce smaller more concentrated paystreaks, whereas gentle turns produce much larger paystreaks but of a generally lower concentration.

In some gold bearing districts all the placer gold is very fine sized - like dust. The Snake River in Idaho is a good example of this, but there are a number of others. Where these rivers bend, they form a somewhat different type of river placer known as a "skim bar". In a manner similar to beach placers, the river flows will accumulate thin layers of heavy mineral sand deposits at the heads of gravel bars on the insides of bends in the river. The deposits are only an inch or two thick, and sometimes show as accumulations of black sand, or at least sands rich in dark minerals. Like beach deposits, these fine gold deposits in rivers re-form regularly during high water periods.

Narrow chutes and deep water spots where the water moves through at high speed during the flood times do not allow gold to find a place to stop. Sometimes however, where there are some deep crevices where coarse gold can lodge even in chute-like sections of a river, but this is an uncommon situation. Normally these chutes end at a point where the river widens out, and groups of boulders drop out there. Under and behind the large rocks in these boulder nests are another place where worthwhile amounts of gold will tend to drop out and accumulate.

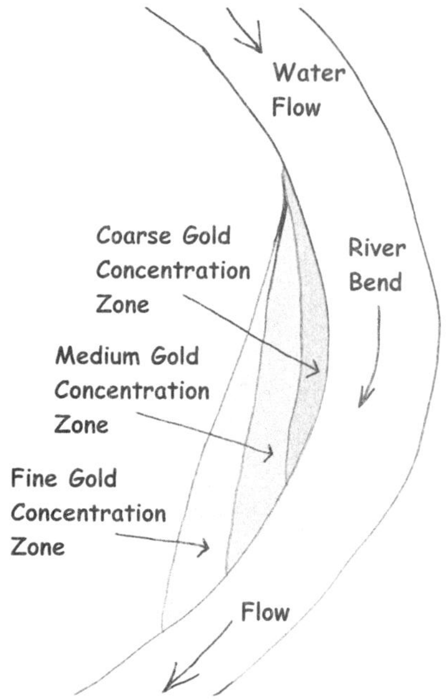

Idealized diagram of how coarse and fine gold deposit on the inside bend of a river.

Another example of a good gold catch is where the current meets a bedrock outcrop, island or large boulder in the middle of the stream. Isolated boulders anchored in the river act as obstructions in a manner similar to the riffles in a sluice. During the flood times there is a quiet water spot behind large boulders that can resist being rolled during the flooding. As the water approaches the obstruction, the current splits and goes around each side and then merges back together on the downstream side. Suction eddies and quiet waters occurring directly behind the boulder cause a velocity shadow, and flow speeds are reduced allowing heavy minerals to drop out of the stream flow. Depending on the situation, a smaller quiet water area can form in front of the boulder as well.

Gold can accumulate in the slow water behind boulders, but under certain circumstances it can also accumulate in front of or underneath them as well. Accumulations of gold underneath a boulder occurs when the rock does not sit flat on bedrock, but allows a certain amount of water to flow underneath it during flood times. This then becomes another quiet water spot where the gold can accumulate. Sometimes gold accumulates better under boulders that it does behind them.

Large bedrock outcrops can act like riffles in a sluice, catching gold behind them.

Similar eddies form both upstream and downstream from obstructions on the sides of streams, around rock outcrops and other obstructions depending on the situation. Bedrock outcrops can also act just like a riffle in a sluice accumulating gold behind them and sometimes under the right kind of circumstances, accumulating gold in front of them as well. The thing about outcrops is that they can be of all sorts of configurations. They can be perpendicular to the river and cut across it or go along in the same direction parallel to it and all possible angles in between. They can also stick up only a little or a lot, and all heights in between. The angle to the river that the bedrock outcrops at is the most critical to how well it captures gold. Outcrops that are more nearly perpendicular to the direction of the stream, and in the line of gold's flow are the best. If the cracks and foliations of the rocks forming the bed of the streams are dipping at steep angles, especially if they strike across the stream, crevices are formed by the uneven decomposition and weathering of the rocks. Any gold that once settles in these crevices is protected from the attack of the stream when the next flood rumbles through. Deposits that have not been so secured are picked up and moved onward when the next flood comes. Crevices in bedrock can really catch and hold gold, especially when they are along the line of gold's travel. The ability of a crevice to catch gold has a lot to do with the direction that the crevice runs, and like outcrops, those which cut perpendicular across a stream, just like the riffles in a sluice. Those which run in parallel to the stream don't stop gold well - the flow tends to scour out the gold. Crevices come in a variety of shapes - some are long, narrow and shaped like a slot, some are V shaped and some are round to oval in shape. These roundish bag-like crevices are variously known as boil holes, potholes or wormholes. Boil holes with rocks and boulders packed into them are a good sign. In fact, almost always where rock and other materials in crevices are loose and easy to remove, the crevice will not normally contain much gold because they probably have already been cleaned out. Where the gravels and rocks are tightly packed and almost hammered into place - those are the ones which are much more likely to contain gold. It's been my experience that iron junk and packed rock are both good signs to productive crevices. Like other gold catches, they are also best when located along the line where most of the gold travels.

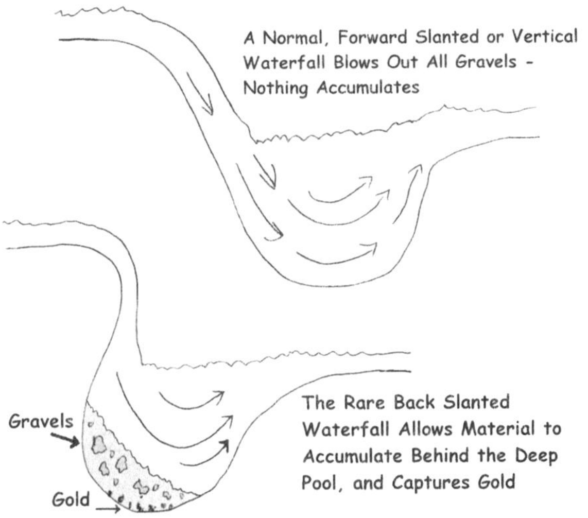

Waterfalls are rarely good places to search, but there are exceptions, and the exception is a rich one. Back slanted waterfalls are a great catch.

A good viewpoint can help you read the river and see all the bends, boulders, bedrock outcrops, etc. that determine where the gold will deposit.

Most waterfalls are vertical or have the base slightly out forward of the top of the fall. This creates a boil hole that does not catch gold, but flushes everything back out into the stream. Where softer rock at the base allows the water to cut backward, behind and underneath the waterfall, a cave is created. This cave like undercut can become a powerful gold catch. However, this exact situation is a very rare circumstance. In a running river, the shallow parts of the stream with outcropping bedrock contain the highest values while the deep water holes are usually barren of gold. These deep pools are normally spots were the river narrows and when the water is deep and fast during flooding these pools become chutes where the water flies through at a high speed, blowing most of the gold right on by. Even within a paystreak, the size of the gold will vary from one point to another within the deposit. Especially on the inside bend of a turn, the head of a paystreak and its outside edge will tend to have the coarsest material. The inner edge and tail of the paystreak to have the finest sized gold pieces. Medium sizes will tend to lie in between the coarse gold and the fine. Because there are a lot more pieces of smaller gold in a river than large nuggets, the coarse side of a paystreak will often be spotty with just a few bigger pieces here and there. While there are always exceptions, the medium and fines portion of a paystreak will normally be more consistent in their production. Remember that no river is a perfectly designed ideal system, and there are always unusual conditions present like boulders, bedrock and other obstructions, so few river paystreaks will meet the perfect ideal, but I have included explanation of the partitioning of coarse, medium and fine gold so that you can get a feel for how paystreaks are typically laid down. Once you have gotten as much as you can about how paystreaks form into your head, reading a river for gold is mostly about choosing a good vantage point and putting your imagination to work. I find that stepping back a bit so I can see a bigger section of the river really helps me get the larger picture of the river's flow at that point. Pick a spot where you can see the river well, and try to envision what that river would look like at flood stage. See all the bends and obstructions in your mind, and consider the likely line of travel for the gold. Using this information, make some predictions about the most likely places for gold to accumulate.

No one is right about reading a river for paystreaks 100% of the time, so it's important to sample the ground to see if your theories about where the gold will be deposited are correct. Pick the spot you think most likely to accumulate gold and dredge a test hole there. Give it a good and thorough testing, but if the paystreak isn't there, move to the next most likely spot for gold to deposit and sample there. It may take a series of test holes, but eventually you will find what you are looking for, and your gold production will increase. If you don't find it on the first hole or two, don't give up. Prospecting can be tough and it's important to stick with it. Many guys who have the skills and knowledge to pick out good spots give up way too easy, and this is a big cause of folks having a lack of success in finding gold. Gold may not drop out where you expect it to for many reasons; the most likely are because 1) Unrecognized forces keep gold from concentrating there; or 2) it's already been dredged up since the last flood. If the gold didn't do as you expected, you just have to keep prospecting around and using your head to select good spots until you find a paystreak with the amounts of gold that you're looking for. I really can't emphasize enough the "if at first you don't succeed, try, try again" aspect of sampling for paystreaks. You need to stick with it. In fact, failure to do enough sampling is one of the biggest reasons for individuals and small operators getting poor results in their mining efforts. If you are not getting enough gold, you need to think on it and look in another spot, even if that spot is only 50 feet away. The same applies if the other spot is miles away. The important thing is to focus your mind on staying in sample mode until you find an acceptable, good deposit. I have a whole chapter talking about sampling a bit later on in this book, so take a good look at the additional concepts there. Once you've found that paystreak, you need to move your dredge (or other operations) to that spot, and sample to determine its downstream end, and side boundaries. If you don't do this at the start, in many cases you will find yourself moving rocks you have already moved and possibly re-dredging your own tailings to get them out of the way. It's important to plan your operations ahead of time. This will allow you to dredge up a paystreak and get the most for your time and effort. Some paystreaks end sharply, but many fade out slowly, grading from good material into poor. It is typical that the fine gold edges of paystreaks often gradually grade from good material into low grade, but sometimes the edges can seem to drop off in grade very sharply. While on the coarse side of the paystreak, the coarser gold often just gets really spotty and you may find only a few large pieces spread here and there on the bedrock. You will have to define for yourself what is an acceptable amount of gold for your daily take, and when you reach points in the paystreak where you cannot recover your minimum acceptable take, and you're not getting enough gold to make it worth your time to continue dredging outward, then you have found a boundary. You have to decide for yourself at what point that occurs.

Bench Placers

As rivers are constantly deepening their channels, through erosion, it is easy to imagine that a gravel bar, having once been formed, may be left stranded as a terrace, as the river drainage channel cuts down through the rock and works its way lower. Because these gravel deposits were once part of flowing streams, if the course of a river is changed, a paystreak that had formed might be left stranded at a higher level than the flowing waters of the stream can reach. These stranded gravel beds are known to miners as bench placers. There are many instances where these old river beds have been profitably worked and yielded more gold than the beds of the present-day streams.

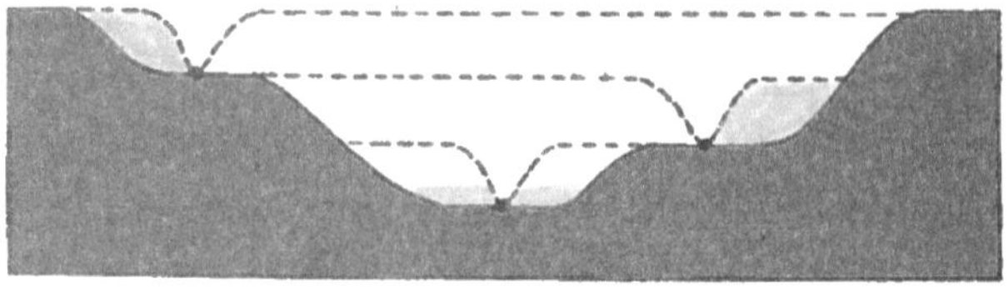

This is a cross section through a river showing how stream erosion cutting down through the mountain ends up stranding patches of older gravels which create bench deposits.

These bench deposits, which are perched just above the more modern drainages, are far more common than true buried ancient channels. Benches can sometimes be located only a few feet above the maximum flood level, but also may be found perched on the hillside up to hundreds of feet above the modern river. These old river gravels can be rich and were commonly mined by hydraulic mining or bedrock tunneling by the early day miners. They are present in many desert and mountain placer districts where a significant geologic uplift has occurred because the uplift increases the forces of erosion. Benches can be as small as only a few hundred square feet on small streams and gullies to large areas covering many acres and hundreds of yards along the length of a large river.

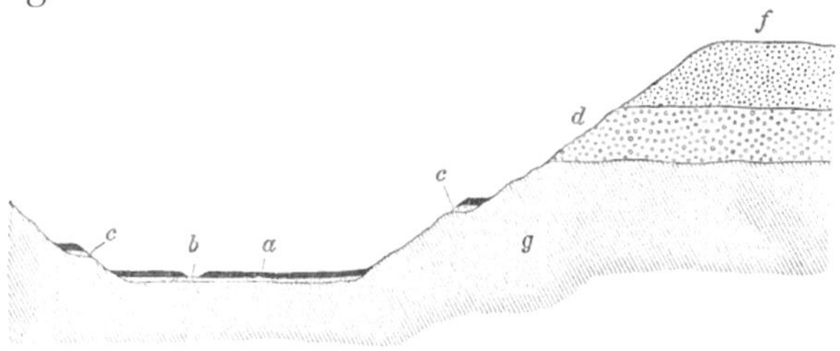

Section across Bonanza Creek, Yukon District, showing modern gravels (a and b); recent bench gravels (shown as c) and much older White Channel gravels (shown as d and f).

Cases of this sort are found on the western slope of the Klamath and Trinity systems in northern California, the Sierra Nevada in California, on Anvil Creek in the Nome district of Alaska, and in the Klondike district of the Yukon. Near Nome on the ridges surrounding Anvil Creek are "high gravels" 600 to 700 feet above the present rivers. These gravels, some of which are rich, are the remnants of an old, now almost wholly eroded system of drainage. In the Klondike high gravels also occur a few hundred feet above the present creeks, the most conspicuous instance being those gravel deposits termed the "White Channel". Portions of these benches that were previously worked by hydraulic mining have been very productive to prospectors armed with metal detectors, as the exposed bedrock of the old benches were not always perfectly cleaned off by the old-time miners and sometimes gold remains.

Ancient and Buried Placers

Not only are placer deposits being formed today, but they have been formed where conditions were favorable throughout geologic history since Precambrian time. However, most commonly the forces of uplift and erosion tend to destroy and reform placer deposits, and as a result, most of the placer deposits being mined today are Cenozoic or younger and occur in unconsolidated gravel materials. There are exceptions which occur when some placers become buried and are thereby geologically preserved. Buried ancient placers, sometimes called "paleo-placers", are found in conglomerate sedimentary rocks as old as Precambrian in age. It is not uncommon that paleo-placers which are eroded by current streams can become the source of present-day placers as the gold becomes reconcentrated in the modern streams. These gold bearing conglomerates occur in a number of areas, and in some places have been very productive. In the US they have been found in the southern Appalachian region, the Bald Mountain region of Wyoming, and in the Black Hills of South Dakota. The South Dakota deposits are famous for their buried placers. The Cambrian basal conglomerate and sandstone rest unconformably above the pre-Cambrian schist series that contains the lodes of the Homestake gold mine. In that conglomerate are found rounded grains of gold.

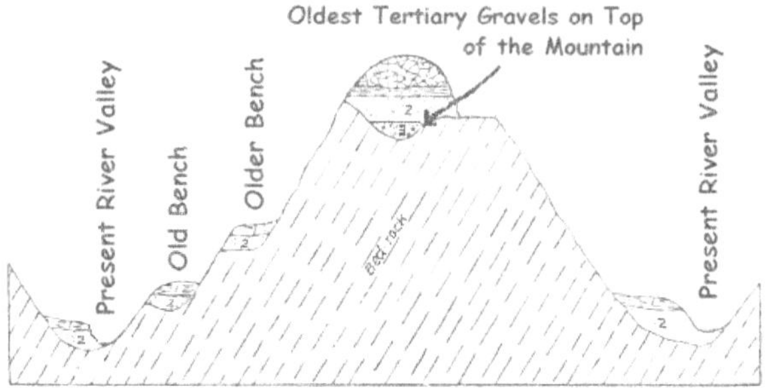

The paleo-placers of the mother lode commonly sit on top of the mountains because of uplift and erosion. Modern streams and benches lie below.

The auriferous conglomerates at Witwatersrand, in South Africa were the most productive gold deposits in the world for many decades and are still very large producers of newly mined gold. From an inconspicuous beginning in 1887, the production of these unique deposits has steadily increased and only in recent years has production begun to taper off significantly. Historical production is so large that there is no question that the Witwatersrand basin has been the world's most important known gold-producing province. It is estimated that roughly 53,600 tons of gold were produced from the conglomerate beds (reefs) of the Witwatersrand basin between 1886 and 2000. This gigantic figure represents close to 40 percent of all the gold mined during the recorded history of mankind.

Perhaps the most productive paleo-placers in the western US are the gold bearing Tertiary River channels of California. They have yielded about 15 million ounces of gold, mostly by the methods of hydraulic and underground drift mining. These channels were created when volcanic eruptions in Tertiary time millions of years ago buried the existing gold bearing rivers and

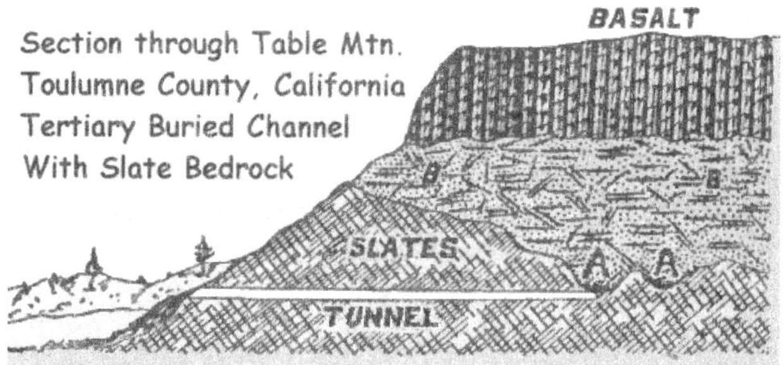

Rich gold bearing gravels (Marked as "A") at Table Mountain lie underneath a bed of gravel and clay, which is in turn covered by basalt. They were mined through bedrock tunnels.

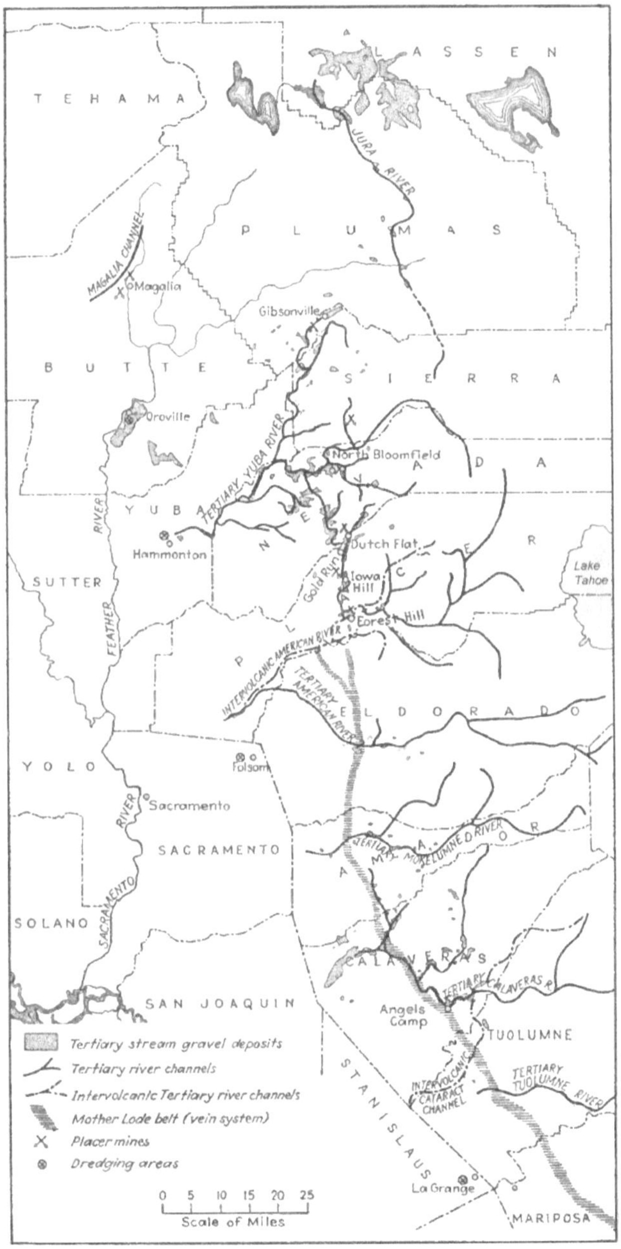

This map, drawn after one prepared by Waldemar Lindgren, shows the locations of ancient gold bearing Tertiary rivers in California.

streams of the Sierra Nevada. The hard, weather resistant lava capped the old channels and protected them. Uplift of the Sierra mountain range forced the water to form new drainages and cut new channels. Because of the great age of the Tertiary channels, they are often heavily cemented and commonly required blasting to excavate and crushing to extract the gold. Where modern streams eroded through the old channels, the old placer was reconcentrated into the modern streams greatly enriching them. Tracing these hillside bleed offs up slope led the old time miners to these rich deposits. In places, these channels yielded very coarse gold, and one of the most famous collections of large California nuggets is the Ruby Mine collection which is on display at the Museum of Natural History in Los Angeles. The Ruby mine extracted coarse gold from an old deep lead Tertiary channel near Downieville in Sierra County.

The ancient channels of Victoria, Australia have also yielded considerable amounts of gold. Like the Tertiary channels of California, they have been buried and preserved by lava flows. The general direction of river drainage has not changed since the deposition of the deep leads, and the channels of modern streams are therefore either superimposed on the deep leads or are parallel to them. No great amount of elevation uplift has taken place in Victoria, and the country has rather been subjected to a general north and south tilting, depressing the lower (northern) portions of the channels and raising the upper (southern) portions. The tilting has occasionally proceeded so far locally as to change the direction of fall of the bedrock. The Victorian deep gravels are late Pliocene in age.

Marine or Beach Placers

While coarser gold will have a strong tendency to hang up before a river reaches the ocean, smaller particles and thin flakes may be washed out on to the beach, creating beach placers. Beach placer environments are formed by the forces of the action of waves upon the shore. While the sand as a whole carries only very small amounts of gold, wave action tends to winnow away light minerals

This early drawing shows how gold was mined on the rich beaches of New Zealand.

and leave heavy minerals to lag behind, similar to the manner in which water currents operate in the stream environment. As each wave retreats it pulls light minerals back towards the ocean or lake. As a result, heavy minerals, along with larger pebbles and cobbles, lag behind on the shore. This sorting action is strongest when the waves are the largest, so the best beach placer deposits are formed at or near low-water mark along the shore at low tide during large storms when waves reach a maximum, and especially where the waves strike the shore at an angle to the shoreline slope.

These "lag deposits" typically form a linear band of visible heavy minerals such as garnet (red) or magnetite (black), which are generally parallel to the shoreline. As noted, the best of these are often found right after major storms. While these bands may stretch several feet up and down the beach slope, they are often very thin in depth. Individuals who live in beach placer areas will often do their prospecting right after large storms during the low tide periods. They will pick up sand from these newly formed bands of heavy material for processing at a later time. If the concentrations are not removed, the next high tide may simply remix them with the sands of the shore. The sand differs greatly in richness at various times: one day, it will be full of golden specks; a few days later, at the same place it will be barren. The concentrated black sand layer in the mean time has been moved by the waves, and replaced by other sand. Generally the layers of black sand that carry significant gold are easily visible. Although these deposits are small and temporary, they can be quite rich at times. In some places, prospectors are able to collect a considerable and worthwhile amount of gold for a short time right after large storms. Small amounts of platinum are also commonly found with the gold in many locations. When it is most perfectly concentrated in the richest deposits, it is said that at times the beach sand in the sunlight is actually streaked with a dazzling golden yellow sheen caused by the shining effect of small flakes of visible gold.

The gold of these beach deposits is invariably small and flaky; that of Oregon is often bi-concave with well-defined rims, due possibly to rolling

Old time miners working the gold rich beaches of Nome, Alaska. Many thousands of ounces were taken from these deposits and yet they are still being mined by prospectors today.

edgewise when moving up and down the beach. The deposits of Oregon are continued north to Washington and south to California. Other notable beach placers are those of Carelmapu and Punta Arenas, Chili; the shores of the Sea of Okhotsk; at Unga Island, Nome, and Cape Yakataga, Alaska; and on the northern coast of New South Wales in Australia. Those of Cape Yakataga are remarkable in that they carry garnet without the ordinary associates of gold in beach sands, like magnetite and ilmenite. In Oregon, New South Wales and at Nome, ancient auriferous beach sands are worked above present high water level. At Nome, Alaska, home of perhaps the world's richest beach sand deposits, elevated beaches have also been mined at a distance from the shore, where there are two old beach lines 37 and 70 feet above the present level of the ocean. In Santa Cruz County, California, a similar elevated beach was mined for some time. Although they are little exploited today, gold-bearing beach sand occurs at various places all along the Pacific coast from San Diego to Alaska, and in many other parts of the world.

Most of the best-known examples of marine placers are confined to the shores of the Pacific Ocean, but they also occur on the coasts of Nova Scotia, Ireland, and Portugal, which shows that any perceived restriction to the Pacific is purely coincidental. Other important beach placers are found on the west coast of the South Island of New Zealand, and below Port Arthur on the Liau-Tung Peninsula in China. Some of the richest beach sands yet found appear to have been those of the Gold Bluff, Klamath County, in Oregon, where the narrow beaches at the foot of

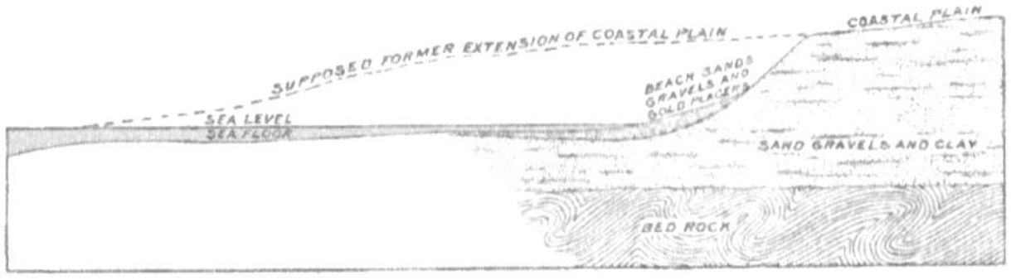

Idealized beach placer area cross section based on the gold bearing beaches of Nome, Alaska.

overhanging bluffs serve as concentrating floors for the auriferous sands thrown within reach of the ocean waves by cliff falls. Geologic forces can move beach placers away from the shoreline over time. Regional subsidence can result in these deposits becoming submerged beneath the water, forming offshore placer deposits as are found in Nome, Alaska. In a similar way, regional uplift can result in these deposits occurring on high benches further inland, which represent ancient shorelines. In California, there are some ancient beach placers around the edges of the central valley, remaining from the time that this valley was an inland sea. Some of these ancient river deltas were prime ground for bucket line dredging.

In the US, there are many laws regarding what a person can or cannot do along the beach. These laws do change from time to time as well. If you'd like to attempt hunting for gold in the black sand deposits along the beach, be sure to find out what the local regulations are and make sure that you comply with all of them.

Placer Deposit Renewal

Weathering, erosion and the formation of valuable placer deposits normally requires eons of time to form. Many folks want to know if placer deposits can re-form, and there are some cases where it does. The potential has a lot to do with the type of placer, and the conditions of erosion present. In a lot of areas there really is not much erosion. In the desert hills, the ground erodes maybe an inch or two every thousand years or so. Out at Rye Patch where it is nearly flat, maybe a few inches get washed away every 10,000 years (probably more due to wind than water erosion). The top of Rich Hill in Arizona was once a rich placer. During long periods of geologic time, yes the ground erodes and the gold deposits re-form. But this will occur in periods of something like 100,000 years or more, not our lifetimes. So if no one mined the top of Rich Hill for 2 million years, probably rich placers would be found there again. Deep bedded placers such as were mined in the past with deep bucket line dredges will only renew in long geologic time as well when uplift washes away the 30 to 100 foot deep beds of now dredged gravels and lays in new material.

Some other types never re-form. This is true for all bench placers where the river or stream can no longer reach and bring in new material. The gold mined out of the ancient Tertiary rivers in California will not be replaced. Those rivers flowed millions of years ago and there is no longer any water flow to bring new gold to them. The gold that has been picked up with metal detectors on the floors of those hydraulic pits - once detected and mined, they will likewise never be renewed.

When the river is at flood stage in modern streams, the gold is stirred up and re-deposited.

Only in the shallow placers of the beds of modern streams, in mountain areas, like in the higher parts of the mother lode country, does the gold in the streams partially renew itself - and then only during the biggest floods, or monsoonal downpours. These placers slowly re-form as the gold on the slopes, in tiny tributary streams and gullies wash new gold into the river bed or wash away old overburden revealing new bedrock. Because they are formed by the actions of heavy waves during storms, beach placers often re-form on a comparatively regular basis. This allows prospectors to return to the same sites over and over again as nature allows, collecting good gold each time the weather creates a new placer deposit along the shore.

The Formation of Large Nuggets

Prospectors have long wondered where large nuggets come from. There were a number of strange and crazy ideas made up by miners long ago. Some prospectors thought nuggets were blown as molten gold out of certain volcanoes and they solidified as they flew through the air. The idea of volcanoes blowing out molten gold which then forms into nuggets is an old one that has never had any science whatsoever to support it. It comes from the appearance of gold nuggets that are blob-like and smooth, because they kind of look like they were once molten and cooled and solidified into those smooth, blob shapes. There are gold districts in California with names like Volcano and Volcanoville because the old time 49ers believed that some giant volcano shot out molten gold and scattered that gold all over the California gold country. There are no real volcanoes anywhere near the towns of Volcano or Volcanoville in California, and volcanoes were certainly not the source of gold there.

Geologists have long realized that the vast majority of the gold in typical quartz veins is in the tiny dust size range, a lot smaller than the gold nuggets found by many placer miners. This is a common thing worldwide in many, many gold districts. Certainly there are some quartz veins with big, coarse gold, but the vast majority of gold in good sized veins is quite small. So if all gold comes from quartz veins (as is normally assumed), how do you resolve that glaring size difference? In addition, many nuggets which have not traveled far show little sign of having come out of a quartz vein. The debate about growth and origins of how big nuggets form has gone on a long time. Some geologists believe large nuggets commonly form under "cold" conditions at the surface through secondary process, while others believe that they form primarily at depth through hydrothermal heated waters. Geologists use the terms hypogene or primary to describe those minerals which are formed within the earth by watery solutions at elevated temperatures. On the other hand, the terms supergene or secondary are used to describe minerals which are formed at or near the surface by the interaction of weathering from air and water acting on those original primary, hypogene materials. More than a century ago, geologists theorized that somehow the gold was dissolved at the surface and reformed into the larger nuggets. Most surface waters have traces of chloride chemicals (like sodium chloride, common table salt), and many rocks contain manganese chemicals. Chlorides and manganese chemicals under acidic conditions come together to form chlorine gas and chlorine can dissolve gold. It has been suggested that this was the process that formed the larger gold nuggets. However, most surface waters are neutral to alkaline and not acidic enough to readily form free chlorine, although in veins which contain sulfides, these minerals can form sulfuric acid from the sulfides, and sulfuric acid is strong enough to produce free chlorine. The problem with this theory has been to identify a mechanism that would make all the gold atoms from a large area come together to form large nuggets. It would seem more likely that instead of a few big nuggets, a large number of little gold particles would form from this type process. In most locations where secondary gold at the surface is unquestionably present, that gold usually tends to be small. Tonopah, Nevada is a great example of this. The veins there contain significant gold, as well as both sulfides and manganese minerals. The district is also located in an arid environment favorable for the formation of secondary minerals. However, all the secondary gold which was formed at Tonopah was tiny in size.

More recently, scientists have discovered bacteria capable of moving atoms of gold. This has been cited by some as a potential explanation of how large gold nuggets could form. Some have theorized that perhaps over huge amounts of time, these bacteria could assemble, a few atoms at a time, a large nugget consisting of trillions upon trillions of gold atoms. The problem with this is that how would the bacteria know to keep working for thousands of generations on the same piece of gold, hauling gold atoms from several feet in distance so that it could be built up into one large piece (several feet being quite a huge distance for a bacteria to travel). While these bacteria can move gold, nothing has been demonstrated in their behavior that they are likely to build up large nuggets, as opposed to many tiny, even microscopic pieces.

So where does that leave us? Geologists still debate where coarse gold really comes from, because until recently there was only a limited amount of hard scientific evidence. Part of the problem is identifying which nuggets might have been formed at depth under hot conditions through heated waters and which may have been formed at near surface conditions by cooler secondary processes. There really is no simple way to just look at a piece of gold and say it was formed by secondary or primary origins. Gold is gold and it all looks similar - so in the past it has been difficult to test and say which process worked to form any particular nugget.

Recently, there has been some brand new research that sought to identify the source of formation for various nuggets. Scientists from Australia (a country well known for its large placer gold) used special analysis techniques to test a number of nuggets which were thought to be formed by secondary processes from a number of locations all across Australia. They tested more than 30 nuggets from different locations in Western Australia, Queensland and Victoria, ranging in size from 1 gram to more than 8 kg. Every single nugget they tested indicated it was formed in deep, high temperature environments, not cold, near surface temperatures. This was a big surprise and exactly the opposite of what might have been expected by many geoscientists.

The Australian scientists looked at a couple of specific features on the nuggets to determine how they formed. One was the crystalline make up of the gold itself. Mineralogists know that under certain conditions, some mineral crystals will sprout new crystals growing in different specific directions. These new crystals are called twins. With quartz, under certain conditions, new crystals will grow exactly perpendicular to the long axis of the quartz crystal, and this is called a Japan law twin. With diamonds, they sometimes have a tendency to grow a new crystal on the octahedral face, forming a triangular shaped diamonds - these are called macles and are also twins. Twinned crystals only grow under certain specific growth conditions, so when twining is present, they tell the geologists studying those locations something about the growth conditions which existed when the crystals formed. The Australian scientists cut through the nuggets and etched their surfaces to observe the crystals that made up the nuggets. They found that all of the gold nuggets they tested had crystallized showing strong twining, indicating that they were formed deep in the earth under hot conditions.

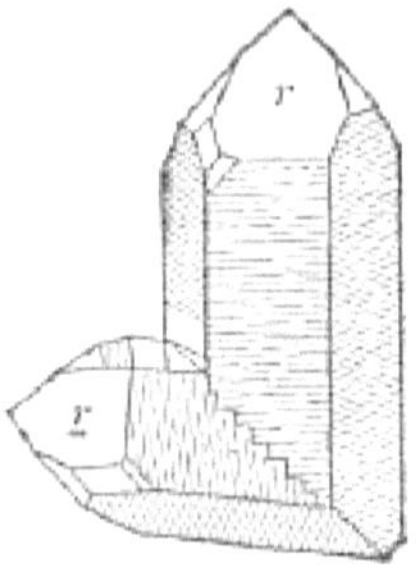

This is a Japan law twin of a quartz crystal. Crystal twins grow only under certain conditions.

Additionally, the Australian scientists also looked at how the impurities were incorporated into the gold itself. For most crystals, the hotter the growth conditions, the more impurities can be incorporated into the crystal without causing growth problems. The presence of significant silver in the gold crystal structure indicated that all the gold nuggets they tested were again, formed under heated conditions, not the cooler conditions that would be found near the surface of the earth.

While more gold nuggets need to be tested in order to say with certainty, the work done so far seems to strongly indicate that the larger gold of so many placer deposits is actually forming under deep, higher temperature environments, and being brought to the surface only under the effects of long term uplift and erosion. This new research argues very strongly against the various theories that most large gold nuggets are secondary in origin, including theories based on both inorganic and bacteriologic processes. Since big quartz veins are not the source of most large placer nuggets as the majority of their gold is for the most part tiny, we are left back were we started, wondering just how coarse gold originates. Tiny veins and stringers can be the source of a good part of the coarse placer gold, but they still do not explain all the coarse gold found near its source with little or no indication of ever having had any contact with quartz. This strongly implies to me that a significant part of the big, coarse gold is somehow coming from somewhere else.

So, the Question is Where?

All of us have been taught how gold is associated with quartz veins - that fact has been drilled into my head many times over the years. While gold truly is commonly found with quartz, that's not always the case, and most lode deposits on the whole yield gold too small for even the most sensitive metal detectors. Over my years of prospecting, I have swung my metal detector over many, many tons of gold bearing quartz and gotten very, very few responses. The truth is that most of the gold I have found with my detector shows little or no sign of quartz. Sure I've found some nice gold-quartz specimens, but the vast majority of the gold I have found has little or no quartz, and shows no sign at all of having come out of a quartz vein.

In many gold bearing regions, the lode and placer deposits don't always go together. The largest placer districts in California have very little lode production, and many of the largest lode districts produced only a little placer. Often even when both are found in the same district, the actual mine locations of the placers and lodes are often separated. This seems to point to the fact that the geology of the two can be different, and in spite of what we have learned, they don't necessarily always go together.

A little known and little studied source of coarse gold is mineralized shear zones in foliated rocks like schist and slate. Fractures suitable for fluid circulation are opened up in these rocks by the actions of parallel faults. Mineralized shear zones within these rocks could be the source of a significant part of those large sized placer nuggets in some areas. These types of deposits are often ignored as they don't obviously crop out and although large, they often contain only sparse nuggets. Some contain only limited amounts of quartz. However, through the process of erosion, these sparse nuggets can be concentrated in gullies and streams and in some places form large placer deposits. These obscure deposits were recognized in some locations in the early days, and were called "seam diggins" by the old timers in California. Where they were deeply weathered in California, they were worked by hydraulic methods. Spanish Dry diggings and the Georgia Slide are both good examples of California "seam diggings" deposits but there are a number of others. There are also deposits like this in other US states and around the world as well. In the southeastern US, they were known as saprolite diggings and were mined hydraulically in a number of places. Worldwide, they have been found all across the planet - Europe, South America, Africa, Asia and even in Australia. A number of these are types of deposits are described in the section of this book on residual placer deposits.

This is an example of slate, a rock that naturally splits into flat slabs. Schist also breaks like this.

As far as the origin of low quartz gold bearing fluids, gold mineralization almost always happens in a series of events. Often in the first events, the fluids are rich in quartz and low in other minerals. As time goes on, subsequent pulses of hydrothermal fluids are normally richer in sulfides and contain less quartz. Often the last pulses of fluid are the richest in gold and leanest in quartz. These seem to be the ones that form those big nuggets. While not every quartz vein forms exactly with that series of events, as a generalization that is the norm and quite common. Often quartz and some other minerals come first, then sulfides, and finally, last and not least is the bulk of the gold. This is a normal situation in both epithermal and mesothermal gold deposits. This fact explains a lot of things. This works out to be a pretty darn good situation because if the gold was laid down first then the quartz afterward, the quartz would encapsulate the gold making it extremely hard to recover from hard rock deposits that produce most of the world's gold. All versions of the cyanide process, from heap leaching to vat processes, would be non-functional with quartz encapsulated gold. Luckily however, the gold is almost always laid down near the last, so the gold is not encapsulated by quartz and cyanide processes do work.

If you look closely at many gold quartz specimens, you can see that the gold has filled up the available void space in the openings - as it was deposited last, after the quartz vein formed. This may well be what happens in the voids and seams in the schist and slate - the gold is deposited last, often after the quartz or at least nearly all of the quartz is gone from the solutions. I think that in the schist, at the end of the hydrothermal process, many of the bigger openings have been filled with the quartz so the gold solutions are forced to go into the little seams and shear zones, and are deposited with some pyrite and maybe calcite or siderite (calcite is calcium carbonate, siderite is iron carbonate). Of course some places just don't have that much quartz to begin with. On weathering, these little gold bearing shear zones mostly have very little quartz and the other minerals just weather to red oxides and clay looking material. If the schist is fairly tight, the gold formed is small and flat. If the openings are large enough, the gold can be larger in size. This is why often the gold from a district or specific area will have a certain normal size range.

Seam Diggings and Little-Known Gold Nugget Sources

So far, we have taken a look at the origins of coarse gold nuggets and the results of some new science about how they form. This new information pointed out that many nuggets once thought to be of secondary origin are formed under hot conditions deep in the earth. This definitely leads us to re-think the importance of some little known and poorly understood sources of placer gold. One possibility that we started to consider were the low quartz - coarse gold deposits that form in shear zones located in foliated rocks like schist and slate.

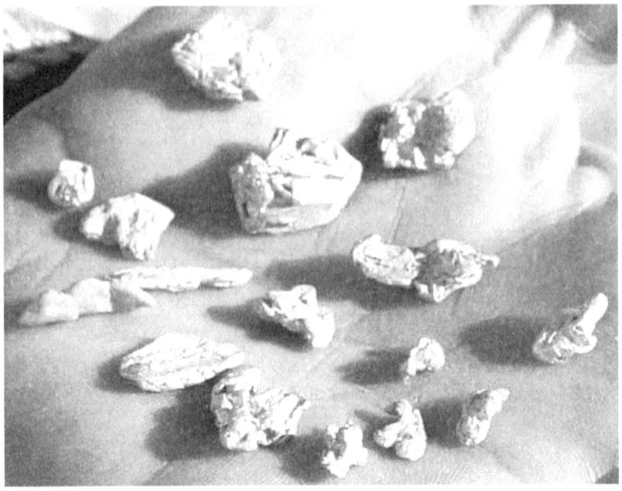

Many nuggets show no sign whatsoever of having grown with quartz. This is especially true for crystalline pieces like these from Nevada.

While this stuff may seem a bit esoteric, it actually is important to the individual prospector in searching for gold. Now I don't want anyone to get the wrong idea here - there is no question that there are gold nuggets that come from the weathering and erosion of quartz veins. There are hundreds of books that tell us this and certainly it's true that some placer gold comes from these sources. That is especially true for small gold pieces less than a grain in size, but what about that larger gold? The more I am involved in prospecting for gold, the more that I think a significant percentage of all nuggets of maybe half a gram and more come from these deposits in foliated rocks (slate, schist, etc.). They are often associated near or in the same general areas as the quartz veins.

I don't know if its half or a third or three quarters of the larger nuggets. Certainly, that would vary from district to district. I think these types of deposits are often ignored because there is no visible outcrop, and they are usually hard to identify. I also think the concept that "all gold comes from quartz veins" has been drilled into most prospectors' heads so many times that it's hard to think outside of that box. However, deposits in schist were not unknown to the old-time prospectors, as I have read century old reports with comments from early day geologists about stringer and shear zones that gave rise to large placer deposits. They were also called seam diggings in some areas. In a few places in California they were mined by large scale hydraulic mining as at Georgia Slide near Georgetown. In most other mother lode locations where they were recognized, they were mined by ground sluicing. I am convinced these kinds of deposits are a lot more common than generally thought and that there are a lot more of them out there to be found. They have also been mined in many other locations across the globe.

This dig hole may not look like much, but it produced a 0.4 ounce nugget that came right out of a little iron seam in weathered and shattered schist – there is no quartz or vein present. The nugget was completely covered with iron oxides.

It took me a while to come to this conclusion, but I just could not deny what my eyes were telling me. A number of times over the last several years, while using a metal detector, I have found and dug nuggets right out of weathered schist that was still "in situ" and had not been moved or disturbed. The rock was fractured and weathered, but tough and nearly solid. I have pulled a number of nuggets right out of rock material where I was almost completely through the weathered material zone and it was getting real hard to dig with my Estwing rock pick. Usually there are quartz veins in the general area (within 15 or 20 feet), but none right around the nugget itself. The metal detector allows the prospector to see the exact nature of the surrounding material as the nugget is dug, since the detector points out the exact location of the metallic target. Good prospectors are always looking for their next new patch and the information that will lead to that next good find, so the existence of mineralized fracture zones in foliated rocks like schist or slate which contain coarse chunks of placer gold is an important concept.

For many years, a lot of prospectors have focused on the bottoms of ravines and other drainages in their search for gold. These certainly have been productive locations as erosion feeds gold off the sides of the hills into the drainages where it is concentrated in the streams, and especially on the bedrock. Sometimes a series of quartz veins may contribute the gold, but in other locations, shear zones where the rocks naturally part into flat sheets (like schist) are important sources as well.

In some spots, it is possible to metal detect or even dry wash these nuggets on the hillside before they get into a drainage, but it depends on their concentration in the schist. If there is only one little nugget in 10 tons of rock, then finding it would be like searching for a needle in the haystack, and a lot of erosion would be needed to make a nice amount of gold in the stream below. On the other hand, you can hit big concentrations of gold in these deposits once in a while, and you might be able to take out 10 ounces from a 20 by 20 foot area on the side of a hill. These are the kind of deposits we all hope to find, and they are a big part of what I want to take a look at in this chapter.

In researching these mineralized fracture zone deposits, I've come to the conclusion that they are a very important source of coarse gold in many parts of the world. I can vouch for having seen and worked this type of deposit in California and Nevada, and other folks I know have worked them in Arizona, Australia, Mexico and other locations as well. I have read of old timers who said that some of the Yukon region deposits of Northern Canada and Alaska were also of this type. A while back I read a report about 17 of the world's largest placer districts, from North America, to Australia to Russia and South America. While many different types of rocks were found in these locations, easily parted rocks like slate, schist and phyllite were important in all 17 of the world's 17 largest placer districts. Not every placer district on the earth has this type of rock, but when 100% of the world's 17 largest placer gold districts have them, you need to take notice of that fact.

The gold nuggets which are found in these mineralized fractures in schist, slates and other similar rocks often occur with little or no quartz. They often seem to be found with red seams of high iron concentrations or in heavily iron stained, weathered rocks of the schist/slate type. Much of the red colored material I've seen in such seams was probably originally iron sulfides, but these have now oxidized to iron oxides and clay

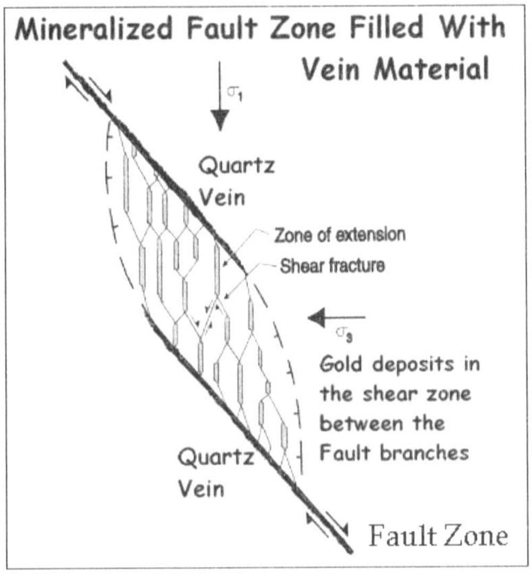

Movement along fault zones through rocks like schist can open up a network of small openings where gold bearing solutions can form nuggets.

through the weathering processes. Often there is some quartz vein material nearby but this gold does not come out of the quartz veins. In some deposits all the gold they produce is small, but in others, they can produce very large (100 ounce+) nuggets (as noted, many individual deposits of this type seem to have a characteristic size range for the gold).

The gold in many of these deposits is sparse, with just a few bits of gold in them spread out over a larger area. Some however, contain much more gold and can produce a pocket or patch with many ounces. However, even the sparse ones, given a long period of geologic time, can contribute some significant gold to the local drainages below them, as erosion and deposition in the drainage allows the sparse gold to accumulate and be concentrated. In total, the sparse deposits can contribute a lot of gold, but it just takes a lot of geologic time to erode and then concentrate them.

A friend recently noted to me that his father, an old time prospector, told him plainly that gold does not ***need*** quartz to occur. That was a great, simplified lesson in gold geochemistry, and so true. The 50 or so years of science since that time have only proven how right this old prospector's observations were. Gold most commonly occurs with quartz, but geochemically speaking, it does not require it, so it can occur apart from it. Another lesson the old timers often taught was that iron is the "mother" of gold. Although this was based only on their own observations, they were more right than they really realized. The hot gold bearing solutions in the earth carry the gold in the form of dissolved gold as sulfur bearing chemicals. Once the solutions cool off enough and come close enough to the surface, the sulfur starts to have a greater affinity for the iron in the surrounding rocks than the gold it is attached to. On encountering iron rich seam zones in the rocks, the sulfur will attach to the iron and form pyrite, leaving the gold behind on its own to grow into crystals and chunks - and eventually nuggets. This is why gold is often found near iron bearing veins, dikes and other iron bearing layers in the country rock. Additionally, small fractures and folds in schist, shale and similar rocks with the same kind of foliation give the gold bearing solutions plenty of chance to contact the rock and react with any iron present in it.

On the other hand, many geoscientists don't always easily accept the idea that coarse placer gold can come from both quartz veins (which everyone agrees on) as well as deposits in fractures and mineralized zones in fissile rocks like schist. Old time prospectors have known it for a long time, but some geologists resist the thought. Here is a comment taken from Waldemar Lindgren, one of the most respected economic geologists on the planet a century ago - his thoughts still reflect what many current geologists would say today:

"It is often stated that gold is distributed in fine particles in schists and massive rocks and that placer gold in certain districts is derived from this source. Most of these statements are not supported by evidence, though it is not denied that gold may in rare instances be distributed in this manner. Even in the Yukon region, concerning

which such statements have often been made, the origin of the gold from veins, lodes and shear zones is beginning to be recognized."

They are still looking for the hard rock source of the gold recovered in the Klondike. It is known that the Yukon gold deposits correlate directly with the presence of the Yukon schist bedrock. There are many different streams in that region, but only certain ones carry gold. In streams where the schist is not present in the upstream drainage, there is no gold. The modern metal detector gives us a tool to see gold in place - in the very spot it was formed. The old-time miners and early geologists could never have dreamed of this. In the end, that's what convinced me - digging the nuggets and seeing they were coming right out of seams in the fractured rock with no surrounding quartz vein, just lots of iron oxides. Rocks that are easily split into flat slabs allow whole areas of rock to be opened up for mineralization along faults and shear zones, because in these rocks the breaking occurs along natural parallel partings. Thin flakes of gold can grow within the flat space of the breaks between the rock sheets. Larger pieces can grow where the openings are larger.

In arid regions with gently sloping to flat terrain, the forces of wind and weathering can concentrate this gold on the surface. Many of the minerals in the schist or slate can weather to form clays and these can be easily washed away by rain water or blown by the wind. In flat terrain, the gold will resist being moved by wind and water and remain behind. Often these areas will leave concentrations of vein quartz scattered around as well. The quartz remains because, like the gold, it is very resistant to weathering. In this way, the gold and quartz of 100 or more feet of rock may be concentrated into a surface layer when the rock is removed through weathering and erosion. Unfortunately, in those types of situations, once the rich surface layer is removed, the material below it contains much less gold. This is what I have described earlier as an eolian or wind-blown placer.

This erosion process shows how a zone that has real sparse gold - with only a nugget or two here and there can still produce a rich deposit. If, after removing the surface gold you scrape the area with a dozer, you could detect that spot and find very little. However, in some of the examples of these mineralized shear zones which have significantly more gold, it is possible to get good nuggets with each dozer pass - basically until you can't dig the rock anymore without blasting. The nuggets would still be in the hard rock, but the rock just gets too hard to dig further.

Sometimes by removing a few inches of dirt off the surface, a bulldozer can reveal considerable gold for recovery by a metal detector.

Hillside patches represent a great potential for the metal detecting prospector, and I really believe there are many undiscovered patches all across the western US, but its just a lot harder to find those hillside deposits than the gulch placers because there is just so much more land to cover. It's not hard to figure out where a gully or drainage is and then test it. On the other hand, hillside placers can be hiding almost anywhere - slopes, ridges, flats, etc. The question is where? One possibility for finding them is to basically walk randomly through an area, swinging across the hill sides with a metal detector. Many of the hillside finds I have heard of actually are made by random swinging over hills, flats and ridges, they were just lucky finds, sometimes made as the prospector was just heading back to his car. However, that's a hard way to prospect, and you need to be both lucky and skilled as you can easily come within 100 feet of a rich deposit and miss it all together. You can do loaming to trace patches, but that does not always work and it also is not easy. There are some visible signs to look for when searching for these types of placers. The hillside placer I am working in California was found by the old timers and I just came in and started working around their old diggings. So perhaps the simplest sign is the presence of old placer workings in hillside locations where there are no ancient stream channels, etc. In mother lode country of California these workings will most likely be ground sluicing areas, but in Arizona and other desert areas they will show up as isolated dry wash areas up and away from a ravine. Jim Straight talks about these types of areas in his book "Follow The Drywashers".

Iron stained outcrops of slate or schist with flat slabby rocks lying around can be a very positive indicator for gold deposits.

Schist and slate can be recognized in the field by their tendency to form flat sheets, and where turned over so the dip of the foliations is near vertical, book like outcrops are often formed. These rocks are normally some shade of gray in color but can be dark gray to nearly black, and sometimes are found in greenish or purplish shades as well. When well mineralized, they are red-brown to deep red in color. Schist areas which weather to create soils which are about brick red in color are a very positive sign. It shows that the rocks have been mineralized with iron. The presence of quartz vein material or silicified schist is also a positive sign that may indicate the presence of gold. Even in places where the gold is not actually in the vein material, the presence of barren quartz vein rock nearby can be a very good sign. Sometimes where the rock is a less weathered and more solid version of this mineralized schist, the rock is dark brown to red brown. Another indicator is the presence of small cubes of pyrite that have been oxidized to a dark brown color. The presence of quartz crystals can also be a positive sign. The physical location makes a difference as well, since gold has a higher tendency to concentrate on the surface in areas with flatter topography such as hill tops, ridges and mesas (in steeper areas, the gold is pulled down into the ravines). There are some common themes in these deposits, such as the red iron mineralized schist; however there are differences in each region and you will have to learn the unique clues for each area. Knowing these typical indicators will certainly help you in your prospecting search.

So what does all this mean to the average prospector? It means that thinking "outside the box" and going beyond the standard ideas that everyone follows can lead to new finds, new patches of gold and more of that valuable yellow stuff in your poke. This is in spite of the large number of prospectors that have come and gone through the gold fields in recent years. With all the new technology and scientific knowledge we have available to us, there are still some great nuggets waiting to be found. Without question, there are a vast number of smaller concentrations of gold in tiny veins and shear zones which may have previously escaped the notice of other prospectors. By learning to think of and recognize those other possibilities, we can go out there and find our share of that gold.

Part IV - Gold Prospecting Techniques

Using Research To Find New Gold Deposits:

Why Research?

During the winter, when the weather is cold and snowy in my part of the country, my mind often wanders to the times when I was out in the hills over the summer and prospecting for gold. Unfortunately, the winter days are short, the temperatures cold and the weather is reliably poor. That's why the middle of winter is the best time for me to be doing research to find new gold deposits to and places to explore when the weather warms and the snow melts.

The hard part of prospecting isn't learning to use your equipment (although that part is important) - the hard part is finding those places with good gold deposits! A lot of guys want to skip the exploration and go right to the digging of the nuggets, but a big part of the whole prospecting adventure is in exploring for and finding those golden spots. Unless you somehow have someone who will constantly give you maps with an "**X**" on it that will take you to those productive spots, you will need to research and find your own productive locations. Being able to conduct research and find your own spots is one of the big things that separate the new guys who are not very successful and the seasoned prospectors who regularly do well.

Yes, you can always go back to the same spots you've been working for years, but if you are like me, you have already worked them hard, and except in a few rivers with upstream sources, the gold doesn't come back (and even then, renewal comes only after a major flood). I find continuing to work the same old spots over and over leads to slim pickings being found all too often. Of course, new productive gold locations are much harder to find, but the rewards are so much greater! Books that tell about locations where gold has been found in the past are just the start of the journey. Good research is so much more! You need to read, collect information and maps and really do your homework! A lot of guys have an aversion to doing research because it makes them feel like they're back in school again - but I will tell you with all honesty, just as I noted above, being able to find your own spots is the difference for many people in being successful in finding gold or failing. The most important prospecting tool you own is the information in your head, and research is a big part of gaining more information and experience. Consider it not as homework, but as a treasure hunt.

Identifying the Type of Deposit You Seek

Perhaps the first thing you want to do in your research is to think about what your target really is. If you are a dredger you want a gold bearing stream, but if you are a metal detector operator you might be looking for old placer diggings high above the water, or areas where the old timers dry washed. The kind of area you are searching for makes a big difference in your research. Setting limits for things that you will look at or not look at in your research will help you become more efficient.

You will be doing your research looking for certain bits of information telling you that this is just the kind of placer you are looking for. Some of those items of information you will be searching for in your reading include past production records telling you how much gold was found, how deep the overburden was, what kinds of methods and techniques the old time miners used, information about how large or coarse the gold they found was, the timeframe in which the majority of the mining in this area was done, why they stopped mining and any other important factors that may give you an edge in finding more gold in your particular location.

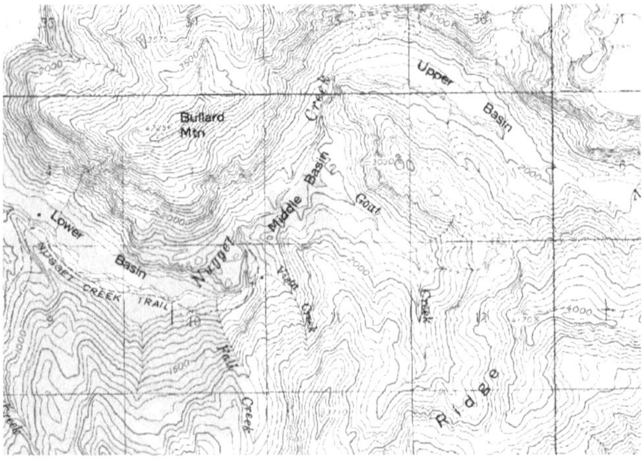

Nugget Creek near Juneau once yielded gold to pioneer prospectors – the name gives it away.

Finding Special Gold Names

One of things that you want to keep an eye out for in both the text and a maps of the areas, are places with special unusual names which indicate that gold was found there - even if the text of the books you are reading does not specifically state that. As an example, these games would include things like Gold Canyon or Nugget Creek. Awhile back I saw an old map with a spot labeled Hard Cash Gulch. Now for the old timers, hard cash was money in the form of gold or silver, so Hard Cash Gulch would be a place with a name which implies gold was found there. Sure enough with little extra research I found out that Hard Cash Gulch was a little-known place where placer gold was found. The trick is that on modern maps it has a different name - so only by seeing that old map, and recognizing a name that implied gold was mined there, was I able to find a little-known spot for gold. Sometimes you need to use your imagination with these names, because they may not be as plain as Gold Canyon or Nugget Creek. Maybe in your area it will be Miners ravine, Bonanza Gulch, or something else. Sometimes miners named their streams after other well known gold areas - a very rich placer gold spot in the Leadville district of Colorado is named California Gulch (the miners named it after the rich placers of California). It may take a little creativity on your part, but just keep an eye open because these names can be great clues to finding little known gold areas.

Identifying the Source of the Gold

Perhaps the most important thing that you'll be looking for in your research is the source of the gold. Gold just doesn't occur randomly, and in pretty much every district the gold is coming from somewhere. In some locations, those sources are fairly well understood. In others, geologists still debate about the source of the gold that was found in certain districts. The important thing to determine is that when there is a known source, that you key in on that and remember what it is. Knowing where the gold is coming from is a very important thing when doing research. Now in some places there will be multiple sources of gold - in some places in California, hard rock quartz veins as well as ancient river placers both contribute important amounts of gold. That's still fine if there are two or three different sources, it's all the better to have more sources of gold. You just need to know in your research what those sources are and keep them in mind. Very often the sources will be limited to one or just a few types of rocks that's going to be critical to know that kind of rock and be able to recognize it in the field because if there are other deposits of similar rocks located nearby there is often an excellent chance that they will have some gold in them as well. We'll take a closer look at this concept in some field examples coming up here shortly in a few pages.

Types of Research Material:

I have obtained good ideas for prospecting locations from a number of different types of sources over the years. When doing your research, it's always best to search in and around locations where gold has been previously found, but still you need some ideas, so here are some sources of information that have worked for me:

US Government Reports

While the federal government has not produced a large number of these types of documents, some of the ones written by employees of the USGS are very useful. Some of the reports best known to individual gold prospectors are the series of books on placer deposits by Maureen Johnson. These USGS publications are a compilation the placer deposits of the states of Nevada, Arizona, Utah and New Mexico. While those reports are of more general nature, the USGS has also produced some detailed reports on specific individual mining districts; one of the excellent ones is report on the Gold Basin district of Arizona, another is an older report about the beach placers in the state of Oregon. Do some research to find out what's available in your state and county. You may be surprised at the number of books and reports that are available for your area. US Geological Survey Publications include various Bulletins; Professional Papers; Open File reports and Geologic Maps. While there are literally thousands of these reports, and some are very worthwhile, the majority are of little interest to the individual prospector - you will have to search through to find the ones of interest to you. Generally the older federal reports have a higher percentage of works that would be of interest to the individual prospector.

State government books can provide great info.

State Government Reports

On a local scale, professional geologists and mining engineers were often employed by the state governments to write official reports on mining activities in their jurisdictions. In those days, mining was seen as a good thing that gave jobs to people and increased the wealth of our nation, not a politically incorrect football that all politicians oppose and love to kick. These old reports often contain a wealth of valuable information, including some very specific details and are not exaggerated. They are some of the sources that I like the best because they can contain information about gold deposits which no one who is still alive may remember.

Many states have created a Geological Survey, or Department of Mines, or some other such arm of the state government. Historically, a major part of the duty of these organizations was to collect data about mineral deposits and mining operations within that state. The reports they have produced over the years can contain many great ideas on where gold was found in the past. As a result they provide great information to the modern prospector on where he ought to go to look for gold. Many of these old reports were written by geologists who were actually out there in the field looking at the mines and talking to their owners during the height of the pioneer gold rush mining operations in the late 1800s. They can be your window into the old days of active gold mining.

An excellent example of this type of publication is Bulletin 193 "Gold Districts of California" by William B. Clark. It was published by the California Division of Mines and Geology, which is now known as the California Geological Survey. Every prospector who is active in prospecting for gold in California ought to have a copy of this book (and you can buy copies at a very reasonable price through the California Geological Survey). It lists hundreds of gold mining districts all over the state of California, including the mines that were worked there, the geology of the location and additional publications where more can be learned about the deposits of that district. It is a great book to start for any research on prospecting in California. It was out of print for many years, but is currently available from the California Geological Survey through their website.

While California may have a great wealth of information which was produced by the state government, they are by no means the only state to do this. For years Nevada, Oregon, Arizona and many other states have also produced worthwhile literature of the gold deposits of their states. These old documents hold a wealth of great information for the modern prospector. Another one of my favorites is "Placer Deposits of Nevada" by William O. Vanderburg. He was a geologist who traveled around Nevada in the early 1930s when individuals were working for gold because they couldn't obtain jobs in the cities during the great depression. It has a lot of great detailed information.

Private Reports

Copies of private geologist property evaluations are sometimes donated to state Bureau of Mines, and usually stored at their offices (commonly on the death or retirement of the author). These private reports were commissioned by mining companies or mine owners and very often they contain some unique information that you just can't obtain elsewhere. Another form of private report is the Graduate Thesis reports done by graduate geologists working to obtain their masters or doctorate degrees. Check with your local University, as they will sometimes have copies of these on file. The Mackey School of Mines in Reno has many hundreds of these thesis reports in their library going back many decades.

Newspapers

In the same vein, old time periodicals, such as mining town newspapers, can also point out locations of important finds, some of which are now little known. Digging through old newspapers can be interesting, and of course they also cover thousands of fascinating things relating to pioneer life in general that have nothing to do with mining. Its important to remember that the mining stories in old newspaper articles are often wildly exaggerated, and a small strike that hasn't yielded $100 to the finder will be built up and extolled in the newspaper as if it were better than the Comstock lode, and contain implications from the writer that the deposit will yield more valuable minerals than all the California mother lode combined. Few of the old-time newspaper reporters were experienced in mining, and they were often just trying to fill up column space in the paper - and the better the story sounds, the more interesting it is to the reader, and the better the paper will sell. So, when you read any of these old newspaper reports, you need to take them with a grain of salt, because serious exaggerations were the norm.

It used to be that finding those old newspaper reports was nearly impossible, but so much old stuff is now online and accessible through the internet. Use Google to see what you can find.

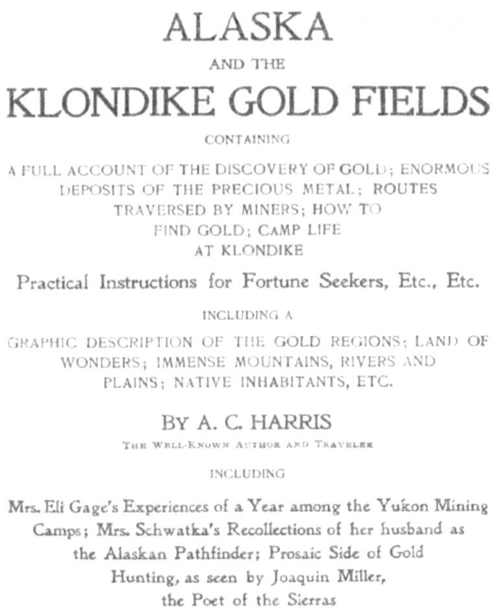
ALASKA
AND THE
KLONDIKE GOLD FIELDS
CONTAINING
A FULL ACCOUNT OF THE DISCOVERY OF GOLD; ENORMOUS DEPOSITS OF THE PRECIOUS METAL; ROUTES TRAVERSED BY MINERS; HOW TO FIND GOLD; CAMP LIFE AT KLONDIKE
Practical Instructions for Fortune Seekers, Etc., Etc.
INCLUDING A
GRAPHIC DESCRIPTION OF THE GOLD REGIONS; LAND OF WONDERS; IMMENSE MOUNTAINS, RIVERS AND PLAINS; NATIVE INHABITANTS, ETC.
BY A. C. HARRIS
THE WELL-KNOWN AUTHOR AND TRAVELER
INCLUDING
Mrs. Eli Gage's Experiences of a Year among the Yukon Mining Camps; Mrs. Schwatka's Recollections of her husband as the Alaskan Pathfinder; Prosaic Side of Gold Hunting, as seen by Joaquin Miller, the Poet of the Sierras

History books can provide some very useful info. This one was written by a guy who was there for the great Klondike Gold Rush.

History Books

I've also gotten some very good ideas out of general history books which were written about the old mining districts. Sometimes it's real hard to find the exact spot which the author is describing as the history books can be fairly general in their descriptions, but the tales of big strikes by the old time miners can really point out good areas that may be interesting to investigate. Sometimes this info can be combined with items you get elsewhere to get a firmer lock on the exact location of a potentially good spot.

Friends

Sometimes I have gotten good advice from friends and long time old prospectors that I know. With friends and partners its important to treat them with the same fairness you'd expect in return - because if they get the impression all you want to do is pick their brain and steal some good ideas, it's a problem. Once that happens, you won't get any more out of them and it won't help your friendship either. Be careful, as gold has split up many friendships. Good friends are their own special kind of gold, and have plenty of value beyond just sources of information. So carefully cultivate friendships with experienced prospectors and treat them right. If you expect others to share with you, you may well be asked to share worthwhile information with them too.

MRDS Records

The US geological survey, in addition to producing a number of reports, also has produced two very interesting computer databases of mines and mineral deposits, one is called the MRDS or Mineral Resource Data System, and the other is the Mineral Availability System, or MAS. The MRDS is a worldwide data base of known mineral occurrences as well as metallic and industrial mineral mine sites which was started in 1969 by the USGS and is still maintained by them. The information in these two data base listings is used in a variety of ways. They were originally set up for US based information, and one of the main activities of the USGS Mineral Resources Program is still the ability to provide mineral-resource assessments of Federal lands, primarily at the request of the Bureau of Land Management (BLM) and the U.S. Forest Service.

Mineral information is stored in these computer records as data fields. Each record can comprise a mineral occurrence, prospect, mine, mining district, region, or processing plant. A numeric record number uniquely identifies each site. The data can be searched and sorted using any of these descriptive fields. New records are continually being added to the database, and existing records updated or upgraded.

These databases contain references to many little known and obscure deposits, including ones that I have never seen mentioned elsewhere. In fact, for the prospector, the chief strength of this information lies mostly in the fact that it contains data on mineral deposits that just have no other published information anywhere besides these two databases. They are a potentially very important resource to the prospector researching for little known gold locations. The weakness of the data lies mostly in the lack of any serious quality assurance done on the information that was used in the original database production. This means that while most of the data is reasonably accurate, there is some information contained in it which is way off the mark. This inaccurate material is spread randomly through the database, and it's not easy to know which information might be inaccurate. Most of the inaccuracy I have noticed is in the location coordinates. Most coordinates point to locations within about 200 yards of the true location, but a small percentage are much farther off, and some can be up to several miles off the correct location. Now some folks might say that problem makes the data unusable, but since the majority of coordinates are reasonably good, it is still a very valuable source of information. I have found that the best way to "use" it - is in conjunction with other data sources, such as maps, old reports, historic newspapers, etc. These other sources can confirm and support the information and help eliminate potential lost time chasing the location of deposits that were entered into the MRDS with bad information.

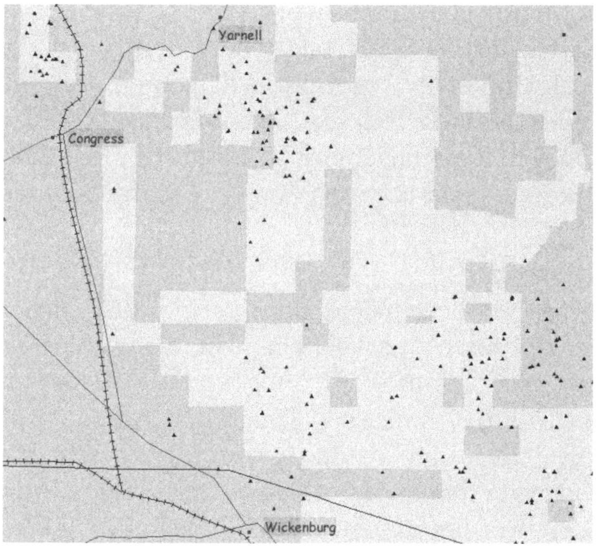

Map of the famous Rich Hill gold mining Area in Arizona from the free USGS site featuring MRDS data. Each black dot is a mine in the database.

As an example of using other information with these database sources to enhance your search, a few months ago I found a spot using the MAS database that I want to investigate further as it sounds like it might be a good source of nuggets for my metal detector. I checked the coordinates from the database against a geology map and the given coordinates were near an area that looks geologically interesting to me. Using a topographic map, I found a spot that looks like the cut of an old hydraulic mine a few hundred yards away from the spot indicated by the coordinates. However the cut out section is small and no mine name is actually noted on the topographic map. When I go up there to investigate, I will concentrate on the spot that looks to be a small hydraulic cut, because it is fairly near to the given coordinates. Because the coordinates are not usually exactly accurate, it is normally necessary to search out the immediate area within a couple hundred yards to find the true location of the deposit mentioned in the database. Both the MRDS and MAS are simply databases - they consist of a bunch of numbers and words organized into fields, and they are not something like an old report that one can simply read and digest. Making sense of either database requires some sort of computer program interpretation system to really make full use of it. Using a computer program can help you sort through the databases. If you are looking for a certain mine name, or perhaps all the gold mines in a certain county, these are the kind of things that the computer program can easily sort through the database and find. Luckily, there are several options available to help the prospector evaluate and interpret these databases. The possibilities include a free internet-based system run by the USGS, as well as some commercial programs that run on your computer.

The USGS has a free map based interactive system on the internet called the USGS Mineral Resources Spatial Data website. The maps allow you to search different areas for mining activity conducted in the past. It can be viewed at: **https://mrdata.usgs.gov/mrds/map-graded.html** It is based on a GIS mapping program and allows various map layers to be shown in conjunction with the MRDS and MAS database information. Searches through the USGS databases are easy with their internet-based program. The program can plot a custom map view of the location of the selected mines and then you can print and take them with you out into the field.

Sources of Research Material:

I have listed a number of different sources of information, the next step is to consider where we might be able to find some of those things.

Libraries

Libraries, especially university libraries, are a great source of information. Any university with a decent program in geology or mining will have quite a few geologic reference resources, especially for the state in which the university resides. Some folks might be intimidated by going up to a university, trying to find a place to park, fighting through all the young men and women rushing off to their classes, and stepping into the large college library and trying to find that hidden spot deep in the bowels of the building where the interesting old mining reports lie sleeping on a dusty shelf. All I can say is, don't let that be you. Nearly all those colleges are funded by your tax dollars, and as citizen you have just as much right as anybody else to look at and access the information found there. Sure, it may have been a long time since you have been on campus, but then it's a very worthwhile visit. If you need to seek the help of one of the librarians in finding the material you're looking for, then by all means take the time and seek them out. Generally, they are friendly and helpful so it's really not a problem. A very large percentage of the old books and reports found in the university library are out of print and cannot be purchased anywhere. In fact, a lot of times I find the copies are marked by the library as "reference only", meaning that because of their rarity, they cannot even be checked out. You can only look at them while you're there at the library. This of course makes for a problem, because I often want to refer to these old reports while doing research at home. What I've done over the years to solve this problem is to make copies. The copyright protection on government documents is not a problem so long as you are not trying to make copies to sell to others, there should be no problem in making personal copies of these old reports for your own use. I actually have been doing this very sort of thing for years, and have

The University of Nevada's Mackay School of Mines Library contains a huge resource of old mining publications and many universities have similar collections which can be searched.

many three-ring binders full of copies of old reports or at least sections of these old reports that I find interesting for my mining and prospecting research. All these libraries have copying machines for use of the students, so there's no reason why you shouldn't make use of those machines as well. Copies cost a few cents apiece and are a bargain when you consider the value of the rare information you're obtaining.

Books For Sale

There are a good many books currently in print about the gold districts in various states and countries. Many of these are great resources and often contain references listing more obscure publications and old reports that you might find in the university libraries as I mentioned above. I have a number of these books in my own library, and there is a list of recommended ones at the back of this book and on my website.

This is the Arizona Bureau of Mines – these offices have considerable research information.

Offices of the State "Bureau of Mines"

Nearly all the Western states and a number of others used to have a Bureau of Mines or other similarly named agency as part of the state government. Many of these have changed names in recent years because of the political incorrectness of mining in nearly any form (almost no politician wants to admit he or she supports mining). As in California, many of these agencies are now known as something like the "Geological Survey". The important thing is however, they still have an awful lot of great information in their files and offices. For many years these agencies have been charged with collecting information about and supporting the mining operations inside their state. During those decades they collected vast quantities of information about past mining operations and ore deposits of all types. Many of them have this information on file and available to the visitor for review and copying if necessary. Some of the reports they may have on file are old maps, hand typed reports, letters and other unpublished documents that you simply cannot find anywhere else. They may have on file a detailed description of a century old gold mining operation that nobody has read in 80 years. These agency offices are great sources of information, and while it may take you a while to sort through all their files, that might make for a very worthwhile project. So check and see where the office of your State mining or geological agency is located, and if it's close enough to your home, be sure to schedule a visit.

Internet

So much information is available on the Internet! There are copies of State Reports, Federal Reports, and more. When I first published this book, only Alaska, had more than just a few reports on line. <u>Placer Deposits of Alaska</u> by E. Cobb is one of several primary "location" type books now available for download. Gold deposits of California by Clark and Placer deposits of Arizona, Nevada and other states by Johnson are all now available for free download on line. More and more is available as the Internet grows, so be sure to search for new stuff from time to time. I am also putting up a lot of useful prospecting information on my YouTube channel, with more than 300 prospecting related videos on line. I cover a variety of topics and I try to put up a new video every week, so be sure to check out my YouTube channel, which is **Chris Ralph, Professional Prospector**.

Using the Many Types of Maps

Maps are a very important source of information to the prospector - information that simply cannot be found anywhere else. There are a number of types of maps, and the four types which are of greatest interest to the prospector are: 1) Topographic maps; 2) Arial Photos (Orotho photo maps and photos that can be found using Google Earth); 3) Geologic Maps; and 4) Land Ownership and Status maps.

TOPOGRAPHIC MAPS:

The USGS took over responsibility for mapping the USA in 1879 and has been the primary civilian mapping agency of the United States ever since. They produce a variety of maps, but the best known USGS maps are the 1 to 24,000-scale topographic maps, also known as 7.5-minute quadrangles. They each cover a four-sided area of 7.5 minutes of latitude and 7.5 minutes of longitude. More than 55,000 different 7.5-minute maps have been created to cover the 48 continental US States. Alaska is covered in larger scale maps with a bit less detail. Maps of this series covering the entire US were completed in 1992, and it is the only series that covers the entire area of the United States in significant detail. The process of producing a topographic map starts with taking a series of aerial photographs of the area being mapped. A pair of aerial photographs, each showing the same ground area taken from a different position along the flight line, is viewed through an instrument called a stereoscope, producing a three-dimensional view of the terrain. Using a set of known control points which have been surveyed on the ground, this setup allows a cartographer to draw a topographic map.

The 7.5-minute topographic map series is available in both paper and computerized digital forms. The feature that most distinguishes topographic maps from maps of other types is the use of contour lines to portray the shape and elevation of the land surface. Elevation lines that are close together indicate steep terrain, while widely separated lines indicate more gentle slopes. They show the names of natural features such as mountains, valleys, plains, lakes, rivers, and vegetation. They also identify man-made features, such as roads, boundaries, transmission lines, campgrounds and of course, mines. Almost 2 million natural and manmade features are identified on the maps of the USGS topographic map series. All this information makes them extremely useful to the prospector. They can tell the prospector where a mine or mining area is located, what roads lead to that spot and what the terrain is going to be like when the prospector arrives there. One of the most important things on topographic maps are the reference grids. These are the systems that help you locate things on the ground. They include things like the cadastral survey, which breaks up the country into sections Townships and ranges. On most topographic maps this grid system is very prominently featured, and easy to see. Each section is 1 mile square in size (this is when they are properly surveyed, sometimes you will see distorted long squares or cut off squares, these happen when the old-time surveyors of the 1800s messed up on their surveying). Each block of sections is 6 miles square, for a total of 36 square miles. They are identified in a grid system marked off from a central identification point, called a baseline Meridian. In northern California and Nevada, this starting point for the grid system is marked off from Mt. Diablo, which is located in the Bay area, not far from San Francisco. Each 6 mile section is marked off, so that range one East is next to Mt. Diablo, range two East is 6 miles away, range three East is 12 miles away, and so on. In eastern Nevada the ranges get up into numbers as high as 70 and 80. The township grid is marked off the same way, but the townships are measured North and South. Other baseline Meridian points are located around the country as reference points. By this system you can quickly identify a parcel of land. As an example; section 29, range 21 East, township 17 North Mt. Diablo baseline identifies a specific 1 mile section of land, which just happens to be Virginia City in the heart of the Comstock lode in Nevada. However, one can identify almost any area in the US using this system. This township survey system is what the US government uses to officially track land blocks and their ownership. It is used by the BLM for tracking mining claim locations in their records. The mile sections can be further broken down into quarter sections and smaller such as the northeast corner of section 29, and even further into parts of that such as the north half of the southwest quarter of the northeast quarter of section 29. A full section is 640 acres in size. Half of a quarter of a quarter section is 20 acres, the typical size of a standard placer mining claim. Along the edges of printed topographic maps there are also a number of other grid systems. Marks and ticks along the edge of the map identify coordinates for things like latitude and longitude as well as the UTM grid system. These systems are useful for identifying or keeping track of locations with your GPS device. For further discussion of these other grids and the advantages of using your GPS device with maps, see the section of this book about using a GPS for prospecting.

In addition to the common paper forms of topographic maps, there are also digital forms which are available for free download over the Internet, or can be purchased as a part of several commercial software programs. The software programs have the collections of maps, but also provide specific software for using the map files and manipulating them. These software programs range from very simple and easy to use systems, to very complex programs capable of a lot of different things. We will briefly discuss one of the more advanced systems known as GIS, here shortly. First however, I want to take a look at digital topographic maps in general. While paper forms are clearly the most easily used in the field (because they don't require the use of a computer to display them), when researching at home or other places where a computer can easily be used, digital maps have some real advantages. The USGS generates its digital maps using the information and data on the original paper maps, commonly through a scanning process. Added into the digital map are the grids I mentioned above as well as other layers of information addressing roads, contours, boundaries, surface cover, and other manmade features. This allows that one can simply place the cursor of your computer over the top of a mine or other location you wish to visit, and the computer will show you the GPS coordinates for that point. This can be very handy for both finding locations in the field, as well as finding road or trail intersections that will help you get to the locations you want to see. We will discuss this more in a bit, but with digital maps you can actually use the computer to make them partly transparent and overlay one across the other allowing you to display two different maps at once. An example of this might be to lay a topographic map over the top of the geologic map allowing you to see the areas of the topographic map which contain certain types of rocks that contain gold deposits.

Topographic maps can be obtained from a large number of sources. Paper maps can be found in certain books (Delorme publishes a series of books

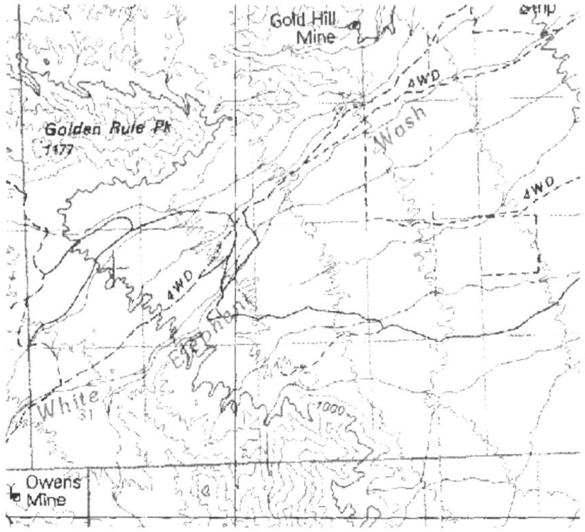

This is a topographic map White Elephant Wash in Arizona, a well-known gold placer area.

covering various US states); many libraries contain a map section, and while you cannot normally check these maps out, you can make copies on the copy machines; many surveying and mapping stores will also stock and sell topographic maps. They are also commonly sold in various sporting goods stores because they are used by hunters and hikers. Digital versions of the topographic maps can be downloaded over the Internet from various sources including state GIS Archives or college libraries. Normally, the map files are in a format known as a TIF file - this means that these images have a computer file name ending with ".tif ". They require a special viewer program to open and look at them. I've done quite a bit of downloading topographic maps and use a program by the name of Global Mapper to view and manipulate the map data files that I've downloaded, but there are others. This piece of map viewing software can be downloaded off the Internet, and while the full commercial version is fairly expensive, the free version is perfectly functional and does the job for handling digital maps. Digital maps can also be purchased as a part of various software packages. Be sure that the software package you purchase has digital maps which are detailed enough for you to use in your prospecting. Some packages contain larger scale digital maps that are just simply not detailed enough for prospecting uses. Partial views of various topographic maps can be seen on some websites which do not require any special program (other than your normal web browser) to view them. Probably the best-known of these free websites is TerraServer which is run partly by Microsoft.

While the 7 ½ minute (24,000 to 1 scale) topographic maps are the ones most commonly used by prospectors, there are a whole range of topographic maps produced by the USGS. I use these larger maps in helping to find roads into areas that I want to see. In this category, the maps I use most commonly are 30 minutes tall by one degree wide, having a scale of (100,000 to 1). These are also available both as paper maps and as files downloadable from the Internet. Many land ownership maps published by the BLM and other federal agencies are produced in this 100,000 to 1 scale size (we will talk more about these maps in a bit).

In our discussion of topographic maps I mentioned a little bit about how they are made, but one of the things I'd like to point out is that these things are made by humans and different makers produce different quality maps. Some guys who make these maps pay incredible attention to detail, and others seem like they are being paid by the piece, so they crank them out as fast as possible. Sometimes there are minor errors on these maps, but generally they are few and far between. The biggest difference that I have noticed is in the stuff that just gets left out and doesn't appear on the map. Not to worry, there are ways of dealing with stuff that is "missing" from the map by using info seen from aerial photos.

With practice you can learn to read some subtle things on topographic maps. Maps are normally marked with prospects and mines but I have learned how to read the steep and flat slopes in and around old hydraulic mines to more accurately see the boundaries of where the diggings start and stop. The steep slopes typically mark the walls of the old pit while the flatter areas immediately adjoining mark the bottom of the pit. Steep slopes in the right locations may mark fault scarp zones; tailings areas or drainage slopes. With practice you will learn to see these less obvious and unmarked things using the topographic lines. There are many advantages to using topographic maps, and every prospector should be making use of them. They are easy to acquire; easy to carry; easy to read; have GPS and other grids; and it's not hard to simply print out copies that you can fold up and put in your pocket for use in the field. While there are many good things about topographic maps there are some disadvantages. As I noted different makers produce different qualities. As a result the maps may not show critical roads and other useful prospector information. I also noted that rarely erroneous information is printed on the maps. Another problem is that you cannot easily distinguish small workings from large. This is because the "X" on a map may indicate a tiny prospect hole that is 10 feet deep, or it may indicate a huge mine with miles of workings which produced gold for 20 years or more. Sometimes the names of places or other natural features change names over time, and you may have an old book which references so-and-so Canyon and you look all over the modern topographic maps for that area and find no place marked as so-and-so Canyon, because now it's now known by a different name. Certain features, sometimes including mines, have been intentionally removed from some maps over the years. However, as I noted before there is a way to deal with things that are just not shown on topographic maps - this is why in addition to topographic maps it is important to make use of aerial photographic maps – like on Google Earth.

Reading and Using Aerial Photos:

Images that are made from the aerial photographs of an area can be amazingly helpful and allow the prospector to effectively look at an area before actually visiting the location in person. When I first was writing this book, a program called Google Earth existed, but the quality of the photos of mining areas was very low and it was hard to see much in the images. The world has completely changed since then. Currently Google earth will allow prospectors to see old working, stacked rocks from old placer mines, roads, rock types, and more. I have grown to be very dependent on Google Earth when I do research on new areas. The views available on Google Earth make a big difference in the success of my research. Often you can see features, mining disturbances and things on google earth from its sky view that are hard to see when one is standing on the ground.

The USGS also produces maps known as a Digital Ortho photo Quadrangle (DOQ or sometimes DOQQ). This type of map is a computer-generated image of an aerial photograph in which the image displacement caused by terrain relief and camera tilt has been removed by computer correction. The DOQ combines the image characteristics of the original photograph with the scaled, geo-referenced qualities of a normal topographic map. These types of maps can be found on the internet and downloaded.

Google Earth images really are great tool to use in tandem with your USGS topographic maps because they contain so much more information! Roads, mines, tailings dumps and other features that the prospector is very much interested in are plainly shown in these aerial photo images. These photo maps will tell you things you just can't learn from a topographic map. In practice I find it usually not hard to see cars on the Google Earth images, and the roads they are traveling as well. In my research, I have found numerous examples of places that have been very productive for me which on a topographic map looked like nothing of any importance, yet on Google Earth images looked highly interesting. A good example are areas that don't look like much on a topo map, but show old timer rock piles from past placer workings.

These aerial photo images are very useful, and in my opinion a critical part of any gold research. Just like with the topographic maps, it will take a little practice for you to begin reading subtle things on aerial photo images, but the practice and time spent is well worth the effort. Aerial photos will also show you the relative size of old workings - you won't have to guess if the "X" on a topographic map is just a tiny ground scratch or if it's a major mine. The photographic map will tell you because you'll be able to see the size of it. You will be able to see the size of the rock piles left by the old timers. Another possibility is to see areas that were scraped with a bulldozer for metal detecting purposes. One of the limitations of Google Earth images is where trees and heavy brush cover the ground. This is why desert areas with their sparse plant growth leave everything out in the open.

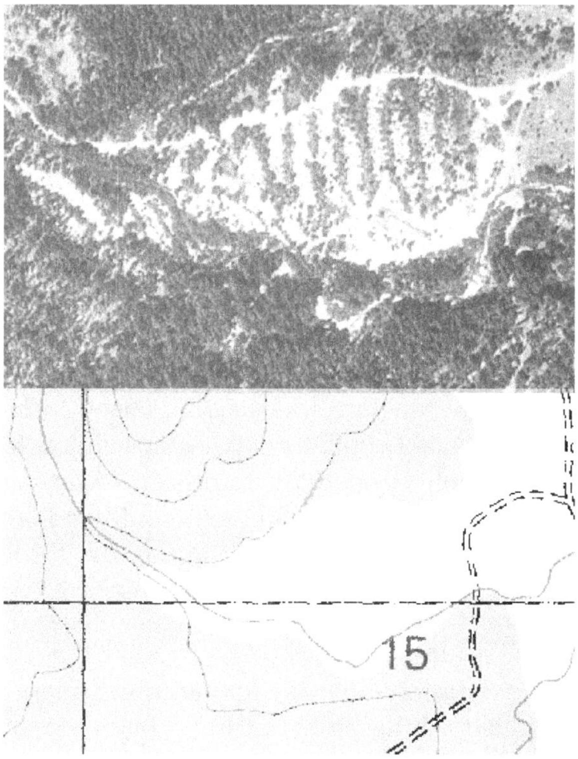

This shows a comparison of an aerial photo with a topographic map. The old dredge tailings not shown on the topographic map are clearly visible

Aerial photo images can also be used to pick out good areas for camping that are not too overgrown. These maps are also a big help with geology - areas of altered rock that has been changed by mineralization, or areas that are heavily impregnated with iron, coloring them brick red will also be visible on these maps. Some types of rock such as serpentine or granite-like rocks can be recognized on these aerial images. For example, serpentine, a favorable gold indicator in the Sierra Nevada will not grow plants well in the soils that form from weathering of this rock type. You can see the sparse vegetation areas of serpentine on Google Earth. You can review aerial photos for mine prospect holes and scrapes, signs of old workings, some of which you can't really see if you are down on the ground, because they are overgrown or otherwise escape the eye. On the aerial map you can see many things that are not marked on the topographic maps, and unless you happened to wander onto the property or get the aerial photo, you'd just never find it. As I mentioned, you can sometimes see faint things with aerial photographic map which are almost impossible to see when you're standing there on the ground. My experience with this is mostly related to faint road tracks going off through the desert. In these kinds of situations, it's actually really good to take your map and figure a GPS coordinate for those faint road intersections you see on Google Earth, especially if you're going into an area where you have not been before. This way you'll have the right GPS coordinate to know when you're supposed to turn from one road onto another and you'll be able to find the same roads that literally sometimes you have to get out of your car and walk around a little bit to see them. Nugget patches that have been raked or were scraped with a bulldozer years ago and now are overgrown can also fall into this category, where you can see them from an aerial photo but when you are standing on the ground they are nearly invisible to the eye. Just like the faint road intersections, having good GPS coordinates will help you find these old hidden nugget patches.

Aerial photo images are just not as common in paper form as your standard topographic map. An interesting thing that can be done with photos of topo maps and google earth images is to combine them. Some photo manipulation programs will allow you to overlay the two so that you can see both the image of the topo map and the Google Earth image at the same time. Doing this may help you to see things on one that is not visible on the other, and train your eyes to better interpret the photo images of Google Earth and recognize old mining spots of favorable rock types for prospecting.

Reading Geologic Maps

Another excellent source of information for the prospector is the geologic map, though some folks are intimidated by them, they are easy to use once you get the hang of it. Some guys just find these maps are too technical, so they don't make use of them. Don't let yourself be intimidated, these maps can be a really great asset! Not many of the individual guys out prospecting use geologic maps as a help and that's a mistake. Some of the new guys may have never seen one before. I will tell you that among the geologists working for large companies and independent folks who are trained in geology and ore deposits, essentially 100% of them use geologic maps, and that ought to tell you something about how important they can be in finding mineral deposits. Geologic maps are not some "magic bullet" that will instantly bring you massive quantities of gold, but they are a useful tool that you can use to help make better use of your time in the field. Believe me, professional exploration geologists make very extensive use of geologic maps in their exploration efforts and if somehow you think you don't need them, I can assure you they know something that you don't. Geologic maps are written to explain the geology of an area, so instead of showing roads, towns, mountains and other features commonly shown on most maps they show the kinds of rocks on the surface. Sometimes they appear a little confusing with all the brightly colored splotches covering all the areas of the map. Reading and understanding a geologic map takes a little practice in a manner similar to getting the practice needed to get the hang of using topographic maps. The secret in understanding

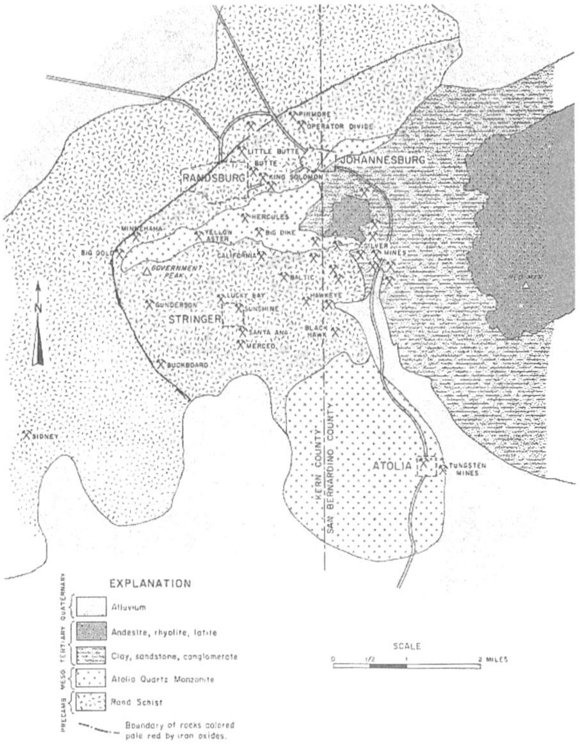

This is a geologic map of Randsburg, Calif.

geologic maps is in understanding the key, which explains what colors and patterns represent as far as each type of rock. When using a geologic map to do research on gold bearing areas, you want to find out from the old reports and other data that you have what the favorable host rock is in that area. In some regions, gold placers are derived from rocks like schist, because they form in the veins and other deposits in these types of rock. In those cases, you would use the geologic map to identify all the areas of schist rock in the region that you are interested in. One of the best ways to use geologic maps is in conjunction with topographic and other maps, including satellite photographs. By comparing all these different types of maps you can see based on roadways, old mines, creek beds and other obvious features were you would want to start prospecting.

Info on How to Read a Geologic Map

There are many types of geologic maps, some covering only a small area in a very detailed way and others which cover regional scales or even an entire state. Some focus in on only one type of rock, others may focus only on faults. The larger scale maps that cover big areas are generally less detailed and less exact compared to the small-scale maps. Regional sized maps like this one are not always perfectly accurate because they have to combine certain different types of rock into one category. The geologists who develop these maps do spend time in the field, but they do not do a field inspection of every inch of the area. Using field notes and aerial photos, they "extrapolate" and project the boundaries of the rocks in places. It does not mean that they are completely inaccurate, just that they are not perfect. Accuracy of Geologic maps is closely related to the scale (the more detail is better). The large-scale maps give you a good general idea of what's going on with the regional area geology. Because of this it's important to consider intent of the geologist who produced the maps. What it comes down to is that maps made at a scale of one to one million are generally not of any use to the prospector, because they just don't have enough detail. What you really need are maps made at one to 100,000 or less. If you can get maps which are produced at a scale of one to 24,000, that's even better.

Geologic maps are available from the State Bureau of Mines and /or Geology in your state. The federal USGS also produces geologic maps. University libraries also have copies of old reports containing geologic maps - and you can use their copy machines to make copies of the maps. If you really want to preserve the colors of the map you might even consider taking a digital color photograph of the map. Some libraries may have a color copier or two – ask the librarians. Detailed maps for all potential mining areas are not always available – check with your state to see what they have. A significant number of geologic maps can also be downloaded from certain websites on the Internet. Check around and do some research to see what geologic maps are available for the area in which are interested.

One of the important things about using a geologic map is having the capability of recognizing certain rock types in the field. Now you don't need to be an expert at recognizing rocks to use a geologic map, you just need to know some basic things about rocks and certain categories of them. The information presented in this book, plus a standard rock and geology handbook with photographs should be more than sufficient to do the job. These types of handbooks are available at larger bookstores and through Amazon on the internet.

After some practice using geologic maps you will learn to recognize certain things. Soil and alluvium generally fill the valley bottoms, so the exposed older rocks correlate generally with the hills and mountain ranges where the uplifting forces have created slopes that are steep enough to allow the sand and soil formed by weathering and erosion to drain off. If you compare a topographic map and a geologic map, you can see this pretty well.

Geologic maps normally have a lot of different colors. This is because normally there are so many rock units in an individual area that you need all these colors to account for all the different kinds of rock that are found. Normally each different rock type gets a name which is explained in the key for the map. The terms used for naming the different rock units normally are chosen to relate to the rock unit being described, so as a result, something like Tba will be used as the name for a Tertiary basalt unit; something like Mzgr will be used for Mesozoic granitic rock; Qa would be used for quaternary alluvium. Sometimes the names are not perfectly aligned, so it is necessary to check the map key to be sure which rock names refer to what type of rock shown on the map. Now we are not just using geologic maps for no particular purpose - there is a specific goal in mind. The *KEY* concept when you are using geologic maps is to determine where the gold is coming from, and look for more occurrences like those sources. Typically you would find out from your study of various old reports what rocks host the source of the gold. It may be coming from certain fault zones; it might be coming from the veins located within a certain rock type; it might be coming from contact zones were two different rocks come together - whatever the source, it's your job to determine what that is during your study. Once you know what that source is, you will then look for similar rock types, similar contacts or extensions of the mineralized fault zones that are producing gold within the district you are looking at, and that's where you want to go to do your prospecting. I'm going to go through a few examples in this chapter of how I do research and put everything together to gain the kind of information I need to find gold out in the field. Take a look at this first example of how I used to geologic map to look at and think about the placers of Humboldt and Pershing County in northern Nevada.

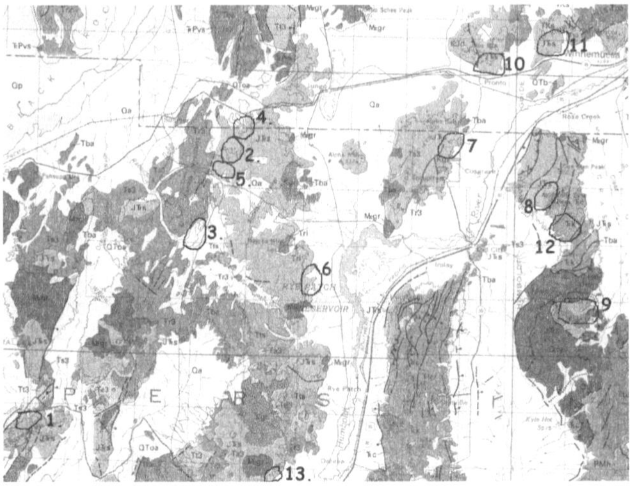

Geologic map of Northern Nevada Gold Districts

Using Geologic Maps to Find Gold - Example: Northern Nevada

I took a map that's produced by the Nevada Bureau of Mines and Geology and altered it with my little photo drawing program on my computer. I have put a copy of the map in with this section, but the original is in color, and this is black and white so it may take a little more effort to read it. I drew onto the map 13 fairly well known placer areas in northern Nevada in the Lovelock-Winnemucca area. Plotting districts out on a regional geology map like this is a real useful technique to see what types of rocks form placers and from that you can take some ideas on where one might explore. Now the ovals that mark the placer areas are generalized and do not perfectly outline exactly where the gold is and is not, but they are roughly correct and do show generally the regions where gold has been found. Because of this, it's probably not the perfect map to the gold of Nevada that some might wish for, but it's very useful for studying the geology of placers in northern Nevada. The 13 districts do not mark all the places where gold has been found in this region, but I picked 13 fairly well known and good-sized areas. There are lots of other smaller spots where folks have found gold, but I have not included them. Also, I specifically left out all the placer areas in the Humboldt Range (of which there are a good number) because the geology is different there. I want to focus on the rest, which are at least generally related geologically. Some are better than others for detecting for various reasons. Some of these have been VERY productive to prospectors armed with metal detectors; others have produced some nice gold for folks with dry washers. Also, just because a site is listed here does not mean that it is unclaimed or open to prospecting - you have to determine that yourself, and that topic is a whole other discussion. The following are the numbered areas I picked:

1. Bluewing Mtns.
2. Rabbit Hole
3. Placerities
4. Sawtooth
5. Rosebud
6. Rye Patch
7. Central / Eugene Mtns / Lunker Hill
8. Dun Glen
9. Willow Creek / Spaulding
10. Blue Mtn.
11. Krum Hills
12. Rockhill / Barber Canyon
13. Trinity

The number for each mining area is marked next to the circle which generally encloses that placer area. I've been to most of these places but not every single one. For those that have a copy of the Maureen Johnson book "Placer Gold Deposits of Nevada" you can look up these locations and see what she wrote about them (history and geology).

They are all in Pershing and Humboldt Counties. Many prospecting dealers sell this book, and its very useful if you plan to prospect for gold in Nevada. On geologic maps, the colors indicate different rock types. The letters in that color area are a description of various regional rocks found there. Geologic maps have keys to explain the different rock types. The key for this map is large, so I didn't try to include it, but a couple different rock types you might note from this map are the rock units marked Qa, JTrs, Qtoa, Mzgr, Kgr, Tr3, Tba and TrK. They are:

Qa = recent alluvium, such as loose soil and sand
Qtoa = Quatenary / Tertiary older alluvium plus older gravels and sedimentary rocks
Tba = Tertiary basalt
JTrs = Jurassic / Triassic Shale & slate rocks
TrK = Triassic Koipato Group - mostly ryholite tuffs and associated sedimentary rocks
Tr3 = Triassic rhyolitic flows and shallow intrusive rocks.
Mzgr = Mesozoic granitic rocks, mostly monzonite and granodiorite
Kgr = Cretaceous granitic rocks, mostly monzonite and granodiorite

The darker lines shown among the rocks in various areas like in placer locations 8 and 12 are fault zones. Not all fault zones are shown, as some are buried in the alluvium, or just are too small to plot. Because this image is so much better in color, I will have it available on my website if you want to check out a larger color version there. So what do you see here? One thing which is important to notice is that nearly all the placer areas marked on the map are associated with outcrops of a specific type of rock listed on the geologic map as JTrs. This group of rocks is a Jurassic / Triassic age shale or slate with similar related rocks. They are rocks which easily part into flat sheets. As was noted in the chapter on placer geology, these rocks are commonly associated with placer deposits, and favorable rocks to be the source of placer gold. It just so happens that a number of smaller deposits that I did not plot on the map are also located in these same shale rocks. Having been out to a number of these locations, it seems that in areas where these rocks show quartz vein materials, gold is found in the same general area. Research from the Maureen Johnson "Placers of Nevada" book suggests the sources of gold are veins in these rocks. So in reviewing all this, you can see that the JTrs rock, as mapped on the geologic map would be a favorable area to prospect. Perhaps all these areas are worth taking a look at as this rock outcrops over a large part of the land on the map and that includes many square miles. Where you see quartz, it's worthwhile to get out your metal detector or dry washer and take some samples to see what, if any, gold is present.

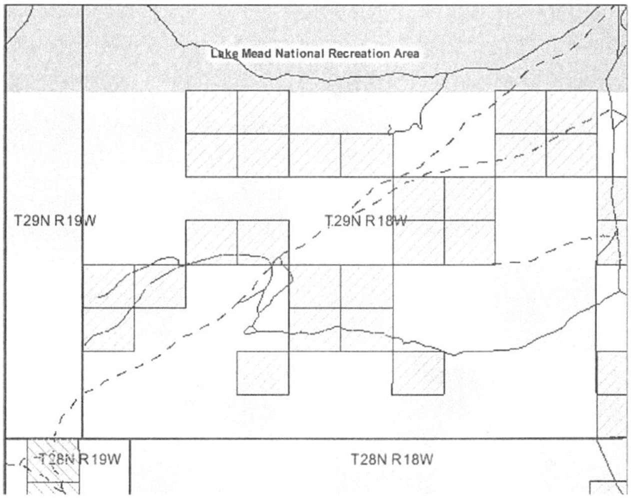

This is a land ownership map for the Gold Basin mining area in Arizona.

Land Ownership and Land Status Maps

One of the more important types of maps that I have are land ownership maps. An important element of doing prospecting research is to make sure that the land you want to prospect on is actually open to your prospecting. We will discuss this subject a lot more in the appendix chapter on mining laws, but there are certain types of land which are open to public prospecting and many others which are not. Lands administered by the BLM or the Forrest Service are generally open for prospecting in the western states where the mining laws apply. However, there are also withdrawn lands such as Indian lands, military reserves, national parks or other federal sites which are not open to prospecting. A series of land ownership maps are prepared by the USGS and the BLM which show the underlying land ownership, specifically private lands vs. public ones and the various categories of public land which may or may not be open to prospecting. I have a set of these maps, some of which are available for downloading over the Internet, and others which can be purchased at BLM or sometimes federal forest service offices. Mostly these maps are produced in scales of one to 100,000 a size which is usually sufficient for rough land ownership research. Detailed land ownership research may require much more detailed maps.

Combining Report Information and Info From Various Maps

Actually, the very best research brings together information from a multitude of sources, perhaps a hint from an experienced prospector, plus some research in old state mining reports and a serious investigation of the topographic and geologic maps of the area. These are then all combined to determine the best possibilities for gold. Be sure that when you are doing your research you make use of all the possible sources of data that you can lay your hands on - the more information you have the better. At this point in my life, I only get to prospect on weekends and vacation time, and the number of hours I have in the field is limited. I need to be efficient with my field exploration time when I get out to the hills. To make the best use of my time, I always have a written plan of the things I want to make sure I take a look at, so I get to see what I need to see. I will normally have the specific GPS coordinates of the spots I want to visit taken from the maps and data base sources in my notes. I typically also make my own maps of the area with notes written on them about what I am thinking about in that area. I also try to write down my thoughts afterwards and make decisions for further actions while that field experience is still fresh in my mind.

Researching and exploring to find new areas is a lot of work, and not every trip to a new area will be successful. However, in the end you will learn a lot about these new locations through your study, and when you find a good new spot with good gold, it will be well worth your time. The end result of your research should be at a number of areas that you want to check out. I strongly prefer to use some sort of handheld GPS device, so I can take coordinates from my research. That way I know when I'm out in the field that I've reached the point that I wanted to see. You will be deciding on those coordinates, and the priority for which you want to investigate them based on your research. Using a GPS isn't always perfect, and sometimes there are difficulties in finding things in the field, especially in heavily forested areas. The tree branches in heavily forested areas can give your older GPS fits, because the GPS needs to be able to see satellites in the sky in order to function. Newer GPS units handle this much better, but with the older units you may need to step away and find some sort of clearing in order to take a reading. In the desert this is usually not a concern. When you do make a worthwhile find, it's also important to use that GPS to mark your finds so you can keep track of them. Some guys are able to keep track of locations in their head, but it's always safer to have a set of coordinates and some notes to remind you about that great find you made, because even after you've cleaned it out you may learn something new, or acquire a better piece of equipment and want to revisit that spot years down the road.

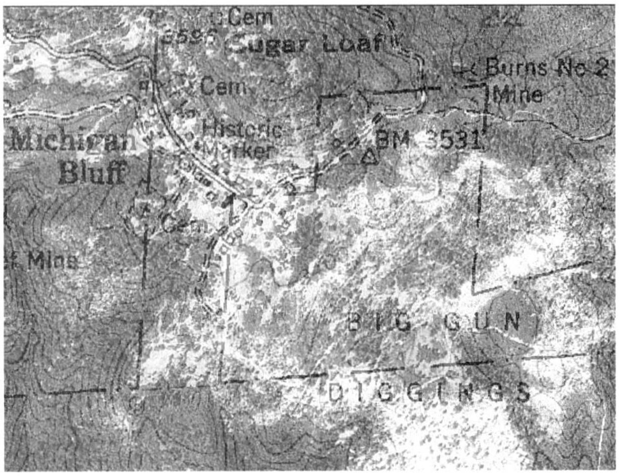

This is the combination of a Topo map and an aereal photograph onto one image using GIS.

Geographic Information Systems

One of the new types of computer software that is on the market is known by its initials GIS. The term stands for geographic information systems and applies to a whole category of computer mapping programs. These programs are basically made for the manipulation and preparation of maps. This category of software can easily plot all sorts of things on the digital versions of topographic maps, geology maps and aerial photos and other types as well. The programs will allow you to switch back and forth or overlay one map with another, so that as an example you could overlay a land ownership map on top of an aerial photograph so that you could see if a certain mine working was located on private property or if it were partly on public property and available to prospecting. It's a really slick method of dealing with maps and information and I know that only a small percentage of prospectors will take advantage of it, but if you have significant computer skills and are interested in incorporating that into your prospecting research, learning about GIS software is the way to go. These programs allow for all sorts of color and image coding, treating each part of the information as a separate layer - so the roads may be one layer, mine locations another layer, property ownership another layer, the geologic map another layer, aerial photographs yet another layer and so on and so forth until you have all the map information you have gathered put together into one package. In the future I think there will be more use of this sort of thing. Some of these programs are free or nearly so (while others can be very expensive), but if you are skilled with a computer and want to push your research to the highest point, look into GIS programs. Many companies actually hire GIS experts and the jobs pay well – and many universities also offer classes in this topic.

Geologic map of Gold Basin NW Arizona, the gold bearing rocks are metamorphics labeled Xm.

Research and Mapping Examples:

As I promised, here are another couple of examples of research I have done on two different areas that I would like to share with you. Through this process, you can essentially look over my shoulder while I explain the steps I went through during the research process on each one of these and how I came up with the conclusions that I reached.

Example 1: Gold Basin:

Gold Basin in Mohave County in northern Arizona is a very interesting district for prospecting with both metal detectors and dry washers. I have been there once prospecting and because a number of prospectors have been very successful there, I thought it might be interesting to do some research about it. I started out collecting up some land maps including topographic, aerial photo and geologic maps. I also got some land ownership map information as well, because there is a mix of both private and public property in this area, as well as a national recreation area that is off limits to prospecting. I acquired all of these by downloading them over the Internet. I then started reading some books I had with information on Arizona. I already had a couple of books on placers in Arizona including the ones by Maureen Johnson and Eldred Wilson. By searching on the Internet I found an interesting report on Gold Basin that was produced by the USGS, with some good information about placer deposits in the region. Now when I read those reports I'm always looking for information about things like: 1) Past gold production; 2) The size of the gold recovered (did they find any larger sized nuggets?); 3) How deep the gold was buried (was the paystreak buried down under a whole lot of gravel or exposed with shallow bedrock); and 4) I always read to find where the gold is coming from (because that's one of the most important things in prospecting research!).

I found out that this area had produced some significant gold for prospectors armed with metal detectors and that it had been pretty good to dry washers as well. Some nice coarse nuggets were produced, and a good amount of the gold was quite shallow. It turns out the gold is coming from veins located in old Precambrian schist rocks which outcrop in a part of the area. I looked out to see what was going on in nearby adjoining old placer districts. It turns out that in Lost Basin and some other nearby areas, the same old Precambrian schist rocks have the same veins which were creating placers in some of those nearby areas as well. Remember how important it is to learn the source of gold in a region, then look for similar geology nearby. The gold bearing source rock was marked on the geologic map as "XM", the "X" being for Precambrian, and the "M" being for metamorphic (schist is a metamorphic rock). In this example I'm looking for locations with the same geology. It turns out that those crystalline continental basement rocks, that is the schist and gneiss labeled as XM on the map, are fairly common in the region and gold veins have been found in them throughout the Northwestern part of northern Arizona and across the Colorado River into southern Nevada at Gold Butte as well. In checking the geologic map for nearby exposures of the "XM" rock, I found that there were quite a few exposures of them in the Black Mountains, an area west of Gold Basin that had no mention in the old reports of any placer gold being found there. It was no coincidence that in the Black Mountains there are also small hard rock gold mines associated with the same exposures of those Precambrian Metamorphic rocks (XM) just the same as there were over in the Gold Basin area. Both areas had locations with the gentle slope topography that is most favorable for accumulations of desert placer gold. I decided that the probability of undiscovered placers at the Black Mountains location was very high! I asked around to some friends of mine who prospect for gold over in the Gold Basin area if they had ever heard of anyone finding gold in the Black Mountains, and sure enough several of them had heard of people finding placer gold to the west of Gold Basin in the Black Mountains. I even did some claim research and found that there are quite a few placer mining claims over there. Nothing can really replace checking a location out in person, but the indications are certainly pretty good for that location. If I lived down in that part of the country, I certainly would check it out. Probably all of the exposures of those Precambrian metamorphic rocks both in northern Arizona and across the Colorado River into Nevada, are worthy of exploration for placer gold especially in around locations where there were small hard rock gold mines. A copy of the geology map for this area printed in color appears on my web site.

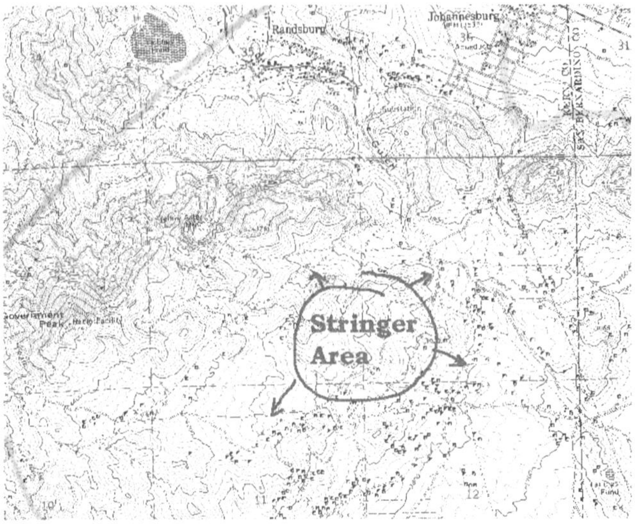

Randsburg area topographic map, including the Stringer placer mining district.

Example No. 2: Randsburg and the El Paso Mountains.

My second example is a well-known southern California desert placer area - the area around Randsburg and the El Paso Mountains to the north. This region has been very productive to metal detectors and dry washers alike. The area produces some very big gold! A 156 ounce nugget, is the largest I know of which was found with a metal detector in California, came from the Stringer area near Randsburg. Many years ago when I was a teenager I used to dry wash in this location so I have a little familiarity with the area. However I was just a kid in those days and knew very little about researching gold deposits. For the most part this would be a brand-new exercise in researching this location. What I did to start my research for this area was to go to my information library, grab my copy of California Bulletin 193 and read what it said about both the Randsburg and El Paso mountains gold districts. I wrote down a number of references from bulletin 193 that seemed interesting, and then I went to the University of Nevada, School of Mines Library and looked up some of these old report references, and made copies and read them. I also made copies of some old geologic maps. I also downloaded topographic maps and aerial photos off the internet.

This is what I found out: The geology of the placer deposits are distinctly different at Randsburg and Stringer as compared to those at the El Paso Mountains, including the Summit diggings and Goler Gulch areas. I read in an old report that there are two very different sources for placer gold in this part of the Northern Mojave Desert: The first type are placers associated with direct erosion of lode deposits in ancient schist associated with granitic intrusions. These are most commonly found near the Randsburg and Stringer areas. Some of this gold was also found in the Atolia area. The gold is sharp and angular - it has not traveled far. The geology of these placers is somewhat similar to many placer deposits in Arizona. Around Randsburg there is an area of schist which is tinted red, and this "Iron Stained" mineralized rock hosts most of the gold vein deposits. It is also nearby to the younger intrusive (granitic) rocks. The schist areas which are in or near to this red colored mineralization are the areas which are most favorable both for placers and vein deposits. These would be the ones to target in your prospecting for the first type of placers. In the second type, the gold is associated with the lowest conglomerate at the base of the Ricardo Formation, a sedimentary rock unit spread across a large portion of the Mojave Desert. There are a number of Southern California placer areas associated with this lowermost conglomerate member of the Ricardo formation. The lowermost conglomerate of the Ricardo Formation is more like a delta with water worn, rounded gravels spread over a very large area as opposed to a single river stream. Erosion has removed large parts of it, leaving isolated patches of gold bearing gravels behind. The conglomerate gravels themselves have only a small to moderate amount of gold, but the processes of erosion have concentrated that gold, forming some decent placers in certain locations. These re-concentrated placers have been found in several places in the El Paso Range. This second type of gold is smoother and shows signs of travel. According to the old report, everywhere the writer examined where modern streams had cut through the Ricardo Sedimentary formation to its base, gold had been found. All the El Paso mountain range placers, plus Coolgardie, and many other southern California desert placers have this same geology (but not Randsburg). Well known and significant placers of this type in the El Paso Mountain range area include Goler, Last Chance and Bonanza Gulches, as well as the Summit Diggings. There are several lesser-known areas as well. It also is the situation that isolated spots of these placers are spread all over and some are very little known by modern day prospectors. A geologic map could be used to determine places where the gold bearing gravels might be found by searching the locations where the Ricardo Formation is exposed near the next formation underneath it in the geologic column. Color copies of the geologic map of this area can be found on my website.

Summary of the Methods of Research

There is no one simple way to do research, and a lot of guys proceed with different methods and techniques and have success. You don't have to do it exactly the way I do to be successful. However, it's good to have set methods because old reports and maps really can give you an awful lot of information. Sometimes experienced prospectors who are friends can give you some great leads and although you may pick up information from a variety of sources, putting it all together is what makes for good research. The way I do it can be summed up in the following steps:

First:

Start with a "location" book, select one or more gold districts for research. Research related area reports to find districts, mines or other areas which seem to be what you are looking for. Build a library of resource information for your area - find out what is available as far as old reports, collect whatever information you can from old timers, etc. Determine, if possible, the source of the gold for this district. This is a key concept.

Second:

In your research, collect as many of the applicable maps for the area as you can, including: topographic, aerial, geologic and land ownership maps. Check with various source reports; make a list of the items you are interested in and visit the local libraries and print out map copies as necessary. Collect as much information as you can on line and study the area on Google Earth and become familiar with the features of the area including any old workings that can be seen.

Compare the different map types for extra information. Read the maps carefully for whatever useful information you can extract.

Third:

Based on your research, select some target sites for field exploration and sampling. Find favorable areas on the outskirts of the main areas; Look for spots with similar geology which are generally nearby to locations where gold has been found in the past. Generate some GPS coordinates for those selected targets. Have a plan and a priority list of different places you want to look at in that area. Sometimes an area that looks good in your research will not pan out when you get there in person. That's why it's important to have a plan B., a plan C, etc.

Fourth:

Actually get out there, put boots on the ground and do that field exploration. If you never take the time to go out into the field, you'll never actually find anything. When you get out there, carefully evaluate your target sites and take good samples so you can really know what's there or what's not there in the way of gold. Do everything possible in the way of sampling and testing to determine if sufficient amounts of gold are present. Don't stop at the first location unless you really hit it big. If you don't strike it rich, keep looking; have the determination that it takes to keep going in the effort. Don't give up on the process; keep looking for new patches, pay streaks or whatever deposits you are seeking.

The basic formula I use for finding gold is to: 1.) Buy good prospecting and gold recovery equipment and then learn to use it properly; 2.) Have the physical ability to get out in the field and the right attitude to keep at it and not get discouraged. Prospecting isn't easy and you need determination and a committed attitude to stick to it when things get tough; 3.) Do your research homework and field exploration. Look for the likely places where gold will deposit and sample or use your equipment to test those locations. Have a plan and a priority list of places you want to investigate. If you don't find what you're looking for the first time, or the second time, or the third time, you need to go back and do more research or find another area to prospect and then repeat the process of prospecting and working hard; 4.) If you keep at it and work hard, then in time, you will find some good deposits of **GOLD!**

Old-time prospectors testing ground by panning.

I've Included Some Maps:

I never intended that this would be a "Where to Find Gold book" with maps and lists of districts that you might want to look at for prospecting. However, I've given some examples of how I do prospecting research, so just to provide some basic information for getting started, I am including some general maps of placer and other gold deposits in the US and elsewhere. These make for a good set of examples and the foundation that you can get started on in your prospecting and see how widespread gold deposits really are. Gold deposits are spread all over the US, although the most important deposits are in the west. I have included full page gold maps of Nevada, California, Arizona, Alaska, and several other states. There is additional gold district information on my website, and in the future I plan to expand the research material found there. I also have a listing of recommended "where to find gold" location type prospecting books on the web site as well. Hopefully this information can help you get started in your prospecting research.

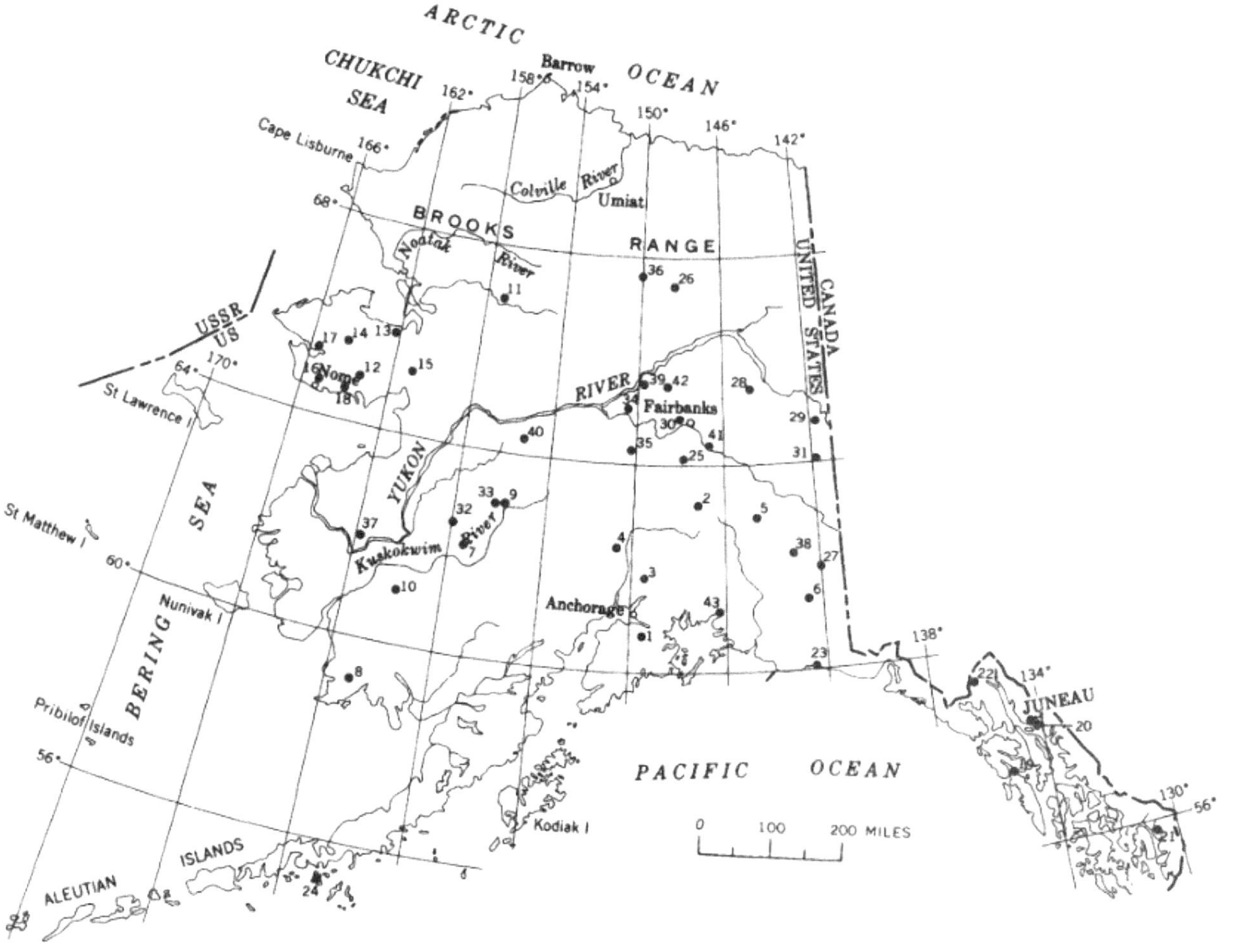

Principal Gold Mining Districts of Alaska

Cook Inlet-Susitna region:
1, Kenai Peninsula; 2, Valdez Creek; 3, Willow Creek; 4, Yentna-Cache Creek.
Copper River region:
5, Chistochina; 6, Nizina.
Kuskokwim region:
7, Georgetown; 8, Goodnews Bay; 9, McKinley; 10, Tuluksak-Aniak.
Northwestern Alaska region:
11, Shungnak.
Seward Peninsula region:
12, Council; 13, Fairhaven; 14, Kougarok; 15, Koyuk; 16, Nome; 17, Port Clarence; 18, Solomon-Bluff.
Southeastern Alaska region:
19, Chichagof; 20, Juneau; 21, Ketchikan-Hyder; 22, Porcupine; 23, Yakataga.
Southwestern Alaska region:
24, Unga.
Yukon region:
25, Bonnifield; 26, Chandalar; 27, Chisana; 28, Circle; 29, Eagle; 30, Fairbanks; 31, Fortymile; 32, Iditarod; 33, Innoko; 34, Hot Springs; 35, Kantishna; 36, Koyukuk; 37, Marshall; 38, Nabesna; 39, Rampart; 40, Ruby; 41, Richardson; 42, Tolovana.
Prince William Sound region:
43, Port Valdez.

Placer gold has a very wide distribution in Alaska, having been found as far south as the southern boundary and at various points northward to the 68th parallel and westward to the Bering Strait. The state is one of the largest producers in the US, having yielded millions of ounces. In addition to the principal gold districts which are listed above, Alaska has produced gold from a large number of smaller mining districts that are not shown on this map. The large majority of all the gold from Alaska has been placer gold. Only a few districts were principally lode producers, with the area around Juneau having the most productive hard rock mines. Only in recent years is modern exploration starting to show the extensive lode gold potential of this very large state as the lode gold in many areas is normally covered over with various forms of plant growth and gravels.

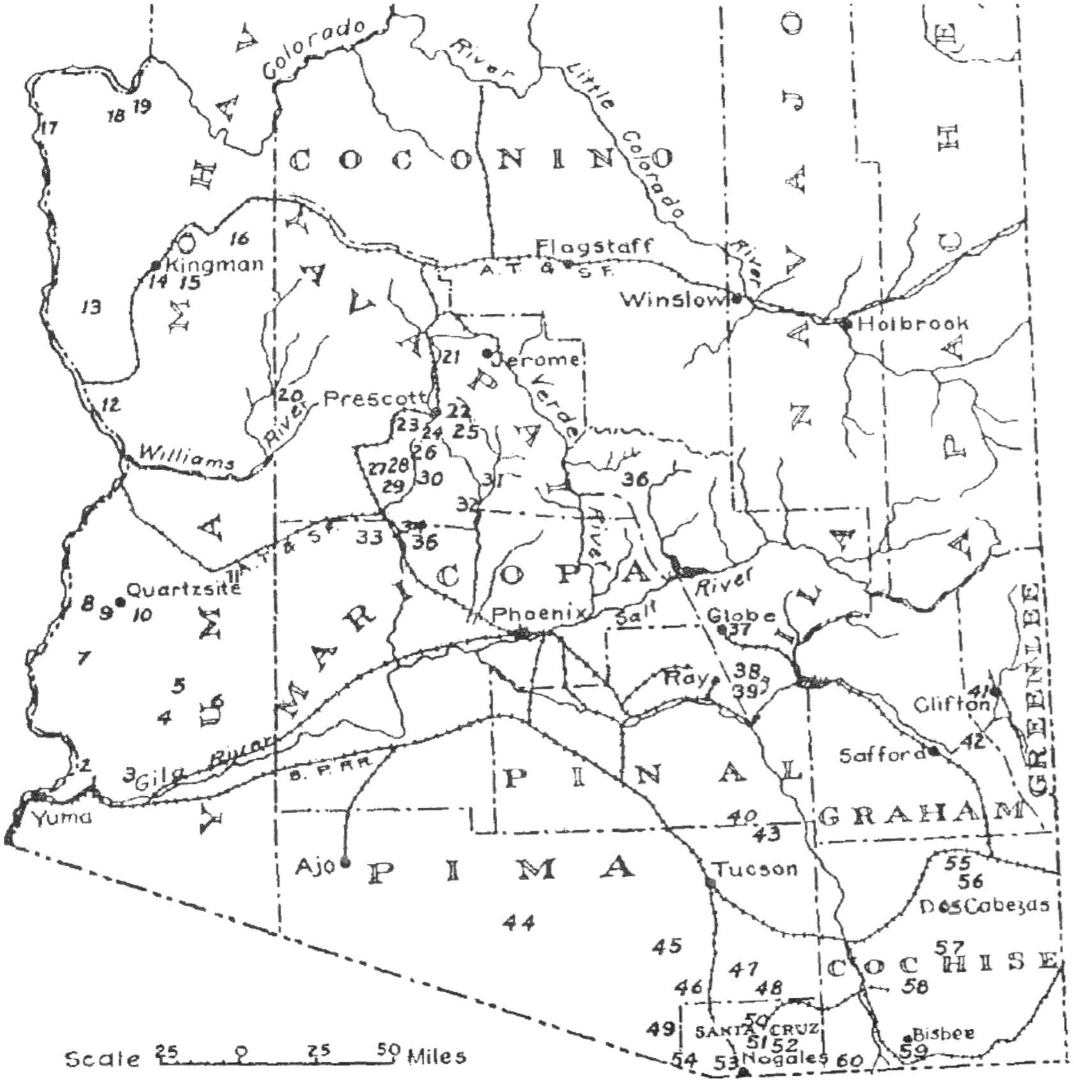

ARIZONA PLACER GOLD DISTRICTS

Arizona has and continues to be an important producer of gold, both from placer and lode. Most of the lode gold comes as a byproduct of copper mining. Small operators still do well in Arizona, using both dry washers and metal detectors. The state is a haven for snow birds who gather here for the moderate winter weather, especially in the lower elevations. Many of the districts shown on the map cover large areas. Look to prospect around the periphery of these districts. In addition to the many districts noted here, there are many other small placer areas which have been mined, and even a few that were never found by the old timers. To find these small placer areas, look for iron stained rocks like schist with chunks of quartz or visible quartz veins. These little known areas are widely distributed across the central, southern and western parts of the state. While not every region underlain by schist or similar rocks will have gold, many of them do.

Index of Arizona placer district map:

1. Gila City (Dome)
2. Laguna
3. Muggins
4. Castle Dome
5. Kofa or S. H.
6. Tank Mountains
7. Trigo
8. La Paz
9. La Cholla, Oro Fino, Middle Camp
10. Plomosa
11. Harquahala
12. Chemehuevis
13. Silver Creek
14. Lewis
15. Lookout
16. Wright Creek
17. Willow Beach
18. Gold Basin
19. King Tut
20. Eureka
21. Granite Creek
22. Lynx Creek
23. Copper Basin
24. Groom Creek
25. Big Bug
26. Hassayampa (Yavapai County)
27. Model
28. Placerita
29. Weaver - Rich Hill
30. Minnehaha
31. Black Canyon
32. Humbug
33. Vulture
34. Hassayampa (Maricopa County)
35. San Domingo
36. Payson
37. Globe - Miami
38. Dripping Spring
39. Barbarossa
40. Canada del Oro
41 Clifton- Morenci
42. Gila River
43. Alder Canyon
44. Quijotoa
45. Papago
46. Armargosa
47. Old Baldy
48. Greaterville
49. Las Guijas or Arivaca
50. Tyndall
51, Harshaw
52. Patagonia or Mowry, Palmetto
53. Nogales
54. 0ro Blanco
55. Teviston
56. Dos Cabezas
57. Pearce
58. Gleeson
59. Gold Gulch (Bisbee)
60. Huachuca

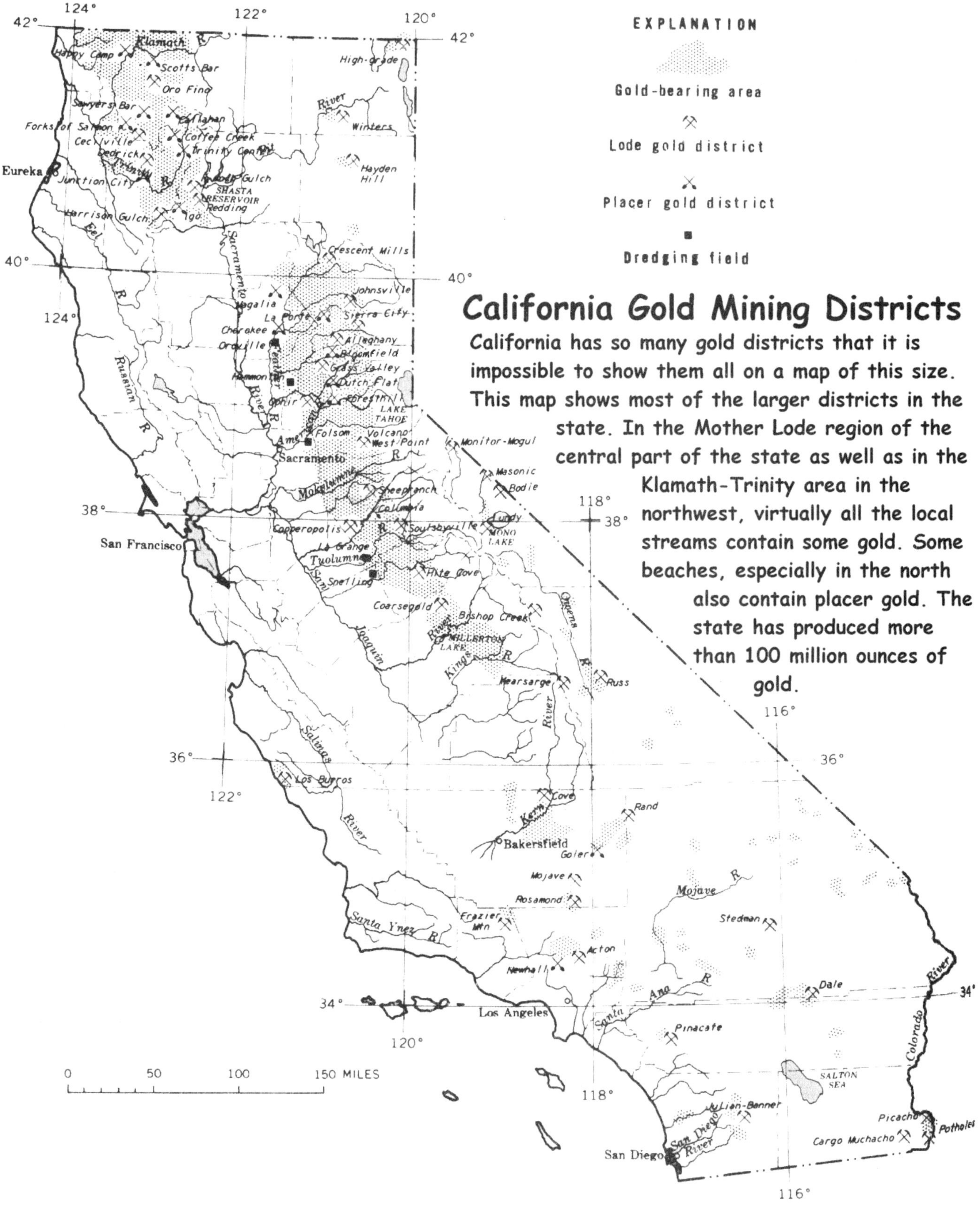

California Gold Mining Districts

California has so many gold districts that it is impossible to show them all on a map of this size. This map shows most of the larger districts in the state. In the Mother Lode region of the central part of the state as well as in the Klamath-Trinity area in the northwest, virtually all the local streams contain some gold. Some beaches, especially in the north also contain placer gold. The state has produced more than 100 million ounces of gold.

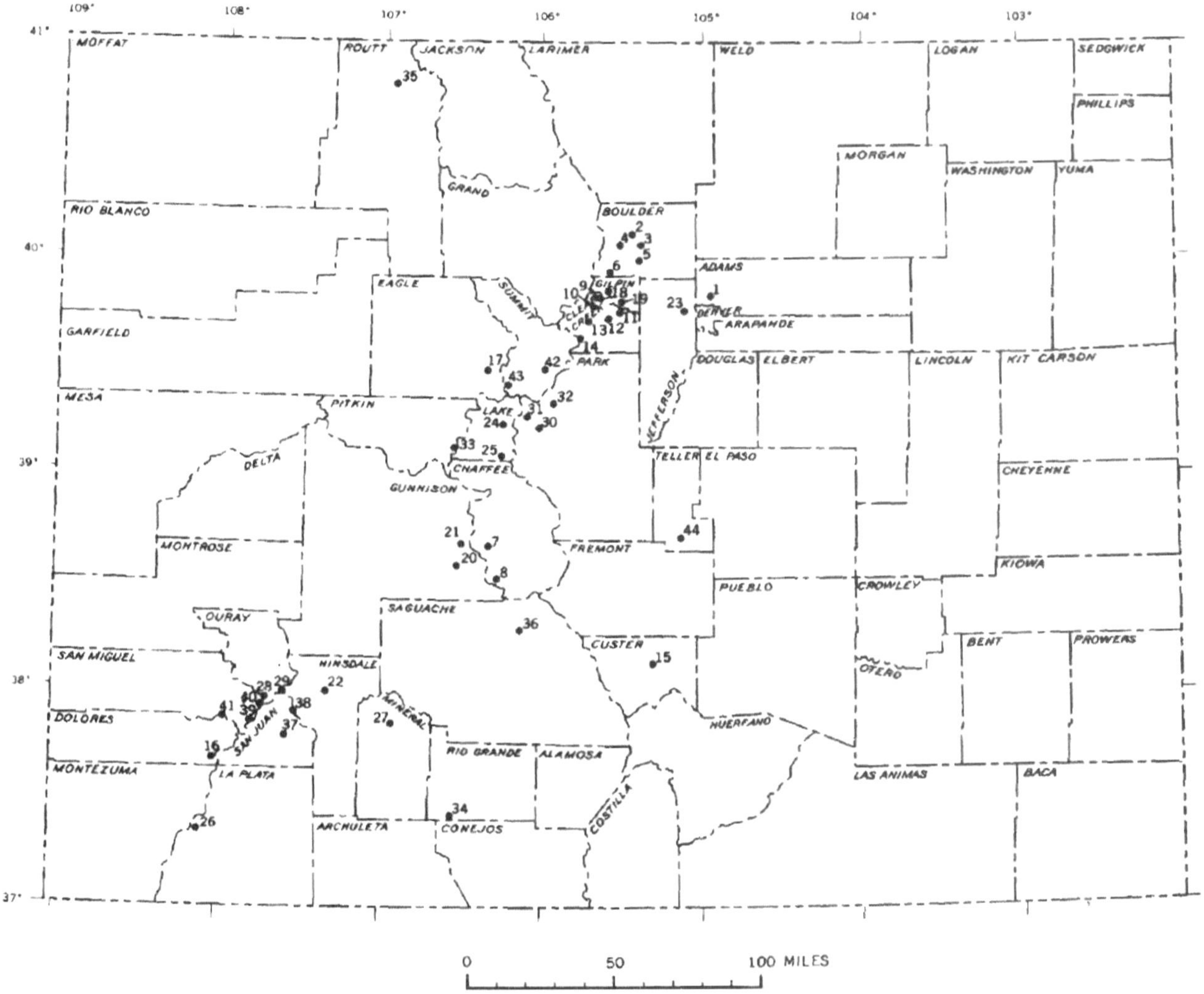

Major Colorado Gold Districts

1) Clear Creek Placers; 2) Jamestown; 3) Gold Hill - Sugarloaf; 4) Ward; 5) Magnolia; 6) Grand Island - Caribou; 7) Chalk Creek; 8) Monarch; 9) Alice; 10) Empire; 11) Idaho Springs; 12) Freeland-Lamartine; 13) Georgetown - Silver Plume; 14) Argentine; 15) Rosita Hills; 16) Rico; 17) Gilman; 18) Northern Gilpin; 19) Central City; 20) Gold Brick - Quartz Creek; 21) Teacup; 22) Lake City; 23) Clear Creek Placers; 24) Leadville; 25) Arkansas River Valley Placers; 26) La Plata; 27) Creede; 28) Sneffels - Red Mountain; 29) Uncompahgre; 30) Alma; 31) Fairplay; 32) Terryall; 33) Independence Pass; 34) Summitville; 35) Hahns Peak; 36) Bonanza; 37) Animas; 38) Eureka; 39) Ophir; 40) Telluride; 41) Mount Wilson; 42) Breckenridge; 43) Tenmile; 44) Cripple Creek

Colorado has a long history of gold mining, and ranks among the most productive states of the US. The gold deposits are found mostly along a northeast-trending zone known as the Colorado mineral belt, which occupies the mountainous western part of the State. Starting near Boulder in the north central part of Colorado, this belt extends southwest to the San Juan Mountains and beyond. The Cripple Creek district, the largest gold producer in Colorado, and several other minor districts are located southeast of the mineral belt. Though considerable gold was been derived from placers, by far the greater part came from mining vein deposits. Among the placer deposits, California Gulch was an import producer, as was Deadwood Diggings in the Clear Creek drainage.

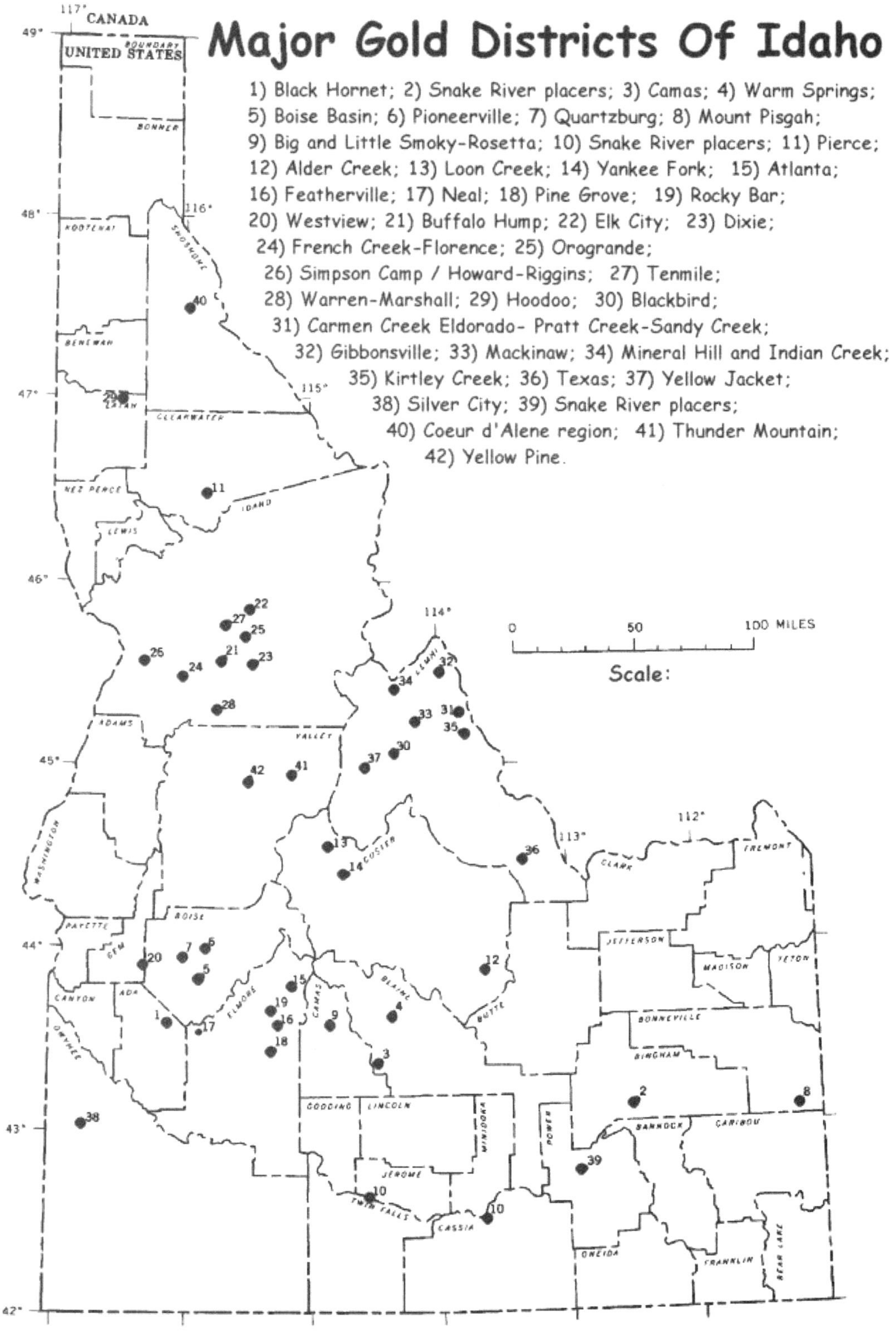

The earliest gold discovery of any consequence in Idaho was made along the Pend Oreille River in 1852. Placers were the major source of the gold mined in Idaho before 1900 and considerable hydraulic mining was done here. However, the best of the placers were exhausted after the period of feverish exploitation from 1860 to 1870. The lode deposits, which had been known and worked on a small scale since 1861, then became more important, though placer mining was rejuvenated in 1897 by the introduction of bucket line dredge operations in several areas. The auriferous region of Idaho lies mostly in the central and southwestern portion of the State.

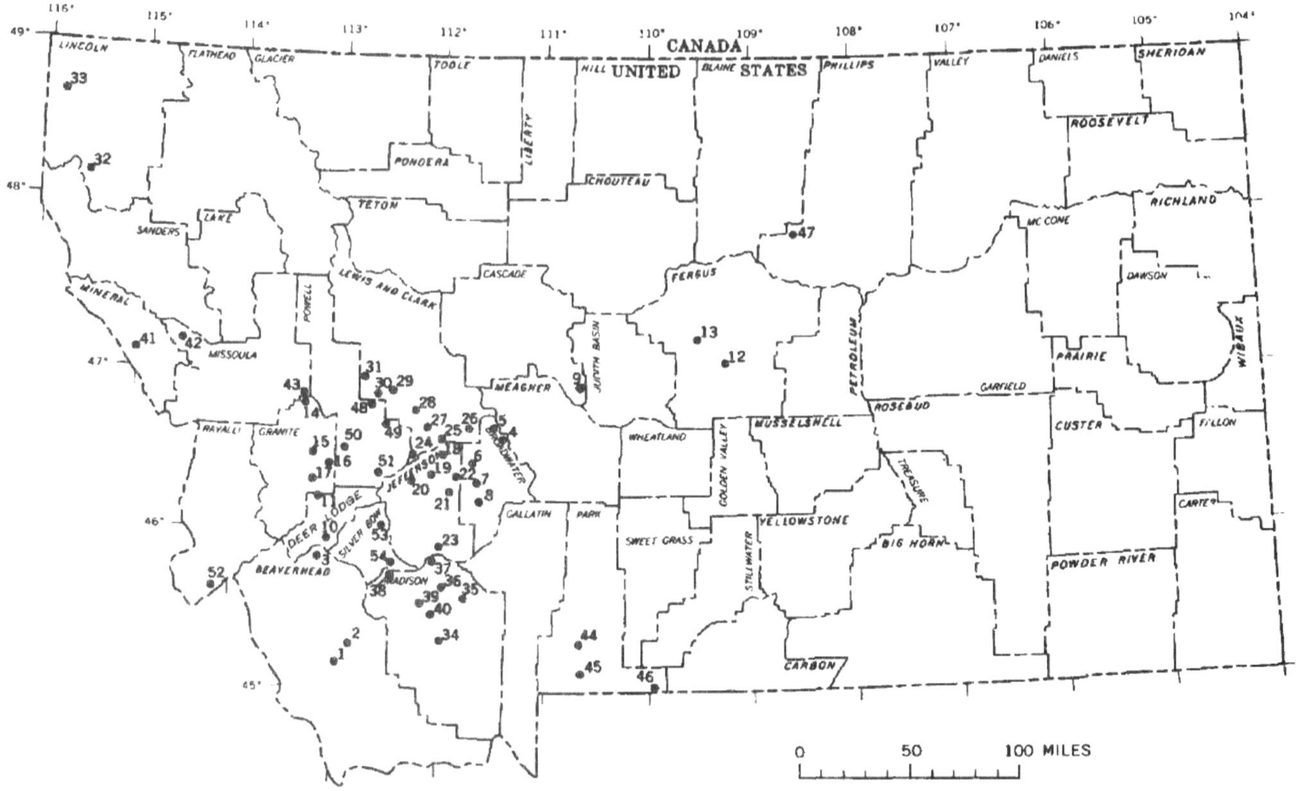

Principal Gold Mining Districts of Montana

1) Bannack; 2) Argenta; 3) Bryant; 4) Confederate Gulch; 5) White Creek; 6) Winston; 7) Park; 8) Radersburg; 9) Montana; 10) French Creek; 11) Georgetown; 12) Warm Springs; 13) North Moccasin; 14) First Chance; 15) Henderson Placers; 16) Boulder Creek; 17) Flint Creek; 18) Clancy; 19) Wickes; 20) Basin and Boulder; 21) Elkhorn; 22) Tizer; 23) Whitehall; 24) Rimini-Tenmile; 25) Helena-Last Chance; 26) Missouri River-York; 27) Sevenmile - Scratehgravel; 28) Marysville-Silver Creek; 29) Stemple-Virginia Creek; 30) McClellan; 31) Lincoln; 32) Libby; 33) Sylvanite; 34) Virginia City-Alder Gulch; 35) Norris; 36) Pony; 37) Renova; 38) Silver Star-Rochester; 39) Tidal Wave; 40) Sheridan; 41) Cedar Creek-Trout Creek; 42) Ninemile Creek; 43) Elk Creek-Coloma; 44) Emigrant Creek; 45) Jardine; 46) Cooke City; 47) Little Rocky Mountains; 48) Finn; 49) Ophir; 50) Pioneer; 51) Zosell; 52) Hughes Creek; 53) Butte; 54) Highland.

In its early yield of placer gold, Montana was second only to California. Its most famous placer deposits were those of Bannack and Alder Gulch. The latter are reported to have produced in the early 1860s some $30,000,000 dollars in gold by simple methods of washing. Gold was obtained from the gravel bars. After the Alder Gulch deposits played out, the rich placers near Helena were discovered and worked. When the principal gulches such as the Last Chance, Confederate, Grizzly and Prickley Pear were exhausted placer mining began to lose its prestige. The Montana Central, the Penn and Cataract placers of Jefferson County were also rich and yielded about half a pennyweight per cubic yard. Hillside placers were worked in Deer Lodge County, but were also soon exhausted. Montana placers which have produced large amounts of gold include: Alder Gulch, Madison County; Bear Gulch, Granite County; Confederate Gulch, Broadwater County and Cedar Creek, in Missoula County. The Cedar Creek placers have a bedrock of loose slate which is easily cleaned.

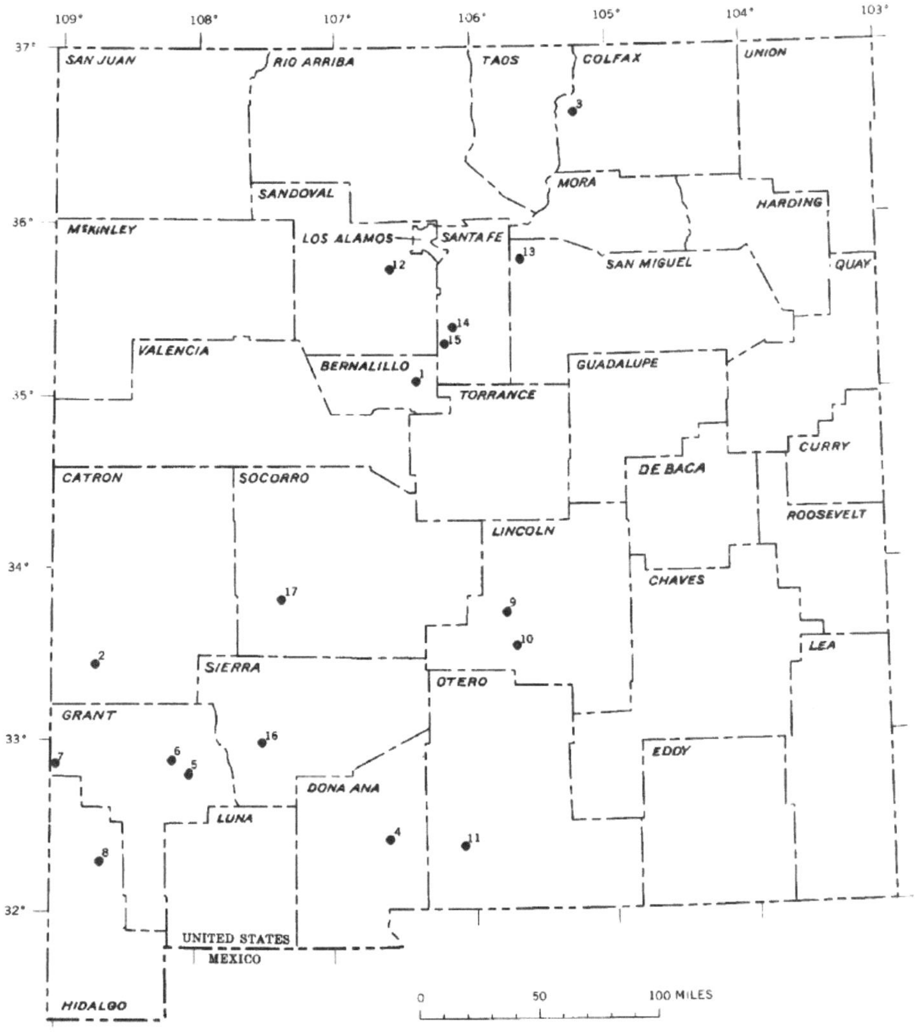

Gold Districts Of New Mexico

1) Tijeras Canyon; 2) Mogollon; 3) Elizabethtown-Baldy; 4) Organ; 5) Central; 6) Pinos Altos; 7) Steeple Rock; 8) Lordsburg; 9) White Oaks; 10) Nogal; 11) Jarilla; 12) Cochiti; 13) Willow Creek; 14) Old Placer; 15) New Placer; 16) Hillsboro; 17) Rosedale

The bulk of the gold production of New Mexico has come from placer mining which has been conducted in this state for centuries. Alluvial gold is said to have been worked by the Indians at San Pedro, Santa Fe County, before the Spanish arrived, but the amounts recovered were probably insignificant. Organized mining in New Mexico began long before discoveries were made in any of the other Western States. The copper deposits at Santa Rita were known and mined late in the 18th century, and placer gold mining began as early as 1828 in the Ortiz Mountains south of Santa Fe. In 1839 placer deposits were discovered farther south along the foot of the San Pedro Mountains. The earliest lode mining, except the work at Santa Rita, dates back to 1833 when a gold-quartz vein was worked in the Ortiz Mountains. The gold producing districts of New Mexico are found along a mineral belt that stretches north-east to Southwest, cutting diagonally across the state. The belt is of variable width and extends from Hidalgo County in the southwest corner to Colfax County along the north-central border. New Mexico's mineral belt is located in mountainous terrain that lies between the Colorado Plateau on the northwest and the Great Plains to the east.

Nevada Placer Districts:

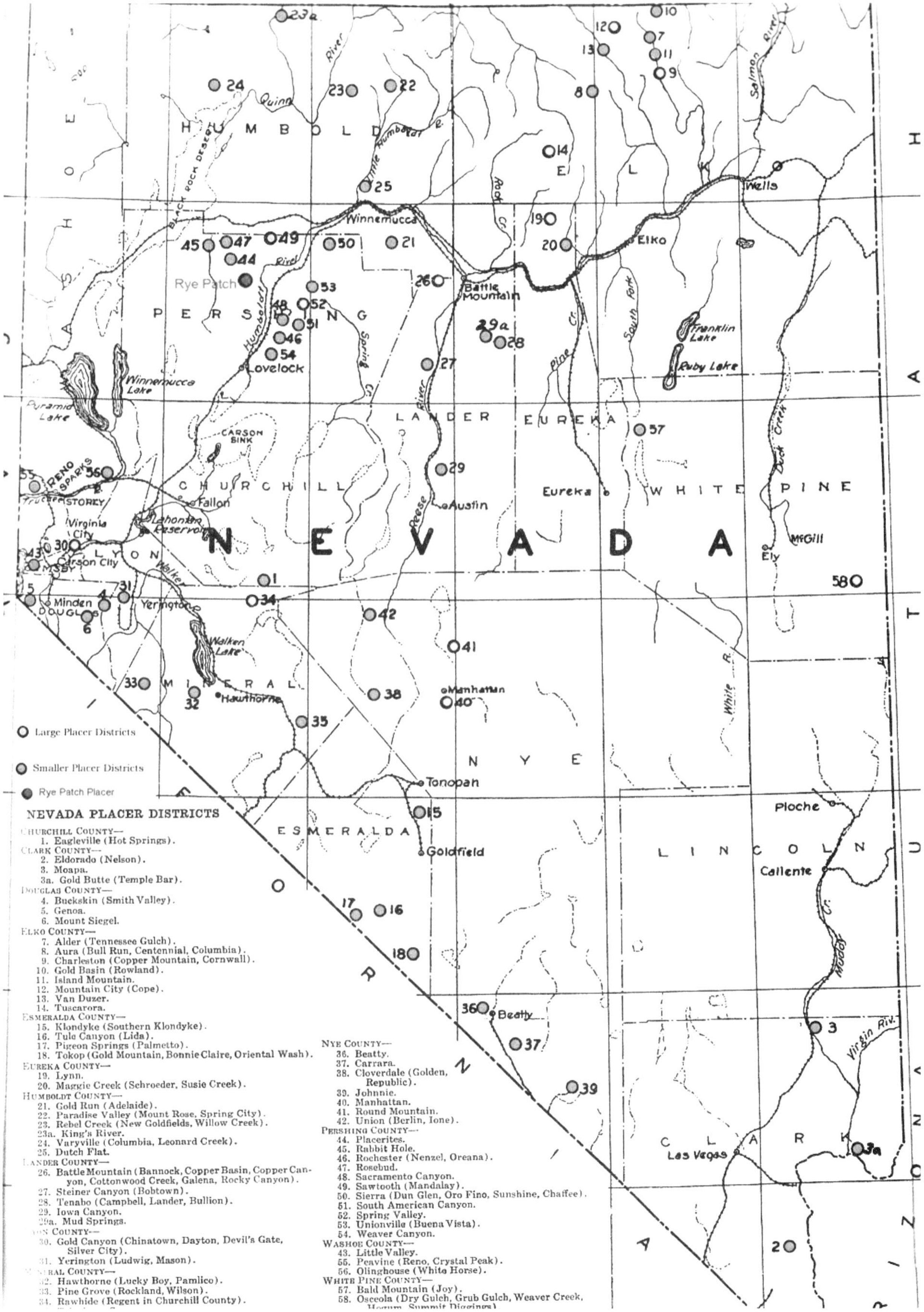

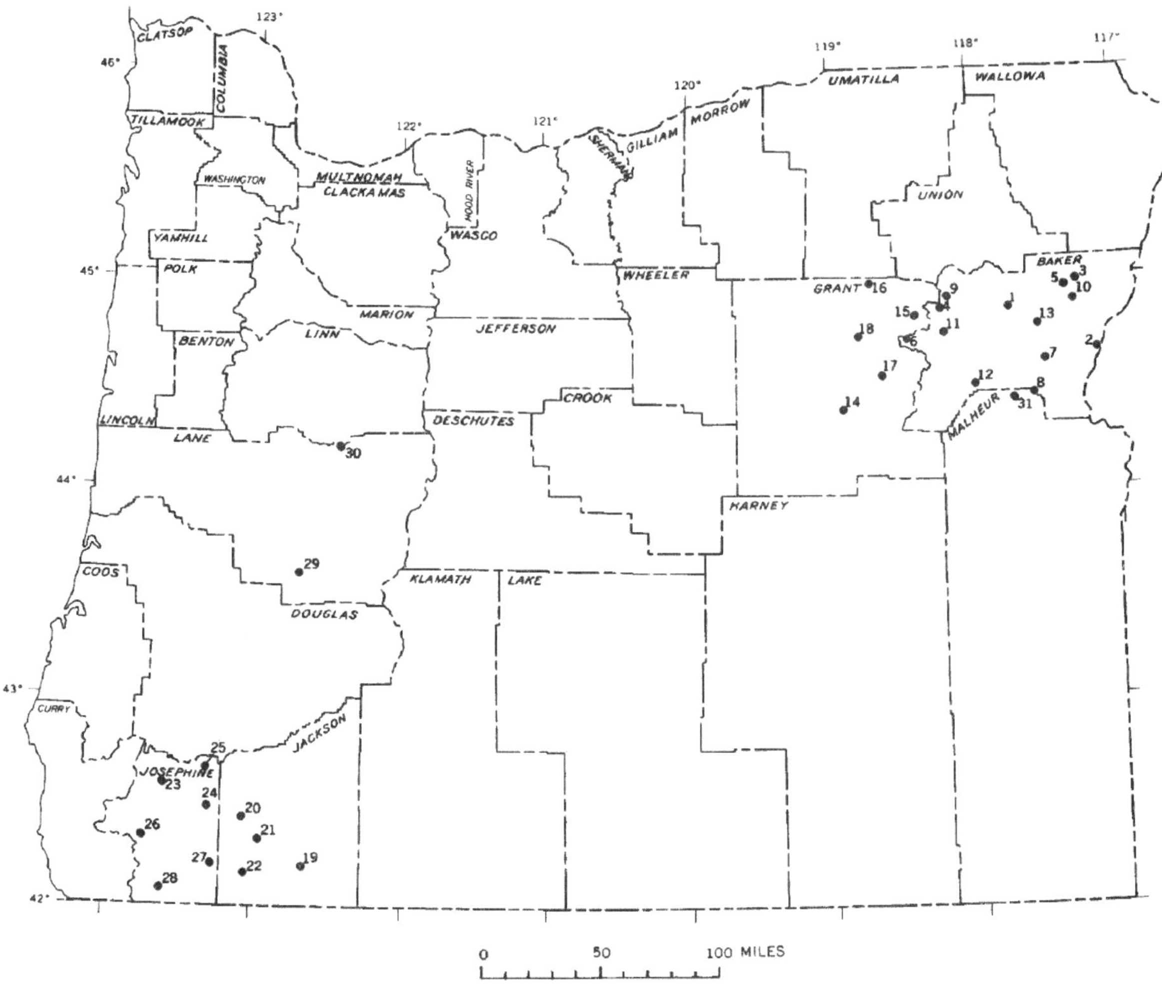

Major Oregon Gold Districts:

1) Baker; 2) Connor Creek; 3) Cornucopia; 4) Cracker Creek; 5) Eagle Creek; 6) Greenhorn; 7) Lower Burnt River valley; 8) Mormon Basin; 9) Rock Creek; 10) Sparta; 11) Sumpter; 12) Upper Burnt River; 13) Virtue; 14) Canyon Creek; 15) Granite; 16) North Fork; 17) Quartzburg; 18) Susanville; 19) Ashland; 20) Gold Hill; 21) Jacksonville; 22) Upper Applegate; 23) Galice; 24) Grants Pass; 25) Greenback; 26) Illinois River; 27) Lower Applegate; 28) Waldo; 29) Bohemia; 30) Blue River; 31) Malheur.

Although gold has been mined in other areas of the state, the most important deposits are found in two distinct and well-separated areas in the State of Oregon, one in the Blue Mountains in the north-east, the other in the south-west, where it forms the northern extension of the Klamath - Trinity gold belt of California. The northeast goldfield covers a considerable area, extending the length of the Blue Mountains, with a width westward from the Snake river of some 130 miles. It is associated with the great central mountain mass of Idaho, and it's placers were first discovered in 1862. By 1870 the richest placers were exhausted and a gradual decline in placer-mining set in. The gold fields of this state are located principally in the mountain ranges, such as the Coast and Cascade ranges, including the Blue, Siskiyou, Umpqua, Calapooya, Snow and Puebla mountains. Beach placers, arising from the action of waves on gravel cliffs, have been worked on many beaches south of Coos Bay, including locations near Gold Beach, Cape Blanco and Port Orford. The beach gold is all very fine in size, but some of these beaches were very productive.

Gold Mining Districts Of Utah

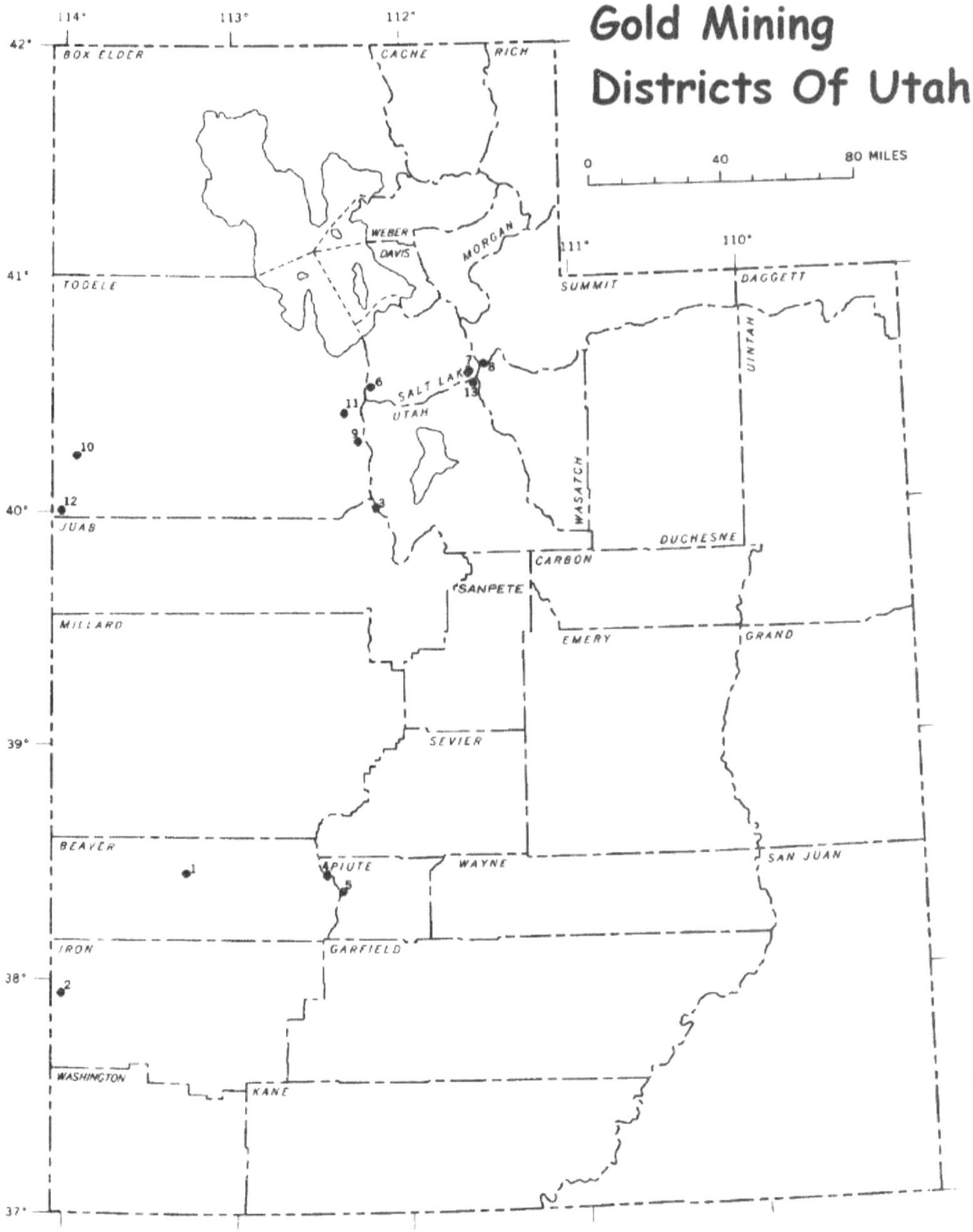

The gold Districts of Utah as shown on the map include: 1) San Francisco; 2) Stateline; 3) Tintic; 4) Gold Mountain; 5) Mount Baldy; 6) Cottonwood; 7) Bingham; 8) Park City; 9) Camp Floyd; 10) Ophir-Rush Valley; 11) Clifton; 12) Willow Springs; 13) American Fork.

Few veins in Utah were worked only for their gold content, and the greater part of the gold produced in Utah has come from copper-gold or lead-silver-gold veins. Most of the gold bearing veins occur at the contacts of sills or dikes of igneous rocks where they meet with sedimentary rocks. The placer deposits of Bingham Canyon were the most productive in the state and have been extensively worked, but much of this area is now covered over by the large dumps from the gigantic Bingham Canyon copper mine. Placer gold was also mined in a number of places along the Colorado, Green and San Juan rivers. These were skim bar surface placers of fine gold which form on inside bends.

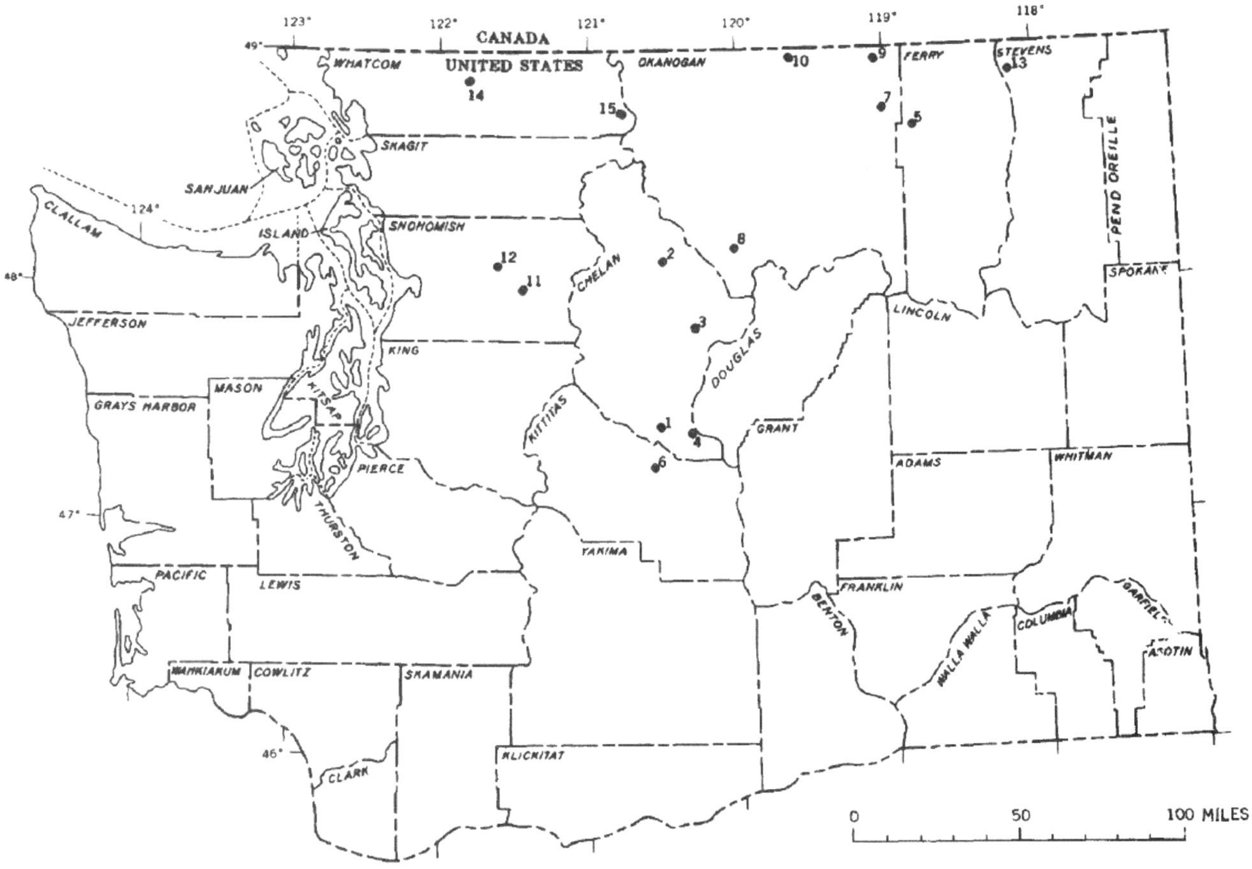

Principal Gold Mines Of Washington State

1) Blewett; 2) Entiat; 3) Chelan Lake; 4) Wenatchee 5) Republic; 6) Swauk; 7) Cascade; 8) Methow; 9) Myers Creek; 10) Oroville-Nighthawk; 11) Monte Cristo; 12) Silverton; 13) Orient; 14) Mount Baker; 15) Slate Creek.

Gold was first discovered in the State in 1853 in the Yakima River valley by a party under the command of Capt. George McClelland exploring for a possible railroad route. In 1855, small placers were found along the Columbia and Pend Oreille Rivers. Two main auriferous belts are found in Washington: a central belt extending along the trend of the Cascade Mountains, and an eastern belt, the southern continuation in Ferry and Stevens counties of the regions of the Rossland and Slocan districts to the north in British Columbia. The geological features are similar to those of the Coast range in British Columbia in the north, and to those of the Sierra Nevada in California in the south. From the 1850's through the 1880's, placers were worked along many streams in the State. Nearly all the terraces and bars of the upper both the Columbia and Snake Rivers in Washington State carry fine gold. The gold tends to concentrate at the heads of bars where the river bends. Most of the richer spots have been discovered and worked by long ago by the Chinese, but because of their shallow and superficial nature, they tend to renew themselves during times of flooding. These placer deposits cover a number of counties including Benton, Franklin, Wala Wala, Garfield, Columbia and others. The Washington Department of Fish and Wildlife publishes a booklet called "Gold and Fish" to provide details and definitions for use by gold prospectors that will be working in the rivers and streams of the state. Be sure to obtain a copy of "Gold and Fish", if you plan on prospecting here as you must carry it with you whenever you are out panning or dredging. The booklet also lists the classes of different streams and seasonal restrictions to certain activities on the various streams.

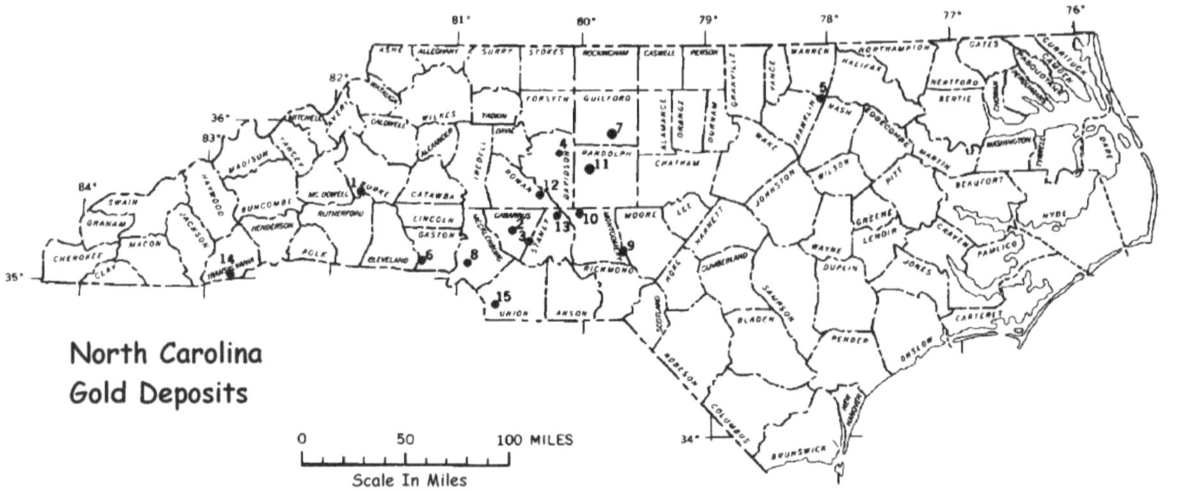

The Major Gold deposits of North Carolina include the following as listed on the map above: 1) Mills Property; 2) Phoenix Mine; 3) Reed Mine; 4) Cid District; 5) Portis Placers; 6) Kings Mountain Mine; 7) Garner Hill, Lindsey, North States and Jacks Hill Mines; 8) Rudisil and St. Catherine Mines; 9) Iola and Uwarra Mines; 10) Russell and Steel Mines; 11) Hoover Hill Mine; 12) Gold Hill District; 13) Parker Mine; 14) Fairfield Valley Placers; 15) Howie mine. More than 500 other properties in North Carolina yielded gold, but most produced only small amounts.

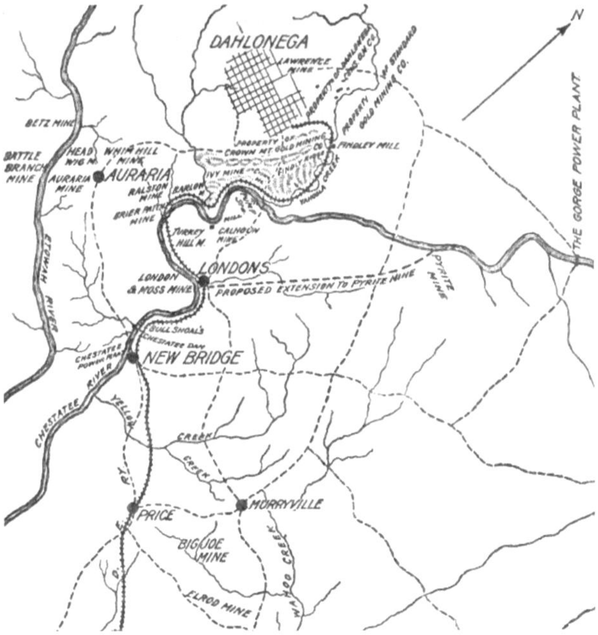

Georgia Gold Mines:

The zone of gold bearing rock begins in the northeastern part of the state and extends southwestward toward the Alabama line. The country rock is mica- and hornblende-gneiss and schist. The area around Dahlongega was the most productive and produced large amounts of gold. A number of old mines are located to the south, southwest and east of town. The mines on Findley Ridge to the south were especially productive. There are some tourist mines which are open to the public currently operating here. Gold was also produced in a number of other locations in Georgia along the mineral trend line.

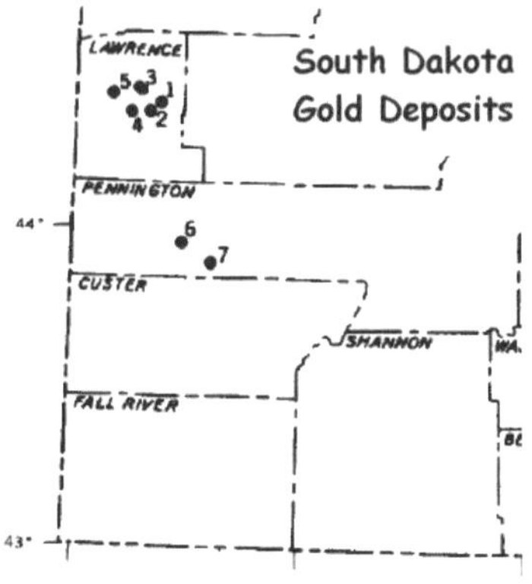

South Dakota Gold Mines:

The southwestern area of the state has produced large amounts of both placer and lode gold. The mines are concentrated around the Black Hills region. The gold districts include: 1) Deadwood - Two Bit; 2) Lead; 3) Garden; 4) Bald Mountain; 5) Squaw Creek; 6) Hill City; 7) Keystone

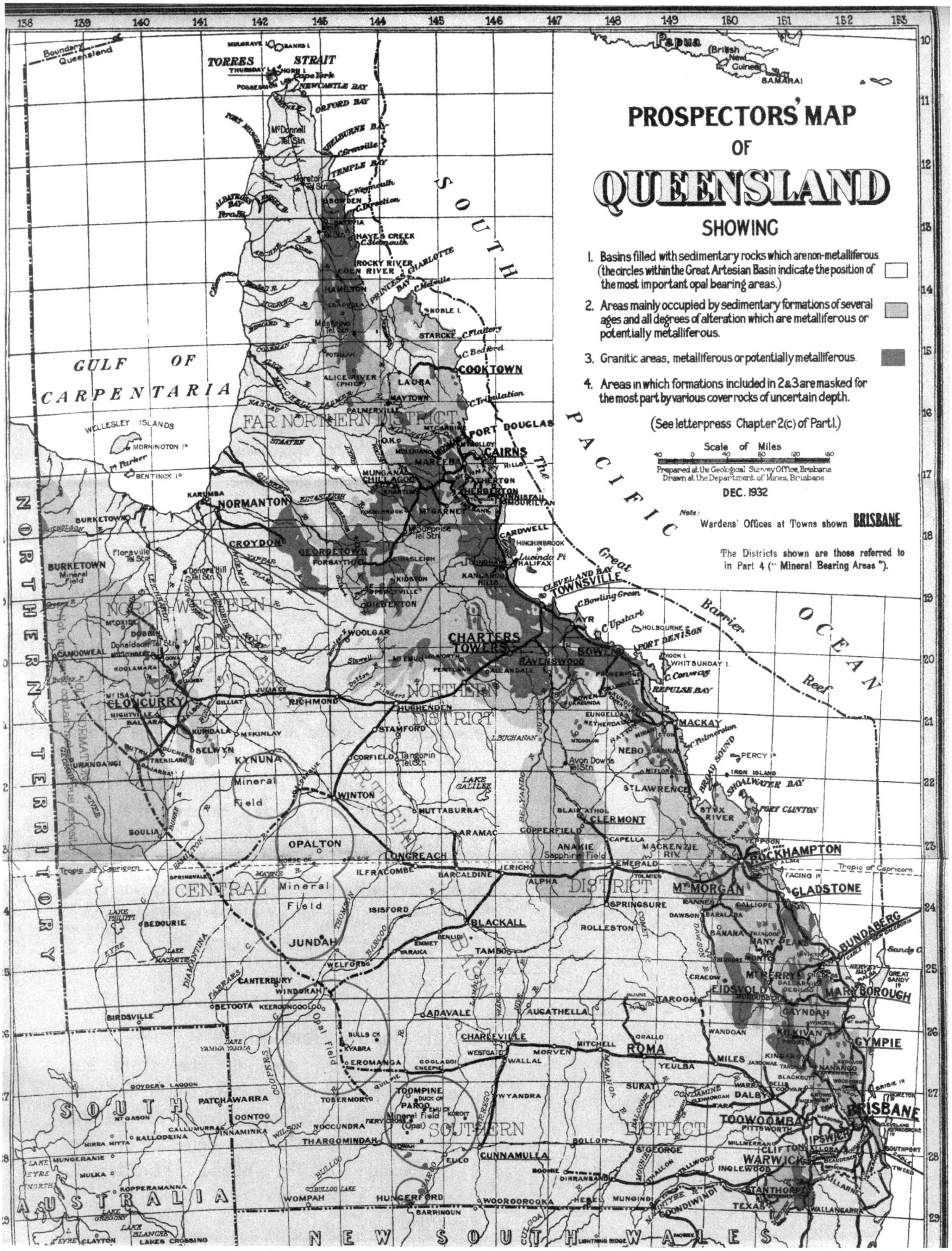

Prospecting maps exist for locations outside of the US – This is a map of Queensland, Australia

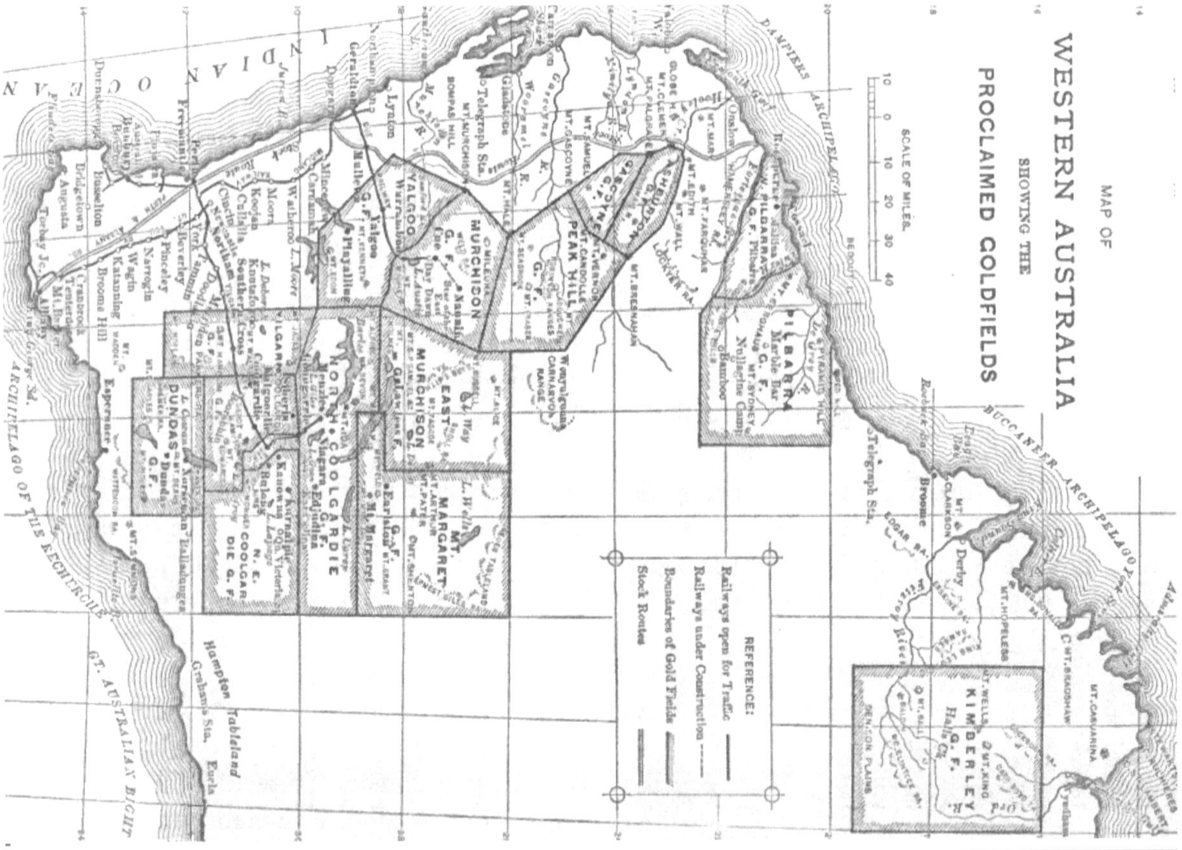

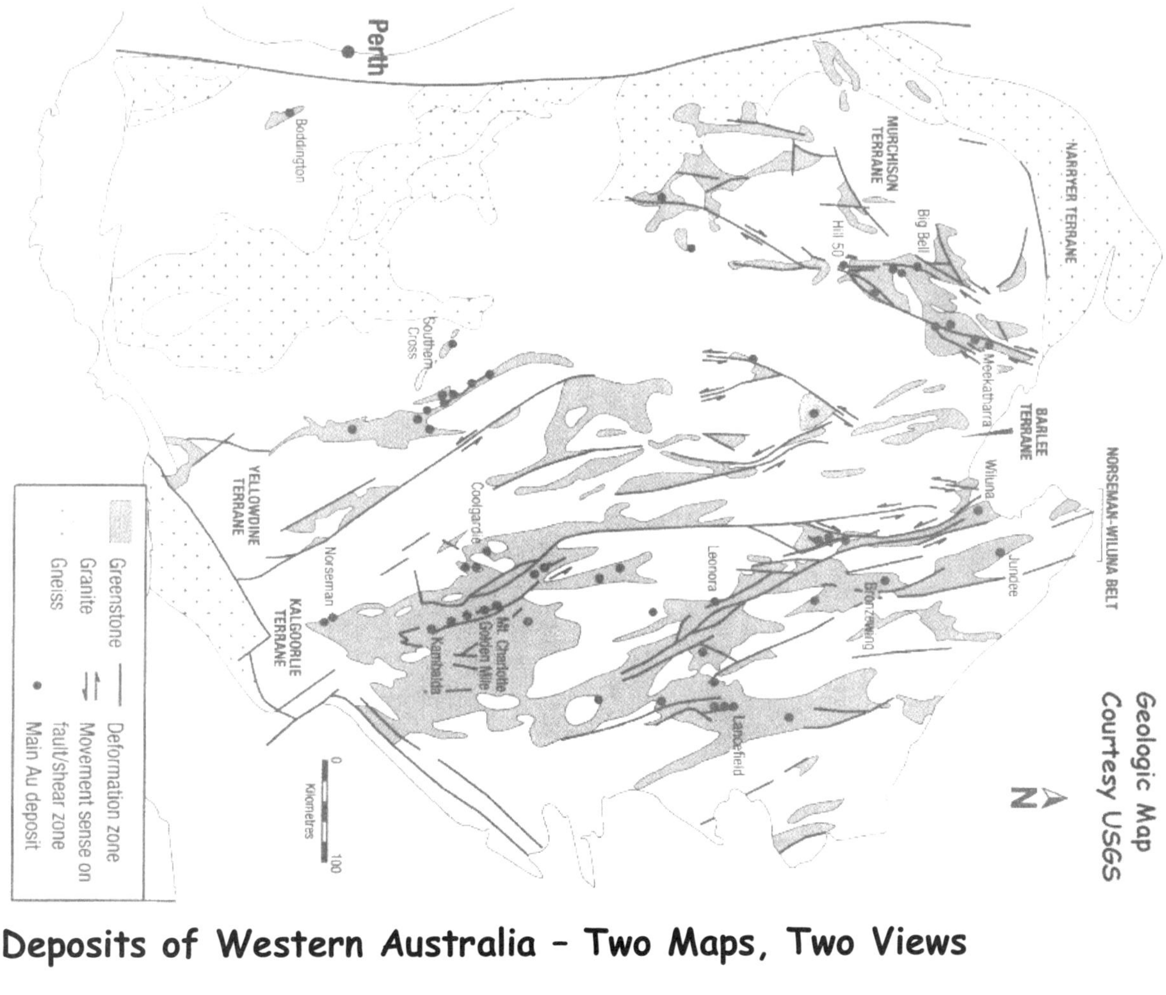

Gold Deposits of Western Australia – Two Maps, Two Views

Techniques for Finding and Working Gold Placers

Large nuggets are in many ways the holy grail of prospecting, so in this section, we will take a look at some hints on finding coarse gold in placer deposits, covering some thoughts, tips and hints not previously addressed in this book. Because of the large difference between the two types, I have divided the discussion here into two parts, first a discussion of residual placers and then a discussion about alluvial gold. I do think that residual placers, because we have tools like the metal detector that no old timer ever had, really do represent a great opportunity for finding some good gold. We will take a look at them first:

Residual placers occur away from streams, even on a mountaintop like this example.

Pocket Hunting for Residual Placers

It is a fact that many of the well-known placer districts of the western US have as their source of gold small veins which contain sparse pockets of coarse gold. This is the main reason why many rich placer districts have seen little or no hard rock mining, because these small veins were difficult for old timers to find and work profitably. While this was frustrating to old time miners hoping for large and profitable lode deposits, it presents a big opportunity to the modern prospector armed with a metal detector. Unlike the old timers, we are able to cover large areas checking for residual placers quickly and when found, work them efficiently. The hills and slopes above the gullies and washes in these small vein districts are excellent targets for the prospector armed with a dry washer or detector, as some of these small veins have formed residual placer deposits which are still sitting on the hill slopes, lying under a thin veneer of soil and clay - just waiting to be found.

Concentrations of coarse gold can occur by water action, but another common source of gold concentrations is the residual placer. These are formed as residual accumulations of decomposed gold veins. Residual placers come about when the original vein has weathered, but the placer gold remains more or less "in place" and at or still within a few feet of the original source. A related type of deposit, called an eluvial placer, occurs where the gold has traveled a short distance down from the source, but has not made it into streams and other drainages - these are often called hillside placers. The line between where you would stop calling a placer residual and start calling it an eluvial hillside placer is somewhat arbitrary, but in general gold at or very close to its source of origin is considered residual. Residual placers can contain from a small number of nuggets comprising less than an ounce, up to accumulations of several hundred ounces. By no means does every source of good nuggets form a valuable residual placer, but they are worth spending a bit of time searching for, especially when one does get into an area that is producing nuggets. This is because many of the really big multi ounce hits found with metal detectors that we hear about from time to time are actually residual placer finds.

The amount of gold retained at a residual placer site in part depends on the amount of gold in the original deposit, and partly on the area topography and climate. Often small broken pieces of quartz will litter the ground around residual placers on flat and gentle slopes. These are the remains of the vein in which the gold formed. Residual placers on flats do not form much of a dispersion pattern as the gold tends to simply remain in place. The gold from residual placers in flat terrain locations really does not have the force of gravity to move or carry it very far. Wind and rain may wash away clay or other light particles leaving a desert pavement of quartz, with gold possibly very near the surface. On moderate terrains a good amount of gold can remain at the original site for a long time, but the gold eventually moves downhill, forming a fan or V shaped dispersion pattern as it progresses down to the drainage below. Coarse gold tends to remain a long time at and very near the source, especially in flat and moderately sloping areas. On steep terrains there is usually little surface material, and it often does not remain at its site of origin very long. Good residual placers generally do not form in areas with very steep slopes. It must be remembered though, that what counts is the terrain at the actual vein location. Even in very steep mountainous regions, there are areas such as ridge tops and mesas where there are locally areas of very flat or moderate terrain. The size of a residual placer is also affected by the amount of rain or snow, which occurs in the area. In arid desert areas, there is often not enough water to quickly wash away materials on the slopes and deep weathering may occur. In these arid regions, wind erosion may move nearly as much material as water erosion - because only a little water falls. These types of situations allow residual placers to accumulate. It is no coincidence that most of the work detecting residual placers is done in dry climates. Placer regions with flat or moderate slopes, minimal erosion by water and sparse plant life can be found only in desert conditions. However, it is true that residual placers can be found in both wet and dry areas.

When working a residual placer, the origin point is the location most likely to have the coarsest gold. This is because it takes far more energy to move a 10-ounce nugget down a hill than a 0.1 ounce nugget. Under most conditions, long after the 0.1 ounce nugget has been washed down a slope into a ravine, the 10 ounce nugget will still be hung up near the original source point. The same concept holds true when comparing force needed to move a 0.1 ounce nugget (that's still a 2 dwt nugget) with the force needed to move a 0.001 ounce flake (roughly ½ grain). The sand and clay formed from weathering of most rocks moves down the hill even more easily because of their low density.

Prospectors with metal detectors are mostly seeking near surface deposits, which can sometimes be a problem in larger drainages, because larger streams often deeply bury the gold - beyond the reach of metal detectors. Luckily, deposits of a residual or eluvial nature often are very close to the surface where the gold has not been transported far from its original source. This makes residual placers not only a good source of coarse gold, but also a good match for the capabilities of a modern metal detector.

Many placer districts in the western US and elsewhere have no known specific source for the gold nuggets found there. In most of these types of locations, the gold has come from mineralized fracture zones and areas where a series of small veins have formed which contain spotty coarse gold as described above. When a number of these small veins erode and release their gold, the accumulation of nuggets can create large and rich placer areas. The old timers worked principally in the rivers, creeks and ravines, but for the prospector with a metal detector seeking to take advantage of the opportunities in locating residual placers, the hill slopes above those productive drainages can present a significant reserve of gold waiting to be tapped. Some prospectors that have sought and found residual placers have had spectacular success in locating significant quantities of gold by locating residual placers overlooked by the old-time miners.

Residual placers have not been river rounded and can produce nice crystal gold specimens.

Prospectors have long known of the existence of residual placers, and methods to find them were practiced during the California gold rush. The old timers realized that when the gold was sharp, coarse and angular that they were probably close to the source. They figured out how to trace the gold to it source using just their gold pans, and a little ingenuity. The method they developed is called tracing or pocket hunting (the equivalent Australian term is loaming). In many desert areas, this method can be difficult as the lack of water may require carrying the samples miles to the nearest water source so the samples could be panned - the process is far easier when the pans can be processed right on site. The modern prospector uses a metal detector to speed up the prospecting process and eliminate the need for large amounts of water.

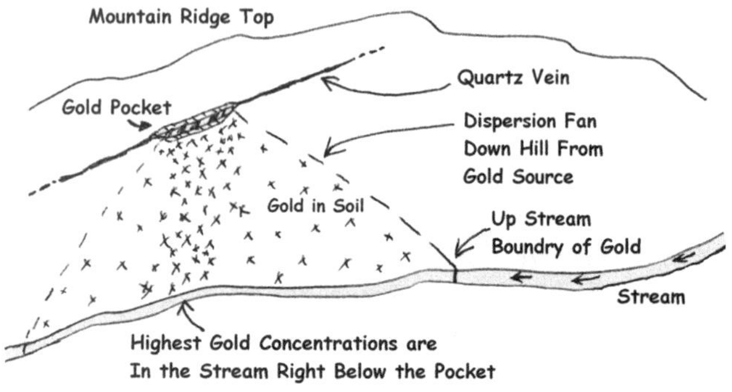

The diagram above shows the dispersion of gold downhill from a pocket on the hillside. Gold will spread out to form a fan shaped pattern below the outcrop of the gold source. Hill side gold can then be traced back to its source.

The methods for pocket hunting are based on the fact that, under ideal conditions, the pieces of gold will form a fan shaped, triangular distribution pattern as they move down hill from the original source. Taking samples from various spots down slope can tell the prospector if he is within the spread of gold particles moving down from its source, or if he has gone too far. The prospector moves uphill, sampling to stay within spread of the gold below the source, continuing upward until eventually the source is located. The down hill dispersion pattern would form a nice fan shaped triangle on a smooth, gentle slope - but real life hills don't work that way. The steeper the hill, the narrower the triangle of the dispersion pattern will be. The number of nuggets will also tend to be more concentrated near the source - as the gold pieces move down the hill they are dispersed and spread out. The number of nuggets in the dispersion trace will also tend to be the greatest along the center line of the triangle. In desert areas where the vegetation is not too heavy, the presence of float (loose) quartz can sometimes be used as an aid in tracing and pocket hunting. There is usually far more quartz than gold in a vein, and visible loose fragments of quartz can sometimes point the way, especially in more arid areas where dense vegetation does not cover the slopes (and the quartz as well).

The first step in pocket hunting is to identify a gold bearing area that may have pockets - usually a smaller drainage in an area known to have produced coarse gold in the past. The goal is to find a drainage with a single gold source, or at least one dominant source that is contributing the majority of the gold - too many different sources contributing gold to the mix makes this method unworkable. The best places to look for residuals are on flat to moderate slopes, with limited vegetation, where coarse gold is known to be present, with drainages that cover only a small area. The larger the drainage area, the more likely that multiple gold sources are contributing to the placer, and this makes finding any single one of them much harder. A great example of the right type of place to look might be in smaller drainages above old placer diggings on a hillside. Information pointing out these kinds of areas can often be found in old geologic reports.

The second step is to take samples along the drainage to determine the limits of the gold bearing area. In the past, this has been done by panning, but it is now possible to do much of the same work with a metal detector. Once the limits have been determined, the pocket hunter then takes some samples up the slope from the drainage to determine which side of the drainage the gold is coming from. If some gold is coming from both sides, the goal is to find one side that is the majority source and has much more gold than the other. If no significant difference is found in one side of the drainage or the other, it may well be that there are too many small sources contributing to easily identify any one source. If a single or dominant side of the drainage can be identified, the next step is to move up the slope of the hill taking samples to determine limits of the gold going uphill. Generally these are taken in a line parallel to the axis of the drainage being tested. Keeping in mind the typical triangular, fan shaped dispersion pattern for the gold, the prospector can trace the gold up slope. Good gold found within a few consecutive samples taken and panned, may indicate you are on the right track and within the downhill dispersion pattern of a residual gold pocket. Another line of sample is then taken further uphill to test the placer. The gold should become more abundant as the samples approach the residual placer "pocket" at the point of the triangle. These methods have remained unchanged for centuries. The techniques used by the California gold rush 49ers were described by Mark Twain in his book "Roughing It". A good example of where this type of technique might work is where you have found a good quantity of placer gold working up a small ravine and the gold suddenly drops off to nearly nothing. This is a strong indication that you should turn and look up the slope of the adjoining hillsides for the source - possibly a rich residual placer deposit.

Unfortunately, tracing (loaming) is limited in its opportunity - only certain types of residual placers can be worked by this method. The system works far better where one or two larger pockets are releasing a good number of gold pieces. Larger pockets of residual gold with a good number of nuggets are required to create the dispersion trace used to follow the gold back to its source. Small residual deposits with perhaps a dozen or so nuggets, don't have the quantity of gold to create a good, strong trace. Multiple sources can also make the trace unclear - and this is a very common situation. As previously mentioned, the small veins that give rise to coarse gold, commonly occur in zones or groups. When a large number of small veins are all releasing a small amount of gold, trying to find any single one of them is like trying to hear just one single voice when 50 people are screaming. In the same way that no single person can be heard in a scream fest, placers that have a large number of small sources cannot be traced by standard pocket hunting methods. If old timers have removed much of the gold - it may be hard to find where the gold deposits start and where they stop. Another frustrating situation can occur when the source has, for the most part, eroded away. There may still be something of a dispersion trace of gold left behind, but the big residual placer payoff at the origin point may be gone. These limitations are a big part of why the old timers had difficulty working residual placers. In spite of this, tracing does sometimes work, and prospectors in California, Australia and elsewhere have used it successfully.

Techniques for Locating Residual Placers with a metal detector:

Although the old timers did their pocket hunting with a pan, it can be done with a metal detector as well. The techniques are a bit different, but for the most part it is the same. There is little doubt that if the old timers had such a tool available to them, they would have placed a far higher emphasis on residual placer pocket hunting.

Small quartz veins like this one contribute a lot of gold to many different types of placers.

When tracing residual placers with a metal detector, smaller drainages are best, as they are less likely to have multiple sources contributing gold, so it will be easier to find a single source. Starting with a smaller drainage that is known to contain gold, if good gold has been found in a shorter section of the ravine, walk a series of sample lines parallel to the drainage up slope. This is similar to the lines of samples taken with a pan and can be used much like standard pocket hunting (loaming) to determine a possible down hill dispersion of gold from a source. The advantage of a metal detector is that it is possible to cover much more ground far faster, and so a number of sample lines can be taken in the time it would take to go through one or two pans full of material. Six to ten passes on each side of the hill perhaps 20 to 40 feet above the bottom of the ravine may be sufficient to indicate if this site might be a promising one for a residual placer. If you find a few rough pieces of gold up slope from the ravine, that is a very positive sign, and it is very important to note exactly where these were found, as they are critical in locating the potential residual placer up above. If you find rough coarse gold on a slope, draw two mental 45-degree lines upslope from the point where the gold was found. The origin most likely occurs within the "V" shape formed by the two 45 degree lines. As you continue prospecting up slope, stay within the two lines. As more gold is found, the 45-degree V shapes formed by each of these pieces limit the area in which the origin spot may occur. In some places, visible quartz can be used to help as a supplement to your detector in determining the dispersion of the gold. Quartz veins are more extensive than gold deposits - and more common, so don't fixate too much on quartz. Quartz is only a help in finding pockets, as gold may not occur right in the quartz vein itself. Your search is for the gold source. Areas on a hill with heavy iron staining can be the same situation. At times, panning can be used in conjunction with a metal detector, and can be helpful in locating the source when you are close. Although you may not find a significant residual placer at many locations, it is worth the time it takes to run a few test lines up slope any time you find a significant placer accumulation in a ravine or smaller drainage, because of the potential for a large residual placer find.

Mark Twain's description of the boom-bust nature and other frustrations of "pocket hunting" is very colorful, and well worth reading. Although he found it "the most fascinating of all the different kinds of mining" he commented that it also provided "a very handsome percentage of victims to the lunatic asylum". It's very true that when you feel like you are close to locating a rich pocket, but try as you may it just can't be found - that frustration can simply drive you crazy. In Twains day, a pan and shovel were the only real tools of the pocket hunter, and when they found a rich pocket they were on top of the world, but then for months they might be searching, close to a big find but never locating it. Pans can still be used, but the modern prospector has available a device which provides a capability the old time miner could never match. Metal detectors provide a very significant tool to use in the discovery and mining of residual deposits of "pocket gold". In the next section we will look at positive signs for locating residual placers using a metal detector and the methods used in mining them.

Field Techniques for Finding and Recovering Coarse Gold From Residual Placers

Residual placers represent an accessible source of coarse gold for modern nugget hunters. So armed with our new knowledge about the nature of gold deposits, let's see how we can take some of this scientific geology stuff, and make use of it in finding gold. Nuggets don't just occur without a reason, so maybe understanding something about the theory of how coarse gold forms can lead to some useful prospecting suggestions. To be honest, many experienced prospectors have already come up with many of the ideas I am about to present just by trial and error, but even the most experienced may find a few new wrinkles, and for many readers, these concepts may be very helpful.

The soil here has been stripped off the surface for its gold, with the rocks piled up by the old timers.

It is a fact that many of the well-known placer districts of the western US have as their source of gold small veins and shear zones which contain sparse pockets of coarse gold. This is the main reason why many rich placer districts have seen little or no hard rock mining, because these small veins and mineral zones were difficult for old timers to find and work profitably. These veins can however, leave residual placers with rich concentrations of coarse gold. The residual placers left behind by these small veins will be the focus of this section. We will go over a variety of techniques for finding and capturing the gold from this type of deposit. In the chapter on placer geology I mentioned seam diggings also known as saprolite diggings in the eastern US. These are deposits of gold in shear and other mineralized fracture zones. These are also important contributors to the formation of residual placers and need to be seriously considered by the prospector looking for residual placer patches. Residual placer patches or pockets can be very rich - it is not unusual that they may contain gold concentrations in the range from ½ to perhaps a few hundred ounces. While the process of locating these residual deposits was frustrating to old time miners hoping for large and profitable lode deposits, they present a big opportunity to the modern prospector armed with a metal detector. The old timers had a tough time prospecting for residual placers as they tend to be widely scattered. Unlike the old timers, we are able to cover large areas checking for residual placers quickly and when found, work them efficiently. The hills and slopes above the gullies and washes in these small vein districts are excellent targets for the detector, as some of these small veins have formed residual placer deposits which are still sitting on the hill slopes, exactly as they were formed - just waiting to be found.

The rocks stacked along this small drainage were carefully placed and set up like a retaining wall.

Finding residual placers in the field:
One of the most difficult problems for the prospector seeking residual placers is the field recognition of these deposits. The first thing to do is to put yourself in or near to a historically productive gold district, but what do you do once you arrive? Many new hands would like a simple rule that - if you see this kind of rock, start prospecting at that point. Unfortunately, it's just not that simple, but there are a number of encouraging and positive signs, which may point toward deposits. There are a variety of things that are both manmade and natural, which provide indications of gold. For the man made indicators, there are signs of previous workings, for the natural markers there are things like: rock contact zones showing areas favorable for mineral deposition; significant iron staining- which often results from the weathering of pyrite deposits and vein quartz accumulations - where quartz veins have been weathering. We will look at a number of these positive indicators, starting with the man made ones first.

Man Made Hints:
The sites of small old time mining operations can be of great interest to the prospector as the early day miners may have found what you are seeking long ago. The four most common types of small placer deposits which may have been worked by the old time miners include:

1) Gravel benches along rivers and streams, especially smaller streams.

2) Tertiary river channels high on mountain tops (these are found mostly in California).

3) Residual placers around hard rock quartz deposits and mineralized zones.

4) Small dry gulches and tributaries that only see water during heavy rains.

Because these old mining sites can become overgrown and hard to see, I want to spell out some of the signs that a prospector might look for in the field when searching for them. Some of the typical workings that mark these old mining sites include:

1) Ground cuts - These are basically the trenches in the ground from which ground sluicing gets its name. These narrow trenches carried the gravels to the sluice box.

2) Stacked rocks - in narrow and steep locations there was little room to move the rocks away. The old timers were forced to stack the rocks into walls alongside the stream they were mining. Sometimes these rocks are nothing more than piles, but either way, they mark the site of past placer workings.

3) Piles of rocks - In many locations the gravels contain rocks that are too large to pass through the sluice. There are no fines mixed in with them and most are dropped right on the bedrock. These cobbles were too coarse to roll through their sluice under the water pressure the miners had, so they were picked out by hand and tossed by the miners. These piles of coarse rocks last for ages and are easy to locate so they make an easy marker for the modern prospector to see. The quartz in these piles should be checked with a metal detector, as one might find gold-quartz specimen chunks in them.

Randomly stacked rocks in some old diggings.

4) **Areas stripped of their ground cover** and top soils are often left behind as the mining operations processed these materials for their gold content. Gold commonly concentrates along bedrock, and miners normally dig down to see what the bedrock holds. Look for barren areas among places with normal growth.

5) **Ponds and Dams** - Small scale ground sluice operations simply could not afford to bring water from long distances. Where creek or river water was not available close at hand, the miners built ponds to collect rainwater and snow run off. They were often constructed near the workings, so check the ground right below them.

6) **Dry Wash Piles** – In the desert where sluicing was not possible, dry washers were used to process the gravels. These leave distinctive piles of coarse and fine screened materials that are right next to each other. Sometimes the fines can get blown away or spread out to the point where they are hard to see, but the coarse piles are resistant to the weather and can be easily seen for a long time. If you see a series of short piles a few feet in diameter that cover an area which seems to have been dug out, you can be reasonably sure you have found an old dry wash site.

7) Potholes and Shallow Diggings:

There are certain wind and water related processes that occur in desert environments which will actually concentrate gold at or near the surface of an area. When the old time miners worked surface enriched placers, the diggings were often very shallow and appeared as dish-shaped potholes or other shallow diggings. In certain desert areas, these potholes mark the site of old timer workings.

8) Deeper Diggings: Adits and Shafts:

Miners sometimes dug deep shafts, adits or other underground workings to access the concentrations of gold along bedrock in gravel deposits that sit well above the water line. These old workings are not difficult to see or recognize. The dumps of these diggings can be recognized as different from those of a lode mine because in a lode mine the dumps will consist of angular broken rock which has been blasted, but for the underground placer workings the dump consists of rounded gravels which have been smoothed by rolling in a stream.

9) Hydraulic Mine Workings:

Where the old miners found large deposits of gravel which had been left behind by the processes of erosion, if there was sufficient water nearby, they would dig trenches to bring it to the gravel. The gravels were mined and run into a sluice box by the use of pressurized water. This kind of operation was known as hydraulic mining. One sign that an area has been worked by hydraulic mining are the steep banks left behind when the work stopped. The sites are also commonly marked by piles of stacked rocks which were too big to go into the sluice.

Natural Geological Hints:

Color Changes: In many districts, acidic mineral solutions have bleached the area rocks to a lighter color. This happens as acidic mineral solutions convert various rock forming minerals to clays and other minerals like sericite. These are the same solutions which can deposit gold if the chemistry is right. This usually results in large areas that show a major color change- almost always to a lighter color. It may appear as if the area has a different rock – but it's the same kind of rock, it's just that mineral fluids have changed the color and makeup of the rock. The same mineralizing action associated with this alteration can also create gold deposits that form placer deposits. In Nevada, it's not uncommon that quartz veins are associated with lighter colored altered zones in rock. These veins may have free gold suitable for location with a metal detector. At times, these bleached areas can be plotted out using satellite photos.

Old Time Yukon shaft access underground placer diggings dump. The dumps were sluiced for gold.

Iron Staining & Gossans:
Not all veins produce much quartz - gold bearing veins can consist of calcite or mostly sulfides - which often weather into iron-stained spots as the pyrites convert to iron oxides. Quartz is often a good indicator, but quartz is not actually necessary for the formation of gold nuggets. It is sulfur that is necessary in most cases. In places where gold occurs with little quartz, there is often quite a bit of iron present, which once (before it was oxidized) was some sort of sulfide like pyrite, etc. Because its sulfur that helps carry gold in solution, some sulfides of one type or another are almost always associated with gold. When the sulfides that are so commonly associated with gold weather from exposure to air and water, iron stained zones are created. Any gold that was formed in these deposits is left in the red soils that remain - or perhaps close by in the non-red soils. These red soils usually have a color ranging from deep orange to red-brown or even brick red color. These high iron areas are termed by geologists as "gossans". Any quartz that formed between the crystals of pyrite and other sulfides often shows the angular nature of the sulfides and the quartz in a gossan often shows what is called a "boxwork" pattern. Not all gossans contain much gold, but this iron stained "good red dirt" is considered as a very positive sign of mineralization in many gold producing districts and can be an important indicator of nearby residual placer deposits. When working a district where iron red soil is an important indicator, remember that with all soils, erosion makes things move downhill. Sometimes this red soil is hard to see as the red soils can be covered over with lighter materials that have moved down the hillside from above. Sometimes you will see the red soil exposed down below the outcrop that is creating it, as it has moved downhill. Always check to see what is up slope on the hill.

Quartz Vein Matter Accumulations:
At times, small accumulations of quartz vein material can indicate mineralization in the area. Gold often forms with quartz, so quartz veining can be a very positive sign. Quartz is very resistant to weathering and no heavier than the typical soil, so it may hang around on the surface long after any nuggets have sunk below the surface of the soil. If the quartz on the ground is showing as a

Accumulations of quartz matter like these white pieces on the ground may indicate gold is nearby.

visible vein outcrop, that can be a good sign as well. The outcrop as well as any areas downhill from it should be checked. On rare occasions, gold can be traced (or loamed in Australian terminology) based mainly on visible exposures of quartz. The fact that not all gold deposits have abundant quartz means you don't need to see hundreds of pieces of quartz littering the ground or an outcrop sticking six feet above the surface to recognize that an area can be attractive for prospecting. Small concentrations may well be worth checking out. As few as a half dozen or perhaps 10 visible quartz chunks in a small area may be an excellent indicator if there is not much quartz in the surrounding area. On the other hand, in places where there are hundreds of pieces of quartz littering the ground, a few pieces may not mean much. The amount of quartz present has to be considered in relation to the general area. Also remember that not all vein quartz is white in color. There are some quartz veins that are darker - some are even black. Quartz heavily stained red-brown by iron is also common. Sometimes veins don't show much of a color difference from their surrounding rock, even where the vein is mostly quartz. In Nevada, there are places where the veins consist of silicified country rock - they are not pure quartz and do not look like pure quartz. There are times where quartz outcroppings may only poke through the surface in places - the vein may be continuous under the soil cover, but it may not crop out above the surface. The non-outcropping places may be good places to check as well.

Productive Rock Types:

The concept of favorable host rocks is an important one. However, the types of rock which constitute "favorable" vary a lot from one location to another and can be significantly different. There are definitely favorable rocks in both California and Nevada, but they are different. Other locations such as Alaska or Australia also have their own favorable host rocks. The prospector or geologist learns what these favorable rock types are for each specific area in which he or she will want to search, and knowing what these favorable types are is very important (searching in areas with very unfavorable rocks can be a big waste of time). In some places it will be schist or slate in other places its volcanic rocks and in some its granitic rocks. Rarely even sedimentary rocks are the favorable host. You will just have to determine in your research which is the favorable host rock in the area for which you want to prospect.

Rock Contact Zones and Faults:

Many quartz veins and other hard rock deposits occur in "zones" along faults or at the contact of two different types of rock. Contact "zones" can be a few yards to a mile or more wide, so don't think of these as being just the actual point where the two different rocks are in contact. Also, at times the mineralized zone related to a rock contact does not occur right on the rock contact, but may be nearby, perhaps hundreds or even thousands of yards on one side or the other.

Different rock contact zones occur for several reasons. The two types most interesting to prospectors occur by intrusion and faulting. Intrusion occurs where igneous or volcanic rocks are squeezed (intruded) into existing formations. The heat from these volcanic or granitic rocks can cause mineralization in rock nearby but not always at the contact. It is not unusual that these rocks intrude along pre-existing fractures and faults. The other example is when movement along faults brings two rock types together. The broken zone of a fault can provide a conduit for geothermal solutions coming upward to the surface. In that case, fault zones may lead to the formation of quartz veins. Faults are a sign of significant geologic activity. They mean that there has been rocks moving and displacement of rocks, often bringing different rocks together. Though they represent significant geologic activity, faults are not always visible on the surface. Sometimes fault zones can be buried and covered by soils eroded down over them. Geologic maps can be very helpful in locating both rock contacts and fault zones. Sometimes the effects of faults can be seen on topographic maps where very steep scarps are present along mountain fronts.

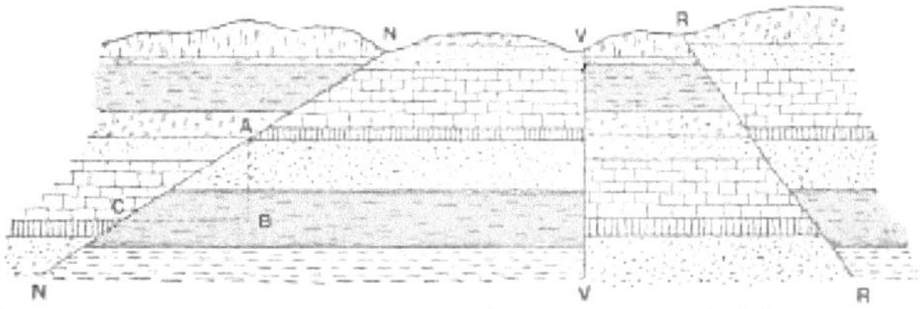

FIG. 2.—Cross-section showing normal fault (*N*), vertical fault (*V*), and reverse fault (*R*). *AB*, downthrow; *BC*, lateral throw.

Fault diagram showing various types of faults

Contacts between two different rock types can be a real good indicator, but the productive types of rock do vary somewhat depending on the geology of the area you are in. For example, the contact between granites and schist can be a real good one in Arizona and elsewhere, but in the mother lode country of California, serpentine/slate has been the most productive rock contact. In California, that combination has produced millions of ounces of gold from both underground veins and placer deposits. You just have to know what is good in your neck of the woods. Its also important to remember that different rocks in contact don't always have a huge color difference, and you may have to look closely to see the two types - geologic maps can be helpful in that type of situation.

Correct Topography:

As a general concept, the coarser gold does tend to hang up farther upstream. But placer geology is just not that simple. If the upstream is too steep, the gold (both large and small) will travel downstream until it can find a place where the water slows a little and it can hang up. Fast water in real steep drainages can push even large nuggets along. Most of the best residual placers form in areas with moderate to flat slopes, again because in places that are real steep, the gold doesn't have a chance to sit and accumulate before its washed down into the drainages.

This shot shows the mild topography around the residual placers at Rye Patch, allowing light materials to be removed and gold to accumulate.

Extensions of Known Mineral or Placer Areas:
Other than pipe shaped bodies, most small scale gold deposits have a linear component. It is fairly common that new deposits can be found along this linear zone of deposition by looking for extensions along the line of known residual patches and similar deposits. The concept is that if a contact or fault zone has produced at canyon A, next door in canyon B, where the same contact is exposed may be good too. Sometimes these may be close to the known deposits, but sometimes they are out a ways, even up to several thousand feet from the known patches. Rock contact zones also commonly have a linear component. When out in the field, be sure to note any specific orientation to the existing diggings. At the Rye Patch placer area in Nevada, the general mineralized zone runs in a NNE-SSW direction. However, within that larger zone, individual residual placers may have a different orientation. One that I have worked (along with others) has a more east-west orientation. A key concept in recognizing this is checking a series of detector dig holes that tend to line up in a specific orientation. If most all of the holes line up in the same general direction, you can check out beyond the ends of this series of holes for extensions of the patch. This happens because a residual placer patch may stop yielding gold for a certain distance, and then pick up again farther out along the line of deposition.

Another good area to search is along quartz outcroppings, including covered areas between exposed croppings and beyond those exposures, along the same general line. Commonly a series of small veins can be found throughout wide contact or fault zones that tend to have a linear direction. If you can figure the trend direction of the mineralized zone, you may be able to find another deposit of gold along the trend of that line. This is not always applicable but worth trying. Good areas and new patches can be found by extending the existing, known placer areas to adjoining sites. Sometimes the extension is not end to end with known deposits, but parallel to them. It can be worthwhile to look for parallel zones near patches and other known deposits.

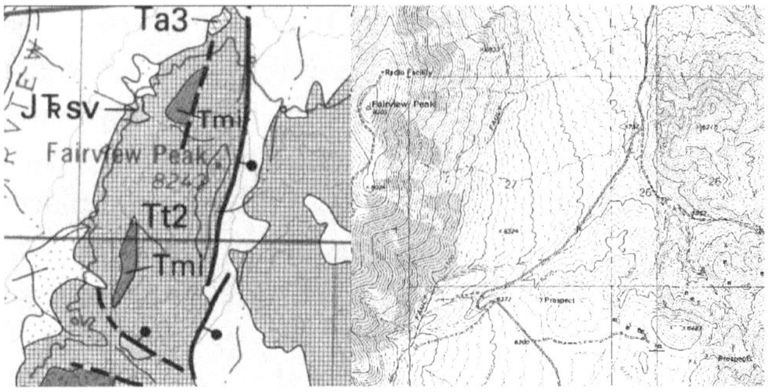

This geologic map of Northern Nevada shows that gold deposits are associated with fault zones.

Similar Geologic Areas Nearby:
If a certain rock type has been productive for gold in one area, if the same rock type occurs a few miles away in the same mountain range, it may well be worthwhile to investigate. Searching for areas with similar geology to productive locations can be a great way to locate new areas. If a contact area between two rock types is productive in one place, and a similar contact between the two rocks occurs nearby, again it would be worthwhile to investigate. This is a case where geologic maps can be very helpful.

Know Your District

The geologic concepts and indications that are the most important for finding gold vary from location to location. What works well in one district may not always work so well in another. In one place a certain type of rock may host all the deposits. In another district the most important may be areas colored red by high iron concentrations. Perhaps the most important thing in prospecting is to know the peculiars of the district you are hunting! Like sports, there is a home court advantage for prospectors who have learned the particulars of the district in which they are working. Studying geologic reports and other information will help you know what conditions are the most productive in your area, but perhaps the best way to gain knowledge of a certain area is to spend significant time prospecting there.

Getting the Gold Out of Residual Placers:

Once you have found a residual placer, the question becomes how best to recover all the gold it contains. When you've made a good strike, it's important to get whatever gold is there. There are a number of techniques for mining residual placers with a detector, but one of the best techniques is called gridding. Rock or other markers are used to lay out a grid on the ground over the patch. Be sure to extend the grid at least a short distance beyond limits of the residual placer patch. The grid sets up imaginary lines that you will walk while swinging the detector. The lines are set up so that the ends of your swings will overlap - leaving no area undetected. Lines are also set up perpendicular to the first set so that if you walked the patch with north-south lines, on the other set of passes you will be walking east-west. As in the first set, the lines need to overlap at the edges of the coil swings, so no areas go undetected. This double pass system is necessary because smaller and deeper nuggets may respond much better when you are swinging the coil one direction vs. another. This is largely because many nuggets are oddly shaped - so the detector sees the nugget much better with the coil head moving in one direction than another. This situation can also be created by nearby hot rocks and other soil effects.

Another important thought is that some nuggets may respond differently to different coils. In a

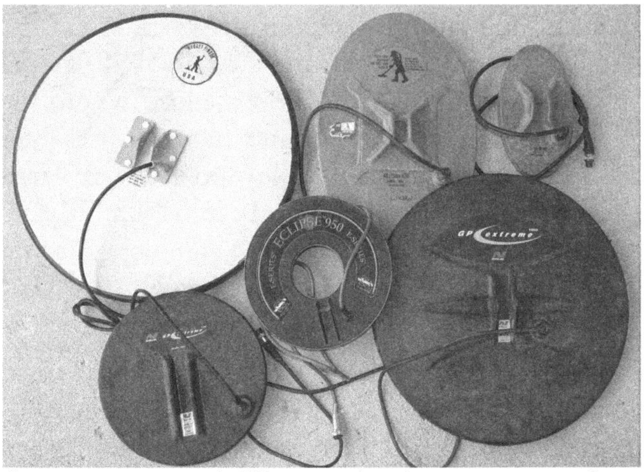

An assortment of different sized accessory coils can be very helpful in getting all the gold present out of a nugget patch area.

very productive patch area it will be worthwhile to conduct the gridding techniques described above with both large and small coils. It may be worthwhile to go over the patch in both directions with a number of different coils, both large and small if a selection is available. Where a good patch is located on a slope, you will want to follow the dispersion of any gold downhill to get the gold that has eroded out of the patch. This technique also can be used to produce gold at the times when you may come across another person's residual placer patch. Usually, these spots that have been worked before are worth going over at least briefly as the discoverer may have picked up the easy gold, but more may remain. Try larger and smaller coils - you may find that the other person did not work the spot with a full selection of coils. Also, if it is located on a hill side, you may be able to pick up nuggets down slope from the patch where they are progressing down into the local drainages by the long process of erosion. Remember that rich residual placers tend to form a triangular shaped dispersion pattern as they move downhill and that the richest concentration of gold is toward the top of the triangle near the source. Start your search within that general triangle, and follow it upward.

Another technique for extracting all the gold possible out of your patches is to scrape and dig down the surface to effectively "renew" the patch. After you have pounded and re-pounded a patch over and over again, eventually even in a really rich patch everything your detector can "see" has been dug. Even the very best of modern metal detectors have a limited depth range, and it is not uncommon that the gold bearing mineralized zone extends well below the depth of that maximum capacity. So where there is still easily moved weathered soil extending down below the depth range of your detector, it may well be worth the work to dig down. Like hard rock deposits, residual placers extend downward into the ground. If you bring out a dozer and scrape off a foot or so of soil from that patch area, the patch is in part renewed, and deeper nuggets that were formerly beyond reach can now be seen by the detector. For small concentrated patches, this could possibly be done by hand with a pick, shovel and wheelbarrow. That's the reason for many of the dozer scrapes that are seen at Rye Patch, Nevada and I've seen pictures of the same type of thing in Arizona as well. This technique works especially well on gentle slopes and flat areas. It is even possible that in places with a good percentage of fine gold that you may wish to dry wash or even pan the materials dug out of a residual placer patch. This would be especially true of very rich placer sections. The concept of a nugget "patch" is commonly used to cover small deposits of many types. Next we will focus on water formed alluvial placers and the patches that are formed in this type of environment.

Recovery of Coarse Gold From Alluvial Placers

The last part of this chapter will look at methods for finding and recovering coarse gold from alluvial (river and stream) placer environments. Unlike the residual and eluvial placers that we have considered previously, the force of flowing water is the most important factor in gold deposition at alluvial placer sites. The alluvial environment is where the gold naturally moves after the forces of erosion takes it down the hills and slopes from the residual and eluvial placer sites. Gold that has been in the alluvial environment for a while becomes rounded and smooth. These are the types of locations which have yielded the most gold to the old time prospectors and they are the first places where the old timers looked and worked the hardest. However alluvial environments still produce good coarse gold for many who prospect them.

Color changes on the ground are often a strong indicator of mineralization, which is often an indication of the presence of gold.

There are a number of different environments that are considered as alluvial placers and they include:

1. **River placers** - True alluvial river placers are found within the high water lines of the river when it is at its maximum flood stage. This means these placers are subject to deposition of new gravels and renewal of gold deposits during flooding.
2. **Desert stream placers** - these are alluvial placers as well, though water does not usually flow year round and in fact, water may be present only for short periods of time after "Flash" floods. These desert streams are often choked with sands and gravels for lack of water to wash these lighter materials downstream. Caliche (limestone or clay hardened gravels) often forms a false bedrock for desert placers. Up in the higher reaches of desert streams the bedrock may be exposed or at least near to the surface and these types of places are popular with detector operators for nugget shooting.
3. **Bench placers**- placer that formed in an alluvial environment, but erosion has left these gravels stranded above the reach of the current river even at its maximum flood stage. Although these gravels are perched on benches that are above the modern river, they still reflect its current drainage and direction. They are commonly formed at inside bends where a river turns, but may occur at other places as well.
4. **Fossil Placers** - These are placers that formed in an alluvial environment, but because of erosion, are stranded high and dry above the water line and no longer in the river system. The difference between fossil placer and bench

gravels is age - bench gravels are old but not nearly as old as fossil river systems which may be millions of years as compared to tens of thousands for bench gravels. The direction of fossil placers is often completely different from the modern river system. These are often mined as underground operations and the gravels may be well cemented, requiring explosives to extract - a well known example is the Ruby Mine in Sierra County, California.

5. Ocean Beach and River Delta placers - these types of deposits don't usually have coarse gold - they are located too far downstream. They are most often known for fine gold flakes and dust. The coarse gold hangs up in the mountains before it gets down to this point. Coarse gold occurs in these types of placers only if the primary hard rock source happens to be fairly near to the coast. Nome, Alaska is an example where fairly coarse gold was found on and near the beach, but even at Nome, most beach gold is tiny in size.

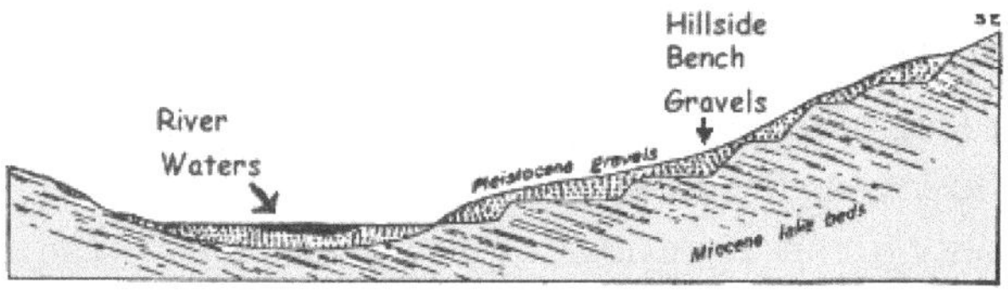

Bench gravels are an important source of alluvial placer gold all over the world. They are gravels left behind during erosion of a stream or river bed. They can be very rich in gold.

One of the most important factors in finding coarse gold is to be prospecting in a location that produces coarse gold. Even within larger areas known for coarse gold, there may be some individual streams, which only produce smaller sized gold, just because of the nature of the local geologic environment found there. In this chapter, we have designated nuggets weighing a pennyweight or larger as qualifying as "coarse" gold. In the search for coarse gold your best chances will be found in places where coarse gold was found previously by the old-time miners. As a result, the type of gold found in the past becomes an important part of any research to find new areas for prospecting.

Where it was physically possible, the early day miners often diverted the river in order to mine the gold which was normally under water.

Because of its high density, gold works its way down through the river gravels quickly in the alluvial placer environment. Because of the effect of buoyancy, under water, gold is more than 10 times denser than typical rock. The watery alluvial environment allows for fast sorting of the heavy coarse gold. Small flakes and gold "dust" may stay in suspension during flooding and be deposited throughout the river gravels from top to bottom, but coarse gold almost always rapidly works it way down to the lowest possible level. On bedrock, this will be the cracks and crevices where it often becomes tightly lodged and stuck. In fact, gold gets wedged in with cemented, compacted gravel that sometimes must be chiseled out. Slide hammers work well to open up packed crevices. Sometimes special care is required to insure that large nuggets are not damaged in the course of extracting them from their bedrock home. Gold does not shake or settle through gravels once they are deposited and in place, it is deposited at the bedrock when the gravel bar forms. Coarse gold nuggets move only when the river is at a high flood stage and the force of water is roaring through the river channel. During these times, the river sounds like a freight train, rumbling and thundering as boulders bounce along the bedrock. While the river is in flood mode, gold moves more slowly than normal gravel as it takes so much more force to move it. This slowness allows the gold to work its way downward to the bedrock or any false bedrock such as a clay layer or caliche in the desert areas.

To get the coarsest gold, you often must be able to get down to the lowest crevices in bedrock. In general, gold does not just sit on top of bedrock, but works its way down into cracks and crevices where it can get out of the flow of the river. In active, flowing rivers, this may well mean that the coarsest gold is located in the places where the water is deep or fast moving. Although normally the coarsest gold will be found on bedrock, at times coarse gold can hang up on a false bedrock that resists the force of the river when it is at flood stage. These layers are often called flood layers, clay layers or caliche, and can be very important to recognize. There are many locations where prospecting by working down through the deep hard packed flood layers simply will not pay for the time it takes to do so, but cleaning off the top of the flood layer (which acts as a false bedrock) will allow the dredger to greatly increase his gold production. Not all streams flow year round, and a lack of water may lead to large amounts of sand and clay that may bury coarse gold deeply. The grade of a river as it flows is also important in determining where gold will deposit - steeper grades generally lead to faster water speeds which may prevent gold from forming good deposits. Gradients too low may lead to gravel filled valleys with the best gold deeply buried under a thick layer of overburden. The deeper the overburden the more likely it will be that there are flood layers part way down acting as false bedrock that hold the gold.

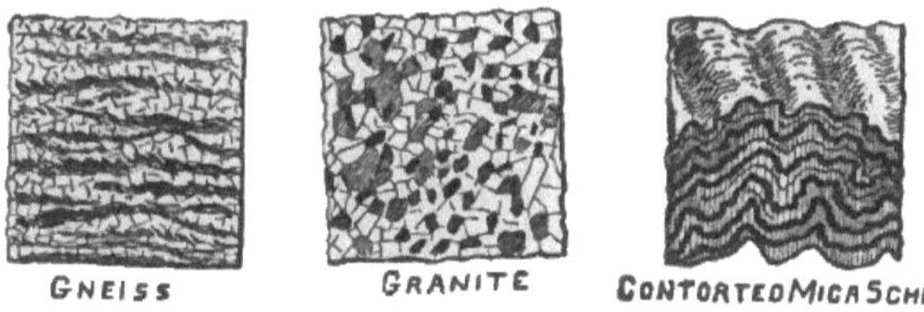

Generally speaking, schist is much better than granite or gneiss in forming the very best gold catching bedrock crevices.

Not all types of rock are created equally when it comes to forming crevices and trapping gold. There are various kinds of rock that give rise to deep crevices and other kinds that do not usually form deep crevices. Limestone often forms deep pockets and other crevices that trap gold. On the other hand, granitic rocks tend to form smooth rounded wear surfaces that have only a few small crevices along the joints in the rock. Other rocks can be poor to great crevice formers - and this may vary along the length of a river. As an example, slates can form poor to fantastic gold trapping bedrock, but it partly depends on the direction of the flow of the river with respect to the foliations in the slate. In one place where the river flows across the slate it may be good, but another where the foliations go parallel with the direction of the river, the bedrock may be smooth and prove to be a poor gold trap. An onsite field examination is important in determining how a certain bedrock may trap gold. Often where good crevices can be seen along the river above the water line, they can be followed down into the river, beneath the gravels and into the parts of the river where coarse nuggets may be found.

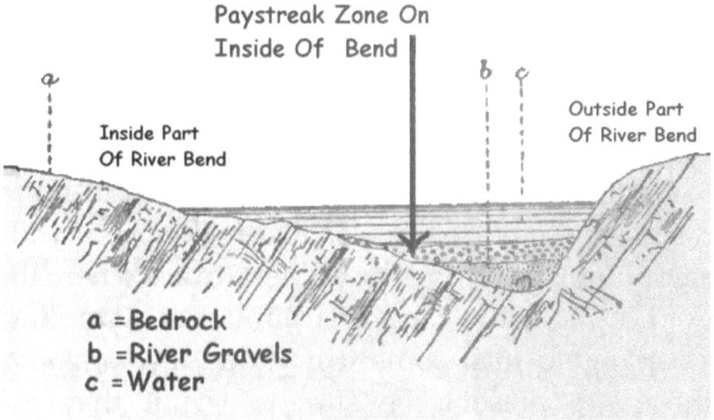

This is a section through a gold bearing River, showing the ideal location for the paystreak zone.

In many places, especially on bends in the river, the majority of any coarse gold tends to follow a certain common line or zone as it travels down the river. Where the speed of the water may slow down, or the gold has a chance to drop out of the flow along this line of travel, it will tend to accumulate. This leads to deposits forming behind large boulders or other outcroppings of bedrock. This effect also forms what is commonly called paystreaks in the river where there are larger areas where the water slows. Gravel bars of coarse cobbles often mark where slowing occurs and are commonly found on inside bends of streams. Where gold does deposit in a gravel bar, it tends to be coarsest at the head and outside parts of the paystreak. This is a tendency and there are many exceptions which can alter the "normal" situation. In wider, long straight sections of a river the gold will commonly spread out, and may have no clear common line of travel. One of the most important things in reading a river for potential placer concentrations is to be able to see the most likely lines of travel for gold as it moves downstream. These paystreak deposits are found not just in the modern rivers but in older bench and fossil placers as well.

Selecting the best spots for alluvial coarse gold deposits involves putting all these factors together, reading the river and then sampling to see what is actually there. Even someone who is very experienced at reading river placers cannot see what all the effects of water and boulders are at high water flow. So it is necessary to sample around to confirm and see what is present - if a spot is not paying, move! It is not unusual that you may find that a really good-looking spot has already been dredged. Sample elsewhere until you find a productive deposit, and don't settle in until a productive deposit has been found. It is also very common that the operator will select what seems to be the best spot and just settle down to go as deep as possible to see what bedrock holds. While sampling, it is important to check to see where the gold is actually coming from when sampling. Coarse gold may come from flood layers, and if those are present, be sure to see if they are producing gold as it is common that it may not be worthwhile to go deeper, especially where the actual bedrock is buried deeply. Sampling is probably the most commonly over looked technique that small miners forget to perform, in spite of the fact that this causes a considerable unproductive work. We have discussed this topic quite a bit and I have gone over these concepts in the placer geology chapter. This is just a review. Even at small hard rock gold mines, failure to sample and find the best part of the deposit is one of the most common causes for failure. If you take nothing else from this chapter - remember that sampling various areas to find the best part of any placer gold deposit is a critical key to finding good gold.

Metal detectors are an important tool for locating coarse gold in alluvial environments, especially where there is exposed bedrock or bedrock that comes close to the surface. They are critical for finding coarse gold in places where water does not flow year round or in prospecting old benches and other gravel deposits that lie where the running water no longer flows. Suction dredges are the other main tool for recovering coarse gold in alluvial environments. Deep holes in the river where the water speed lessens and gold can drop out are sometimes favorable places to look with a dredge, but these deep holes require diving equipment and special techniques for deeper water when one exceeds 30 feet in depth. It is not always necessary that the dredger work down through large amounts of over burden, in many gold bearing streams only a few feet of overburden may lie on top of bedrock. A dredge allows the prospector to work down through the gravel overburden, but a common and unprofitable mistake is to try to work excessive depths with a dredge that is too small for that deposit. An old saying is that the maximum practical overburden depth that can be worked by a dredge is roughly its hose diameter in inches converted to feet. On that basis, a 2.5 inch dredge can reasonably work through about 2.5 feet of gravel over burden, a 4 inch could work through about 4 feet of over burden, etc.

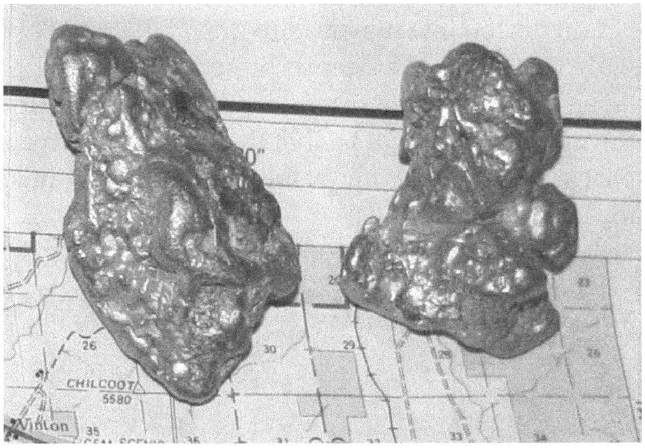

Big coarse nuggets like there are out there to be found, but they are few and far between and not at all easy for anyone to locate.

Finally, seeking coarse gold is never an easy target, and even armed with loads of good knowledge, it's still a difficult quest. Keeping a positive mind set and a good attitude will keep the prospector working at it, and unless you are very lucky, one does have to put in the time to make the finds. Keeping a good focus, good attitude and working hard at it are the keys to making regular finds of large coarse nuggets in whatever environment you will be prospecting.

I've often heard the old phrase "gold is where you find it" - it's a classic statement passed down from the old timers of many years ago. How can you disagree with something as plain and simple as that! However, I think too many folks (both beginners and even some more experienced) take that old line to mean things like:
1) You can never figure out where the gold might be - even if you guess correctly it's an accident;
2) There's just no explaining where gold might be found - it's a total mystery; or
3) Basic rules for prospecting might help a little, but most gold is just a random occurrence.
Consider this: It's a fact that a small percentage of prospectors find a majority of the gold. Some say that perhaps as few as 10% of the folks who own metal detectors find 90% of the gold. Although there is no scientific survey, there is little doubt that a small percentage do far better than the majority. These successful few do put in more hours prospecting, but that's not the only explanation. The truth is they are not "just totally guessing" as to where the gold is. It may well be that a good part of the unsuccessful 90% that only finding 10% of the gold are (as implied above) pretty much "mostly just guessing" as to where the gold is.

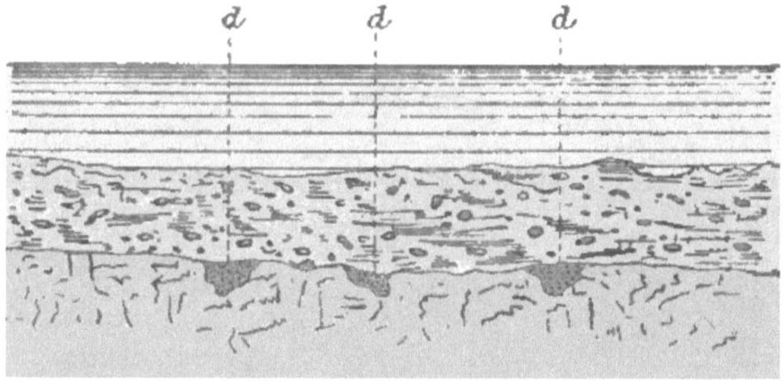

Coarse gold will very commonly get hung up in the deepest parts of any cracks and crevices in the bedrock (marked as d), especially along the line of travel for the most of the gold.

My point is that the resting location for the vast majority of all placer gold is not a big mystery. Sometimes gold occurs in seemingly unexpected places, but most of the time when that happens, it's because we lack some important pieces of information we need to really understand why this gold is here in what would otherwise be considered an unexpected place. As an example, it's also a well-known old saw that: "Gold is found on the inside part of the bends in a stream" Even more information in that same vein is that the coarsest gold tends to be near the head and toward the outside edges of the paystreak gravel that accumulates on the inside bend. In general those are true statements. However, there are almost always exceptions. There are many things going on in a river during the flood stages when gold is being moved around. Sources of gold on the hillside above can contribute gold to the outside of the bend - because it takes a bit of distance in the stream before gold works its way into the common line of flow. What the stream really looks like during true high water flood conditions may be unexpected.

Obstructions in a strong flow can throw gravel one way or another and make deposits that otherwise don't look "right" when we see them in the spring and summer when there is no flow or just a trickle by comparison. Streams can change their beds and leave behind gold deposits where the flow used to be but is no longer. There are similar "exceptions" to other standard prospecting rules like shallow and exposed bedrock is a good place; or that iron red dirt means gold is nearby.

So the next time you find gold in what seems to be an "unexpected" place - think hard about it -there is an explanation, and the explanation may help you find more of that "unexpected" gold. Don't give up and simply chalk it up to "gold is where you find it". Be tenacious and think about what force or cause acted to put that gold in what seems (at least on the face of things) to be an unusual spot. Who knows, you may be on to something. Is there a hidden patch up slope in the hill that is waiting to be found that is leaking gold down slope? Is there some hidden shallow bedrock just below your feet that can't be seen? Does a nearby rock outcropping cause the force of water to throw the gravel an unusual way? Was this gold left high and dry by erosion or a change of flow in the stream? If you work at it, a lot of times you can eventually figure out the cause. You basically turn yourself into a gold detective. There is no question that the most important piece of prospecting equipment you need to take into the field is your brain! Knowledge and experience are even more important than a good detector; because a good detector can only find the gold if the coil is over a detectable nugget. Your brain has to do all the necessary research to find a good spot and move the detector coil to the place where the gold is located. The same is true of all the other forms of prospecting, including dredging - your knowledge and skill must guide the dredge to sample the likely paystreak locations. Great equipment can only take you so far in this game. With every nugget found you need to ask - why is it here, and where are its brothers?

All prospectors that are successful on an ongoing basis have learned certain rules about the likely locations for gold that they use. Normally these are based on past experience. Consciously or unconsciously, you have to have some concepts in your head as to where the most likely places are to search. It's when you find something unexpected and just say "gold is where you find it" and then forget to think any more about it, then you have failed. You have slipped just a bit too far into your comfort zone and are not really learning what you could. You may be missing out by not doing a little detective work to follow up on all the geologic clues and seeing what the real causes of that "unexpected" gold are. None of us knows it all - there are always curveballs and exceptions that nature throws our way in the quest for gold. Keep a positive attitude even when things seem difficult and you will enjoy the "quest" much more. Successful prospecting is one part tenacity, one part knowledge and one part plain old work.

Good prospectors learn to look for and find all the most likely places when they arrive at a site. The more you know about geology, gold deposition, and placer deposits the better you can read your surroundings, and the more accurate your guesses on likely gold locations will be. Some of this can be gained through study, other elements are gained through experience. Either way, that level of prospecting skill takes time and practice to acquire. Knowledge, attitude and experience are very important for your long-term success. Now please don't take this to mean that somehow I think I already know it all. No way, not even close - and the truth is that no one does. But I do think that the more you learn, the better off you are and more successful you are likely to be. Coarse gold is not easy to find even once you raise your level of learning. However, the more you learn, the more you know, and the more experience you have, the better off you are. If you have taken the time to read this whole book up to this point, you have probably learned at least a few new things and will be better equipped in your search. So, when you get out in the field to seek your share of those coarse gold nuggets that are still out there to be found, remember to keep a good attitude, keep your eyes open, keep learning, keep exploring and keep seeking. Enjoy your time in the quest for gold.

One of the great secrets of successful prospectors is that they know what to look for and then get out into the field and work hard at searching. There is a great big beautiful world out there to explore and prospect. Get yourself out there and enjoy it to the fullest!

The Importance of Sampling in Gold Prospecting and Mining

There are a lot of differences between the operations of successful mining companies and ones that become financial failures, but one of the consistent key differences between successful mining companies and unsuccessful ones is their sampling programs. One is reminded of the old saw about the three most important things to remember about investing in real estate - they are location, location, location. For mining it is similar, and perhaps the three most important things in designing a successful mining project are sampling, sampling, sampling. No mineral deposit is uniformly rich; in fact most are extremely variable. If the property is a good one, some spots may be rich, some will be lower grade but still rich enough to be profitable, but much will be too low in grade to work profitably. When rock is too low in grade to make money, it is waste and should not be processed. The big operators spend HUGE amounts of money to sample so they know in advance what they have to work with at their mine. They have a good idea before mining begins how much waste and how much ore they have. Successful mining companies know that the investment necessary for a good sampling program is well worth the time and dollars expended. Even when actively mining, they keep a tight control on feed grades. Far too often small operators, in a mistaken attempt to reduce costs, conduct little or no sampling before beginning to mine and not nearly enough even after mining starts.

I cannot tell you how many times I have been told crazy things like "We can't afford the costs and delays of waiting to sample" or "Old Tom worked this place in the '30s and he told us it was rich." or "We know it's rich, so there is no need to sample." I previously told the story of a placer miner I knew that lost his shirt because he didn't bother to sample - but it's not an unusual situation. It's amazing to me that many small operators who are willing to spend hundreds of thousands of dollars on a mining venture, won't take the time to spend a few thousand to properly sample their deposit. One company I worked for spent many hundreds of thousands and planned to mine old dumps while they got into full production. They took a few grab samples off the dump but when other samples didn't confirm, they just went with the higher numbers from the picked grab samples. They never were able to produce what they thought they would and so from the start they couldn't make expenses, and a few months later they were out of business and all their investment lost. Something similar to this sad tale is the normal fate of commercial mining operations that don't do their sampling "homework" beforehand.

If you are running a commercial operation and your total operating costs are $10 per yard to process gravel, how much $3 gravel should you process? None, because you are losing $7 per yard. How much $50 gravel should you process? As much as possible! Unfortunately, without good sampling, you have no idea what it is you are putting in the hopper. Is it $50 material or $3 per yard material? You might have some kind of guess, but who really knows if you don't have samples! You will find out eventually, but if you ran money loosing rock, the money will be already gone before you find out! This is a lesson that even small scale individual prospectors should keep in mind.

Taking samples should be a well thought out process done with all the right tools on hand.

Designing a Sampling Program

Probably the first thing one needs to determine is what the goal of your sampling program is. Mark Twain tells a funny story of miners seeking to sell their property who would search for hours through the mine to find some small piece of ore that was high grade. That tiny fragment, perhaps the size of a walnut, would be taken in for assaying. The results would show extremely rich ore, with perhaps hundreds of ounces of silver and many ounces of gold per ton. In selling his claim, the miner would use the assay as if it reflected the average grade of all the rock the mine produced, even though there might not be even half a pound of that type of rock in 100 tons of ore. That's how Twain got his saying that the definition of a mine was "a hole in the ground with a liar standing next to it".

Using poorly planned sampling methods is a technique guaranteed to give bad results. When beginning a sampling project, it's important to plan what you are doing in advance. When designing a sampling program the first thing that has to be considered is: What are you trying to determine? Are you exploring for new deposits or trying to characterize the average grade of a known vein or other deposit? The goals of a prospecting or exploration sampling program are not the same as the goals in ore characterization sampling. In prospecting or exploration, one is looking for ore deposits, concentrations of mineralization that would be worthy of further investigation. In an exploration project, the miner is looking to find new ore bodies, pay streaks or patches of nuggets. The goal in this type of operation is to identify likely locations for deposits and prospect there to see if some valuable deposits can be found. In sampling for ore characterization, one is taking a more or less known deposit and testing it to describe the extent and grade of deposit, determine its size, typical grade and other characteristics. This is usually done when considering a commercial mining operation in order to make an economic analysis or feasibility study to determine if it is economic to mine that deposit. It is important to recognize that most of the gravel in many placer mines is waste and should not be processed through the recovery plant (the same is true for vein rock in most hard rock mines). When prospecting, you are searching for the valuable parts of those deposits and in sampling, you are deciding the typical grade and size - what gavels are profitable ore and what gravels are not worth processing.

The Nugget Effect

Designing a good sampling program requires an understanding of the deposit itself. The famous "nugget effect" occurs when a sample includes a nugget or other piece of high grade material, giving an overall value much higher than the true average. Some deposits are like chocolate chip cookies with a few "chips" of high grade spread thinly through the "dough" of the whole deposit. Many placers tend to be like this and as a result these types of deposits are very difficult to sample as you will get a few samples with very high values where the good stuff was included, and many more that are very low in grade where the high grade was not included in the assay. Larger volume bulk samples and a combination of both the high and the low results need to be considered to give a true average. There are hard rock deposits with characteristics like this, and they are among the most difficult to sample properly. Dealing with these kinds of deposits requires a good understanding the size and dispersion of the mineralization.

Sampling a placer with many layers and false bedrock or flood layer surfaces can be difficult.

Sampling Placers

Sampling of placers is difficult, as placer deposits are typically stratified with increased concentration of values in layers, especially on bedrock. Often the paying layer of bedrock may extend one or more feet into bedrock because of crevices. If the placer values are coarse in size, the nugget effect can be especially bad, making the determination of a true average especially difficult. Checking placers by fire assay almost always gives mis-leading, higher than expected results. This is because the fire assay captures gold within fragments of quartz and other mineralization which normal placer operations simply cannot recover. Good placer sampling requires larger sized bulk samples and accurate collection of materials from the bedrock or other paying strata such as false bedrock or flood layers. It is also much better to test a placer using recovery methods similar to that which will likely be used when the deposit is mined.

Sampling Mine Dumps

Sometimes old dumps contain enough mineralization that they can now be economically mined and processed. Special techniques, which consider the layered structure of mine dumps must be used since mine dumps are created in a way far different from natural hard rock deposits. Channels cut across the surface of the dump need to be arranged perpendicular to the layers of the dump - a direction which is usually parallel to the haulage track on which the dump was built.

Sampling Hard Rock Deposits

Effective hard rock ore body sampling requires some understanding of the structure of the ore deposit. Sometimes this can be difficult when the geologist is considering a new deposit - what direction does the deposit go? Some deposits have sharp contacts with the country rock, and some just slowly grade from ore into the waste. Different techniques are required for surface or sub-surface deposits. In drilling, the angle and depth of the sample are designed to go (as much as possible) across the structure of the deposit, so that a true sample across the deposit can be obtained. In sampling hard rock deposits, there are a number of types of samples a geologist might take. They include:

Grab Samples: which are samples of rock material from a small area, often just a few pieces or even a single piece of rock "grabbed" from a face or dump or outcrop. These are the most common types of samples collected when surveying an area or beginning exploration for hard rock deposits in the field. The sample usually consists of material which is taken to be representative of a specific type of rock or mineralization. It is more associated with exploration programs.

Composite Chip Samples: which consist of a series of small chips collected over a larger area. The sample source may be a line or a grid laid out on the face of an exposed outcrop. Composite samples are normally taken from similar sorts of material but are specifically intended as representative samples, so the balance of the material they represent is taken into consideration. The procedure is to collect small chips and pieces of rock over a large area and in doing so, to make the sample as representative of the material as a whole as is possible. A composite sample is typically collected to determine the average grade of the mineralization present. If the area covered is too large, the information as to what is present (i.e., small high grade portions) will be lost to the averaging effect of this sample method.

High Grade Samples: This is more like the type of sample Mark Twain mentioned in the story referred to previously. A high grade sample consists of selected pieces of the most highly mineralized material, specifically excluding any less mineralized rock. As a result, a high grade sample is not representative of the overall mineralization found at that location. A high grade sample might be collected to get an idea what the richest possible values are, or to provide material for certain types of trace element analyses. If a very selective sample like this does not return the expected high grade results, then it is unlikely that a significant amount of valuable mineralization is present, so it is sometimes used as a "worst case" type of test. When a high-grade sample is collected it is important to label it as a high-grade sample, so its values will not be misinterpreted as representing the "average" mineral values contained in an ore deposit.

Cutting a channel sample

Channel Samples: consist of small chips of rock collected over a specified linear interval. Channel sampling is used for veins and other structures exposed on the surface. The objective is to cut a linear channel across the vein or ore body in order to obtain the most representative sample possible for the designated interval. Most of the time chip channel samples are collected in succession along a sample line which is laid out in advance using a tape measure, and is designed to parallel the true width of the deposit. Often the chip channel samples are collected along surface trenches or the floors or walls of various underground workings. When chip channel sampling is conducted along a rock face, often a piece of canvas or plastic is laid out for the material to fall on so as to avoid contamination and make the collection of the sample easier. Sample intervals are set at a specified width, usually ranging from 1 to 20 feet, commonly 5 feet. Similarly, with rotary drilling, the geologist can capture a certain length of the chips collected along the drill hole length, and produce a composited chip sample out of a drill hole.

Exploration Sampling Methods:

A century ago, prospectors in the field sampled hard rock deposits by crushing and carefully panning out the heavy materials, hopefully including gold. The method is known as pan sampling. The old timers tested veins and other deposits by collecting a sample, crushing it to a fine sand, then carefully panning the results to see what heavy minerals and gold would be found. With experience, they learned to judge the amount of color they saw after processing a pan full of crushed vein material. They could then roughly estimate the ounces per ton reading one would get from a normal fire assay. A very large percentage of the hard rock gold mines across the world were originally discovered using some variation of this method, especially in combination with techniques for pocket tracing (the method for following placer fragments up a drainage up to its source - AKA loaming). Many small mines were run for years based mostly on "panned" assays of channel samples. The advantage of this method is that it is a cheap and quick, and it gives a good evaluation of coarse free gold, but also shows up any sulfides or other heavy metallic minerals. Because sample results are available in only a few minutes, many samples can be taken and many veins can be tested in a short period. The disadvantage to this type of analysis is that it won't capture extremely fine gold or any gold tied up in sulfides or tellurides. Sometimes crushed sulfides are hard to recognize and rich tellurides are hard to distinguish from things like pyrite. Some very rich deposits (including those at the Carlin trend in Nevada) have only very microscopic gold, which cannot be panned by this method. The other part of this method is that you still need to take multiple samples to really characterize the vein. You never can really tell the whole story from one pan because even if you get a pennyweight of gold in one pan, until you've taken a few pans you don't know what is typical.

Most Commercial sampling is done by drilling.

Geochemical Sampling

Geochemical sampling involves collecting and analyzing things like soils, stream sediments and rocks and testing for certain chemical elements which are associated with valuable mineral deposits. These methods are extremely efficient exploration tools and a few well-located sediment samples can be used to test vast areas for potential mineral deposits. The presence of these trace elements can be nearly impossible to recognize in a hand specimen, but lab analysis by methods such as atomic absorption (AA) can reveal elevated levels that are only in the parts per million range. Many well-known ore deposits were discovered using these methods. Geochemical sampling works because it is fairly normal that large hard rock gold and silver deposits have a surrounding envelope of weak mineralization, also known as a dispersion halo. They are formed by "leakage" of mineralizing fluids into the surrounding rocks, and are best developed around large bulk type deposits and those deposits which form in rocks that are porous enough to allow for fluid flow. These halo areas have elevated concentrations of metals commonly found in association with gold deposits - the concentrations are low, but significantly higher than the normal background levels found in unmineralized rock. Looking for mineralization halos through geochemical sampling helps exploration geologists locate valuable deposits. A variety of elements can be analyzed, typically focusing on expected elements, which commonly associate with known deposits. As an example, typical elevated elements that may be found around epithermal gold or silver deposits include mercury, antimony and arsenic. Knowledge of these pathfinder elements and which elements are associated with certain types of deposits is critical for interpreting geochemical test results.

Fire assaying is the standard method for testing gold and silver content of ores and has been used for hundreds of years with little change.

Even after the discovery of a possible new deposit, geochemical sampling is important in delineating the areas of mineralization. For example, geochemical sampling of soils is often employed to outline the general area of mineralization at shallow depths where outcrops of bedrock are minimal or nonexistent. This can then be used in designing drilling programs. Sample results for certain elements are often plotted on a map to locate regions of elevated concentration.

Drilling Operations:
Various types of drilling, mostly by core and reverse circulation, are the main methods of sampling ore bodies below their surface outcrops. Core drilling yields a solid rock cylinder sample of the ore deposit, while reverse circulation grinds and chips the rock, resulting in a sample that consists of small chips and sand. Drill holes are often set in regular grids that help the mining company identify areas where ore occurs. When good ore is found, the space between the holes is shortened and the miner gets more accurate information from the closely spaced holes.

The Assay and Analysis Process:
Hard rock samples must be assayed to determine the metals they contain. The first step in any assay is sample preparation. If the sample is wet, it is first dried, and then crushed. The crushed material is split into a reasonable fraction, and then pulverized to a powder. Splits of the pulverized powder are then set aside for various analysis methods, depending on what analysis is to be done. Normally, only a small part of the rock sample is actually analyzed.

The fire assay process works by fusing or dissolving a known amount of the rock sample in a crucible using a lead glass flux. Fluxes of various types are used on various types of ores. When the reactions are complete, the crucible is removed from the oven and the molten glass formed from the flux and ore sample is poured off. The remaining lead metal at the bottom of the crucible is poured into a mold where the lead settles to the bottom and the whole mass cools and solidifies. Once cool, the molds are turned over and the lead buttons removed. This mass of lead metal contains all the gold and silver that was present in the sample. The mass of lead metal is then placed on a cupel made of bone ash. This material is such that it readily absorbs lead oxide which has properties like glass, but it does not absorb metal. The lead button is placed in the cupel and heated. What is left afterward is a tiny bead containing all of the gold and silver that was in the original sample. The bead is then weighed, and the total weight of the gold and silver together is known. After this, the gold and silver are parted by flattening the bead and putting it in a solution of dilute nitric acid. Dilute nitric will dissolve the silver but not the gold. The remaining purified gold is then weighed. The weight of the silver in the assay is calculated by subtracting the weight of the gold from the weight of the bead that contained both the gold and the silver. Using the known original weight of sample that was used in the assay, the weight of the gold and the weight of the silver, the content of the ore sample in ounces per ton (or other units) can be calculated.

The AA assay is the modern alternative to the ancient fire assay technique. To perform an "atomic adsorption" or AA assay, the assayers will take a weighed sample; then crush it to a fine powder. They add nitric and hydrochloric (aqua regia) in a flask then put the flask on a heating plate and let it sit until the acid is nearly gone and the sample is nearly dry. It is cooled and distilled water is added. The salts of the metals in the sample (gold, silver, copper, etc.) dissolve in the water. The Atomic Adsorption unit can then analyze for dozens of different elements - not just gold, silver, copper, etc., but also halo and other trace elements like mercury, arsenic or antimony that may indicate a large hard rock deposit is nearby.

For The Individual Prospector:
So what does all this talk about sampling mean to the small operator or weekend prospector? The individual needs to know the difference between exploring and mining, and that exploring always comes first. Take your cue from the pros. Explore and prospect around to find the best parts of your claim, and mine there. Very few small operators spend enough time prospecting and sampling - they want to move too quickly to mining. The best pay streaks are not always exactly where you expect to find them, so if you are dredging, punch a few holes around to see where bedrock is, and what flood layers are present, how deep bedrock is, etc. If you don't find good gold, move on and prospect further - keep sampling and exploring. Far too many simply find a likely spot and start working down to bedrock. Once they hit bedrock, if there's not much there, they just spread the hole in an attempt to find something. It would be much better to move up or down stream a ways, and then punch a series of sample holes until you get into some good gold.

The message to the individual is to sample, explore and prospect to find a valuable deposit first, then test to determine its extent and value second, then mine it third! The most successful individual prospectors think like snipers - prospecting around an area to find a paying deposit first. Depending on your situation, it might even be best to evaluate a property with a pan and crevice tools first to determine the best place to bring in a dredge and sample on a larger scale by punching holes. When you do find a paystreak, punch a few more holes to see how far the paystreak extends, both up and down stream, and the inside and outside edges. Start dredging seriously for production at the lower downstream end of that paystreak and mine upstream so you don't dump your tailings on top of the un-mined part of the paystreak. Don't start out mining. Start out prospecting, then go to sampling, and finally on to mining.

Exploration sampling and prospecting can be the most difficult part of an operation for small operators and individuals, but done correctly, it will have the most rewarding results. Until you find the paystreak, you are uncertain as to what you might find, whether the property you are testing will work out, and if you will be able to find the gold if it's there. All these uncertainties can be tough on the prospector mentally. Don't let those uncertainties and doubts play with your head. If you stick to good procedures and use what you know and the experience you have, you will eventually get into good gold. Remember that when you are sampling to find a deposit, you gotta follow that gold! The gold in your sluice box or pan never lies - it tells you exactly where the paystreak is. You don't want to spend any more time digging barren material than absolutely necessary. A bad sample has still useful information - what is it telling you? It tells you where the gold isn't! I always do better when I stop and think a bit about things a bit. I ponder what I'm doing, what the material is telling me, what the amounts of gold I am finding is telling me, where the paystreak is seeming to go, where the source of the gold is located, bends in the stream, changes in grade, etc.

Procedures For Continued Success at Gold Prospecting

Back at the end of the chapter on metal detecting, I started to get into a discussion about keeping a positive attitude and not giving up. I also noted that this was true for pretty much all types of prospecting. The problem is that long term, ongoing success in prospecting for gold is never easy. I often tell folks that if finding gold were easy, gold would not be so valuable! Successful prospecting and finding nuggets truly requires keeping a positive focus and putting in the necessary time to explore. This is equally true for all types of prospecting. In fact, having the right attitude is so important that I decided to spend a whole chapter in this book about the related topics of keeping your confidence, persevering, being dedicated and not allowing yourself to get discouraged.

You need confidence and determination to hang in there during times of drought. Learn to enjoy being out in the field for its own sake, even if you don't find any gold that day. Doubt and uncertainty are perhaps the most difficult obstacles faced by the independent prospector. Prospecting isn't easy and a good dose of confidence will help get you through those difficult times of drought. Just stick with what works and hang in there and in time you will get into good gold. That's how prospecting works.

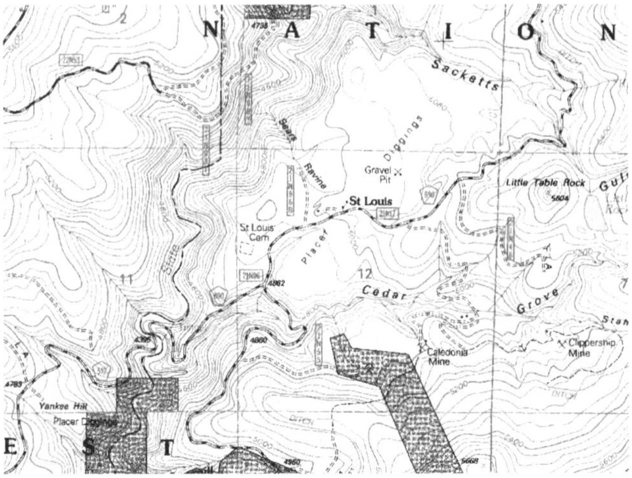

Exploring and sampling for gold deposits is critical for ongoing success. Sometimes you have to get out and hit the road to find new deposits.

One of the things you have to keep focused on is the fact that perseverance pays off. I have a little story I'd like to share that really illustrates this point. A few years ago, a couple friends of mine decided to team up for the year and work together. They both live in California and come up to dredge almost every weekend in the summer time (plus a few vacation days here and there). Because they were going to be working on the North Fork of the Yuba, the same river where I dredge in the summer, I knew I would see them periodically so I could ask how things were going. They were prospecting around various locations on the North Yuba where they had access, but for most of that summer things did not go especially well. They were hoping to find something nice - some nuggets beyond the ordinary run of the mill flakes and pieces running about a pennyweight or less.

Early in the season, they had spent several weeks making a deep hole in the river, trying to reach bedrock. The walls of the pit were loose and material was continually sliding down on them, so the hole got very wide as it reached depths of 8 or 9 feet deep in the overburden, yet any signs of bedrock remained out of sight. How much farther would they need to go? Was it another foot or another 20 feet? It was impossible to say. They had the good sense to pull out of that hole and move on because they realized bedrock would not be easy to reach at that location. Pulling out of a hole where you have invested considerable time is always difficult, but successful dredgers know that sometimes you just have to call it quits and move elsewhere. They moved on and continued to sample other locations where they thought some good sized gold might have gotten caught up. They worked a riffle of exposed bedrock on an inside bend, and found a few positive indicators, picking up a little more gold. They worked in and around this river bend, sampling various locations and did recover a lone quarter ounce nugget, but after a few more weekends' worth of effort, this area also failed to yield the expected results.

By the end of August the weather starts to cool and many prospectors leave the area, and I had also pulled my dredge out of the water. My friends had collected about 2 ounces of gold by that time, and they had moved a lot of boulders and dredged a lot of gravel in various spots to get that. However, they also knew that sampling was the key to finding that rich spot - so they were persistent, and continued on in their sampling efforts, still trying to find a paystreak of coarse gold that would make their efforts really pay off. In September, they decided to try one last location. It was only now accessible at the end of the season, with the water levels down, as the flows had been a bit too fast earlier. They decided to give it three additional weekends working there before they would pull out for the season.

Not long after moving to the new location, they hit an area of coarse gold - just the kind they were looking for. In the course of their operations, they picked up a one ounce nugget. Now that was more like it! Shortly afterward, in winching a boulder out of their hole the bedrock they had tied off on broke. What seemed like a minor inconvenience turned to more excitement as they found out a bit later that the crack along which the bedrock broke contained a second one-ounce nugget. Clearly their luck had strongly changed for the better!

The water was cooling rapidly from the summer highs and they could not spend as much time working in the water as they had to come out regularly to get warm. As good as their luck now was, it then got even better, as they were shortly to make the biggest find that either had ever made. While continuing to expand their dredge hole, they came upon another large nugget. It was visible in the clear water long before the nozzle would have (maybe) sucked it up. It was in water about 5 feet deep and under what was initially about 3 feet of gravel cover. It originally showed only on one end, so they thought it was another nugget of around one ounce. But it was lodged tightly between the bedrock and a boulder, and as they worked to gently free it (without scratching or marring it), it just got bigger and bigger as the sand and gravels were moved away from it to free it. They could see it was not just another "ouncer" but was something much larger. When the nugget was brought back to camp it was too large for any of their scales to measure, so they took it to a local gold shop. It was too large for the scales there as well! Finally they ended up taking it to a local butcher shop, where it was first weighed. Final calculations show it comes in right at 7.5 troy ounces. Their last three weekends of dredging yielded a total of 11 ounces of gold, which includes that big 7.5 ounce nugget. Together with the 2 ounces from earlier in the season, the total take was about 13 ounces. Their experience proves that persistence really does pay off!

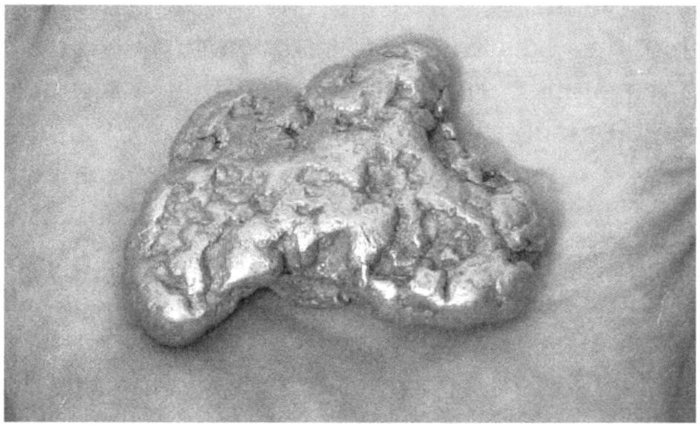

The 7.5 ounce nugget found by friends with patience and perseverance.

I have certainly faced dry streaks myself. I remember a few years back when I went on a prospecting trip to a new spot and in the first 10 minutes, I found a 4.2 dwt nugget, nearly a quarter-ouncer. That was one nice nugget, a genuine prize and I was real pleased with it, and I was real optimistic after finding it so quickly in a spot that had little or no sign of any other modern prospecting. However, I was there for 3 more days after that and found nothing but trash the rest of that trip. Although it didn't seem too bad at the time, that was the start of it. Several weeks later I took off to Alaska for a week of prospecting. I had a great time and hung around with some great guys, but I got the skunk for an entire week. I ran my detector all over, but never found a single nugget. Now I did run a highbanker for a while on that trip and picked up 7 dwt of beautiful Alaskan gold - and that was great. Some may say that wasn't a skunk, but as far as gold found with my detector, I drew a big zero. Then in mid-July, I returned to the Sierra Nevada high country - my own backyard - and spent 4 days prospecting and again got nothing for the whole trip. I dug lots of trash, including small, good-sounding items like heavily rusted boot tacks, small bullets and even an old nickel from the 1940s, but no gold. That worked out to be a total of 14 days detecting without a single nugget (although it's true I was not detecting all day long for each of those days).

Still, I have to say that's one of my longest streaks that I can remember without detecting any gold. When you are in a dry streak, there may well be gold where you are looking, and others may even be finding it, but you can't seem to get your coil over a nugget- that's basically the definition of a dry streak. So I was officially stuck on the Skunkville express and I needed to find a way out.

Here are the nice nuggets that broke my streak of bad luck. The larger one is 7 pennyweights.

Many weeks later my bad luck ended when I returned to a spot that I had been successful at the year before. The previous year I had found a 3.6 pennyweight piece at a certain place, and was hoping that I might be able to find a couple more - maybe a small one or two that I had missed. I took a friend with me and started exploring around the area where I had found gold. It didn't happen right away, but I hit a real nice nugget. It later weighed in at seven pennyweights, and was just the kind of find to turn around a dry streak. I found it because I just wandered through the area I had prospected previously in a bit different direction than the year before. The nugget was plenty loud on my detector, and as soon as I got through the top layer of soil and into the hard pack, I knew my chances for gold were pretty good. It was quite a sight for me when I pulled it out of the ground - I knew it was well over a quarter ounce. My new found good luck continued when only about 2 feet away there was another nugget just a bit under a pennyweight. I was real excited and all the failure of the previous weeks was forgotten. Like the story of my dredging friends, my perseverance paid off and got me some good gold. Its important to remember that gold is so very valuable because it is so rare and difficult to find.

Breaking That Bum Luck Streak

Look, we all hit periods of time where we don't do as well as we had expected. We find ourselves coming home with little or no gold to show for our efforts, and that's not the position we want to be in. It's common, even for guys who have tasted success that they will start to doubt themselves and get discouraged. When you get in a slump, you have to step back and make a plan. Explore, change how you're doing things, check your attitude, whatever, shake things up and try something new - but most important of all: DON'T GIVE UP!

Sometimes things go well, and sometimes they do not work out so great. There is no doubt the more you know and the more experience you have prospecting in the field the better off you are, but in no way does that make you immune to the skunk. So how do you get out of Skunkville and get back in a productive mode? Here are a few of my thoughts; a kind of checklist to go back to if you are getting discouraged to see if you are on the right track.

1. Make sure you are not doing something systematically wrong.

Probably the first thing you need to do is to make sure you are not doing something systematically wrong that will prevent you from getting gold. Take a close look at your procedures. If you are metal detecting you might ask questions like: Are you digging those weak signals? Is your detector set up right? Are you using the right coils? If you are dredging you might ask: Am I reading the river properly, Am I doing enough sampling to find the paystreaks? Don't overdo it by focusing on the minutia, but step back take a hard look at what you are doing to see if there is something seriously wrong, then make any changes as necessary.

Dry spells can make you wonder and really doubt your skills at finding gold. I know that I am doing OK as long as I am following the proper procedures and methods that I know will lead me to find gold. When I am confident that I am doing things right, I know I just need to keep at it and I'll soon be back into the gold nuggets.

2. Keep up your enthusiasm - never give up, never surrender.
It's easy to get discouraged when you are not finding anything, but you'll be stowing away your prospecting equipment for good if you lose your optimism. Prospecting for gold requires perseverance - you need to keep at it just like my friends did in the story I mentioned. Dry streaks happen, and they can last a while, but you need to have the enthusiasm, perseverance and confidence to ride them out and find those new patches and paystreaks.

Sometimes when you are not doing well you have to hit the hills and explore, sample new areas and find new paystreaks and nugget patches.

3. Get out - explore new places and sample new locations.
It's easy to get in a rut and just go back to the same old spots - trying to eek out a few last crumbs. Eventually things can get pretty well cleaned out and the gold just stops coming. One of the best ways to break out of a rut and make a new find is to get out and do a lot more exploring. Try some new places! Do some research, check the claims and land status for the area, then get out and hit the field. Exploring new locations, ones that hopefully have not been pounded previously by others, can lead to new finds.

4. Take a break and try something different.
Sometimes you just need to take a short break from the type of prospecting you've been working on - just try a little change of pace. If you have been detecting, maybe to take a break you should try panning or sniping. Maybe if you've been dredging you might try detecting. Shake things up a little. Sometimes a short break will recharge the confidence and enthusiasm, especially if one is successful with the change.

5. Enjoy your time, no matter what gold you find.
Getting out in the field is a great thing on its own. Whether you are out alone, with friends or family, you have to remember to enjoy the time. The mountains and the deserts are beautiful, so take time to notice and appreciate their majesty. None of us will be out prospecting forever, so enjoy the time you have. If you can enjoy your prospecting trips even when you have not found any gold, you are well on your way to keeping the right attitude and frame of mind no matter what you are finding.

6. Keep a positive attitude, no matter what gold you find.
Keeping a positive attitude and refusing to allow yourself to become discouraged is one of the most important keys to long term success. When I first started prospecting, I was ecstatic over finding just a few tiny flakes of gold and a dozen little specs. Now days it takes a whole lot more to get me fired up. Sometimes when I am a little discouraged, it's nice to think of all the fun I have had searching over the years, and remember that excitement I felt when I saw the first gold I had ever panned out of a creek on my own. A positive attitude and enduring perseverance are what it takes to be successful in the long run when prospecting for gold.

DRY BLOWERS AT WORK

Prospecting for Commercial Scale Gold Deposits

While most individual prospectors out there are searching for placer gold, there is a different type of prospecting being done by a handful of individuals - the search for commercially viable large scale gold and silver deposits. The techniques are different, and it's a lot more difficult, with the odds being plainly against the prospector, but the rewards are potentially much greater. Participation requires knowledge and skills far beyond the basic level, but there are plenty of individual prospectors out there who have most if not all of what it takes to hunt for commercial scale mines. These are properties that no individual would be able to tackle, properties that contain hundreds of thousands of ounces of gold and require many millions of dollars to develop into producing mines.

Big mining companies are always looking for potential new deposits to explore and work.

With the high price of gold holding steady and projected to rise in the coming years, mining corporations need suitable exploration drilling targets to test for precious metals, and as a result there is a definite demand for good properties. How this process normally works is that if you control a desirable property, the exploration company will lease the ground for an annual fee and the prospector will also retain a certain percentage royalty on any future production. If the property is ever put into production, the royalty percentage is paid directly to the prospector. These numbers are typically in the 2% to 5% range, so if a mine produces a million ounces, and the prospector collects a 3% payment then he receives 30,000 ounces. None of that is paid, however, until the mine is put into operation and produces the gold. Another common requirement in these agreements is for some minimal level of exploration expenditure on the property to insure the company conducts a serious exploration effort.

The annual lease fees, which can add up to a handsome amount in and of themselves, range from 5 or 10 thousand per year for raw, wildcat properties with potential but no proven ore, to more than 100 thousand per year for proven properties - it just depends on how much testing has been done and how likely the property is to become a valuable, producing mine. One can see in thinking about these numbers that if the prospector really does come up with a valuable property, there is a real potential for riches! There are some variations in these agreements, and as an example, sometimes part of the annual lease will be paid in stock, or there may be a provision to allow the company to purchase back all or part of the royalty, but overall the general structure of these agreements is fairly standard.

Now because the prospector will be marketing these properties to commercial exploration companies, it is important to be offering the kinds of properties that they are interested in acquiring (you have to give the customer what he wants to buy, right?). So the prospector has to keep in mind what kinds of mining properties are marketable. What constitutes a suitable exploration target is something that could contain several hundred thousand ounces of gold or more, or perhaps an equivalent dollar value in combined gold and silver. This means that they are generally searching for bulk deposits of large size, which contain a huge tonnage. They are not looking for a single narrow vein that may have a few thousand ounces of gold - there is just not enough value in a small deposit to pay the cost of a big mining company to come in and develop it. That's why "big" is perhaps the most necessary feature of these commercial properties. The key to coming up with potential properties of this type is research and time in the field. Old reports, geologic maps, aerial photos and all the trappings of researching gold deposits are very important, not to mention the ability to do claim research and determine what is still unclaimed and who has the rights to what property. You'll need to have the skills required for staking claims and keeping them valid. Sometimes to create a suitable exploration target, groups of properties will need to be assembled among an association of owners.

A big part of getting an exploration company interested in a property is preparing a report explaining the potential of the deposit and why it has great potential for significant gold. Normally it is worthwhile to take samples of the rock outcrops on the claims, analyzing them not just for precious metals, but the associated trace metals that often accompany large scale deposits. Those results, together with any geologic maps and a written explanation of the potential of the property can be assembled into a report that could be distributed to the various exploration companies. Negotiations with interested companies will hopefully result in a contract for exploration of the property.

It's not easy, coming up with a good potential property, because they are not common and finding one still open to claim would be even rarer. However the prospector searching for commercial deposits is up against long odds from the start and must be willing to accept that. Geologically, in order to create a large deposit, you need large amounts of mineralized water. Wide fault zones and rocks that are permeable like tuffs or breccias, or brittle rocks that have been structurally prepared by faulting and fracturing are needed to conduct those large amounts of mineralized waters. A source of heat from the earth is also necessary to drive the circulation systems and move the mineralized solutions. You will be looking in your research for geologically favorable areas meeting these requirements.

The gold in many commercial sized deposits is finely disseminated throughout the rock and of a size that it is too small to recover in a gold pan or see without significant magnification. That's why the old timers missed so many of these deposits - they panned out the gravels in nearby streams and came up with nothing, so they figured it was barren. That's why assays are important when you are doing this type of prospecting - a good fire assay will show the gold and silver values that cannot be seen with the naked eye.

There are a lot of different types of geologic situations that can form desirable commercial scale precious metal deposits. In the rest of this chapter we will take a brief look at some of the various types of larger bulk scale gold deposits, so you might get a feel for what you would be looking for.

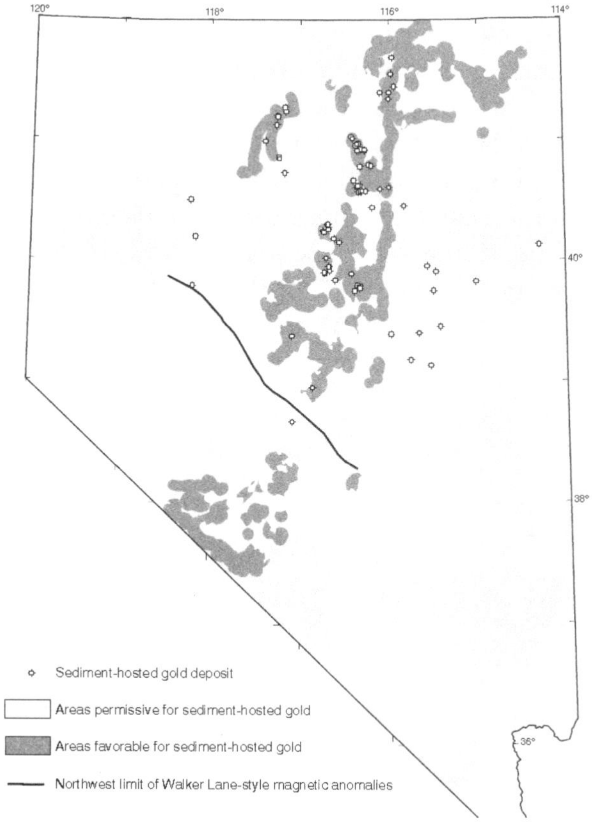

This map shows areas which are permissible (light gray) and most favorable (dark gray) for Carlin type deposits in Nevada.

Sediment Hosted Gold deposits (Carlin type)

"Carlin type" deposits are strata-bound disseminated gold mineralization, occurring in carbonate rocks that have been metamorphosed to varying extents. Their location is usually structurally controlled by steep faults, but may also be controlled by strata of favorable rocks. Acidic mineralizing fluids dissolve away the carbonate strata, leaving behind residual dolomite, collapse breccias and bodies of silicified rock (jasperiod). While the majority of this class of deposits has been found in Nevada, they have been located elsewhere as well. The normal host rocks are thin-bedded silty or argillaceous, carbonaceous limestones or dolomites, commonly associated with carbonaceous shale. A lesser percentage of the host rocks are intrusive rocks such as small stocks and felsic dikes. The typical mineralogy of these gigantic deposits include: native gold (very fine grained), pyrite, realgar, orpiment, arsenopyrite, cinnabar, fluorite, barite, stibnite with quartz, calcite and carbonaceous organic matter. The gold is localized along thin fractures and disseminated throughout sedimentary host rocks. The deposits contain generally less than 1 percent of fine-grained sulfides. They often feature selective replacement of carbonaceous carbonate rocks adjacent to and along high-angle faults, or regional thrust faults or bedding with silica replacing the carbonate rocks with a jasperoid quartz material. These fault zones serve as an important conduit for hydrothermal mineralizing solutions. Depending on the location of the ore body, mineralization may be predominantly oxides, sulfides, including refractory or carbonaceous sulfides depending on how weathered and oxidized the rocks are and how close they are to the surface.

Hot spring Gold-Silver:

Hot-spring gold-silver deposits are sparsely distributed throughout the western US, and include several very productive mines. Geologic studies of the processes active at hot-springs form the basis for the descriptive model of these deposits. Hot-spring deposits form at rhyolitic volcanic centers, rhyolite domes, and shallow parts of geothermal systems related to volcanism above subduction zones and rifted continental margins. Regional fracture and fault systems allow the flow of large amounts of mineralized fluids. These hydrothermal systems are active within a few hundred meters of the surface and are characterized by siliceous sinter, fumarolic mineral precipitates, and hydrothermal eruption breccias and massive area wide silicification. Deposits typically have a high gold-silver ratio and contain precious-metals and sparse sulfides disseminated throughout a thick blanket-like section of permeable tuffs or as stockworks in brittle rocks. The sulfides may be very fine grained and hard to see in a hand specimen. The mineralogy of these deposits includes native gold, pyrite, stibnite and realgar. Sometimes arsenopyrite, sphalerite, chalcopyrite and fluorite may also be present. The veins generally contain quartz in the form of chalcedony or even opal.

Most commercial exploration is done using either core or reverse circulation deep drilling. There are a lot of uncertainties in a drilling program like this as no one knows for sure what drilling a new deposit will reveal.

Hot spring gold-silver deposits are sometimes related to hot-spring mercury deposits which typically form in siliceous sinter near the paleo-groundwater table in areas of fossil hot-springs. Gold and silver will sometimes be deposited a few hundred feet below the mercury systems that form at or on the surface. Deeper levels of hot-spring gold-silver deposits have geologic characteristics in common with epithermal veins, such as fault control, open space textures, and wall rock alteration. Round Mountain and Sleeper in Nevada are examples of volcanic hosted hot-spring deposits in which metals are deposited over a wide vertical range and which have many characteristics of quartz-adularia systems. Other well-known hot spring gold locations include McLaughlin, in California and Delamar in Idaho.

Comstock Type Epithermal

Epithermal quartz-adularia vein deposits of the Comstock type are related to hot springs but are formed deeper, hundreds of feet below the surface. They are distinguished from hot-spring deposits by the lack of evidence suggesting deposition at or very near the paleosurface, by the presence of complex sulfosalts in the ore mineralogy, by a higher silver-gold ratio, and by vein form and structure suggesting the deposition of metals occurred primarily in fissures. The mineralogy features acanthite (argentite), gold or electrum, silver sulfosalts and naumannite (silver selenide). Galena, sphalerite, chalcopyrite, tellurides, hematite, and arsenopyrite are also present in moderate to sparse amounts. However ore minerals typically constitute only a few percent of vein. Gangue minerals include quartz, pyrite, adularia, calcite, sericite and chlorite. Sometimes other gangue minerals like barite, fluorite and rhodochrosite may also be present.

Small single vein deposits are not normally of great interest to large mining companies, even if they were productive in the past. However, near to this mine in Randsburg, the veins splayed out into a stock work system that has been mined on a large scale in recent years.

Productive deposits are characterized by a complex ore mineralogy, with a variety of precious and base metal sulfide minerals, extensive and gradational alteration, especially propylitic alteration, vuggy, comb, and crustiform vein structures with banded sulfide layers. Quartz, calcite, and pyrite are the most important gangue minerals. The host rocks are normally felsic to intermediate volcanic rocks like andesite, dacite, quartz latite, rhyodacite and rhyolite. These volcanic rocks commonly overlie sedimentary rocks, or their metamorphic equivalents. Epithermal vein deposits are found in the Tonopah, Comstock, and Seven Troughs districts in Nevada, among many others. Many of these deposits are too small to constitute commercial scale deposits, but the larger ones may be suitable for commercial mining.

Mother Lode Type Mesothermal in Stockworks / Mineralized Intrusives

These deposits consist of gold in massive persistent quartz veins mainly in regionally metamorphosed volcanic rocks and volcanic sediments. Vein structures include saddle reefs (in folds), ribbon quartz and massive quartz veins. The veins are persistent along regional high-angle faults and joint sets. The best deposits overall are found in areas with greenstone. High-grade ore shoots form locally at metasediment-serpentine contacts. Larger more commercial disseminated ore bodies are found where veins fray out into stock work zones or cut into granitic rocks. In the stockwork zones, the better ore grades are usually found within the thicker quartz pyrite veins, but the mineralization is also disseminated along micro-fissures in between the thicker quartz veins. Where the wall rocks are permeable like tuffs or altered intrusives, mineralization will extend into the wall rock.

Host rock types include greenstone belts; oceanic metasediments like slate and schist; regionally metamorphosed volcanic rocks, graywacke, chert, shale, and quartzite. The typical mineralogy includes quartz, native gold, pyrite, galena, sphalerite, chalcopyrite and arsenopyrite. Locally tellurides are important in some of these deposits. Productive quartz veins are often grayish or bluish in color in many instances because of the fine-grained sulfides contained within them. Weathering of these deposits produces abundant quartz chips in soil. Placer gold may commonly be recovered from the soil by panning. Smaller single mesothermal quartz veins are generally not of commercial interest, but if there are enough veins in a small area or the veins break up into a series of small stock works in a fracture zone, then a commercial type deposit may be formed.

Distal Disseminated

Distal-disseminated silver-gold deposits are a variety of deposits that contain silver and gold in stockworks of thin sulfide bearing quartz veins in sedimentary rock, and contain trace elements indicative of a plutonic-related suite. They are similar to sediment-hosted gold deposits but differ in some important ways, the most important of which is that these deposits contain a considerably higher silver grade than the Carlin type sediment-hosted gold deposits. They are found in or near districts with major skarn, replacement, and polymetallic vein base-metal ores, but are localized in the most distal parts of these districts.

They are found in carbonate and clastic sedimentary rocks that have been folded and fractured along shear zones. Fracture generated permeability is the most important ore control. The ore occurs in irregular bodies locally conformable to the bedding of the sedimentary rocks. Intrusive rocks are always found nearby. Ore minerals include native gold, native silver, electrum, argentite, tetrahedrite, stibnite, sphalerite, galena, chalcopyrite and arsenopyrite. Leaching and redeposition of silver minerals can create bonanza silver deposits as at Treasure Hill in Nevada and Tombstone in Arizona.

Large scale commercial mining operations produce many thousands of ounces of gold and silver each year. The operations are conducted on a gigantic scale and require big deposits.

Gold Bearing Skarn:

Skarns are a specific type of rock that forms at relatively high-temperatures and consist largely of minerals like garnet, epidote, calcite, pyroxenes, ampiboles and lesser amounts of idocrase, wolllastonite and chlorite. In the western US, they have been the largest source of tungsten deposits. These special mineral deposits are related to hydrothermal activity associated with the contact of hot granitic bodies with sedimentary strata that include limestone or other carbonate-rich rocks. While most of the valuable skarn deposits in the US have been worked for tungsten, there are some that are valuable for other minerals including copper and gold. Gold skarn deposits are found in a limestone host near major fault zones. Native gold (or electrum) and chalcopyrite are the most valuable minerals in these deposits. Gold-bearing skarns have been found in a number of locations worldwide including several in Nevada and Montana, British Columbia, and Australia.

In weathered terrains, skarns may be indicated by exposures of limonite, iron hydroxide, iron sulfate gossan, clay, or carbonate rocks that are associated with these deposits. In areas that have been historically mined, there are often multiple adits, pits, and tailings piles because skarn deposits were often explored or worked at different sites, by various operators in the past, along the limestone-granitic rock contacts. Unlike some of the other large scale commercial gold deposits, not all of the gold in skarn deposits is microscopic, and placer gold workings may be found downstream from many gold-bearing skarn deposits. Locations upstream from small placers in and around copper deposits where limestone and granite are nearby would be favorable locations to prospect for gold skarns. This includes some skarns that were worked for copper in the past where the small amounts of free gold present in the deposit may not have been previously recognized.

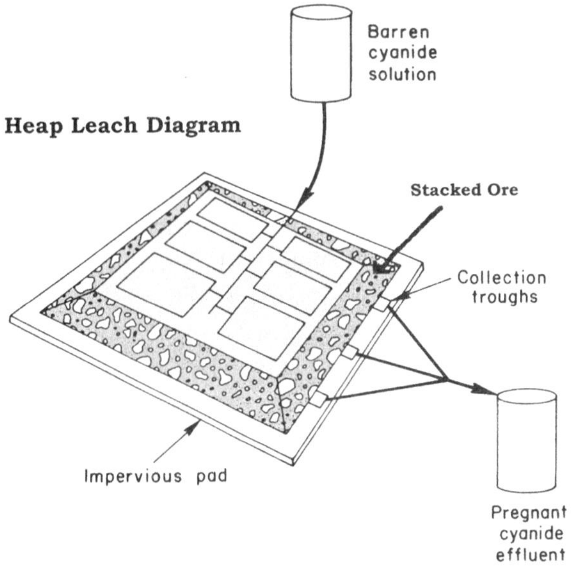

Many large commercial operations employ heap leach designs similar to that pictured here.

Part V: Appendices Of Information

Definitions: Hard Rock Mining

While most small scale prospectors and miners focus their energies on placer deposits, most large mining companies put their efforts primarily into hard rock mining. Although many miners may see these as two separate worlds that can never meet, a better understanding of hard rock mining for small operators can be very helpful in many ways. This is true from both a standpoint of hard rock geology as well as historic hard rock mining operations. As an example, research of hard rock operations can lead to more opportunities in placers as many hard rock deposits with free milling gold have eroded to form placers. Perhaps even some placer prospectors will decide to make the jump over to hard rock operations.

The first step in gaining a better understanding of the hard rock end of things is to decipher the terminology and special "lingo" that is used to describe these operations. A number of the articles published in books and magazines like the GPAA Prospecting magazine contain references to shafts, portals, adits and other mine workings. Though most experienced placer miners have some basic idea about most of these terms, this chapter will explain the full detail of these terms to help readers better understand hard rock mining including the methods used by both the old timers as well as present day operations. Also, to better understand the complete process of hard rock mine operations, it will be necessary to look at terms related to milling and extraction of metals. Because the Mining Journal focuses mainly on precious metal mining, this chapter will also be focused in that direction, though many of the terms would apply equally well to other types of mining operations. This chapter will cover both underground and surface mining processes, and although this chapter will not make anyone into an expert on mining, hopefully it can give the reader a better picture of how hard rock mines work. Also, it is important to remember that the inside of old underground mines, however interesting, can be very dangerous places. An expert with many years of experience is often necessary to determine if a mine can be entered safely - be careful around old mines.

The history of mining terminology reflects the mineral development of the western US. For the early years after the first discovery of gold in California, the miners were interested almost exclusively in placer gold (placer being a term with Spanish roots). In spite of the abundant richness of their finds, the earliest quartz miners had a difficult time mining and processing the hard rock ores at a profit because they were so inexperienced. When experienced hard rock miners arrived from Europe, they brought with them not only better mining techniques, but European mining terms. As a result, European terms were commonly in use throughout the US in the 1850s and 1860s - and it is normal to see them used in American mining publications and reports of that time period. However, as American mining developed on its own throughout the 1870s and 1880s, the terminology changed as we developed our own unique nomenclature. Remembering this little bit of history can help when interpreting old mining reports.

Knowing the meanings of hard rock terms and a little about how these mines work can help the prospector to understand reports and stories about both modern and historic mine operations. Researching old mining reports is important to the individual prospector for finding new deposits. Knowing a little about mining terms and methods can aid the prospector in interpreting those old mine reports and understanding where on a mining property rich ores were removed in the past. We will look at terms and methods for both underground and surface hard rock mining. Then we will look at geologic terms related to hard rock mining and at the milling methods used to extract the precious metals.

Underground Mining Terms and Methods

Historically, the first hard rock mining was conducted when placer miners began to follow the outcropping veins they had discovered underground. All blasting had to be done with hand held steel drills, with a stout hammer pounding on the drills to make the holes for the explosives. All this was done with only candles to light their work. How long does it take to hand drill all the holes for a blast? Many hours, even though the old time miners usually only went about 18 inches deep when they were hand drilling. It's hard to even imagine the amount of hand steel work that went into driving the hundreds of miles of underground workings in Nevada alone. In modern times, underground mining has revived in many places as companies follow rich ore bodies down below existing open pit mines. This happens when the depth and stability of the wall rock make surface mining too costly, however in some of these situations, ores below the pit can still be profitably mined by underground methods. Underground mining on a large scale now allows some of the rich gold mines of Nevada to continue in production when they would otherwise be forced to close.

Shaft - A shaft is a vertical (or near vertical) access way to an underground mine. This opening is used to bring in men or supplies or remove ore and provide ventilation. In European terms, this might be called a well. The entrance to a shaft at the surface is called a collar.

Headframe - The head frame is a structure, which sits over a shaft at the collar. A large pulley, called a sheave wheel, sits on top of the head frame. This wheel functions as a giant pulley to direct the cable, which goes over this wheel, and connects the mine hoist to a skip (container for ore and supplies) or a cage (for men). The skip or cage is raised or lowered by the hoist much like an elevator, but it usually moves much faster.

Adit - An adit is a horizontal access way that is used to bring in men or supplies, remove ore and provide ventilation. If an adit is inclined downward a small amount, it is called a decline. If steeply sloping it is called an incline or inclined shaft. The entrance to an adit or decline is called a portal.

The horizontal entrance to a mine is an adit. These men are standing at the adit portal.

Drift - A drift is a horizontal working which does not connect to the surface but which runs parallel to the vein or mineralized structure. If the working runs perpendicular to the vein or structure but does not connect to the surface is called a crosscut.

Winze - A winze is a vertical working driven downward that does not connect to the surface. A vertical working that is driven upward is called a raise.

Level - A level is a group of workings, which are on a single elevation. As miners move downward in a mine, levels are opened up to access the vein and remove any ore that is present. It is fairly common that these levels are established about every 100 feet or so vertically. Levels which are used principally to move broken ore out of a mine are called haulage levels.

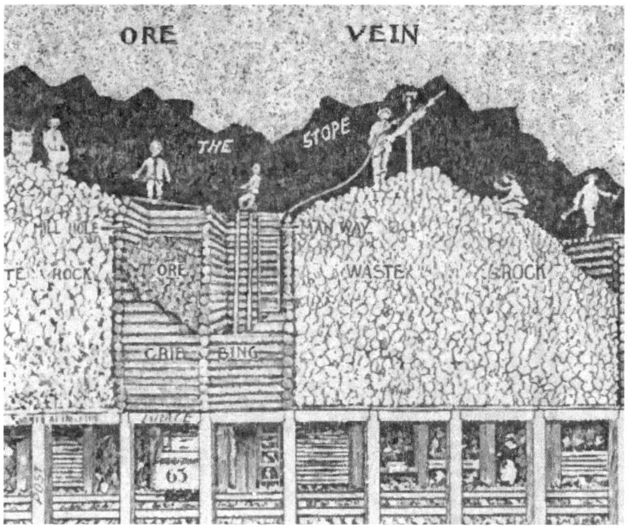

Old time miners extracted ore from a working called a stope, where mining was done standing on timber support stulls or broken rock

Stope - A stope is an excavation in the area of the mine where ore is removed. At times, when large bodies of ore are removed, enormous rooms are opened up, hence the European term for stope, which is gallery. When miners work from the top of an ore body downward, the method is called underhand stoping. When the miners work from the bottom upward, it is called overhand stoping. Overhand methods are much preferred as they allow the miner to use the force of gravity to assist them in moving the ore once it is broken by blasting. When working a vein, the part above the miner at the top of the opening is called the hanging wall or the back, the lower part below the miners feet is the footwall or sill. Stopes which begin at the surface and go downward are often called grass roots stopes. Historically, many old stopes are at least partly filled with waste rock as miners often dump waste rock from underground workings into the old, mined out stopes. This saves them the work of hauling the rock all the way to the surface and also helps to stabilize the walls of the stope. In places, where the miners hand sorted the ore underground, much of the material once considered as waste which was dumped into the old stopes would now constitute good ore.

Jackleg - Hard rock mining requires drilling and blasting. Before 1900, most drilling work was done using a set of steel chisels, called hand steel. Holes of about 18 inches were drilled by pounding on these steel drills and turning them as the hole went down. When the hammering was done by two people - one holding the drill and the other holding the drill steel - it was called a double Jack operation. If one man held both the hammer and the drill, it was a single Jack operation. A tool called a spoon was used to clean out the hole prior to inserting the dynamite. The jackleg is the drill of the modern miner. It is a drill mounted on a support cylinder which is powered by compressed air. These drills are heavy and the support cylinder allows the operator to hold the drill level while working. The leg part is set at an angle so that part of the force of the cylinder also pushes the drill into the rock, helping to speed the drilling process. The difficult part for new operators is getting the hole started - called collaring the hole. Although these drills are powered by compressed air, they also have a water supply, which feeds into the drill bit to suppress dust. Without the water, the miner is exposed to harmful levels of dust, which is usually high in silica.

Timbering - When ore is blasted out and removed, an underground void is created and because such voids are unstable, some sort of support is usually required to prevent cave-ins. Depending on the strength of the ore and surrounding rock, and number of different techniques can be used to support the mine openings. Timbering is used both for support in access openings and in stopes where ore is being removed. When driving access ways through broken ground (said to be ground which "runs") special techniques are used, including special sets called spiling. In stope areas, methods are different, and these can range from simple wooden posts called stulls spanning short distances across a vein, to more complex timbering systems such as square sets to support large openings. Another common method of wall support in stope areas is the use of support pillars, by leaving behind areas of the ore to support the backs above. Removing pillars is sometimes attempted to recover ores left behind by the old timers, but it can be highly dangerous to remove the supports, and may cause caving.

Working face - the end of a current working that is in the process of being extended. This is where the miners are actively drilling and blasting rock to extend the mine workings.

Surface Mining Terms and Methods

Much of the large scale mining currently being done by modern mining companies is based on recovering lower grade values from large bulk deposits near the surface. Large surface excavations to remove these ore bodies are called open pit mines. In the first large open pit mines, trains were used to bring the rock up out of the pit, but that changed as large haulage trucks were developed to replace the trains because they could handle steeper slopes and required less maintenance. Haul trucks at modern mines may carry hundreds of tons of ore or waste in a single load. Drills for blasting are usually track mounted, larger units and ANFO (Ammonium Nitrate - Fuel Oil mix) is used as a blasting agent because of its low cost.

At the beginning of any mining operations, the soils that may cover the ore body are removed and set aside for reclamation after the project is complete. Bulldozers clear off pads for the mill location, access roads, mine dumps, and other similar facilities such as heap leach pads. This initial work is established according to the mining plan for haul roads, bench width and height, drilling and other characteristics of the pit. After the initial stripping of waste overburden, mining and the separation of waste from ore begins. Waste materials are sent to a waste dump, and ore grade rock is sent to the mill or other facilities to be processed. As the pit gradually deepens, the problems of water encroachment, haulage road maintenance, pit slope, and increasing waste-to-ore ratio result in increased mining costs. In the end, a predetermined closure point is reached or the ore body bottoms out so that it is no longer profitable to keep mining. In the modern operation, reclamation of the property begins at that point.

Stripping Ratio - In order to maintain a stable pit wall, surface mines must remove both ore and waste as they work deeper and deeper. The stripping ratio is the ratio of waste to ore that must be mined. A ratio of 2 to 1 means that two tons of waste must be mined to produce one ton of ore.

Bench - As a surface mine works downward from the surface, the operations are done a level at a time. As the mine works downward the walls of the pit come closer together and the bottom of the pit becomes smaller. The walls slope down in a stair step method. The step-levels are called benches, and their remainders can be seen on the pit walls. Benches are also a safety feature, as they keep rocks from rolling down off the high walls into the pit bottom where the miners are working.

Angle of Repose - There is a maximum angle of slope where the rock will no longer remain stable - and wall failures (landslides) are likely when that angle is exceeded. Most rock in mining areas is broken, weathered or altered - so that a good part of its strength is already lost. Miners do not wish to remove any more waste rock than they absolutely must, but the slope of the open pit wall must be engineered to remain stable for the life of the mine. The angle of repose, minus a reasonable safety factor, is used to figure that safe pit wall angle.

This glory hole excavation on a quartz vein was made to take out a small high grade pocket.

Glory Hole - Small surface excavations where rich ore was present and removed directly from the surface in small pits are called glory holes. This is an older term and is generally not used to describe full sized surface mines like open pits but is reserved for smaller excavations.

Definitions: Geologic Terms

Although placer deposits can have their own complexities, there are a vast number of possible geologic situations that can be found in a hard rock mine. Understanding the basic geologic terms can allow the individual prospector to be more successful in researching mineral deposits, including finding locations along the projected line of existing, known deposits. Old reports can contain a wealth of information, but the geologic terms can make them difficult to decipher. The following are a few geologic terms that commonly appear in both new and old reports on hard rock mining operations:

Vein - Veins are comparatively narrow deposits of mineralized rock. They are often formed where geothermal hot water solutions circulate within pre-existing natural fault zones. This is why most veins are shaped like fault zones - that is a long and narrow plane. This is the process that forms nearly all gold-quartz veins. Veins can commonly form at the meeting of two different rock types, also for the reason that on the contact where the two rock types meet is a zone where water circulates more freely than through the solid rock itself. Sometimes a zone of broken or fractured rock develops with no strong single fault shear, and in that case a series of small parallel veins may develop. These are known as stockwork zones. Large regional mineralization may result in deposits which are not vein-like in any way. Most of the Carlin type gold deposits of Nevada are not really vein-like. Lode is an older term which functionally means the same as vein.

Ore - Ore is mineral matter that contains concentrations of desirable materials. To be truly considered as ore, the material must be rich enough to be mined and processed at a profit, producing some valuable commodity. If the valuable product cannot be produced at a profit, it may be mineralized rock, but technically, it is not ore.

Gangue - These are the minerals found mixed in with the ore, which are not valuable. After processing, these are usually sent to a tailings pile. Quartz is the most common gangue mineral found in many gold bearing veins, but a variety of other minerals such as calcite, adularia and barite are common gangue minerals as well.

Host Rock - the host or country rock is the enclosing rock that surrounds the vein or other mineralized structure.

Strike - The strike describes the horizontal direction that a vein runs with respect to North, South, East and West. Sometimes on the slope of a hill the outcrop may appear to be somewhat misleading as outcrops usually show a combination effect of both the strike and dip at the same time. Where referred to in geologic reports, the strike is the true strike of the structure, as if the strike were measured on a horizontal plane.

Dip - The dip describes the vertical direction of the vein or deposit in comparison to the horizontal. In this system, a vein with a 90-degree dip is perfectly vertical. A vein with a 0-degree dip is horizontal.

Quartz vein outcrops, like this one pictured here, do not always stick out much above ground level.

Outcrop - Where a vein or other deposit reaches the surface of the ground. Though at times an outcrop can stick out boldly and be very conspicuous, many outcrops are partly hidden and not easy to see, even where the vein is strong and continuous underground.

Apex - The highest part of a vein or other type of deposit. This can be an outcrop where a vein reaches the surface, but this is not necessarily the case where a deposit does not actually outcrop on the surface. US mineral law allows claimants to follow veins from the apex downward through adjoining claims. Where deposits are not present as veins, or veins branch and recombine regularly, this has led to considerable litigation to decide who actually owns the mineral rights to certain ore bodies.

Fault - A plane where geologic forces cause the rock on one side of the fault to move relative to the rock on the other side of the fault. The fault is the line or zone of breakage. Faults with dip-slip movement have a movement that is vertical, - those with strike slip displacement have a movement that is horizontal. The well-known San Andreas Fault system has a horizontal movement, so it is a strike slip fault. Most faults in mining areas in the western US are dip-slip in nature. Faults can be anything from very tiny - only a few feet long and less than an inch of movement - to huge regional faults like the San Andreas Fault system. Faults can bring two different types of rock together, or take part of a productive vein and move it to another location. Many productive veins have been lost in the past due to movement caused by faulting. Movement after the vein is formed is called post mineral faulting, and that before is pre-mineral faulting. Pre-mineral faults often become the conduits for mineralizing geothermal fluids, which are the source that form veins and other deposits. The broken rock material between the walls of the fault is called fault gouge or fault breccia.

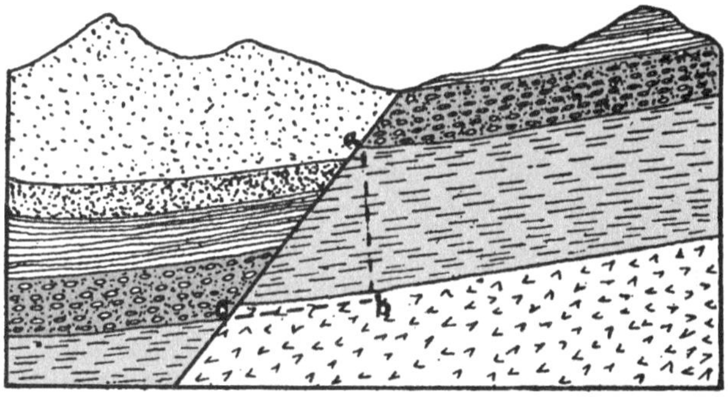

Faults move various rock types around. Many contacts of different rocks occur on faults.

Exploration Drilling - Large scale mining requires a considerable investment of capital, and a full understanding of the ore body and the metals it contains must be obtained before a financial commitment to the project can be made. Most modern mining companies do their exploration by drilling, putting down holes into the potential ore body and pulling up the rock chips or cores that result. These are tested to determine their values. Both core and rotary drilling may be used depending on the type of information the geologists are seeking to obtain. Rotary drilling is usually less expensive, but core drilling brings up intact sections of rock that may be examined for information beyond simple assay results. A series of small roads located fairly close together, each usually ending in a small pad, often indicates the site of a past exploration drilling project.

Definitions: Milling Terms

The one of the biggest differences between placer and hard rock mining is that the rock has to be processed (usually involving crushing) to extract the valuable minerals in the ore. In placer mining, natural weathering has released the gold, but in hard rock, the precious metals values must be pulled out by the miner. There are many types of equipment and techniques that are used, depending on the types of ore to be treated. We will cover some of the most common types of processes.

Mill - A mill is the place where ore is taken to be processed to extract the valuable minerals it contains. For most ores, this means the rock must first be crushed to a small size before putting it through the process to separate the valuable minerals from the waste materials. Most pre-1900 mills used stamps to crush the rock (stamps are large rods of heavy iron which are raised and dropped to crush the ore. Modern processing mills usually use a multi-stage process with a jaw or cones for coarse crushing, and a ball or rod mill to finely grind the rock. The stamp type mill is an all-in-one crushing system, but it is prone to over grinding and that was the main reason for its demise. No mill recovers 100% of the ore values, so every operation has a recovery rate for the ore it treats. Percentages recovered vary widely depending on the type of rock, the values to be recovered and the milling methods used. Old timers recovered as little as 50% of the values on refractory ores, which may seem rather low, but modern heap leaching typically recovers only about 60% of silver values and 70 to 75% on gold. Modern, well-designed cyanide mills will often recover about 90 to 95% of the gold in the ore.

Free Milling Gold – In many ores, gold is closely associated with sulfide minerals like pyrite and galena, to the point where these other minerals enclose and hide the gold. Where weathering or other processes have removed or oxidized the sulfides so that the gold is exposed and easily recovered, the ores are said to be "free milling". Ores with free milling gold can be hand crushed in small lots, and careful panning will often reveal the visible free gold once the gangue minerals such as quartz have been panned away. Ores where the sulfides and metal values are closely mixed make the values harder to recover, and such ore are called "refractory". In refractory ores, the values are hidden in the ores and special techniques are needed to recover the values, such as smelting, where the ores are fused in a furnace, or autoclave processes where the ores are chemically oxidized to make them free milling. Many mines have a mixture of both free milling and refractory values, so that only certain percentage of the gold in an ore may be said to be free milling.

Mine Tailings – Are the piles of unprocessed country rock and ores that are found at the entry portals and collars of old mine workings. The material will be a combination of coarse and fine rock that is usually well mixed. At times these tailings can yield valuable specimens. Many old-time small-scale mine operations sorted ores on the dump and shipped only the richer materials – sometimes these yield only coarse rock. Sometimes very valuable specimens or pieces of ore were overlooked. These types of tailings are of more interest to the individual prospector and many may be profitably prospected with a metal detector, especially if the mine produced free milling gold ore.

Mill Tailings – These are the remains of ores that have already been crushed and processed, with the valuable materials removed. They usually have the consistency of fine sand. For the most part, these are not of interest to the prospector, except certain older tailings from rich mines where the ores were not treated by cyanide, and these may still contain significant values. Even then the valuable minerals have been crushed very fine – often dust sized. Mill tailings are usually found below an old mill site and are generally uniform in color.

Mill Extraction Processes - Once the ore is brought up from the mine, its first destination is most often some sort of feed or surge bin. This bin allows the mill to take in a constant feed. After crushing, the extraction processes vary greatly, depending on the metal or mineral to be extracted. However, since we are focusing on precious metals, we will look at gold and silver extraction methods. Commonly used methods include the following:

Gravity Concentration - These are systems which use gold's high density. Gravity based equipment includes jiggs, spirals, sluices and tables. Gravity systems are often the lowest cost system, but only a certain part of the gold is usually coarse enough to be recovered with gravity systems. It is common that ores be processed by two methods - gravity is used to remove the coarser gold from an ore and afterward the ore is processed by cyanide to recover the fine particles of gold and silver. Gravity concentration methods are the ones most commonly used by the individual prospector in working placers.

Vibrating tables, like the one shown above are great at capturing fine sized free gold particles. They are a gravity based process.

Froth Flotation - this is another common method to concentrate gold and silver minerals. This works because most sulfides (the most common class of gold and silver bearing minerals) are attracted to the surface of air bubbles. The crushed ore is mixed with water into a thin slurry, and then frothing chemicals and air bubbles are added. The sulfide bearing bubble froth is collected off the top of the ore slurry. Sulfide concentrates from froth flotation often need to go to a smelter for additional processing to convert the sulfides to metals.

<u>**Cyanide Extraction**</u> - this is the most common method used on low grade ores by modern companies. It has been the method of choice for processing gold ores since it was first commercialized in the 1890's. The process is based on the fact that extremely dilute solutions of cyanide salts actually dissolve gold and silver right out of the ore. Only a pound or two of Sodium or Calcium Cyanide is added per ton of water (roughly 240 gallons). The cyanide reacts with oxygen from the air to dissolve both gold and silver. Although this method does not work well with coarse gold (because it takes too long to dissolve) it does allow recovery of precious metal particles so small that they cannot be recovered by gravity separation or other similar methods. Once the precious metals are dissolved, the liquid is recovered and the dissolved metals are removed from it. There are two main versions of the cyanide process, vat leaching and heap leaching. In vat leaching, the finely crushed ore and cyanide solutions are mixed and processed in vats - this allows for an efficient and speedy processing of the ore. In heap leaching, coarsely crushed ore is piled up on an impermeable layer and cyanide solution sprinkled over the heap. The heap leach pad is constructed on a gentle slope so that the solutions percolate down and collect on the impermeable layer and then drain down slope into a collection pond. The heap leach method takes much longer and generally does not recover as high a percentage of the gold and silver in the ore, but it is extremely low cost, which is its main benefit. The heap leach method has revolutionized the processing of low-grade ores that have been discovered worldwide. The disadvantage of cyanide is that it is a deadly poison, and considerable safety precautions are required to handle it safely. All solutions must have a protective base added to keep the Ph high enough to prevent the solution from going acidic. Under acid conditions cyanide gas is created and released. In both heap and vat leaching, dissolved gold and silver is accumulated in the cyanide leach solution. The gold and silver bearing liquid, called pregnant solution, is separated from the ore by various drainage and filtration processes. The most common methods to remove the gold and silver involve activated carbon, or zinc precipitation. The carbon method requires an additional step where the metals are removed from the carbon. The end result of both methods is a finely divided black metal powder of gold and silver. This impure mix of gold and silver is often melted and poured into bars before it is shipped. The resulting gold and silver alloy is called a dore. The metal recovered from these processes is impure and is sent to a refinery for further purification. At the refinery, the dore alloy is re-dissolved in a water solution and purified by electrolytic plating methods. Once purified, the metals are used to create various products such as jewelry alloys like 14 Karat gold or sterling silver, coin alloys as well as things like high purity investment bars.

Definitions: Rock Types

Rocks actually have their own definitions spelling out their nature based on the minerals they contain. The following are some definitions of certain common rock types:

Igneous Rocks (Intrusives):

Granite A rock formed at considerable depth by crystallization of magma and/or by chemical alteration. It is characteristically medium- to coarse-grained, of granitoid texture. A plutonic rock defined as having quartz between 20 and 60% and feldspars between 10 and 65%.

Diorite A rock formed at considerable depth by crystallization of magma and/or by chemical alteration. It is characteristically medium- to coarse-grained, of granitoid texture. A plutonic rock having quartz between 0 and 5% and large amounts of plagioclase more sodic than An50.

Gabbro: A rock formed at considerable depth by crystallization of magma and/or by chemical alteration. It is characteristically medium- to coarse-grained, of granitoid texture. A plutonic rock with less quartz and more iron than diorite.

This piece of light colored granite has an inclusion of dark colored diabase (like gabbro).

Igneous Rocks (Extrusives / Volcanics):

Rhyolite: A generally finely crystalline or glassy igneous rock resulting from volcanic action at or near the Earth's surface, either ejected explosively or extruded as a lava. The term includes near-surface intrusions that form a part of the volcanic structure. A volcanic rock defined as having a quartz and feldspar content the same as granite.

Andesite: A generally finely crystalline or glassy igneous rock resulting from volcanic action at or near the Earth's surface, either ejected explosively or extruded as a lava. The term includes near-surface intrusions that form a part of the volcanic structure. A volcanic rock defined as having a quartz and feldspar content the same as diorite.

Basalt: A generally finely crystalline or glassy igneous rock resulting from volcanic action at or near the Earth's surface, either ejected explosively or extruded as a lava. The term includes near-surface intrusions that form a part of the volcanic structure. A volcanic rock defined as having a quartz and feldspar content the same as gabbro.

Metamorphic Rocks:

Metavolcanic rock: A volcanic rock that shows evidence of having been subjected to strong heat and pressure.

Greenstone: A field term applied to any compact, dark-green, altered or metamorphosed basic igneous rock (e.g. spilite, basalt, gabbro, diabase) that owes its color to the presence of chlorite, actinolite, or epidote.

Greenschist: A schistose metamorphic rock whose green color is due to the presence of chlorite, epidote, or actinolite; a common product of low-grade regional metamorphism of iron rich igneous rocks like basalt.

Slate: A compact, fine-grained metamorphic rock that possesses slaty cleavage and hence can be split into slabs and thin plates.

Phyllite: A metamorphosed rock, intermediate in grade between slate and mica schist. Minute crystals of graphite, sericite, or chlorite give a sheen to the surfaces cleavage or schistosity.

Schist: A strongly foliated crystalline rock, formed by dynamic metamorphism, that can be readily split into thin flakes or slabs due to the well developed parallelism of more than 50% of the minerals present, particularly those of the lamellar or elongate prismatic habit, like mica.

Mica schist: A schist whose essential constituents are mica and quartz, and whose schistosity is mainly due to the parallel arrangement of mica flakes.

Gneiss: A foliated rock formed by regional metamorphism, in which bands or lenticles of granular minerals alternate with bands in which minerals having flaky or elongate prismatic habits predominate. Generally less than 50% of the minerals show preferred orientation.

Quartzite A granoblastic metamorphic rock consisting mainly of quartz and formed by recrystallization of sandstone or chert by either regional or thermal metamorphism.

Sedimentary Rocks:

Sandstone A medium-grained clastic sedimentary rock composed of abundant sand-sized fragments, which may have a finer-grained matrix (silt or clay), and which is more or less indurated by a cementing material.

Limestone A sedimentary rock consisting chiefly (more than 50% by weight or volume percentages under the microscope) of calcium carbonate, primarily in the form of the mineral calcite.

Shale A laminated, indurated rock having more than 67% clay-sized minerals.

Conglomerate A coarse-grained clastic sedimentary rock, composed of rounded to subangular fragments larger than 2 mm in diameter typically containing fine-grained particles in the interstices, and commonly cemented by calcium carbonate, iron oxide, silica, or hardened clay.

Researching and Staking Mining Claims

One of the greatest opportunities still available to Americans is the opportunity to stake and own your own mining claim. Much of the western US consists of lands which are open to claim by any US citizen (or anyone who has declared their intent to become a citizen). It does involve following a certain series of legal steps and filing paperwork with both the local county and the federal government, so the process is not as simple as it could be. Although one does not have to be a lawyer to stake a claim, a certain amount of knowledge of the law is needed to complete the process. In addition, every prospector ought to know a little about mining law in order to keep out of trouble.

As an example, a while back I made a trip to the BLM office in Sacramento where all the California mining claim information is kept. The reason for my visit was to do a little research on some placer claims located in an area I was interested in claiming. I wanted to see if any of the land there was open. I looked the spot up in the BLM's Mining claim database on the Internet (MLRS), and there was not enough information to really determine the actual location of the claims in the area - that's why I went to Sacramento. Every claim has a government file with all the paperwork ever submitted in relation to that claim. The BLM keeps the files at the state BLM headquarters. For Nevada, it's on Financial Blvd. in Reno, for California it's on Cottage Way in Sacramento. The Arizona office is located on Central Ave. in Phoenix. These files are public record and can be accessed and read by anyone. An examination of the records at the BLM office made clear why the MLRS claim database seemed confused. The original records were wrong. Two claimant groups had claimed over the same area, but because all the maps and claim location descriptions were erroneous, it was impossible to sort out who had the legitimate claim to the area or even exactly where the claims were located.

One claimant, with 4 signatures, claimed 80 acres as an association placer claim (that's OK) but his description of the claim was 400 acres in size (and that's not OK) the pieces of the claim did not adjoin or even touch (again, not OK). The map he filed with his claim paperwork showed several claims in the area. None of the claims on the map was his! (That is not OK - the map is supposed to show where YOUR claim is!). The other claim, which had been filed by a different claimant, had 2 signatures and was about 340 acres in size (again, this is not OK - the max is 160 with 8 signatures). He did not even bother to submit a map (violation of the code of Federal regulations on Mining claims). BLM had contacted him and was asking him to fix his filing as it was not legal. Although it's hard to say exactly as the two claim filings are so messed up, it's likely that the two claims overlap, and probably are both covering virtually the same area.

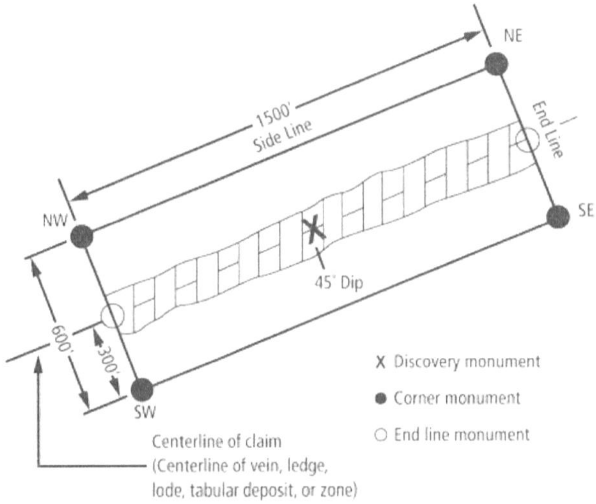

Claim maps must show how your claim is situated and how the various corner markers are laid out. It is supposed to tie into the federal land survey to show where it is located.

Failure to properly locate and describe a claim and complete the necessary paperwork can lead to the claim being invalidated. Although the majority of claims are okay, a surprisingly large number look something like the ones I mentioned above and have such major problems that they are technically invalid. Anyone who files a claim wants to make sure his claim is right and so it's important to take the time to do all the paperwork properly and understand what you're doing. Even in the best of cases, bad paperwork may encourage and lead to other prospectors staking over your claim and creating considerable confusion and frustration as to who the real owner actually is.

Some of the details regarding procedures for locating mining claims on Federal lands have been left to the individual states to determine, however state laws may not contradict or alter the underlying Federal laws (30 USC 28; 43 CFR 3831.1). Because of this, some of the requirements to stake a claim vary a bit from state to state, so the reader must in the end, do some research to find the exact details of mining claim law established within his state to insure his claim will be valid. However, because all state mining laws are established upon the federal law they are very similar, and many of the steps involved in locating a mining claim are the same within all states. This chapter will emphasize the requirements of the federal law, which apply to all the states.

It is important to remember that not every state signed on to participate in the original 1872 mining law. It is not a situation where the mining law rights have been stripped away from certain states. The law never applied in the 31 states not covered by the law. While we are used to federal laws that apply exactly the same in all of the 50 states, such is not always the case. Perhaps the most infamous of all laws which applied unevenly among the states are those laws which led to the civil war: laws which allowed slavery in certain southern states. The 19 States in which a citizen may locate mining claims or sites include Alaska, Alabama, Arizona, Arkansas, California, Colorado, Florida, Idaho, Mississippi, Montana,

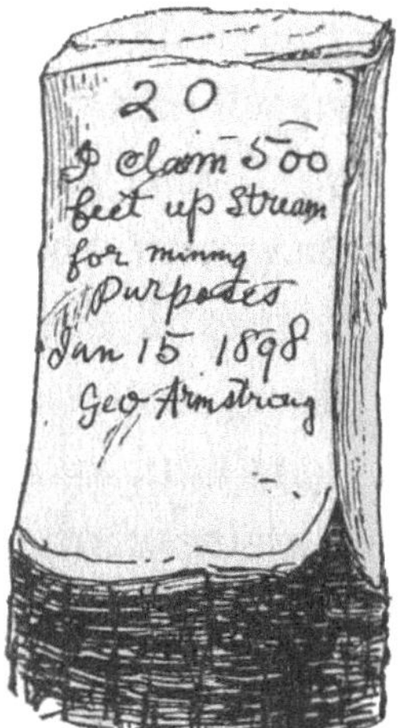

In the past, claim notices were posted in many ways. This note was made on a tree cut down to become a monument during the great Klondike gold rush of the late 1890s.

Nebraska, Nevada, New Mexico, North Dakota, Oregon, South Dakota, Utah, Washington, and Wyoming. Many of these states have large and significant amounts of potentially locatable federal land within their borders. The BLM manages the surface of public lands in these States and the Forest Service manages the surface of National Forest System lands. In the other 31 states, the amount of federal land which is potentially locatable is generally much smaller.

Many miners and prospectors know that the granddaddy of all US mining law is the 1872 Federal mining law originally titled "An Act to Promote the Development of Mineral Resources of the United States." All mining claim procedures are still based on this law, but the original scope of the law has been changed by a number of legislative and regulatory actions. Most of the Federal laws concerning mining claims on public land can be found in the United States Code (USC) under Title 30 "Mineral Lands and Mining" and Title 43, Chapter 35 "Federal Land Policy and Management" (FLPMA), and in the Code of Federal Regulations (CFR) under Title 43 "Public Lands." These laws have been interpreted and refined through numerous court cases. As a result, even though mining claim law is established by laws passed by Congress, it is also governed through regulations established by land management agencies like the BLM, and precedents established by the result of court cases previously decided.

The General Mining Law of 1872, as amended, is the major Federal law governing mineral acquisition by the location and maintenance of mining claims. This law opened the public lands of the United States and allows US citizens the opportunity to explore for, discover, and purchase certain valuable mineral deposits on those Federal lands that are open for mining claim location. Its intent was to accelerate development of the mineral resources in poorly populated portions of the US in the late 19th century, and this law has given many opportunities to the citizens that have availed themselves of its rights. Even in the states where the mining law applies, certain federal land areas are withdrawn from any mining claims or mineral entry. These include National Parks, National Monuments, American Indian reservations, most reclamation projects, military reservations, scientific testing areas, most wildlife protection areas (such as Federal wildlife refuges), and lands withdrawn from mineral entry for other reasons. Lands withdrawn for power development are subject to mining location and entry only under certain conditions. Mining claims may not be located on lands that have been: 1) Designated by Congress as part of the National Wilderness Preservation System; 2) Designated as a wild portion of a wild and scenic river; 3) Withdrawn by Congress for study as a wild and scenic river. The purpose of an unpatented mining claim is for mineral prospecting, mining or processing operations, and uses reasonably related to such mining operations. A claim holder does have the right to prevent others from prospecting and mining on the claim and as well as the right and duty to safely secure his operation from trespassers. However, the claim holder cannot prevent others from crossing the claim for uses recognized under the Multiple Surface Use Act of 1955 (30 USC 611-615) or engaging in lawful recreational activities provided that they do not interfere with the claim holder's operations. Deposits of gold, silver and other metals are locatable under US mining claim law, as well as most gemstone deposits. Materials which are not locatable include things like deposits of oil, gas, salts, phosphates and coal. Some deposits like building stone or sand and gravel cannot be located but may be leased. Other non-locatable items include things such as fossils or meteorites.

When doing claim research, normally the first step in considering a location to be claimed is to determine the land status of the spot. Large amounts of land in the Western US are controlled by the Bureau of Land Management or the Forest Service. With certain exceptions like wilderness areas, most of the land controlled by these agencies is open to prospecting and the filing of mining claims. As I noted, certain lands are not open to prospecting, including those controlled by the Department of Energy, Department of Defense, or Park Service are generally not open to claim. There are also patented mining claims where the government has transferred property ownership to the mining claim owner. Patented mining claims are private property, and like other private property lands, these areas are not open to claim. Topographic and other maps are available to determine the section, township, and range of the location according to the cadastral survey of United States. This special survey locates lands across the US in both remote and rural areas and is the basis used by the government for surveys of land both public and private. In the cadastral survey grid, the land is divided into blocks of 36 sections with each one a square mile in size. These blocks are identified using a townships and range number measured off in a north - south and east-west direction from a central baseline and meridian point. T29 N means that township is 29th one north of the meridian point. The same is true for R 10 E being 10 ranges east of the baseline point. There are a number of these location points across the US and one or more may be used in each state. Special mineral rights ownership maps and records are available from the BLM which show the exact status of mineral rights for government lands. These maps can be consulted to determine if the area of interest is located on claimable lands. These maps also show areas that have been withdrawn from claim activity for various reasons such as wilderness areas or wildlife refuges. There are also some areas where the BLM or Forest Service control the surface rights, but do not control or own the underlying mineral rights. Be sure to check all the right maps and carefully

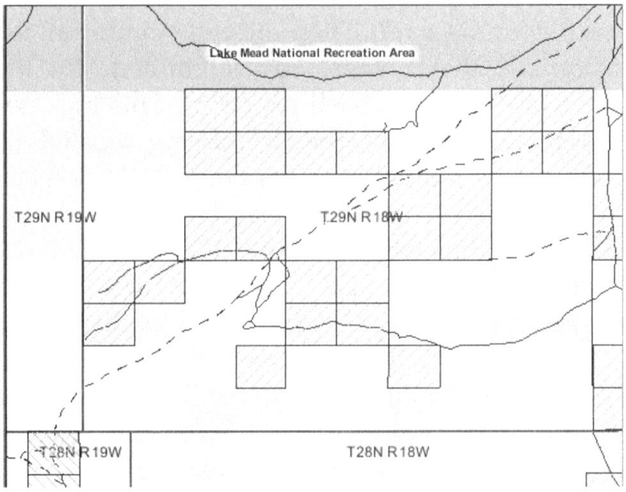

An example of the claim status products which can be generated with the BLM's internet based MLRS mapping program. (**https://mlrs.blm.gov/s/**) *The map shows a part of the Gold Basin District in Arizona. Hashed portions denote the presence of existing claims. Blank white is private property.*

determine the true status of the mineral rights on the land you may be intending to prospect or locate a claim on.

The next step is to determine if the land has already been claimed by a previous locator. The BLM has an online database which I mentioned earlier which they call the MLRS (Mineral and Lands Record System). The new web-based system is a considerable improvement on the previous LR2000 system. The on-line version can be found at **https://mlrs.blm.gov/s/** . Because 60 to 90 days are allowed to file paperwork with the BLM, claim records are a bit dated sometimes. The on-line records are usually kept fairly current, but it is certainly not perfect as there can be times when staffing problems cause new claims to take a bit of time to get entered into the system. BLM does offer a geographical map read out that is a GIS map based system, called the MLRS Research map. Be patient as it takes a while to load onto your computer. This handy tool had been off line for many years since early in the Obama administration, but it available once again. It is great for rough research and it quick to employ, but it is not 100 percent accurate partly doe to the delay in getting the records into the system. If you really want to know exactly where a claim is located, and learn for certain if the claim is valid or not, you need to go to the state BLM office and pull the file for the claim. Then you can examine all the materials the owner has submitted to the government to establish his claim. This includes the map of where the claims are located as required by federal law. A lot of information can be determined including both old invalid and current claims once you have determined the Section, Township and Range of the location you're interested in. If a claim had formerly been staked at the location you are interested in but it has become invalid - such as the claim owner did not file his annual assessment work, etc., then the land becomes open again, and it is open for prospecting and claim staking. It is important to follow the rules and keep your claim current and valid, because if you do not, then folks can and will come in and stake new claims on the area. If the land is owned by the government and it's status is such that it's available to claim, and no one has previously filed a claim on it, then you have the right to both prospect and file a claim.

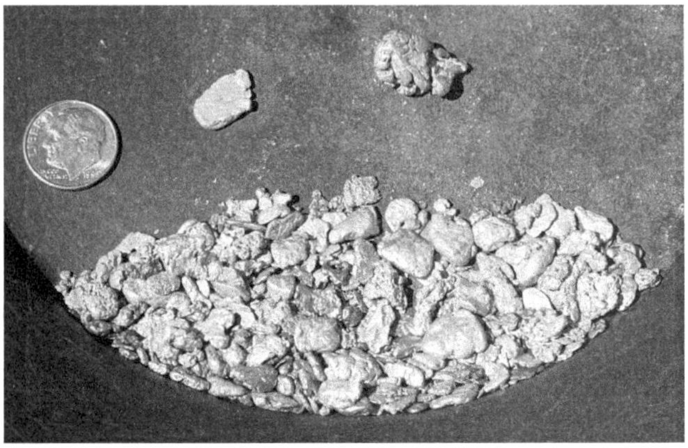

Placer gold is material like this – loose nuggets, flakes and chunks of natural gold.

on that site. Federal mining law (30 USC 26; 43 CFR 3831.1) actually requires that a discovery be made before a claim can be considered valid. However, the courts generally have recognized that a certain amount of time is necessary to make a discovery. The discovery can be made either at the surface in an outcrop or in some sort of excavation made below the surface. At the present time, because the government is using both the "prudent man" and "marketability" tests, most discoveries are made after the date of location, and many claims technically do not have valid discoveries. Sometimes prospectors are confused by the difference between Placer claims and Lode claims. Under US law, mineral deposits must be categorized as either lode or placer and the appropriate type of claim filed (43 CFR 3840). Placer claims are taken on "all forms of deposit except veins of quartz or other rock in place" (30 USC 35 - citation from US Federal law). This includes tailings from lode claims, gravels, ancient and buried gravels, and other loose materials derived from rocks or veins. A standard placer claim is 20 acres in size but with placer claims you have the option of filing an association claim. An association placer claim allows groups of individuals to claim up to 20 acres per person, up to a maximum of eight people with 160 acres. If you take the maximum size which is 160 acres, that is a full quarter section of land.

Lode claims on the other hand, are taken on deposits in the form of veins, lodes or other rock in place (30 USC 23). The maximum size for a lode claim is 1500 feet along the length of the vein or deposit with up to 300 feet on either side from the center line of the deposit (this gives a total lode claim size of 1500 x 600). There is no provision for larger claims taken by groups of individuals such as there is with association placer claims.

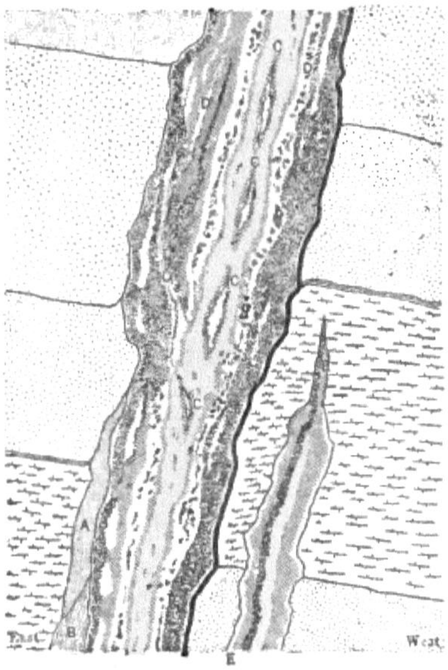

Vein or lode deposits are mineral matter in place, fixed between two walls of solid rock material.

With the differences between placer and lode claims, some prospectors will wonder if it is legal to file a new claim over an old one where the old claim is a lode and they are interested in the placer deposits on that claim. In US case law (the results of previous court cases that set a precedent to interpret US law); the courts have decided that it is illegal to file a placer over a lode or a lode over a placer unless one has the permission of the owner of the valid claim. There fore you cannot legally file a lode claim to prospect on someone else's placer claim. In the same way, you cannot (without permission) enter onto a lode claim and stake a placer claim to mine the placer material on that claim. (Clipper Mining vs. Eli Mining and Land Co Case). Sometimes folks are unsure what the nature of the deposit they are exploring actually is, so they will think about filing both types of claims on the same ground. That is not illegal - but I think it's a waste of money. I always advise folks to decide what they are truly interested in and stake accordingly. If you are really interested in panning, dredging or metal detecting, you are interested in the placer and should file your claims as placer. If you plan on leasing your claims to a big mining company for hard rock exploration, then stake hard rock claims.

There is no limit to the number of claims an individual or company can stake, but the federal government charges $140 per year per claim as a rental fee, and if you're going to be staking hundreds of claims that can really add up to a large amount of money (the $140 is the current annual fee as I write this, but the BLM adjusts the amount periodically for inflation). Large mining companies who do own thousands of claims pay hundreds of thousands of dollars each year in federal fees. If an individual holds 10 claims or less he can file with the federal government for a lease fee waiver. Although some other fees still apply, the $140 per year fee can be waived, which can save a lot of money for the individual prospector and other small scale operators. To actually file and legally record our claim, we need to follow certain procedures.

Each state has slightly different requirements for claim markers. Here are examples of some that are legal in the state of Nevada.

All claims require some form of monument to be placed on the ground to physically mark the location of the claim. The exact requirements for staking and posting mining claim monuments are one of the things that does vary a bit from state to state. It is also a bit different depending on if you are locating a placer or a lode claim. As an example, Nevada will allow a wooden stake one and a half inches square as a legal claim monument. However, in California, a post that small in diameter would be too small to legally count as a claim marker. You may need to research your own state's requirements a bit, but a 4 inch square wooden post, long enough to stick out of the ground 4 feet above the surface is sufficient enough in nearly all states. A mound of stones at least 3 feet in diameter and 3 feet high is also a sufficient marker in most states. At one time in Nevada and elsewhere, 4 inch PVC pipes perforated for septic systems were commonly used for claim monuments. Because of environmental concerns, these have been outlawed in the federal rules and they can no longer serve as claim markers in Nevada. PVC pipe of any type should not be used to mark new claims. In addition, they are easily blown away in harsh weather. The exact positions on the claim which are required to be marked by a monument also varies from state to state.

Mining records can be researched at state BLM offices. This is the Nevada claim records office.

Most states require marking the corners of any lode claim, but some also require side or end line markers. Again, you will have to research the specific requirements of the state you're interested in to be certain that you are properly marking the monuments on your new claim. Each claim corner or sideline must have some sort of mark indicating which claim corner it is, such as the "northeast corner of the Big Nugget claim". In general, lode claims are taken at the maximum size which is 1500 feet long and 600 feet wide. This is supposed to be structured such that 1500 feet are taken along the length of a lode or mineralized geologic structure, with 300 feet on either side from the center line of the lode or structure. Normally this makes for a rectangular shaped claim, but that shape is not required by a law and sometimes claims are taken in other forms like triangle and squares. Claims can be taken in smaller sizes where that is desirable, including fractional areas that might exist between claims.

Placer claims are located somewhat differently than lode claims in that they can be located according to the boundaries of the cadastral public land survey. This allows the locator to claim up to 20 acres described as a small part of the public land survey. In fact, this is the most common way placer claims are made, simply because it is so much easier than marking all the claim corners. In addition, federal law allows groups of prospectors to form an association and claim up to 20 acres per person with a maximum of eight people per claim taking 160 acres for the group. In most states, when taking a claim described as a part of the public land survey, it is not necessary to mark or monument each corner. Usually, only the location monument is required. This means that the location monument may be the only monument required for the staking of a claim which could be up to 160 acres in size. When describing your claim using the public land survey, it is important to understand that the description is the actual claim you're making. If you location paperwork says your 20 acre claim is the northeast quarter of section 17, it does not mean that your claim is 20 acres inside the northeast quarter of 17 (as is common with lode claims), it means that you are claiming the entire northeast quarter of section 17 which is 160 acres in size. I've seen many folks say they are taking up a placer claim by the public land survey, but they're paperwork will say something like the claim is the northeast and southeast corners of a certain section. Thus, their paperwork is saying they are claiming 320 acres, which is not a legal claim. So when making a placer claim based on the public land survey, it is important to be very accurate and precise in describing your claim or your claim may be an invalid one. A 20 acre claim might be described as the S ½ of the NE ¼ of the SW ¼ of section 12. Twenty acres is half of a quarter of a quarter section. In fact, if you are artful and precise in designing your claim according to the public land survey you can sometimes take in a considerable length of stream or river - this summer some family members and I staked an association placer claim that takes in almost a mile and a half of stream yet is only 160 acres in size. Placer claims can also be taken without using the public land survey by marking the corners in a manner similar to lode claims.

Claims like this are sometimes staked to follow a stream or other type of placer deposit.

All state laws require the marking of a location monument within the claim both for placer and lode claims. This monument will contain the claim location notice, signed by the locators and other information concerning when the claim was filed, its general course or direction, the address of the locator, the county the claim is located in, etc. Copies of various forms for claim staking with blanks to be filled in by the locator are available on the Internet. The forms vary slightly depending on the state in which the claim is to be located because the differences in mining claim law from state to state. Once the claim has been properly staked, and all monuments posted and marked, it is necessary to file paperwork with both the BLM and the County recorder for the county in which the claim is located. The period of time in which the locator must get this paperwork filed with the local County recorder does vary somewhat from state to state - in Nevada the time allowed is 90 days. For the BLM all necessary paperwork must be submitted within 90 days after the date of location. Both the BLM and the local County recorder will charge fees for processing the claim paperwork. At this time, the BLM fee for new claims is currently $189 per claim - it is adjusted periodically for inflation. For local County recorders, the fees vary wildly depending on the location and your fee may be anywhere from as little as $10 to as much as to $369 (this is the current cost for 160 acre association placer in the state of Nevada). If you mail in your claim paperwork and forget to include the necessary fees, you will not have a valid claim. Because of this, it is worthwhile to call the County recorder and perhaps the BLM office to confirm exactly what your fees will be, especially if you mail them in, so you can send the correct amount. I strongly recommend that you don't wait until the last second to submit your paperwork because if there are problems which you need to resolve, it can make it real difficult to resolve them before the required filing date expires and you then have to start all over and re-stake your claim.

The forms for the location notice paperwork you will need to prepare and file with both the BLM and the local County recorder can be found at some mining supply stores as well as on the internet. Federal laws, as well as some state laws require the filing of a map of the claims. It is not necessary to hire a draftsman or an engineering firm to prepare the map, but it must be legible and accurately show the location of the claims. Some state claim laws, such as those of Nevada specify the map scale, size of paper and other specifics of the map. Having an accurate map is to the benefit of the claimant as it can help to ensure that others will not be unintentionally filing claims over yours. Once your claim is filed, you have to maintain it. Each year work must be done and additional paperwork filed in order to keep your claim valid. There is a fee of $140 per claim which is charged by the BLM. However, if

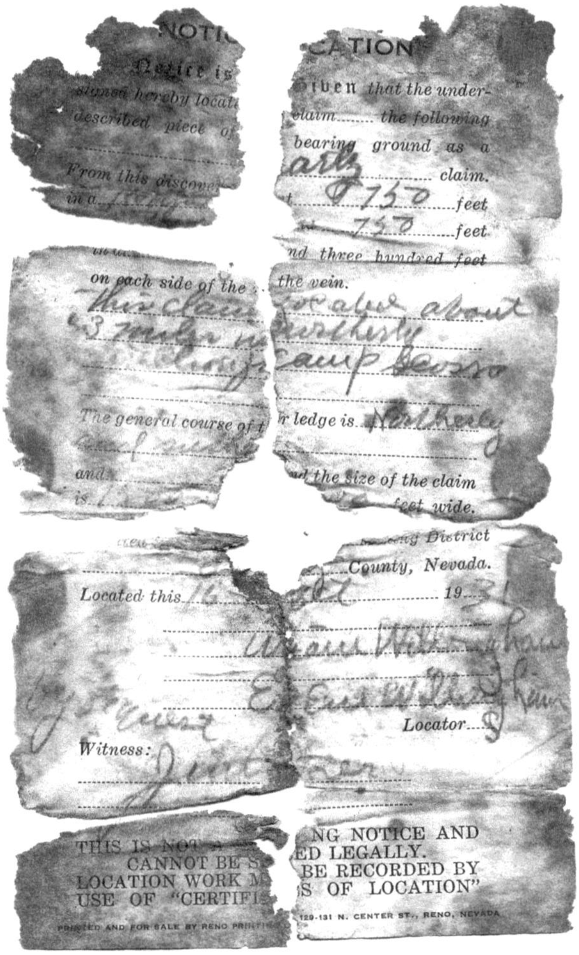

Legal mining rights may hang on the validity of old mining claim locations like this one, staked in 1931. Assembled like an ancient document, this location was made during the Scossa gold rush.

the prospector owns 10 or fewer claims he may file for a "small miner's" exemption to the $140 per year claims fee. The 10 claim exemption is actually figured per household and to get the exemption, you have to sign a paper certifying that you and your household have 10 or fewer claims. Payment of the fee or filing of the waiver paperwork must be received by the BLM by September 1 of each year. If you file for a claim fee waiver, you must perform at least $140 worth of assessment work per claim per year to hold them (the amount changes and is adjusted for inflation from time to time). Even if you pay the $140 claim rental fee, you must still file assessment work, both with the BLM and the County. The "year" allowed for assessment work runs from September 1 through August 31. Assessment work is labor and other work that you perform which benefits and develops the claim for production. Examples include repair of claim markers and access roads, geologic mapping and testing, collection of ore samples, and mining a small test lots. On blocks of claims which adjoin each other, work done on one claim may count as work done for the benefit of all the claims in the group. Forms and paperwork documenting the assessment work and the findings or accomplishments of that work must be filed with the BLM before December 1. The BLM still charges a fee for this, but it is small. Assessment work papers must also be filed with the county recorder and any required local fees paid. The timeframe allowed for filing paperwork with the County varies from state to state, ranging from 30 to 90 days.

To put yourself in the best position legally, you need to be doing all your assessment work properly and filing the necessary paperwork in a timely manner. Not doing the right type of assessment work or failing to follow the legal requirements for proper assessment work can put you in a deep hole if someone wants to make a legal challenge to your mining claim. For your assessment work to be valid you must put in at least the required amount of effort working on each claim group that you hold. Your annual assessment work can actually be any type of labor or improvement that will benefit the claim or group of claims you hold. While this is a very general definition, your assessment work must be something that tends to disclose or develop the valuable minerals present on the claim. It must be useful and productive for your claim. If it can be shown that the work you did was not beneficial for development and future mining, your claim may be declared invalid by a court of law. Some examples of beneficial work that would qualify include the cost of digging and earth moving on the claim: this includes cutting trenches, sinking shafts, running adits or dewatering and cleaning out old workings. Most specifically, assessment work includes the activity of mining or extracting ores. For many small placer miners who work their claims, this is the most important form of assessment work which they conduct. The law also allows that geological, geophysical, and geochemical mapping or surveys may qualify as assessment work, but only for a limited period, and only when performed by a qualified expert who possesses the background and education to conduct that type of survey. The various forms of geological surveys may not be used as the annual assessment work on a claim or claim group for a period of more than two consecutive years, and they cannot be used for more than 5 years in total. The survey work cannot be of a repetitive nature or duplicate the work of a previous survey done on the same claim. In order to use such surveys as your assessment work requires that you file a detailed report on the results of the work. If you do this type of work for too many years in a row, then it will be considered repetitive and may not qualify as valid assessment work.

Valid assessment costs include not just the labor costs, but also the value of any materials which were consumed in these activities (such as explosives or fuel). Costs of labor calculated for the work must be reasonable and figured using the prevailing wage rates for the area - you cannot figure your time at $500 an hour unless you can demonstrate to a court that you really are worth that amount. A common mistake which a lot of small miners make is to fail to properly describe the work they have done in the assessment paperwork. While they may have done a legal amount and type of work, paperwork can be a pain and they may not take the time to properly describe in writing a full and complete account of what was done in the assessment paperwork. Neglecting the proper description of your work may lead directly to legal challenges and problems. If the claim owner holds a group of claims, the full amount of assessment work labor and expense does not have to be spent on each individual claim. So long as the claims are contiguous (that means the group of claims all share at least one common sideline, not just touching at a claim corner), then work done on one claim is generally held to be done to the benefit the entire block of claims. In this way, the situation of a placer miner who holds three contiguous claims and mines on one them, the mining he does on that one claim is taken to benefit the development of all three of his claims.

Assessment work can even be done outside the claim block in cases where the effort is done clearly to benefit the claim group. Federal law requires that all assessment work be performed within a certain specific period of time referred to in the law as the assessment year. Work done either before or after the allowable deadline must be counted only for those other years. The legal assessment year begins at noon of each September 1, and ends at noon on the next September 1 the following year. It's important to plan ahead because if you miss doing your work within the proper period, you cannot go back in time and fix the problem. Additionally, paperwork must be filed at the correct office. One cannot just file paperwork at any BLM office - claim paperwork must go to the appropriate BLM state claims office in the correct state.

While the right to own a mining claim really is a great opportunity, there are sometimes disadvantages to staking a mining claim. A lot of individual prospectors choose to have no claims, while others decide to stake and hold claims. It is not a simple thing that one way or the other has all the benefits. There are benefits of having a claim, but there is a downside, too. Probably the most important thing for the individual prospector is that any secrecy about your discovery goes out the window forever if you file a claim. All claims are part of the public record. In the BLM's MLRS claim database, you can search by location, but you can also do a search on the last name of the prospector who may own claims and find out if he has any claims, and then their exact locations. It is possible to go to the main BLM office in any state and pull the files for his claims and check out all his paperwork. No secrecy is one of the biggest disadvantages of filing a claim. People can check if

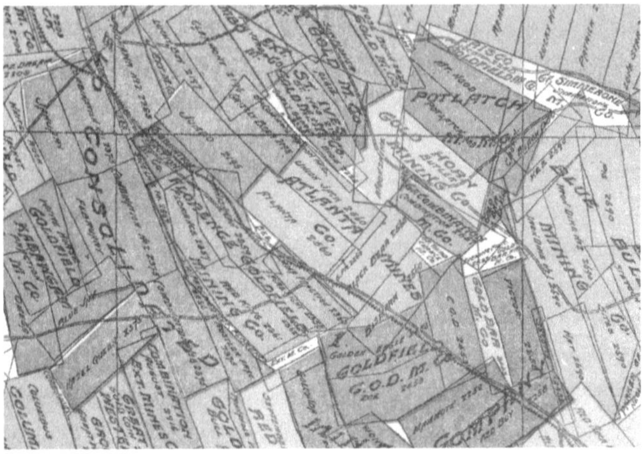

There is a tangled mass of overlapping claims in the heart of the Goldfield district in Nevada.

you have a mining claim and where it is located. Another of the big disadvantages of having a claim is that you may find that you have to actively defend it. Many prospectors, who understand the claim system and respect the rights of others, will know where existing claims are and respect their boundaries and stay off them. This is not the case for everyone. Many will go in to an area to prospect and have no idea where existing valid claims are located. Others, even if they know where the boundaries of existing claims are, will not respect them. In addition to finding others working your claim without permission, you may find other prospectors staking new claims right over the top of your existing valid claim. I know that in the early 1980s, when the price of gold was high and the interest level in prospecting was at a maximum, there were places on the North Fork of the Yuba where 4 or 5 layers of new mining claims were staked on top of older, valid claims.

The BLM will generally not step in to try to settle any dispute between claim owners. It makes sense to notify both the BLM and any adverse claimant of the conflict with your claim, however any paperwork you file with the BLM about another claim conflicting with yours is mostly intended for setting up a legal action. If the person who has claimed over you has only made an honest mistake, a notice of the problem should resolve the issue. Unfortunately, it does not always work that way. When you file a new claim, the BLM will usually send you notice if your claim conflicts with another (older) claim or if it conflicts with private ground (I won't guarantee they always do this, but they usually do). BLM will also notify you if they find you have failed to file your annual paperwork, and they will eventually declare your claim null and void if you don't file your annual paperwork. That's about all they will do. They will not jump in to settle a dispute – normally you need lawyers and a judge and 99.9% of the time it just becomes a big, very expensive can of worms. You can see with the expenses involved how if the conflicting claim owners are hot heads and both believe they own the claim, where such a situation could even lead to violence. As a claim owner, you need to know that campers, fishermen or other folks may come onto your claim legally as they have a right to camp, hunt, hike, shoot target practice or any other non-mining related activities on the federal land that constitutes your claim. The mining claim only gives you rights to the minerals and deposits on the land you claimed. Unless you have an active, permitted operation which could be a hazard, you cannot keep them from accessing the land. However, if they leave behind trash and junk or otherwise mess up your claim, the BLM or forest service may well blame you for their mess and try to order you to clean it up. There can also be liability issues that may be associated with old mining operations - contaminated ground or even old holes that people could fall into. In the end, there are a lot of benefits to owning a good mining claim, but the claim owner has to keep his wits about him, know his legal obligations and requirements, be aware of people coming on your claim both legally (to camp, hike, etc.) and illegally (to prospect or extract your minerals), and take the appropriate actions. There is no doubt that as the price of gold and other mineral commodities continues to rise there will be more and more claims staked.

One thing I need to talk about is the possibility of buying a mining claim. Many new prospectors, when presented with a number of claims already out there in the hills, decide that their best alternative is to buy their own mining claim. There is a certain romance about it - owning your own mining claim just sounds really good. Typically, these claims are purchased via ebay, the Internet or some other source like a magazine ad. While the idea of owning your own gold mining claim does sound really great, there are a huge number of drawbacks if you don't have the skills to sample and evaluate them yourself. Begining with the earliest days of mining claims, there have always been dishonest folks who sought to pawn off worthless claims. In fact, more than a century ago the famous author Mark Twain wrote that the definition of a mine was a hole in the ground with a liar standing next to it. He wrote how a mining claim owner would search high and low on a claim to find even one tiny piece of good ore; he would then have that tiny piece assayed. It would show rich values, and then he would claim that the entire mining property was filled with this rich high-grade ore. Other tales were told of claim owners who would fill shotgun shells with small bits of placer gold, stand back and shoot selected areas of the claim with gold. The dishonest claim owner would then invite the prospective buyer to sample for himself the rich gravels of the claim (of course the rich spots were very small, but if the buyer only took a few samples he would find that the claim seemed to be exceptionally rich in gold). Things have changed only a little during the last century. One can purchase a perfectly valid claim (in so far as the paperwork is concerned) which contains little or no mineral value. I've seen dozens of these claims staked all over the West where a perspective claim seller will locate land which contains little or no mineral value that is in the general vicinity of an area which has produced good gold or silver. He will then seek to sell you claims which are located in the well-known so-and-so mining district, even though they are not in a part of the area which has ever produced any significant minerals. So, what is the value of a claim to the minerals on a parcel of land that has no gold or other valuable minerals? The value is nothing. So unless you have the skills and knowledge to do your own independent evaluating and know well enough not to be fooled, I think you are much better off joining a club than you will be buying your own claim.

Some charlatan mining claim sellers suggest you might convert your mining claim to private property. It is true that in the past it was possible to patent a mining claim or mill site so that the Federal Government conveyed the full property title to you, making it private land. A century ago this was fairly easy. The US kept making it harder and harder until on October 1, 1994, the BLM was prohibited by an Act of Congress from accepting any new mineral patent applications. It is unknown exactly how long this moratorium will continue, but it will likely continue always, and the chance to patent a claim is almost certainly gone forever. Certain extremist liberal politicians are pressing to change the US mining laws and take away all our rights to prospect and hunt for gold and minerals on public lands. I do not know what will happen in the future, but I urge you to keep up with this ongoing battle to preserve our rights and consider participating, either through clubs or other public land rights organizations like Public Lands For The People (PLP - see **https://www.publiclandsforthepeople.org/**). The GPAA Prospecting magazine covers all the aspects of these issues and can keep you up to date on the most current developments.

Finally, because of the lawsuit happy nature of our society, I need to offer some important disclaimers: I am not a lawyer and nothing in any part of this entire book should be construed in any way to constitute legal advice. I have not and do not offer legal advice or opinions because I'm not an attorney. I am not qualified to offer legal advice, so that's why I don't do it. If you need legal advice or other legal opinions, please consult a competent attorney. Mining Law is a complex topic, and many lengthy books have been written on that subject alone. My comments here are only of a general nature and do not cover the topic in a complete and detailed way, which is far beyond the scope of this book

Addresses For Some Western US BLM offices:

BLM Alaska
222 West 7th Avenue #13
Anchorage, AK 99513-7599 Phone: 907-271-5960

BLM Arizona
One North Central Avenue
Phoenix, AZ 85004-2203 Phone: 602-417-9200

BLM California
2800 Cottage Way, W-1834
Sacramento, CA 95825-1886 Phone: 916-978-4400

BLM Colorado
2850 Youngfield Street
Lakewood, CO 80215-7093 Phone: 303-239-3600

BLM Idaho
1387 S. Vinnell Way
Boise, ID 83709-1657 Phone: 208-373-4000

BLM Montana, North Dakota and South Dakota
P.O. Box 36800 (59107-6800)
5001 Southgate Drive
Billings, MT 59101 Phone: 406-896-5000

BLM Nevada
P.O. Box 12000 (89520-0006)
1340 Financial Blvd.
Reno, NV 89502-7147 Phone: 775-861-6400

BLM New Mexico, Kansas, OK and TX
P.O. Box 27115 (87502-0115)
1474 Rodeo Road
Santa Fe, NM 87505 Phone: 505-438-7400

BLM Oregon and Washington
P.O. Box 2965 (92708-3420)
333 SW 1st Avenue
Portland, OR 97204 Phone: 503-808-6001

BLM Utah
P.O. Box 45155 (84145-0155)
440 West 200 South, Suite 500
Salt Lake City, UT 84101 Phone: 801-539-4001

The Unsuccessful Prospector

Joining A Prospecting Club

Are you interested in getting started in prospecting for gold? Joining a club is an idea that has a lot to offer (and it's something I suggest to nearly all new prospectors). If you are new to prospecting for gold and just getting started, I have some suggestions to help get you going in the right direction. Perhaps the best way for a new prospector to jump in and get started is by joining a prospecting club. When I first started out looking for gold thirty years ago, there were very few prospecting oriented clubs or organizations, but all that has changed. I never heard of such a thing back then and I sure wish I would have been able to take advantage of that kind of opportunity back then. Today, there are both national organizations like the GPAA (the Gold Prospectors Association of America) which has many local affiliate clubs, as well as a large number of well-known independent regional or local clubs. The cost to join most clubs is comparatively minimal, especially when you consider the benefits they offer. From the chance to learn from more experienced prospectors, to the opportunity to prospect on claims for which the club has access, and even the chance to see the latest types of prospecting equipment in use in the field, being a part of the prospecting club is really a great opportunity. Even though I couldn't take advantage of club membership when I was just starting out, I am a member now and certainly will remain a part of one club or another for many years to come. While there are certainly plenty of benefits in being part of a club even for experienced prospectors, for new prospectors there is virtually no downside.

One of the biggest hurdles for most new prospectors is just finding a good place to hunt for gold where you will not be trespassing on someone's mining claims, and where you won't have to worry about stirring up any angry property owners. Most prospecting clubs have mining claims that are available to their members to use, and some of these are staked in some very good locations. Being part of a club that owns their own claims will give you access to places to prospect where you can have confidence that you can find your own gold without causing problems or breaking any laws. Some clubs even have group digging operations where the members band together to operate some larger equipment and then share the gold that is produced. If there are several different clubs in your area with claims, it might not be such a bad idea to join all of them, as that would give you the greatest amount of choice in your prospecting opportunities.

This is the AZO get together on the LDMA property at Stanton, Arizona. Club outings other or organized get-togethers like this one allow for training and education as well as making and renewing friendships.

The regular club meetings will provide you a chance to talk to other prospectors about the best local places to search, state and government regulations, as well as what really works to get gold in your neck of the woods. These things are all critically important for the new prospector trying to learn the ropes. It will also give you a chance to meet some folks with more experience in prospecting for gold. Attend the group outings and hang around with some of those more experienced guys. Don't be bashful - most will willingly share information about how to find gold, and their experiences of how they have found gold in the past. The friendships you will gain are worth far more than the cost of the membership dues. Even many experienced prospectors are members of more than one club, both because of the friendships as well as the prospecting opportunities they present from the different club claims. Because very few of us will ever really strike it rich, in the end, the fun and enjoyment we have when we are out prospecting is what it's all about, and having friends along can make a big difference in improving your prospecting experience.

In addition to regular meetings, many clubs have special field outings where the members get together to work on club claims or visit other claims on which the members have been invited to prospect. These can be a great learning opportunity for hands on experience with other prospectors, including those with more experience. Success in prospecting is very dependent on what you know and your ability to recognize likely spots where gold has been deposited. These outing events often turn out to be great opportunities to ask questions and learn from the more skilled members of the club. Additionally, club outings will often include things like pot luck cook outs, so be sure to join in and enjoy all the opportunities to make friends and socialize with your fellow gold fever victims. Every club has its own organizational structure and there are usually plenty of opportunities to get involved and make a contribution. In fact, many clubs suffer from the fact that only a few individuals are willing to volunteer to put in the necessary time and effort to keep the club running. Keeping a prospecting club functioning does take a good bit of planning and effort, and in my opinion, if you are going to partake of the good things the club has to offer, it's only right that you volunteer some of your time to contribute to the operations and upkeep of the club as well. Helping hands are usually very welcome and appreciated, so be sure to offer to do your part.

The Gold Prospectors Assn. of America (the GPAA) is a nationwide organization that puts on various events including the Gold Shows and various expeditions to places like California or Alaska. They have many affiliated local clubs.

My personal experiences with prospecting clubs have actually been very good. I've met a number of folks that really enhanced my experience and knowledge of prospecting; some of them have become friends for the long term. Through club membership, I've enjoyed opportunities to dredge, snipe, dry wash and metal detect for gold on a variety of club claims, and while I have yet to really strike it rich on any of those club claims, I've consistently been able to find some nice gold. Several of the club claims that I have been on have been among the most beautiful, wild and remote places I have ever prospected and were worthwhile visiting if only for the scenery. I expect that your experiences would likely be similar to mine.

First off, I'd suggest you join a prospecting club like GPAA (Gold Prospectors Association of America) or one of the other well-known regional/local clubs. Start by doing some research and seeing what clubs are active in your region. Many prospectors are actually members of more than one club, and some are members of several different clubs. If there are several different overlapping clubs in your area it might well be a good idea to check them all out and see which ones seem friendly and interesting and join all the ones that seem good to you. Most have very low cost to join, but a few are higher cost. Be careful with the higher cost ones because while most of them are fine, there are a few higher cost organizations I have heard of that are not really set up as prospecting clubs, but are more investment type organizations. Before you put a lot of money into a "little known club" ask around to the more experienced guys and get their opinions of it.

For those you who have never prospected before, and have never found any of their own gold in the wild, I'd suggest starting small by purchasing a gold pan and some basic digging tools. Start out slow and let your equipment inventory grow with your experience. Spend some time in the field and learn some of the basic skills and techniques of prospecting for gold first. Then once you have the confidence that you can succeed in finding gold, consider bigger prospecting equipment like a detector or a dredge. The experience you gain learning how and where to find gold will help you later when you decide to buy that expensive equipment. You don't need to wait forever, but don't get overly excited and rush things either. If you really have the gold fever, you'll be genuinely thrilled even by those first few colors in your pan that you found yourself in the wilds through your own skill and effort (trust me, I know).

Prospecting has long been an activity that is best shared with friends. Although there are new organizations and types of equipment, friends still remain the truest form of gold you can find.

There is a lot to learn about finding gold and it's unlikely you will strike it rich right off the bat. Be patient and stick with it - the gold nuggets are still out there and your finds will come in time. Dogged persistence is an important asset for all prospectors - but is especially true for those who are just getting started. For your convenience, I have assembled a listing of links to some of the prospecting clubs across the United States. There are FAR more clubs than the few listed here. This is just a selection for you to start with. Further research will show what clubs exist in your area, if your area is not listed in any of these links.

The GPAA (Gold Prospectors Association of America) is a nationwide organization that has many, many affiliated prospecting clubs located throughout the United States. For a complete listing of all the local chapters including the necessary contact information of the GPAA clubs, see this website (Nationwide, USA)
http://www.goldprospectors.org/membership/chapters/Index.aspx

There are also many independent clubs, especially in the western US. Here is a partial listing of just a few of them:

Havasu Gold Seekers (Lake Havasu Area, Arizona)
http://www.havasugoldseekers.com/

Roadrunners Prospectors Club (Arizona)
http://www.roadrunnergold.com/

Gold Searchers of Southern Nevada (Southern Nevada)
http://www.goldsearcher.com/

Comstock Gold Prospectors (Northern Nevada, Mother Lode in Calif.)
http://www.cgpgold.org/

Golden Caribou Mining Assn. (Northern Mother Lode Country)
http://golden-caribou.com/

Mother Lode Gold Hounds (Mother Lode Country, California)
http://www.goldhounds.com/

New 49ers Prospecting Club (Klamath River Area, California)
http://www.goldgold.com/

West End Prospectors Club (California)
http://www.wepcgold.com/

Valley Prospectors Inc.
www.valleyprospectors.net

Nye Gold Seekers (Nye County, Central Nevada)
http://www.nyegoldseekers.com/

Northwest Mineral Prospectors Club (Washington and Oregon)
http://www.nwmpc.com/

If the links above do not provide what you are looking for, do some research on Google looking for prospecting clubs in your state. There are both independent and GPAA affiliated clubs all across the US and there are various organizations for prospectors in Canada and Australia.
Lots of folks will ignore my advice to join a club to get access to prospecting ground but after years of talking to new folks wanting to get involved, I stand by it as sound advice.

How to Figure the Amount of Gold in a Gold - Quartz Specimen

Many prospectors who find specimens of rich quartz with visible gold often wonder about the gold content of their specimens. Jewelry rock types of pieces with significant visible gold can sometimes be found at hard rock mines, by prospecting with a metal detector, or sometimes they can be found even in rivers by dredging. The value of these specimens is dependent on the amount of gold contained in the piece, so it is important to know how much gold a specimen carries, especially when the prospector is planning to offer it for sale. Often the gold is largely encased in quartz, so it is impossible to see the majority of the gold without crushing the specimen. Since crushing the specimen might well destroy its value, the prospector needs some reliable method to figure how much gold is inside. This chapter presents a handy and easy method to estimate the gold content of these specimens without damaging the piece. The method gives a good ballpark number, but it is not perfect and gives only a rough estimate, not an exact measurement of the gold. It also only works with gold - quartz pieces such as "jewelry rock" which contain some significant visible gold. It also does not work well with specimens that contain only very small amounts of gold, since the weight of only a few gold specks is not within the accuracy of the results this technique produces. It will not work as an alternative to an assay for standard grade ores with little or no visible gold present. The method basically assumes that the specimen is made only of gold and quartz. This is why it also does not work where there are lots of other minerals besides gold and quartz present in the specimen, such as calcite, sulfides like pyrite or even iron oxides.

The method is based on the fact that gold is much denser than quartz, and the more gold that is present in a specimen, the higher the density of the piece will be as a whole. There are a number of versions of this same equation out there, but this one is a simplified version of the equation and works just as well as the more complex versions to calculate the gold content.

The first step in the process is to find out the overall density of the whole specimen. For this formula, we will call the overall density of the gold - quartz specimen "D". The density (D) is simply the dry weight divided by the volume of the piece. Weighing the dry sample is no problem, but the trick then is determining the volume of the piece, as most are very irregular in shape. The best way to figure D is by weighing the piece both in air (dry) and then weighing it again when it is submerged under water. The submerged weight will be less than the dry weight, as the buoyancy effect of the water will reduce the weight. The volume is then the dry weight minus the submerged weight. Because of this, the density, D, is equal to:

(Dry weight) / (dry wt. - submerged wt.)

This means that you take the dry weight and divide it by the result of the dry weight minus the submerged weight.

Gold bearing quartz specimens can be valuable, depending on the amount of gold present.

There is also rough and dirty way to figure an estimated volume by submerging the piece in a measuring cup or other measuring container. Most glass measuring cups for cooking have a volume measurement in milliliters or ml (as well as the normal cups, ounces, etc.). A crude estimate of the volume of the piece can be made by partly filling a measuring cup then seeing how much the water level goes up when the specimen is immersed in the water. If one could get a very accurate reading of the water displacement, it would be just as accurate as the procedure of weighing the piece dry and submerged as noted above, but usually measuring water displacement directly is simply not as accurate. The water displacement method, however may be the only way to get a volume measurement for some very large chunks of gold quartz that weigh many pounds, or even tens of pounds. Using measured displacement volume, D or density is then:

D = Dry weight in grams / volume displaced in ml

If the density, D, calculates out as 2.7 or less, there is not enough gold in the sample to use this method, as its just not accurate enough to measure a few very small bits of gold in a piece that is nearly all quartz. Once the volume of the specimen has been determined, the percentage of gold in a gold - quartz specimen is estimated as:

Decimal fraction Au = ((1.15 X D) -3.03) / D

This means first multiply 1.15 times D, and then subtract 3.03 from that number. Then take that result and divide it by D. That is the part of the sample that is gold (the number is less than 1). Take the fraction Au and multiply it by the dry weight of the sample and that is the weight of gold you have in the specimen.

To see how this method can work, let's go through an example gold quartz specimen. In the first step, we determine the density. In testing, we find that our specimen weighs 128.5 grams dry, and 96.9 grams when submerged in water. The density, D, for this piece would be:

128.5 / (128.5 - 96.9); this would be the same as 128.5 / 31.6, or 4.07 grams per ml or cc.

The decimal fraction for the gold would be ((1.15 X 4.07) -3.03)/ 4.07, which would be (4.68 – 3.03) / 4.07, or 1.65 / 4.07, which is 0.405. This number means that the specimen is 40.5 percent gold by weight. The amount of gold is 128.5 grams x 0.405, or 52.0 grams. Since there are 31.1 grams in a troy ounce, that means the specimen contains roughly 1.67 ounces of gold.

Another method for taking a look at specimens that contain gold encased in quartz without destroying them is conductivity testing. A voltmeter (available at many electronics shops) can be used to see if gold on one side of a specimen rock is continuous to the other. Since gold is a fairly good conductor, a voltmeter can be used to measure the electrical resistance across the piece. Where the gold is continuous, resistance will be low, perhaps only a few ohms, but if it is not continuous, resistance will be essentially infinite. If a piece has multiple spots of gold exposed, a meter probe can be used to test various combinations as only some may be electrically continuous.

Knowing how to estimate the amount of gold present in a quartz specimen can be very handy.

Etching Gold Quartz Pieces:

Some prospectors use a special solution of Hydrofluoric acid to etch some of the quartz off of their specimens to improve their appearance and value. The increase in value can be very dramatic. Often a combination of chipping, sawing and masking off parts of the specimen are involved in the conversion procedure. In many ways, this etching process is more of an art than a science and it is not always a simple matter to convert raw specimen gold-quartz pieces into spectacular gold quartz mineral specimens. Knowing the amount of gold that is present in a piece and whether or not it is continuous from one side to the other can help in determining if it is worthwhile to try to etch the specimen. As a reminder, acid products that etch quartz or glass (such as Hydrofluoric acid) and the gasses which they produce when reacting with quartz can be hazardous. Because of these dangers, before any such attempt should be made, the user needs to fully understand the nature of these chemicals and take all the necessary safety precautions when using them.

Platinum Placers and Nuggets

Sometimes placer miners will find silver-colored heavy materials in their concentrates. These small bits and nuggets of metal are associated with gold and other heavy minerals in the concentrates, but they exhibit a silver-gray color rather than the normal tone of gold. The question asked is, what is this? Well, there are a lot of possibilities such as steel or other junk metal, or even gold that has become coated by mercury. However, the most common occurrence by far is lead. Shotgun pellets, bullets, fishing weights and other bits of lead will often find their way into a stream and become concentrated along with heavy materials such as gold. In most California streams, lead is far more abundant than gold. However, in certain regions of the US, especially in parts of California and Alaska, there are gold placers which also contain platinum. Platinum is a white metal with a specific gravity slightly more than gold, and is found with the concentrates of the placers in various locations, including the basins of the Klamath and Trinity Rivers. Their particles are usually fine scales, only rarely reaching a quarter of an ounce in weight in California. The largest platinum nugget ever found in that state was a little more than two ounces. In Russia and some other locations, the pieces can be larger. Because it is actually heavier than gold, platinum will concentrate with gold in most placer deposits. The platinum metals cannot be separated from gold by washing, but they do not easily unite with mercury, and therefore they can be separated from gold by amalgamation (leaving the metallic platinum behind). Although finding a platinum nugget is an uncommon occurrence, it's worthwhile for the prospector to know a little bit about platinum and how it occurs. Platinum normally is found alloyed with several other metals, including Palladium, Rhodium, Ruthenium, Osmium, and Iridium. These closely related elements are known as the platinum group metals (PGMs).

Platinum is the rarest of the three precious metals, because deposits of the platinum group metals themselves are fairly rare. Mankind has known and used gold and silver from pre-historic times, but platinum was only identified as a unique material in 1735. It has a specific gravity of 21.46, and is ductile and malleable, but the presence of minute amounts of impurities render platinum hard and brittle. It is silver white with a grayish tinge with a metallic luster less brilliant than that of silver. Its melting point is 1780 C, much higher than gold which melts at 1063 C. It is used in many forms of chemical apparatus in which a high melting-point is necessary. It also has strong catalytic properties, accelerating the reactions of many chemicals.

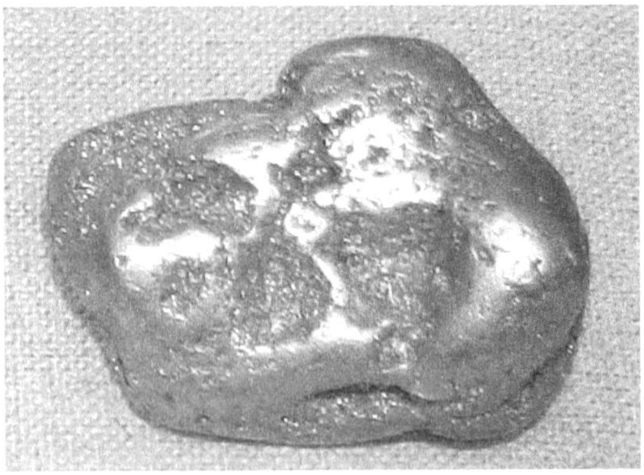

This large multi ounce nugget has a shape and wear pattern typical of gold, but its steel gray color gives it away as platinum.

The members of the platinum group occur intimately associated with each other and are found alloyed together in placer deposits. Where one is present, the other members of the family are usually also present in varying quantities. They are all grayish-white, lustrous, ductile and malleable metals. Native placer platinum typically contains from 16 to 40 percent of the other platinum metals. The other platinum group metals can be described as follows:

Palladium has the color, luster and appearance of platinum, but takes a finer polish. It is also malleable and ductile, and is the most easily melted of the platinum group metals. It is usually found in the metallic state, sometimes with gold and silver, and also associated with platinum in its ores. It has many of the same catalytic properties of platinum and is used that way industrially. Palladium can scarcely be distinguished from platinum by its color and both are used in jewelry. Palladium is noted for its capacity to absorb large amounts of hydrogen which is why it was used in the "cold fusion" experiments. It is soluble in nitric acid.

Osmium is a bluish metal, harder than glass and infusible in the oxyhydrogen flame. It crystallizes in cubes or rhombohedrons and is the heaviest of all known solids. It burns in the air to form the tetroxide, which has a peculiar penetrating odor. It is toxic and injurious to the eyes. It alloys with metals, notably iridium. It is found associated with platinum in the Ural Mountains, Brazil, California, Borneo and Australia.

Iridium is a hard, white, lustrous metal which resembles steel. It is malleable at red heat. It melts only in a high temperature flame. It is brittle and very hard. It is a powerful catalytic agent when finely divided. It is obtained by purification from platinum ores by a complicated process. It is found with placer platinum in Brazil, the Ural Mountains, California, Borneo and Australia.

Rhodium is a white, malleable, ductile metal. It is very resistant to chemical reaction and is insoluble in acids and unchanged in air. It is recovered from platinum ores. It is the most costly of all the metals, being worth several times as much as platinum. It is plated on silver jewelry as a protective agent. It occurs in small quantities in platinum ores and in trace amounts in some native gold. It is found in the Ural Mountains and in Brazil.

This rounded half ounce platinum nugget was mined in Trinity County in northern California.

The geology of the platinum group metals is quite different from the geology of gold and silver. Economically interesting concentrations of platinum group metals are most commonly are found in mafic and ultramafic igneous intrusive rocks such as dunite, pyroxenite, peridotite or layered complexes of such rocks. This also includes serpentine, which is an altered from of some of these rocks. Ultramafic rocks are presumed to originate in the upper mantle, and represent what is left behind after the differentiation process separated out basaltic magma at rift spreading centers. These types of rocks are called mafic because they contain less silica (less than 45%), and usually greater than 90% dark colored minerals with a high percentage of iron and magnesium. They are generally associated with deep-seated magmatic processes and the platinum group metals contained in them have been concentrated by high temperature separation in the magma during the slow process of cooling.

Because of their high density and resistance to chemical change, platinum minerals that form in igneous host rocks also become concentrated in placer deposits when the rocks are eroded away. These placer deposits were the first sources of platinum to be found and exploited by mankind. While placer deposits still remain a source of platinum, most newly produced platinum is now mined from hard rock deposits in layered ultramafic complexes associated with either copper or nickel sulfides, including chalcopyrite, pyrrhotite and pentlandite. In some deposits, platinum metals are also associated with chromite, which is an ore of chromium. Mines in the Bushveld layered ultramafic complex in South Africa, and the Norilsk layered complex in northern Siberia account for 93% of the world's Platinum group metals production.

Historically, the world's most productive platinum placers have been the deposits on the eastern slope of the Ural Mountains. In 1819, platinum was discovered in serpentine and other ultra basic rocks on the Siberian side of the Urals. Until 1823, the world's commercial supply of platinum came solely from South America. After that time, Russia became the largest producer until the production from the lode deposits of Africa took over. Platinum, with the metals mentioned above and some gold, has been extracted from the more basic parts of peridotite, dunite, and serpentine masses occurring in the Ural Mountains at Nizhne Tagilsk, Mount Solovief, and other districts, and placers have been worked in the valleys of the rivers Issa, Wyja, Tura, and Njassma, which drain the peridotite and serpentine region. The two principal centers of production were Blagodat on the eastern slope, and Nizhni-Tagilsk, on the western slope. The placers of the Nizhne district are derived from the peridotite masses of Solovsaia. Although the Ural placers taken as a group are the richest in the world, platinum has been discovered in other regions of Russia, such as in the auriferous gravels of the Miask district. In the Ural Mountains, platinum nuggets occur in the gravels of many streams, and all of the productive streams have their headwaters in regions of serpentine, peridotite and pyroxenite bedrock. It is associated with iridosmine, iridium, chromite, and often also with some gold, but the platinum is present in a greater quantity than the gold. The natural platinum forms small rounded grains, but larger pieces are found, occasionally as nuggets of up to 25 pounds in weight. The fineness of the placer platinum ranges from 750 to 850, with the remainder being iron, copper, and various metals of the platinum group, particularly iridium.

Second in importance for the historical production of platinum placer material has been Colombia, South America, where placers were worked along the Choco River and its tributaries. It was in 1735 that the platinum from these deposits was first recognized as an independent and unique element, as before that time it was thought to be some odd form of silver. Like the Russian placers, the platinum is obtained there in a greater proportion than gold. Platinum-bearing gravels occur in modern riverbeds and Tertiary conglomerates. Other platinum placers worldwide include the alluvium of certain rivers in Borneo, which are derived from serpentines and gabbros, also yield platinum. At the Congo Soco mines in Brazil, platinum occurs in the decomposed schistose rocks associated with gold. In New Zealand the placers of the Takaka and Cluthe Rivers on the Middle Island are also very similar to those of the Urals; they contain osmium and iridium as well as platinum, and are derived directly from peridotites. At the Ruwe gold mine, in the Tanganyika district of Africa, recent alluvial deposits and certain sandstones carry gold, platinum, and palladium, but so far only the placers are worked. The Tulameen district, in British Columbia, contains some platinum that is associated with chromite and formerly yielded some production as well.

In the United States, placer platinum occurs in small quantities together with gold in many of the gold-bearing districts of Alaska, California and in southwestern Oregon, where serpentine or peridotite bedrock is found. Historically, the most productive platinum deposits in the US have been the placers at Goodnews Bay in Alaska. These placers were worked, mostly by bucket line dredge, from 1928 to 1975. According to government records, from 1928 to 1975 they yielded at least 650,000 troy ounces of platinum, osmium and iridium. While Goodnews Bay has been by far the largest placer platinum producer in the US, there have been a large number of gold placers in California, Oregon and other parts of Alaska that have produced significant platinum as a byproduct. Platinum also occurs in certain beach sands along the coast of northern California and southern Oregon, mixed together with more or less gold. Small amounts of platinum have been recovered from these deposits. I have even seen some small platinum nuggets produced from Missouri River placers in the state of Montana.

Bucket line dredges produced the majority of the platinum mined in California and Alaska.

In California, small amounts of placer platinum were recovered from the large bucket line dredge fields along the western side of the mother lode. Butte, Humboldt, Siskiou, Trinity, Calaveras, Sacramento, and Del Norte have been the most productive of the California counties. While most of the California production was made by large commercial dredges, some platinum was also recovered from the numerous hydraulic mines in the Trinity and Klamath River regions of northern California. The largest nuggets recovered from this area have been in the one to two ounce range, although most have been much smaller in size. Suction dredge operators working on the Trinity and Klamath River systems and their tributaries still recover small amounts of placer platinum. Metal detector operators working the old hydraulic bench mines in that part of California have potential to find platinum nuggets as well. At all the known California platinum locations, the quantity of placer platinum in the deposits is very low, and PGMs are recovered only as a byproduct of gold mining. Most California deposits yielded platinum in amounts less than 1/100th of the gold value. All California platinum is associated with bodies of Serpentine and other ultramafic rocks, which belong to the Ural - Alaska type of deposit. No mining of any primary platinum deposit has taken place in California.

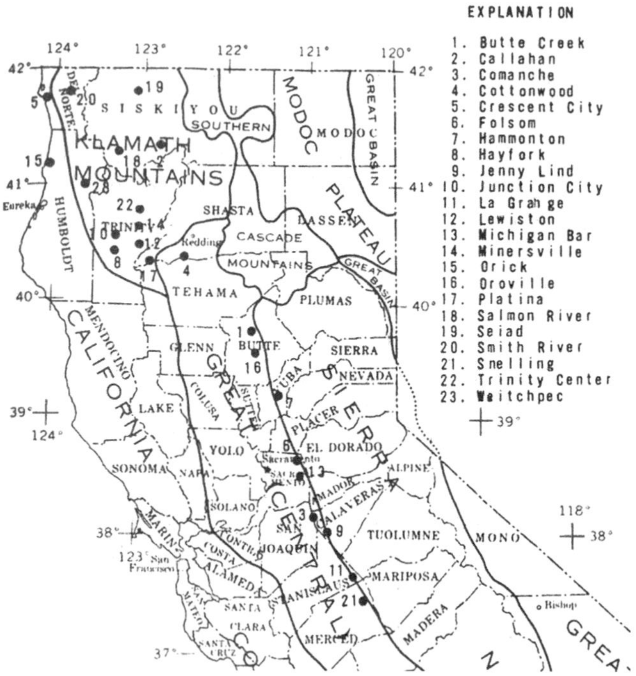

This map shows some of the many locations in Northern California where placer platinum has been mined in the past.

The primary types of rocks that give rise to these types of placers are known as Ural - Alaska type complexes. The platinum comes from bodies of plutonic rocks which are mafic to ultramafic in composition, are commonly associated with volcanic arc terrains and are thought to be the result of interaction between mantle and crustal melts. Because of their relationship to tectonic subduction, these zones of mafic to ultramafic rocks are commonly associated with linear orogenic belts which are also are related to plate subduction. Ural-Alaska type plutonic complexes are typically concentrically zoned, with dunite or peridotite cores surrounded by pyroxene-bearing rocks which are in turn rimmed by more differentiated rocks including gabbro and norite. The complexes may form sills, stocks or intrusive bodies often with poorly expressed external geometry. Generally, these types of rocks are not nearly as well segregated as the layered mafic complexes, and as a result, few if any "in place" or lode deposits have been worked within Ural - Alaska type complexes. This is because the platinum tends to be too widely dispersed to form good continuous economically viable ore grade concentrations. As a result, they are important only as the source rock for platinum placer formation. In California and Oregon, most of the ultramafic rocks of this type have been altered to serpentine, but these bodies still generate platinum placers, even though their original composition has been changed. Deep erosion of the Ural - Alaska type complex rocks at Goodnews Bay in southwestern Alaska led to the formation of that large and significant placer PGM deposit. Most of the placer PGM nuggets that arise from Ural-Alaska type plutonic complexes tend to be very small in size, and although larger nuggets are uncommon, they do occur occasionally. As I noted, the deposits in the Ural Mountains in Russia have yielded some exceptional specimens up to nearly 25 pounds in weight. Most geologists believe that the gold in these placers, which are commonly found with the platinum deposits, are formed in separate, geologically different processes.

The Stillwater platinum mine, located in the Beartooth Mountains of south-central Montana is the largest currently active platinum mining operation in the US. Like the Buchveld and Norlisk deposits, the Stillwater mine is located in a layered complex of strongly basic igneous rocks. The ultramafic to mafic rock layers found here are derived from a large magma body which was emplaced deep in the Earth's crust. The molten mass was sufficiently large and fluid enough to allow its chemical constituents to cool and crystallize very slowly. The minerals formed in certain sequences as the chemistry of the melt changed and as the crystals formed, they settled out to the bottom of the melt, producing extensive and uniform layers of varied mineral concentrations. Exploration of the layers within this complex body of rocks disclosed a zone of rock which is significantly enriched in platinum and palladium. This layer has been named the Johns Manville (J-M) Reef. In addition to platinum group metals, the reef also contains significant amounts of copper, nickel, and trace amounts of gold and silver (palladium predominates over platinum in the ores here). The Johns Manville reef extends over a very considerable distance, with surface outcrops that have been mapped and sampled for approximately 28 miles along its east-southeasterly course and over a known expression of over 8,200 feet vertically in drill holes and underground workings.

There are a few unusual in-place nickel-copper deposits in Clark County, Nevada that contain significant platinum and palladium. These include the Boss Mine at Goodsprings and the Key West mine at Bunkerville. The PGMs here are dispersed through the ores and no platinum placers are known in Nevada. Platinum is also recovered as a byproduct of copper mining in Arizona. The amount of platinum in the copper ore is infinitesimally small, but it is recovered from the sludge during the copper refining process. Such trace amounts of PGM are typical of copper porphyry ores found in mines elsewhere as well. According to the State of Arizona Department of Mines and Mineral Resources, they are not aware of any placer platinum has ever been produced from any deposits in that state.

While a considerable amount of platinum is used for jewelry purposes, unlike gold, the platinum group metals are essential to a good number of the industrial operations of modern society. Platinum group metals are important catalysts for chemical reactions, and essential for many industrial processes, as in many cases no other metals can substitute for them. Critical industrial uses include pollution control catalysts, catalysts needed for crude oil refining, electronic contacts and anodes, fuel cells, and corrosion resistant lab ware. Industrial demand and limited supply has led to very high platinum prices in recent years. Vaulting up to over $2000 per ounce in a recent price spike, gold seems cheap by comparison to platinum. These high prices and continuing demand seem to indicate that there will certainly be continued exploration for additional PGM deposits in the future. It is also important to note that for many years, platinum deposits have been the object of numerous mining scams and con games, as few people know much about platinum deposits. For individuals considering investment in any platinum mining operation, extra care and investigation is warranted because legitimate platinum deposits are so rare.

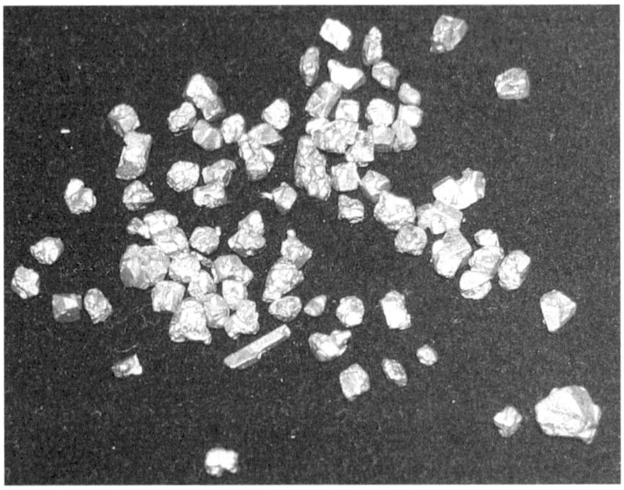

Like gold, platinum can also be formed naturally into crystals. These platinum crystals were mined out of platinum placers in Russia.

For the individual prospector wanting to explore for platinum deposits, the search should begin with the identification of bodies of mafic and ultramafic igneous intrusive rocks - geologic maps can be useful in doing this type of research. Learning to recognize these rocks in the field, with their high percentage of dark colored minerals would be a high priority. Recognizing the waxy green-gray appearance of serpentine, a rock formed from the metamorphic alteration of mafic and ultramafic igneous intrusive rocks would also be helpful. Regions where platinum has been found in the past would be a likely place to start a search for platinum metals. Checking local creeks and drainages which cut through bodies of ultramafic igneous intrusive rocks by panning the stream gravels would probably be the quickest and most efficient method to identify the platinum potential of an area. Any black sands collected by this method should be examined under a microscope to check for their platinum content as a visual check for larger platinum nuggets would not be a sufficient examination. Any promising locations could be followed up with assays for PGM content.

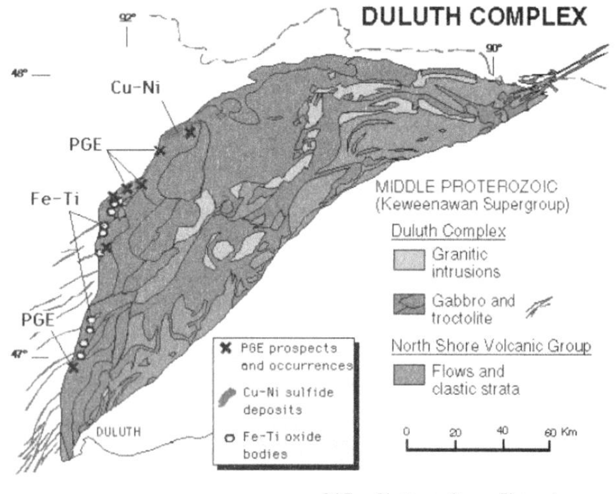

Source: Minnesota Geological Survey Documents PGE = Platinum Group Elements

Most commercially mined platinum comes from layered igneous rocks like the Duluth complex.

It is important to the individual prospector to be able to identify platinum nuggets in the field. Platinum is less malleable than gold, and placer platinum nuggets will remain angular further from their original source. Lead chunks often show something of the original bullet, fishing weight or shotgun pellet shape, and those are easily rejected. Natural PGM nuggets, of course, will not show this. In addition, after long exposure to moisture, lead will develop a white coating of lead hydroxide-carbonate. Platinum does not oxidize and will show a metallic silver-gray coloration. Perhaps the simplest way of testing a suspect nugget to determine if its lead or to remove a coating, is to immerse it in nitric acid. Nitric acid will react with and dissolve many common metals including lead, copper, iron, zinc and silver. It will also dissolve many coatings found on gold, including most manganese and iron oxides. It will not dissolve platinum or gold unless hydrochloric acid is added, which forms aqua regia. Suspect nuggets can be immersed in diluted nitric acid and allowed to sit overnight. Platinum will not dissolve in diluted nitric acid. The acid should be diluted with equal amounts of water for this test - never add water to acid - always add acid into the water. Some lead alloys dissolve only very slowly, so the time to sit overnight is needed. Lead and many other metals will release bubbles and dissolve. Platinum will be unaffected. Nitric acid can be harmful if mishandled, which is why it's a lot harder to buy than it used to be. Most prospecting shops used to sell it, but that is no longer true. Anyone who uses nitric should take the time to learn its potential hazards and be fully prepared to use it safely with all necessary gloves and other protective gear.

As I noted before, the common parent rock that gives rise to platinum placers is the ultra-basic ferro-magnesian igneous rock known as peridotite. It is normally a dark gray to near black color, but in many locations worldwide, this rock (or a related rock, dunite) has been altered to a greenish gray rock known as serpentine, because it is so easily converted to this material. This alteration however, does not change the platinum if it is present. Whether it is changed from its original form or not, it is known that platinum occurs as a primary constituent of peridotite and dunite rocks, and specimens showing its intergrowth with olivine and chromite have been described. Nearly all platinum placers have been traced back directly to the decomposition and erosion of this type of rock. However, not all occurrences of peridotite or serpentine rocks contain platinum, so not all bodies of such rock will lead to the creation of platinum placers. Placer platinum is almost always associated with segregations of chromite, and, although not all chromite segregations will carry platinum, this mineral is an important indicator of the platinum group metals. Early day prospectors looking for gold often considered platinum and the platinum group metals as nearly worthless or merely as a curiosity. Historically, prospectors and miners never made them the chief object of their search, as they had no regular or reliable market for sale, and platinum was worth significantly less than gold in the 1800s. Rich deposits of platinum that did not contain significant gold may well have been ignored by the old timers as not worth working. In more recent times however, platinum is easily saleable, and worth much more than gold. Because of this, the importance of prospecting all alluvial deposits draining from regions of basic igneous rocks that contain chromite segregations cannot be too strongly urged. If you carefully prospect for platinum in these types of locations, you could well make an important find.

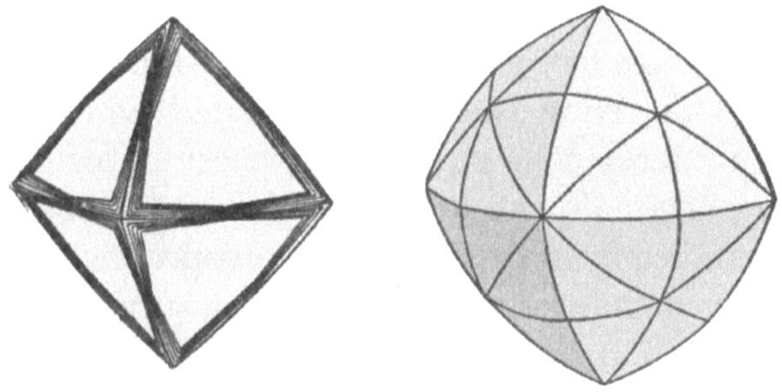

Although natural diamond crystals come in a wide variety of shapes, these are two of the most common crystal forms – an octahedral and what is essentially a rounded octahedral shape.

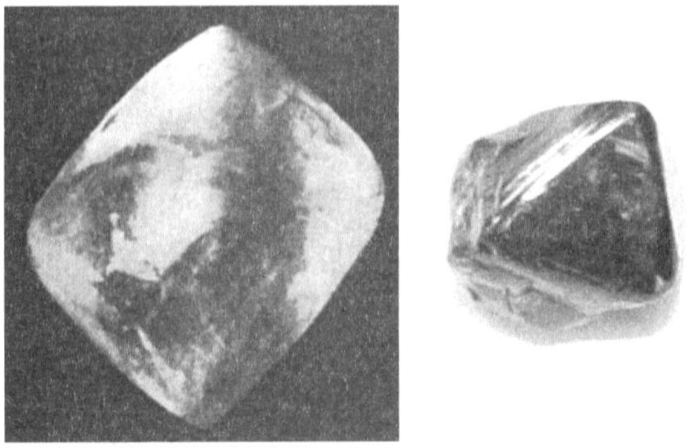

Many diamond crystals, like these pictured here, show some form of an octahedral shape. These are slightly rounded and distorted. The cleavage is parallel to the octahedral face.

Diamonds

Although they are far rarer than platinum, alluvial diamonds do turn up occasionally in some placer gold deposits. Worldwide, there are actually diamond placers which are worked only for their gem content. In the US, diamonds have been found with placer gold in California, Colorado, Wyoming, and several of the Southern states including Tennessee and Georgia. They are only a little heavier than regular gravel, having a density of 3.5 as compared to 2.7 for most rock (gold is 19.3). As a result, you are much more likely to lose a diamond out of your sluice or pan than you would be to lose gold. Most prospectors who think they have found a diamond are typically confused by quartz crystals or other transparent minerals. Most diamonds are of industrial quality and therefore dark and opaque, or, at most, translucent. However, purer stones are also found, and these include the gem quality stones. These are transparent and pale to colorless in tone. Gem quality stones of gray, brown, green, yellow, blue and red tints are also known, and, with the exception of the blue and red diamonds, these other colors are all more common and less valuable than the colorless or so-called white stones. Bort, an industrial form of diamond, is usually gray or black, translucent or opaque and typically comes in rounded masses, with a rough exterior and the structure of a crystalline aggregate as opposed to that of a single crystal. Bort diamond is normally crushed as an abrasive. Diamonds can be distinguished by their great hardness, their sparking adamantine luster (they appear to be slightly oily even when clean and dry) and their octahedral cleavage.

The minerals with which diamond is liable to be confused are much softer, and, so there is little difficulty in distinguishing between them in a hardness test. Sometimes prospectors will try to scratch glass with a suspected diamond, and while diamonds do scratch glass, so do many other rocks and minerals, including quartz. A hardness test for diamond needs to be done on something that is itself very hard, like a sapphire or ruby (both are corundum of hardness 9). If you have a suspected diamond which can easily scratch sapphire or ruby, then you likely do have a real diamond.

Kimberlite rock is the most important host - the ore rock for diamonds. It's a kind of volcanic rock, but a very rare and unusual one. Geologists figure there are many kimberlite pipes out there which remain to be discovered because they are covered over with soils. Kimberlite is a dark-colored, heavy, brecciated (fragmented) igneous rock that contains diamonds in its rock matrix. It starts out dark colored, but is often altered and/or weathered to various lighter colors.

Normally diamonds recovered from placer deposits show little or no signs of abrasion or wear as diamonds are so extremely hard. They may be broken or cleaved by impacts from larger rocks, but they are rarely worn. In addition to their very extreme hardness, diamonds are also often recognized by their crystal shape. Those crystal shapes are many and varied, but very commonly are some form of distorted octahedron or cube, with the octahedron shape being by far the more common of the two.

Hazards of Prospecting:

While normally a whole lot of fun, and an activity well suited for the whole family, under certain circumstances prospecting can be a hazardous occupation. Most of the dangers associated with prospecting are just the standard dangers of being outdoors in a wild and uncivilized environment, but the prospector needs to be aware of his surroundings and to take the necessary steps to be safe. He needs to be prepared for a variety of hazards and other unexpected emergencies. Its way beyond the scope of this book to cover and fully delve into each possible hazard that one might encounter out in the field. However, in order that you can be better prepared and at least aware of some of the more common hazards that you might face, we will spend some time considering the potential hazards of prospecting and how you might deal with them.

Dangerous Animals:

Encounters with dangerous animals are one of the most common hazards associated with prospecting. Poisonous snakes, mountain lions, bear, javelina and wolves present dangers ranging from moderate to extreme. In nearly all cases, the best strategy for dealing with dangerous animals is to avoid them if at all possible. Give them a wide berth and steer clear of their path where you can. Many prospectors sometimes choose to carry hand guns in defense or even to frighten off animals.

Rattlesnakes are poisonous and various species are common in the western US. They are a common hazard faced by many prospectors. This one is coiled up and ready to strike its target.

Rattlesnakes are well-known hazards that many prospectors face in the desert. Many folks that do metal detecting wear "snake chaps" which are hard exterior coverings which protect the leg from the knee down to the ankle. They are commercially available in some prospecting shops and from many outdoors equipment suppliers. In combination with sturdy boots or shoes, these will protect the prospector from most snakebites which occur below the knee. Bites in places above the knee are less common, and occur when folks put their hands in places they can't clearly see. So watch where you are stepping and be careful what you grab when climbing. The important thing to remember is that if you receive a snake bite, you need to seek help at a hospital emergency room as soon as possible. Rattlers do not always inject venom in every bite they make but you can never know. The swelling and symptoms can show fast or take a while to develop. That's why it's important to get to the hospital quickly.

Mountain lions are rare, but they are out there. In several decades of prospecting, I have only seen a few bobcats, never a full-size mountain lion (American cougar). Unlike bear however, cougar make a significant effort to remain unseen. They are potentially dangerous to prospectors especially men of smaller stature, women and children. A mountain lion sometimes will take on a full-size man, but generally they prefer a smaller prey. Attacks on humans are rare, because for cougar prey recognition is a learned behavior and they do not generally recognize humans as prey. Attacks on people may occur when the cat establishes a home near to humans or is in a condition of severe starvation. Attacks are most frequent during late spring and summer, when juvenile cougars leave their mothers and search for new territory. They are wild and unpredictable, but exaggerating the threat you present to the animal through intense eye contact, loud but calm shouting, and any other action which may make you appear larger and more menacing, may make the animal retreat. Fighting back with sticks and rocks, or even bare hands, is often effective in persuading an attacking cougar to disengage.

Bear on the other hand, I have encountered several times. My first choice is to steer a wide berth around them and let them do their thing and ignore me if possible. Often bear will ignore you if you present nothing interesting to them. If necessary, they normally can be frightened away by yelling or waving your hands, firing a pistol up in the air, blasting on an air horn or otherwise making a lot of racket. I have also had a bear stare me down which is a somewhat unnerving experience. In that particular case I had a friend nearby who heard me yelling to try and scare off the bear. When he called down to me asking if I was okay, the bear heard the second voice from a human he could not see, and not wanting to mess with two humans, took off running. Grizzly are by far the most dangerous, but their populations are few in most of the lower 48 states. Alaska is different and there are a number of grizzly in certain parts of that state. Black bear are much more common than grizzly and also much less dangerous in most situations. Still a mother with cubs should always be left alone - these are among the most dangerous situations - when the mother feels threatened she will attack. I like air horns, but experts say that the best defense against bear is the large pepper spray cans made for that purpose and available at many larger sporting goods stores. Bear have a sense of smell that is extremely powerful - they are more than 10 times better at sensing smells than the best dogs around (bloodhounds). Because of this they are very sensitive to things like pepper spray in the face (and up the nose). Pepper spray has been known to turn back even crazed mother grizzlies who thought their cubs were threatened.

While many types of wild animals roam the remote forests and deserts, perhaps the most dangerous of them possesses no threatening claws or fangs. Sometimes the most dangerous animals in the woods or out on the desert are other humans. I know prospectors who have been shot at, and I even know of some others who have been killed during disputes, etc. Additionally, some California goldfields are also fine locations for growing dope. Years ago, I was hiking to a remote location on the New River in Trinity County, CA. In the middle of the trail as I walked was a big rock. I went to kick it out of the way and noticed there was a note underneath the stone. Being a nosy individual, I read it - the note was from one

Unless you hit just the right spot, a shot or two will not bring down a bear even at point blank range. Many old timers were mauled or even killed in exactly this type of situation.

dope grower to another inviting him to come over for a smoke (I guess he figured the other guy would be on this trail going to his "location"). I carefully replaced the stone and looked around to see if I was being watched. The coast was clear and I beat it. I had been warned by other prospectors in that area of the dangers of booby traps and the possibility of being shot by dope growers in traveling on that trail. Another episode occurred a while back when I was prospecting in southern Arizona down near the border. I looked up and couple hundred yards away were a group of men looking at me - all dressed in black and carrying very large packs on their backs. I dove into the bushes and disappeared up a ravine. If you are near the border areas, drug runners can be every bit as dangerous as drug growers. Additionally, if you find a nice bit of gold, be careful who you tell about it - gold has a way of attracting thieves, and if you are out in the back country working a spot everyone knows is rich, you just may get followed, and possibly robbed.

Dangerous Insects:

In recent years, one of the things that seriously concerns prospectors out in the desert areas of the southwest are Africanized Bees, more commonly known as "killer bees". Wild bee hives, especially in AZ, NM, and the Southern Calif. locations are a danger. The majority of wild hives in those locations south of latitude N 34 are generally assumed to be Africanized. I am a former beekeeper, and I have taken up to 75 stings at one time, so I have a few hints.

The first and most obvious is stay away from the hive. The sound of many buzzing bees arriving at the hive is a recognizable sound, and you should avoid the hive where possible. If you should accidently get into a hive and be attacked, keep your wits about you. In a straight race, they can run you down until you drop - they can fly long distances at 30+ mph - so in a flat-out race, you will lose. The good thing is that bees are dumb - they live on instinct. If you ever have to run from a hive, pick a direction where there are some big bushes or small trees (you need something 6 foot tall or more). Pick a bush and run a circle around it a couple times (staying in tight and up close to the bush or tree) then go 5 or ten yards to the next bush and run a circle around it a few times, pick another bush or tree and repeat. Each time you circle around the bush about 75% of the bees will be confused and stay with the bush when you move on. Weaving in between bushes helps also, but the circling works best. Circle around enough trees fast enough and you can quickly turn 300 bees chasing you into zero - but once you loose them, keep moving away from the hive as they will be looking for you if you head back.

Bees defend their hive - brushing up against a few while they are collecting nectar from flowers is no big deal - at worst you might have one sting you and that's not a big deal unless you are allergic. It's being stung by several hundred at once that can send even a normal person into shock. So the danger is at the hive where there are many thousands of bees and they are very territorial of their home. Bees sound the danger alarm against intruders by odor - the smell gets spread around and then they get all worked up and ready to attack. Also, when one bee stings, she leaves an odor mark that tells other bees - "sting here - sting this guy!" The venom is a histamine reaction inducer - so anti-histamines may be useful to reduce your reactions to any stings you would receive.

Spider and scorpion bites are normally avoidable. Typically they are a case of one sticking his hand in a place where it should not be. Prospectors using metal detectors can dig a target hole and accidently get into an insect nest, and if not paying attention, grab a handful of angry insects. Scorpions come to mind, but fire ants can also be a problem in this category. I recommend using heavy gloves and a plastic spoon for target digging, and pay attention to the spot you are digging in. Scorpions also have a tendency to get into shoes and boots, so if you camp out and leave your shoes on the ground, shake them out in the morning - it might save your foot from a scorpion.

Dangerous Plants:

One does not normally think of dangerous plants in the same way you would think of dangerous animals or even insects, but plants can be a problem in their own right. The various types of cactus and other thorn bearing plants can be an annoyance, especially the Cholla or jumping Cholla as it sometimes known. These things just have an amazing capacity to latch on when you brush against them and then puncture when they get the chance. Catclaw in Arizona and Blackberry bush in the West coast mountains can grab hold and scratch out their share of blood.

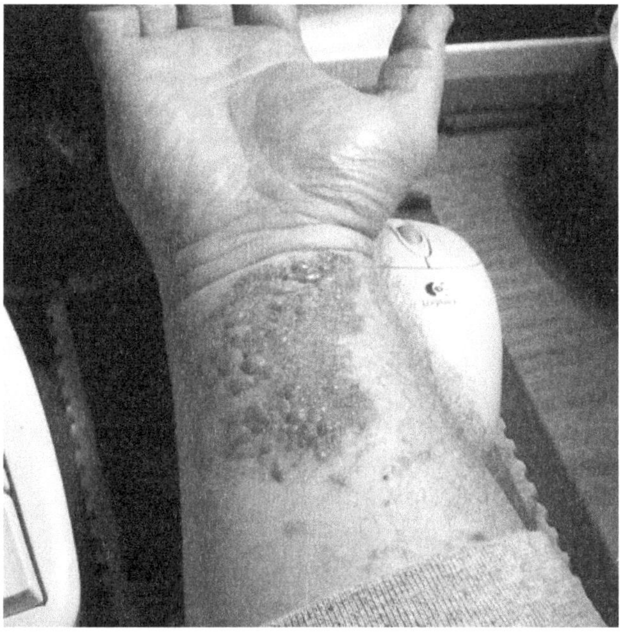

This patch of blisters on the author's arm is the result of a close encounter with Poison Oak.

Another common annoyance from plants is the itching and blisters caused from poison oak and poison ivy. These two are closely related species that can cause problems even when the leaves are off the bushes. I had a bad experience some time ago when I went to visit a friend to do some prospecting with a metal detector. My friend gave me fair warning about the poison oak when I was down there, but I had never had much problem with it (some folks are a lot more sensitive than others). Unfortunately for me, it was mid October and all the leaves had fallen. I know to recognize the plant but it's hard to tell one kind of bush from another when the branches are all bare. I did a lot of pushing through the brush trying to find spots that had not been previously detected. I wasn't wearing gloves but did have long sleeves. I got scratched up by the brush but its not thorns, so it's no big deal, right? It was about 4 or 5 days until the rash and blisters started to show.

Out on the desert and in other places as well, you may find yourself many miles from the nearest potential help. Be prepared for unexpected emergencies with a cell phone, tools, supplies, water and anything else you might need.

I'd never had a case anything like that before, and never have since, but that one was pretty bad. A product named Technu works well if you put it on right after you are exposed. The point is to wash it off quickly with solvents. If you wait too long, the oils from the Poison Oak will chemically bond to the proteins in your skin and after that no amount of washing can remove it. By the time you start to see the symptoms showing on your skin, it's too late. Once you have the symptoms, antihistamines pills and cortisone (steroids) on the skin are the only treatments (and don't scratch!). Its an allergic / histamine reaction with the skin. Basically mammals other than humans don't react to it and deer can even eat it with no problems. I did a lot of research when I had that case! Some form or another of Poison Oak or Poison Ivy (AKA Poison Sumac - they are all very closely related plants) grow in every state except Nevada (too dang dry), Alaska (too dang cold) and Hawaii (too far offshore). Stinging nettles are another well-known plant which causes skin irritation and itching when you brush against it, however the problems do not last nearly as long as the case with poison oak or poison ivy.

Dangerous Impacts of the Outdoors:

There are dangers of just being outdoors for a long period of time. Some of these include heat stroke, sunburn, hypothermia, dehydration. Most are preventable if you take precautions such as sun block lotion, taking along sufficient water, warm clothes, etc. The point of this is that when you go out in the field, you need to think about potential problems, plan ahead and be properly prepared for the situations you will encounter.

Another great challenge is the problem of physical injury out in the field. Minor scratches, bumps and bruises are common, especially if you are a little clumsy like me! I always carry basic first aid items with me into the field, from bandages to ibuprofen to antiseptic materials, it's good to be prepared. The real problems come with more serious injuries including things like deep wounds, broken bones, compound fractures, etc. The potential dangers of these and other major physical problems like heart attack or stroke are greatly amplified in the wild. I get first aid training at work, and it's important to learn how to help yourself and others during emergencies. When you go into the field, know where the nearest sources of medical help are located. Certainly, there are many reasons why it's good to have a partner out in the field, someone who can help in the case of major medical situations. Sometimes just having someone around who can drive you to the nearest medical center can make all the difference in the world. Cell phones and other communication equipment can also make a huge difference, and service coverage in remote areas gets better every year.

Dangerous Situations:

Vehicle failure in the desert or other wild regions becomes a major danger - small breakdowns which would present only a minor inconvenience in the city can be life threatening if one is not prepared. It is also possible to get stuck in the road itself - in soft spots, mud, or get high centered on a rocks, etc. If you are venturing out into the field, always be sure to have sufficient water to last for a few days. That same water might also serve as emergency source of fluid if you spring a leak in your radiator system. Some extra food would be good as well, but you can go longer without food than you can without water. Figure one gallon per person per day, and take enough for several days in case of emergency. Also take the tools and other equipment needed to fix some simple car problems - if you can help yourself, you may be able to resolve the problem without assistance. If you do get stuck, it is normally best to stay with your car. In some situations where you can see nearby help it may be worth it to hike to help. Always let someone know where you are going - give them maps, etc. so that they can tell the local authorities where you can be found if you don't come home on time. This is another case where having a buddy with you can make a big difference, especially if they have their own car. Cell phones now cover a huge area of this country, and the more you can do to help yourself, the better. Two-way radios can also be used to call for help in some situations. One of the new things on the market is the GPS satellite transponder. There are more than one on the market, but the product I use is the Zoleo Satellite Messenger. It allows you to send a text message to any phone or call for help in the case of an emergency. The Zoleo Company monitors the units and can send responders where to find you using GPS accuracy even in areas with no cellular phone service. It works off satellites rather than the cellular antennae grid, and can send a message in areas that are very remote. I go to some pretty remote places looking for gold, so it is important to have a way to communicate if there is a problem. It is also good just to stay in touch with my wife and family too. There is a monthly service fee, but it is worth it for safety.

Getting Lost in the Wilderness

Once when I was on a prospecting trip in the sierras with some friends, it started to snow. We had been out camping and prospecting for the weekend and snow was not expected. The snow storm obscured all the mountains around and I lost my bearings. Hiking back to our cars we took a wrong turn on the trail and I soon realized something was wrong. I was able to figure the problem out, but many times when lost it is best to hunker down for a bit - the snow would have cleared and then I could have then seen the mountains and been properly oriented. If you are lost in a car, stay at the location with your car. Cars are nearly always easier to spot from the air by searchers because of their larger size and colors. Better yet, take good maps, a GPS and a compass and don't get lost in the first place!

Underground Mines

Underground mines are dangerous places, and I urge you very strongly to stay out of them. For some reason a lot of guys think they are going to go inside and find a treasure. They often contain many unseen dangers, and most guys go in completely unprepared and without the knowledge or equipment they need to be safe in these environments. Prospecting should be about having fun, not getting killed or seeing your friends suffer or die. As a serious warning, let me offer you some stories about just how dangerous exploring old mines can be.

Old Mines like this one are extremely dangerous and should not be entered! They hold hazards which are both visible and invisible to the eye.

Years ago, I worked in a mine at Virginia City, Nevada. Our workings inter-connected to the miles of old underground tunnels that weave through the Comstock Lode underneath the surface. Thousands of cords of wood lie rotting in these old workings from the 1800s, and the fungus in the rotting wood consumes all the oxygen from the air. When I worked there and the mine was operational, we had a 6 foot diameter fan that flushed the bad air out of the mine, and kept the air underground good and clean. We also tested the air regularly with hand held devices to make sure it was breathable. When the mine shut down, the fan was shut down, and the entrance locked with heavy gates and chained shut. Several years later, a teacher up at the local school and a friend of his decided to explore our old mine. They took a bolt cutter and cut the chains and went in with flashlights. Not realizing that bad air was in front of them, they entered and walked down the decline. Not more than a few hundred feet in, both men were overcome with a lack of oxygen and fainted. Unfortunately, the air on the floor had even less oxygen. There was no bad odor or other warning associated with the low oxygen air - they were probably both overcome quite quickly. They laid there and died after a few minutes. Their bodies were recovered several days later when someone figured out where they had gone (they didn't tell anyone else). They were both well-liked young men and it was a terrible tragedy.

A word from the state of Nevada, the truth about abandoned mines: ***Stay Out and Stay Alive!***

Here not long ago in Northern Nevada, several friends entered an old underground copper mine to explore it for fun. This mine did not have a problem with bad air, but they had no knowledge of that when they went in. As they were going through some of the old workings, one of the friends felt the floor give a little underneath him and whoosh - he suddenly disappeared out of sight. Unknown to him, he had been walking on some boards placed over an old winze dug when the mine was still operating, nearly 100 years ago. The wood was covered in gravel so it could not be seen. Those boards, now partly rotted, gave way underneath his weight and he fell crashing down some 95 feet to his death. There was nothing his

Dangerous open stopes like this can be common in mining country. They go down hundreds of feet

friends could do but report the accident to the authorities and wait for them to recover the body.

A former co-worker ran a program for the state of Nevada with the following slogan about old mines "Stay out and stay alive". In my opinion it's a very valid sentiment. Old mines are very dangerous places and many of the dangers are, as I have described in my stories here, difficult or impossible to see until it is too late. Most folks who enter into these mines have no experience working underground, no ability to recognize the dangers, and none of the equipment that would be needed to stay safe. When you enter an old mine you are asking to receive disastrous consequences. No amount of gold is worth dying for. Please stay out of old mines.

Plan Ahead and Stay Safe

Much of dealing with hazards is about knowing the dangers, being prepared and keeping your wits when emergency situations happen. When you are packing your gear, don't rush - be sure to include that extra water, first aid kit, GPS and other items that will help you stay safe and allow you and your family to enjoy prospecting for gold and silver for many years to come.

Finding Locations to Prospect: Pay To Prospect Opportunities

Finding locations to prospect that have good gold for the digging have always been a central focus of most prospectors. Once a prospector has enough experience and has learned all the techniques, in the West there are places on public land that can be accessed for most prospecting opportunities. There are however, problems and exceptions to this rule. First is that the prospector must have enough experience to be able to do research, find areas and accurately check the land status to know if it's private property or under claim. Not all new prospectors have those skills. Second, in the eastern areas of the US there are limited amounts of public land, and in a lot of those states the US mining laws do not apply. Because of this, eastern prospectors need to find other options.

The first option I recommend for most new prospectors, and those looking for places to go prospecting, is to join a club. There are many clubs out there of all types and organizational structures. Clubs often have access to property that would otherwise be closed or sometimes they have their own claims. I've gone into detail about joining a club and given my comments about all the benefits that are involved in club membership in a separate chapter, so I won't go into those details here.

The other option for those wanting to prospect for gold on their own are the locations where one can pay a fee to hunt for gold. Because of the interest in small-scale prospecting, a number of landowners have chosen to offer opportunities to prospectors to use their property. For those prospecting on the East Coast of the US, these types of opportunities may be some of the very few that are available to prospectors. The costs for these opportunities vary quite a bit from small amounts to more significant outlays. Fees are charged sometimes by the day, or sometimes for longer periods. These really do represent a great opportunity for the prospector. In the past, I visited the Moore Creek and Gold King Creek mines in Alaska, both pay to prospect operations. The total costs of some of these trips, including the travel to places like Alaska or Australia, can be quite expensive. For most people, however, it's a once or twice in a lifetime type of trip, and well worth the expense if one has the money to put into this sort of adventure. Advertisements by the property owners offering these sorts of opportunities appear in magazines like the GPAA Prospecting magazine. One can also hear of them discussed on some of the internet prospecting forums I have mentioned. Another advantage to the prospecting forums is that one can ask for opinions from those who have already been on these trips, and get honest input and information about the adventures they experienced, the gold they found, etc. I can certainly testify that my own trip to Alaska was a wonderful experience through some beautiful and highly remote country, and a very worthwhile adventure.

Finding locations on your own is always a possibility, especially for those who have access to mineralized public lands in the Western US. I've gone into quite a bit of detail on how to conduct research, both with respect to finding likely districts as well as researching land ownership and mining claim status. For those with the proper experience and knowledge, there are many opportunities along these lines. It just takes time and experience to be able to do that sort of research work. For new prospectors just learning the ropes, joining a club or accessing gold bearing lands by a pay to prospect type of arrangement can represent an excellent opportunity. As far as the research part of it, I plan to add a considerable amount of historic information on the gold districts of the United States and even across the world onto my website in the coming months. Check on my site regularly to see that information.

Internet sources of Information

One of the greatest opportunities available to the individual prospector is the exchange of information over the Internet. A number of web sites and forums that are dedicated to various aspects of prospecting for gold have sprung up in recent years. They offer a very unique chance for prospectors worldwide to converse about their favorite topic. Some sites specialize in various aspects of prospecting such as detecting for nuggets or dredging. The internet allows prospectors from Australia, New Zealand, and all over the western states including Alaska, and elsewhere across the globe to share their insight and knowledge on finding the elusive yellow metal. I often participate in these forms myself, and have made friends with a number of the other participants. Log on sometime and check it out and maybe we can talk about prospecting for gold. The following is a list of some of my favorite sites (including my own):

Arizona Gold Prospectors info and forums
http://www.arizonagoldprospectors.com/

49Mike Higbee's Forums
http://www.tomashworth.com/

Nuggethunting Prospecting info and forums
http://www.nuggethunting.com/

Arizona Outback Prospecting info and forums
http://www.arizonaoutback.com/

Nugget Shooter Prospecting info and forums
http://www.nuggetshooter.com/

Virginia Gold Prospectors (East Coast)
http://va-gold.com

The Prospecting Channel (Prospecting Videos)
http://www.prospectingchannel.com/

Chris' Gold Prospecting Encyclopedia (The Author's Prospecting website):
http://nevada-outback-gems.com/prospect/chris_prospect.htm

Gold Prospectors Assn. of America
http://www.goldprospectors.org/

Some websites are valuable for research. Here are a few of my favorite informational websites with data useful for prospectors:

BLM Map Based Mining Claim Info
https://mlrs.blm.gov/s/
The new system has a map-based research tool. Once you have reached the website, you can click on the research tab on the MLRS home page to access the map.

BLM MLRS Claim Info Website
https://mlrs.blm.gov/s/
or you can just Google BLM MLRS

USGS Mine Database Mapping System http://
https://mrdata.usgs.gov/mrds/map-graded.html

USGS Publication Warehouse
http://pubs.er.usgs.gov/

Google Earth:
http://earth.google.com/

California GIS Information Clearing House
https://gis.data.cnra.ca.gov/

Alaska DNR Mineral Resources Info
http://dnr.alaska.gov/mlw/mining/

Nevada Bureau of Mines
http://www.nbmg.unr.edu/

There is a lot of information on the internet on prospecting of all types, including dredging, metal detecting, dry washing high banking and just panning for gold. More information is being added all the time, so any summary of websites that I publish here will be somewhat dated as new sites will pop up. So do your own research and check up to see what is available in the way of forums and other informational websites.

Some Recommended Books For Further Study:

I have long kept a home library of books on prospecting related topics. Nearly all prospectors have some reference materials. On this page I'm going to give you a few recommendations for books you might think of downloading or purchasing. Many of these can be downloaded for free on the internet. A lot of what you might be interested in purchasing for your own library depends on what part of the country you live in, but here are a few of my thoughts. These book recommendations cover various different categories. For a much more extensive listing of gold, rockhounding and geology book recommendations, take a look at my website, where I have more than one hundred different book recommendations in a wide variety of categories, together with my comments on each book.

Prospecting Related Books:

Gold Prospectors Handbook
by Jack Black

Where to Find Gold In Southern California
by James Klein

Where to Find Gold in the Mother Lode
by James Klein

Where to Find Gold and Gems in Nevada
by James Klein

Dry Washing For Gold
by James Klein

Advanced Prospecting & Detecting for Hardrock Gold
by Jim Straight

Advanced Dredging Techniques (all volumes)
by Dave McCracken

Gold Mining in the 21st Century
by Dave McCracken

A Guide To Gold Panning In Utah
by Alan J. Chenworth

The Elusive Pocket Gold Of Southwestern Oregon
by Tom Bohmker

Placer Deposits Of Alaska, USGS Bulletin 1374 by Edward H. Cobb

Placer Gold Deposits of Arizona
by Maureen G. Johnson

Placer Gold Deposits of Nevada
by Maureen G. Johnson

Placer Gold Deposits of New Mexico
by Maureen G. Johnson

Placer Gold Deposits of Utah
by Maureen G. Johnson

Gold Districts of California
Calif. Geological Survey, Bulletin 193
by William B. Clark

Geology Related Books:

The Practical Geologist
by Dougal Dixon

Rocks & Minerals
by ROBERT W. JONES

A Field Guide to Rocks and Minerals
by Frederick H. Pough, Jeffrey Scovil

Simon & Schuster's Guide to Rocks and Minerals
by Simon & Schuster

Gold - The Noble metal
Published by Lapis International

Manual of Mineralogy (after James D. Dana)
by Cornelis Klein, Cornelius S. Hurlbut, et al

Roadside Geology of Northern and Central California
by David Alt
(I like all of the Roadside Geology series books)

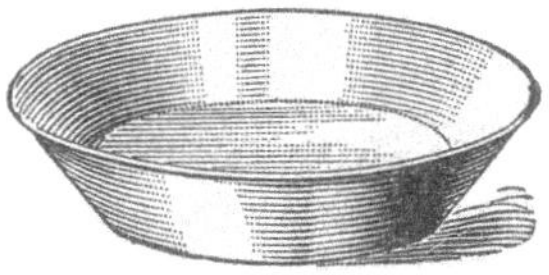

Recommended Books, Part 2: Old Time Texts and Reports

Toward the end of the 19th century, over 100 years ago, a series of important gold strikes were made around the world and these discoveries continued into the early years of the 1900s. These great finds fired the imaginations of the men of that time and many great books on geology, prospecting and mining were written. While some of the information in these great old texts is still usable, other parts of them have become a bit dated. They mention nothing about the important technology advances and new scientific findings of our time. In many ways, this book is an attempt to basically compile and update the materials presented in some of those great old works. Most of these classic texts are so old and long out of print that they are now public domain and no longer covered under copyright law. For those books no longer protected under copyright law, I have made extensive use of their illustrations, diagrams and charts because many of these charts and illustrations depict well the concepts I have wanted to convey in this book. Nearly all were specially restored, shaded and annotated by me for their use here. I also used a number of diagrams, maps and charts from various reports prepared by the US government (USGS) and various state geological surveys which are also in the public domain. This listing is then both a recommendation for your research as well as an acknowledgment that these classic texts have been the source of many of the illustrations used in this book.

US GEOLOGICAL SURVEY PROFESSIONAL PAPER 610: Principal Gold-Producing Districts of the United States By A. H. KOSCHMANN *and* M. H. BERGENDAHL, 1968

PRINCIPALS OF ECONOMIC GEOLOGY
William Harvey Emmons, PH. D. 1918

MINERAL DEPOSITS
BY WALDEMAR LINDGREN 1919

GOLD: Its Occurrence And Extraction
ALFRED G. LOCK, F.R.G.S., 1882.

CALIFORNIA MINES AND MINERALS
Published By The California Miners' Association
Edward H. Benjamin, SAN FRANCISCO, CA. 1899

The Idle And Industrious Miner
JAMES ANTHONY & Co, PUBLISHERS, SACRAMENTO 1854.

GOLD: ITS GEOLOGICAL OCCURRENCE AND GEOGRAPHICAL DISTRIBUTION
BY J. MALCOLM MACLAREN, D.Sc.
The Mining Journal: LONDON, 1908

Professional Papers on Geology: AIME
By Thomas A. Rickard, 1902

GOLD AND SILVER: A History of Mining
BY WALTER R. CRANE, PH.D, 1908

DANA'S MANUAL OF MINERALOGY
BY WILLIAM E. FORD, 1912

THE PROSPECTOR'S HANDBOOK
J. W. ANDERSON, M.A, F.R.G.S. 1911

PROSPECTING FOR GOLD AND SILVER
By ARTHUR LAKES, 1895.

A TEXTBOOK ON METAL MINING
Colliery Engineering Co., 1899

THE MINER'S HANDBOOK
By JOHN MILNE, F.R.S. 1893

PROSPECTING FOR MINERALS
By S. HERBERT COX, 1921.

UNDERGROUND TREASURES: How and Where to Find Them. BY JAMES ORTON, A.M., 1881.

Practical Stamp Milling and Amalgamation
By H. W. Mac FARREN
THE MINING MAGAZINE, 1914

Geologic Guidebook Along Highway 49
Calif. Geological Survey, Bulletin 141
By Olaf P. Jenkins, et al., 1948

GOLD DISTRICTS OF CALIFORNIA
Calif. Geological Survey, Bulletin 193
By William B. Clark, 1963

Meet The Author:

I got into prospecting as a teenager when my grandparents, who were light duty rock hounds that enjoyed camping and the outdoors, took me on several trips. I got into rockhounding and it wasn't long before I was planning rockhound trips for them to join me on. One of the trips they took me on was a tour of an underground gold mine in Julian, California. The mine owner told me about prospecting for gold and that I could learn to do it for myself. The idea of prospecting for my own gold really sounded great, so I bought a pan, made a sluice box out of wood, and read some books. I hadn't even held my driver's license very long when I took my younger brother and we went out to a place where the books said gold had been found. I dug for most of the day and ran it through the sluice box - I found nothing, not even one speck. I tried a couple other places mentioned in the book, and still nothing. It wasn't until I took a trip with my family up to the mother lode country on the North Fork of the Yuba River that I found a little 3/16 inch flake in my pan. Not very large, but plenty large enough to see it was clearly gold. After that, I kept exploring different locations closer to my home in the LA area and ended up doing better, finding spots where I always got at least some color and sometimes a few very small nuggets (little pickers). I spent a lot of time around Randsburg and the East fork of the San Gabriel River back then.

Out dry washing and prospecting near Randsburg, CA in the mid-1970s.

By the time I graduated from High School, I had decided I wanted to go into geology, so I enrolled at Long Beach State (in the LA area). I moved up in my prospecting as well, and within a couple years I bought a small puffer type dry washer and a 3 inch dredge (both from Keene) and I started processing more material and getting a little more gold. In those days (30 years ago) metal detectors could not do what they do now in finding nuggets, so I did not own one back then.

After a couple years as a geology major at Long Beach State, in 1979 I decided to make a switch, and I transferred to the Mackey School of Mines in Reno and changed my major to Mining Engineering. Reno is a great area and I was now close to some great prospecting in the mother lode country, as well as the northern Nevada deserts. I spent a lot of weekends in those days kicking around the mother lode and especially the North fork of the Yuba River. I had met and started dating a very special lady before I left the LA area and we were married in 1980. Soon we were both attending college in Reno, and I finally finished up and obtained my Mine Engineering degree in 1982.

Here I am surveying some old mine workings underground in the 1980s.

The first mining company I worked for was Candelaria Metals, where I helped set up and run a small heap leach project in central Nevada about 70 miles from Tonopah. We mined some ore underground, but got most of our material from the huge old dumps of the early day mining. Cyanide solution was sprinkled over ore heaped up on heavy plastic sheeting, and the solution dissolved out the gold and silver which we captured. I really enjoyed working outdoors, the excitement of producing our own gold and the friendships of the other guys on staff there. At Candelaria, I really was involved with pretty much everything from initial construction of the heap leach facility, to running the mining equipment, through the smelting and shipment of the metal. We produced quite a bit of gold and silver, but the company had huge debt and overhead that were established long before I arrived, and in spite of the mine's general success, the company went bankrupt.

At Virginia City, we mined a number of the old dumps from the 1800s for their gold and silver.

Within a few months, I was able to get on at the New Savage mine in Virginia City, a district with a rich and famous mining history, and not so far from my home in Reno. I worked mostly underground, surveying out where the working miners would drill. At Virginia City I discovered an important near surface ore body - I put together our modern maps with some old surveys from the 1800's and showed there should be a valuable ore body in a certain spot on our claims. I convinced the head geologist to drill there and sure enough, there was a nice little area of high grade rock, with about 20,000 ounces of gold. Not too bad! Unfortunately, this small mining company also had serious financial problems, including too much overhead, and it too went under as prices of gold sank after the early 1980s highs and new sources of funding dried up. Except for the corporate instability of the small mining companies, the mining industry was all I had hoped it would be while I was in school - I really enjoyed it.

Declining metals prices meant fewer jobs were available in the US, and many of my former classmates were going overseas, some even to South Africa - then a very politically unstable place. That just seemed way too risky. Instead, I chose to take a job as an engineer with the local county Health Department, and unlike the little mining companies, I knew my county paychecks would not bounce! The county job was a safe haven, and my wife and I built a house and started a family, and I stayed at the same county job that I told myself was "temporary" until I retired in 2011. In spite of how much I enjoyed it, I never did return to employment in the mining industry.

I enjoyed dredging for gold until it was outlawed in California in 2009. These clear waters are the North Fork of the Yuba River near Sierra City in the California mother lode country.

On the other hand, I have never left the love of hitting the hills prospecting that I had when I was a kid. For the last 20 years I have spent vacations prospecting in the hills for placer, hard rock and even large-scale bulk mineable gold deposits. From Ely to Denio to Tonopah and beyond in Nevada and the Klamath River to the Cargo Muchachos in California, I've been all over Nevada and California at various times. I've also traveled to other states like Alaska and Arizona to search for gold. I've even spent time is Australia and West Africa prospecting and finding gold. In the winter I focus on research, and when the nice weather arrives, I do my prospecting. Dredging, panning, nugget detecting, I enjoy them all, and sometimes I do pretty good with finding the gold as well. Now that I am retired, I have a lot more time available for these activities.

Here I am doing a GPAA gold show for the Prospecting and Mining Journal in about 2010. I always enjoyed doing these shows and talking about prospecting, but sadly the GPAA is no longer holding them (at least for now).

A few years before I retired, I began to write occasional articles for the ICMJ Prospecting and Mining Journal, a magazine I had first known as the California Mining Journal and later as the International California Mining Journal. I enjoyed writing about prospecting things worked out so well that in December of 2007, I joined the staff as the Associate Editor. In addition to writing, I began to represent the magazine at various gold prospecting shows (like those put on by the GPAA), and do some other things like speaking at club meetings and other events and appearing on video materials. Sadly, the Prospecting and Mining Journal went out of business in 2024.

A few years ago, I began to work on a website, partly to sell jewelry made from turquoise I dig from my Nevada mining claims, but I also wanted to share with friends about my prospecting experiences. The website has continued to grow and I have added to it materials about mining and prospecting as well as related topics I am interested in. It now has a tremendous amount of information, and I have plans to continue on adding to and improving it in the future. As I have been writing this book, I have collected a lot of information. I could have easily made this book one or two hundred pages longer - but hey, at 360+ pages, it's long enough! So, I have put the bulk of this extra material on the website.

In 2019, I finally got on board with YouTube and now have an active and vibrant YouTube channel. The name of the YouTube channel is "**Chris Ralph, Professional Prospector**." At the time of this 2025 update, I have over 300 videos on the channel, with more being added every week. I make a concerted effort to answer all the comments and questions on the videos I post on my YouTube channel. Please check it out regularly in the coming years as I will be adding more and more video content, with my adventures, maps, reference material and other useful information. I also have become active on Facebook, as "The Professional Prospector." I'd love to have any of you participate in my social media activities.

I hope that you have enjoyed this book, as it's been a lot of work, but a labor of love at the same time. The book has been a long time under development. I've spent years writing the text, collecting the images, laying out the pages, etc. - working off and on as time was available. Although finished in 2009, I drafted up the first outline of the contents for this book way back in 2001, so you can see it's been a very long-term project. With this latest 6th printing in 2025, I've done a major update to the information in the book. I've poured a lot into it over the years, in an attempt to make it a very useful reference. If you would like to get in touch with me about this book, the website or even my YouTube channel, please feel free to shoot me an email at:

Chrisralph218@gmail.com